Horticultural Plants Facing Stressful Conditions - Ways of Stress Mitigation

Horticultural Plants Facing Stressful Conditions - Ways of Stress Mitigation

Editors

Agnieszka Hanaka
Małgorzata Majewska
Barbara Hawrylak-Nowak

MDPI

Basel • Beijing • Wuhan • Barcelona • Belgrade • Novi Sad • Cluj • Manchester

Editors

Agnieszka Hanaka
Maria Curie-Skłodowska
University in Lublin
Lublin
Poland

Małgorzata Majewska
Maria Curie-Skłodowska
University in Lublin
Lublin
Poland

Barbara Hawrylak-Nowak
University of Life Sciences
in Lublin
Lublin
Poland

Editorial Office
MDPI
St. Alban-Anlage 66
4052 Basel, Switzerland

This is a reprint of articles from the Special Issue published online in the open access journal *Horticulturae* (ISSN 2311-7524) (available at: https://www.mdpi.com/journal/horticulturae/special_issues/Horticultural_Plants_Stressful_Conditions).

For citation purposes, cite each article independently as indicated on the article page online and as indicated below:

Lastname, A.A.; Lastname, B.B. Article Title. *Journal Name* **Year**, *Volume Number*, Page Range.

ISBN 978-3-0365-9444-6 (Hbk)
ISBN 978-3-0365-9445-3 (PDF)
doi.org/10.3390/books978-3-0365-9445-3

Cover image courtesy of Małgorzata Majewska

Contents

About the Editors

Agnieszka Hanaka

Agnieszka Hanaka works as an Associate Professor in the Department of Plant Physiology and Biophysics, Faculty of Biology and Biotechnology at the Maria Curie-Sklodowska University (UMCS) in Lublin, Poland. She earned a Master of Sciences degree in Biology (specializing in biochemistry) in 1997 and a Ph.D. in Biological Sciences (specializing in plant physiology) in 2007. She obtained a post-doctoral degree (habilitation) in biological sciences in biology in 2019. The aim of the study is to examine the physiological and phytochemical responses of plants to abiotic stressors (e.g., excess metals and temperature), which are altered by the presence of signaling molecules and plant growth-promoting microorganisms. This research focuses on exploring changes in the synthesis of primary and secondary metabolites of plants treated with various environmental factors.

Małgorzata Majewska

Małgorzata Majewska works as an Assistant Professor in the Department of Industrial and Environmental Microbiology, Faculty of Biology and Biotechnology at the Maria Curie-Skłodowska University (UMCS) in Lublin, Poland. She earned a Master of Sciences degree in Biology (specializing in microbiology) in 1996 and a Ph.D. degree in Biological Sciences (specializing in microbiology) in 2003. She obtained a post-doctoral degree (habilitation) in biological sciences in biology in 2018. The aim of the study is to identify the rate and efficiency of the mobilization of cadmium previously immobilized in the root biomass and Cd immobilization via components of the soil. This research focuses on the bioremediation of the soil contaminated with heavy metals, microbial interactions in the rhizosphere and plant growth-promoting microorganisms in metal-induced stress.

Barbara Hawrylak-Nowak

Barbara Hawrylak-Nowak works as an Associate Professor in the Sub-department of Plant Physiology and Biochemistry, Department of Botany and Plant Physiology at the University of Life Sciences in Lublin, Poland. She earned a Master of Sciences degree in Biology (specializing in environmental biology) in 2000, as well as in Biotechnology in 2002 and a Ph.D. degree in Agricultural Sciences (specializing in plant physiology) in 2004. She obtained a post-doctoral degree (habilitation) in agricultural sciences in horticulture in 2017. The aim of the study is the application of trace elements and other phytoprotectants in the mitigation of the impact of unfavorable environmental factors on plants. This research focuses on the biological activity of trace elements in plants, especially grown under stressful conditions, as well as the mineral nutrition of plants, enhancement of their secondary metabolism and ecological investigations.

Preface

In recent years, numerous studies have investigated the effects of various stressors on crop plants and their ability to tolerate biotic and abiotic stresses using different substances. This Special Issue provides the latest research and information on stress-induced changes in horticultural plants, as well as the biological actions, effectiveness and principles of mitigating factors in different cropping systems. A deeper understanding of the traits associated with improved plant stress resistance can accelerate the practical implementation of these advancements in enhancing crop yields.

We sincerely appreciate the authors for sharing their expertise, experience and critical discussion in preparing and revising their interesting articles in a timely manner. Moreover, we are thankful to the reviewers for their valuable suggestions and insightful remarks that significantly contributed to enhancing the quality of each manuscript. We trust that this Special Issue will be useful to those readers who are interested in stress physiology and resistance of horticultural plants.

Agnieszka Hanaka, Małgorzata Majewska, and Barbara Hawrylak-Nowak
Editors

horticulturae

Editorial

Horticultural Plants Facing Stressful Conditions—Ways of Stress Mitigation

Agnieszka Hanaka [1,*], Małgorzata Majewska [2] and Barbara Hawrylak-Nowak [3]

1 Department of Plant Physiology and Biophysics, Institute of Biological Sciences, Faculty of Biology and Biotechnology, Maria Curie-Skłodowska University, Akademicka 19, 20-033 Lublin, Poland
2 Department of Industrial and Environmental Microbiology, Institute of Biological Sciences, Faculty of Biology and Biotechnology, Maria Curie-Skłodowska University, Akademicka 19, 20-033 Lublin, Poland; malgorzata.majewska@mail.umcs.pl
3 Department of Botany and Plant Physiology, Faculty of Environmental Biology, University of Life Sciences in Lublin, Akademicka 15, 20-950 Lublin, Poland; barbara.nowak@up.lublin.pl
* Correspondence: agnieszka.hanaka@mail.umcs.pl

1. Introduction

Rapidly progressing climate change is increasing the frequency and severity of drought and salinity stresses, which are the major factors affecting crop production and the quality of ornamental plants, fruits, and vegetables. Natural acclimation and adaptation mechanisms of plants may not be sufficient to cope with these swift climate changes, and plant yield may be reduced and have a lower quality. For this reason, it is necessary to look for ways to reduce the negative effects of stress to provide constituents of the human diet in sufficient quantity and quality. Scientists are investing efforts toward the development of new cultivation techniques and improvements in the nutritional and pro-health value of fruits and vegetables.

The present treatise examines selected topics relating to stressful conditions and ways of its mitigation. The findings of these studies is especially important in understanding the molecular, metabolic, and genetic basis upon which to increase the plant resistance to environmental stresses. This Special Issue comprises twelve research papers and one review. The review article focused on the influence of salt stress on plant growth and methods of its reduction. The presented research studies were conducted at the morphological, anatomical, physiological, and molecular levels. The authors described the impact of stress caused by salinity [1–7], high temperature [8,9], drought [9], spring frosts [10], copper [11], asbestos [12], and biotic stress [13]. Different plant species were studied, i.e., sunflower, zinnia, orchid, buttercup, tomato, snap bean, lettuce, jojoba, grapevine, strawberry, and apple. This Special Issue is dedicated to the following topics: (1) the application of exogenous compounds and bioinoculants to increase stress tolerance [1–7,9,13], (2) the improvement of fruits and vegetables quality and their nutritional value [2,7,10], and (3) the planting of ornamental plants in polluted areas [1,11,12]. Interestingly, Markulj Kulundžić et al. [8] showed that daily changes in temperature and light intensity can induce unfavorable changes in the process of photosynthesis. They observed that morning environmental conditions were optimal for photosynthetic responses of sunflower in the flowering stage, while the afternoon was characterized by elevated temperatures and excess light intensity.

Citation: Hanaka, A.; Majewska, M.; Hawrylak-Nowak, B. Horticultural Plants Facing Stressful Conditions—Ways of Stress Mitigation. *Horticulturae* **2023**, *9*, 1146. https://doi.org/10.3390/horticulturae9101146

Received: 27 July 2023
Accepted: 7 August 2023
Published: 19 October 2023

2. Exogenous Compounds and Bioinoculants Increasing Stress Tolerance

Various studies indicated that application of exogenous compounds can serve as an alternative to mitigate salt stress in commercial cultivars. Eisa et al. [1], El-Beltagi et al. [5], and Annadurai et al. [9] found that melatonin application decreased the negative effect of salt, drought, and high-temperature stresses, respectively. They recognized that melatonin, as a novel biostimulator, has potential in scavenging reactive oxygen species through increased antioxidant activity, which shields the photosynthetic membrane from damage and therefore helps toward stress mitigation. Melatonin was found to have a positive effect at the morphological, physiological, and biochemical levels. In accordance with expectations, plant growth [1,9], fruit yield [9], photosynthetic rate [9], the activity of antioxidant enzymes, such as superoxide dismutase, catalase, peroxidase, ascorbate peroxidase, and glutathione reductase [1,5,9], and the contents of free proline [1,5], chlorophyll, carotenoids [1,5,9], and total soluble sugars [5] were increased in stress- and melatonin-treated plants. Moreover, El-Beltagi et al. [5] demonstrated that a combined treatment of melatonin and putrescine could enhance resistance to salinity in snap bean seedlings.

Aboryia et al. [4] evaluated the effect of foliar-applied proline on the enhancement of jojoba tolerance to salt stress. The physiological responses to proline treatment were manifested by an increase in the content of chlorophylls, carotenoids, minerals (N, P, K, Na, and Cl), and phenols in leaves, as well as the promotion of the antioxidant system compared with the control plants. Also, the exogenous application of proline improved most morphological (e.g., shoot number, stem diameter, number, area, and weight of leaves) and anatomical characteristic of jojoba leaves in saline conditions. In turn, Sayed et al. [2] used nano-silicon fertilizer to mitigate salinity stress. They observed that nano-silicon application on the leaves of tomato plants (*Solanum lycopersicum* L. grafted on *S. pimpinellifolium* or Edkawy) enhanced shoot and root growth, fruit yield, and fruit quality. Higher levels of mineral content, gibberellic acid, abscisic acid, and proline were detected in shoots that were grafted and subjected to nano-silicon application compared to the control treatment.

Polyhydroxy steroids phytohormones, such as brassinosteroids, can also be used to improve crop salt tolerance [6,7]. Furio et al. [7] found that strawberry treated with brassinosteroids was characterized by a higher shoot and root weight as well as a higher total weight of fruits per plant. Additionally, the quality of fruits obtained in salt stress conditions after brassinosteroids application was significantly higher than in non-treated individuals. Also, El-Banna et al. [6] indicated that grapevine seedlings sprayed with brassinosteroids tolerated salt stress better by sustaining higher photosynthetic pigment concentrations, maintaining ion homeostasis and water status, and stimulating antioxidant capacity, as well as affecting the preservation of the proper leaf anatomical attributes compared to untreated plants.

Stressful conditions can also be alleviated by using bioinoculants, which can offer a crucial alternative for organic farming [13]. Bioinoculants are a group of microorganisms which promote plant growth by making essential nutrients more bioavailable, control phytopathogens and disease development (biotic stress) as biocontrol agents, or increase tolerance to pollutants (abiotic stress). Pacheco-Trejo et al. [13] described the possibility of using *Trichoderma* spp. to protect and stimulate the growth of horticultural crops. They proved that the plant signaling events triggered by *Trichoderma* spp. are of high importance in order to understand the molecular basis involving plant protection against stresses. Also, the signaling elements of the plants from *Trichoderma* perception through late defensive responses were described. The authors explained that the activation of defense in plants stimulated by *Trichoderma* spp. will lead to an increase in crop production with consequent benefits for human health and the environment.

3. Improvement in Quality and Nutritional Value of Fruits and Vegetables

Plants exposed to various stresses deliver a lower crop production and quality of edible plants. Furio et al. [7], Ozherelieva et al. [10], and Maglione et al. [3] were looking for ways to prevent the negative effects of stress and maintain the good quality of harvested fruits and vegetables. They proposed brassinosteriods [7], iodine [3], and adaptogenic preparations [10] as protective factors against stress with the ability to improve the quality of potential yields.

Furio et al. [7] demonstrated that a pre-harvest treatment with brassinosteriods promoted the growth of strawberry in normal conditions and under exposition to water or saline stresses, as well as evaluated the percentage of fruits of commercial quality. In turn, the results of Maglione et al. [3] showed that positive stress (eustress) or essential and non-essential elements can improve the nutritive values (biofortification) of lettuce. Iodine application under moderate salinity increased the amount of the bioactive compounds acting as antioxidants, e.g., polyphenols and anthocyanins, thus exercising functional effects on human health.

Ozherelieva et al. [10] estimated the suitability of the adaptogenic preparations (phytomodulator "White Pearl Universal Antifreeze" and phytocorrector "White Pearl Drip Ca + Mg") to increase the yield and quality of apple cultivar "Sinap Orlovsky". The foliar application of these products in the early spring period significantly reduced the effects caused by scald in fruits, decreased bitter pitting, increased the average fruit weight, as well as increased the amount of sucrose and L-ascorbic acid, thus improving the taste qualities of fruits.

The conducted research including brassinosteriods [7], iodine [3], and adaptogenic preparations [10] proved their validity as additional techniques in traditional cultivation technologies, thus inducing a higher production and better quality of crops.

4. Planting Ornamental Plants on Polluted Areas

Heavily polluted areas lose their suitability both as agricultural lands and as ecological lands. The accumulation of toxins limits the growth of plants and hinders their revitalization. The constant increase in contaminated areas forces us to seek new opportunities for their cleanup and re-use.

Tugbaeva et al. [11] proposed to use *Zinnia elegans* (Jacq.) in landscaping because they observed that it performs well in copper-polluted growth medium (200 μM $CuSO_4$). The adaptation of zinnia to the excess Cu was associated with the metabolic changes in the phenylpropanoid pathway. The intensified lignification in the roots led to the accumulation of Cu in this organ and limited translocation to the shoots, which provided plant growth. An increase in the content of H_2O_2 and intensity of lipid peroxidation, as indicators of stress, an elevated amount of phenolic compounds, as well as the level of expression of genes encoding 4-coumarate-CoA ligase, cinnamoyl alcohol dehydrogenase, and class III peroxidase were observed under Cu treatment.

The possibility of *Epipactis atrorubens* (Hoffm.) cultivation on serpentine dumps post asbestos mining was analyzed by Maleva et al. [12]. The aim was the naturalization of regionally rare orchid species, very beautiful bright flowers, in contaminated areas. The authors observed that the leaves of orchids colonizing these dumps accumulated lipid peroxidation products, ascorbate, free proline, soluble phenolic compounds (including flavonoids), and non-protein thiols. For this reason, they recognized that non-enzymatic antioxidants increased the adaptive potential of *E. atrorubens*. They also attributed an important role to the increased mycorrhization of orchid plants growing under adverse conditions.

In addition to studying the adaptation of plants to growth in a saline environment and the response to this stress, Eisa et al. [1] also studied the possibility of alleviating stress using melatonin. The subject of the study was buttercup (*Ranunculus asiaticus* L.), highly prized as a cut flower due to its very decorative inflorescences. It was found that the foliar application of melatonin decreased the salt-induced symptoms of retarded vegetative growth, improved physiological parameters, and soothed oxidative stress, thereby enhancing stress resistance, which can be considered as an effective practice for production under stress conditions.

Author Contributions: Conceptualization, A.H. and M.M.; writing—original draft preparation, A.H., M.M. and B.H.-N.; writing—review and editing, A.H., M.M. and B.H.-N. All authors have read and agreed to the published version of the manuscript.

Acknowledgments: We gratefully acknowledge all the authors that participated in this Special Issue.

Conflicts of Interest: The authors declare no conflict of interest.

References

1. Eisa, E.; Honfi, P.; Tilly-Mándy, A.; Mirmazloum, I. Exogenous Melatonin Application Induced Morpho-Physiological and Biochemical Regulations Conferring Salt Tolerance in *Ranunculus asiaticus* L. *Horticulturae* **2023**, *9*, 228. [CrossRef]
2. Sayed, E.; Mahmoud, A.; El-Mogy, M.; Ali, M.; Fahmy, M.; Tawfic, G. The Effective Role of Nano-Silicon Application in Improving the Productivity and Quality of Grafted Tomato Grown under Salinity Stress. *Horticulturae* **2022**, *8*, 293. [CrossRef]
3. Maglione, G.; Vitale, E.; Costanzo, G.; Polimeno, F.; Arena, C.; Vitale, L. Iodine Enhances the Nutritional Value but Not the Tolerance of Lettuce to NaCl. *Horticulturae* **2022**, *8*, 662. [CrossRef]
4. Aboryia, M.; El-Dengawy, E.; El-Banna, M.; El-Gobba, M.; Kasem, M.; Hegazy, A.; Hassan, H.; El-Yazied, A.; El-Gawad, H.; Al-Qahtani, S.; et al. Anatomical and Physiological Performance of Jojoba Treated with Proline under Salinity Stress Condition. *Horticulturae* **2022**, *8*, 716. [CrossRef]
5. El-Beltagi, H.; El-Yazied, A.; El-Gawad, H.; Kandeel, M.; Shalaby, T.; Mansour, A.; Al-Harbi, N.; Al-Qahtani, S.; Alkhateeb, A.; Ibrahim, M. Synergistic Impact of Melatonin and Putrescine Interaction in Mitigating Salinity Stress in Snap Bean Seedlings: Reduction of Oxidative Damage and Inhibition of Polyamine Catabolism. *Horticulturae* **2023**, *9*, 285. [CrossRef]
6. El-Banna, M.; AL-Huqail, A.; Farouk, S.; Belal, B.; El-Kenawy, M.; Abd El-Khalek, A. Morpho-Physiological and Anatomical Alterations of Salt-Affected Thompson Seedless Grapevine (*Vitis vinifera* L.) to Brassinolide Spraying. *Horticulturae* **2022**, *8*, 568. [CrossRef]
7. Furio, R.; Salazar, S.; Mariotti-Martínez, J.; Martínez-Zamora, G.; Coll, Y.; Díaz-Ricci, J. Brassinosteroid Applications Enhance the Tolerance to Abiotic Stresses, Production and Quality of Strawberry Fruits. *Horticulturae* **2022**, *8*, 572. [CrossRef]
8. Markulj Kulundžić, A.; Viljevac Vuletić, M.; Matoša Kočar, M.; Antunović Dunić, J.; Varga, I.; Zdunić, Z.; Sudarić, A.; Cesar, V.; Lepeduš, H. Effect of Elevated Temperature and Excess Light on Photosynthetic Efficiency, Pigments, and Proteins in the Field-Grown Sunflower during Afternoon. *Horticulturae* **2022**, *8*, 392. [CrossRef]
9. Annadurai, M.; Alagarsamy, S.; Karuppasami, K.; Ramakrishnan, S.; Subramanian, M.; Venugopal, P.; Muthurajan, R.; Vellingiri, G.; Dhashnamurthi, V.; Veerasamy, R.; et al. Melatonin Decreases Negative Effects of Combined Drought and High Temperature Stresses through Enhanced Antioxidant Defense System in Tomato Leaves. *Horticulturae* **2023**, *9*, 673. [CrossRef]
10. Ozherelieva, Z.; Prudnikov, P.; Nikitin, A.; Androsova, A.; Bolgova, A.; Stupina, A.; Vetrova, O. Adaptogenic Preparations Enhance the Tolerance to Spring Frosts, Yield and Quality of Apple Fruits. *Horticulturae* **2023**, *9*, 591. [CrossRef]
11. Tugbaeva, A.; Ermoshin, A.; Wuriyanghan, H.; Maleva, M.; Borisova, G.; Kiseleva, I. Copper Stress Enhances the Lignification of Axial Organs in *Zinnia elegans*. *Horticulturae* **2022**, *8*, 558. [CrossRef]
12. Maleva, M.; Borisova, G.; Filimonova, E.; Lukina, N.; Chukina, N.; Ermoshin, A.; Tugbaeva, A.; Voropaeva, O. Adaptive Redox Reactions Promote Naturalization of Rare Orchid *Epipactis atrorubens* on Serpentine Dumps Post Asbestos Mining. *Horticulturae* **2022**, *8*, 603. [CrossRef]
13. Pacheco-Trejo, J.; Aquino-Torres, E.; Reyes-Santamaría, M.; Islas-Pelcastre, M.; Pérez-Ríos, S.; Madariaga-Navarrete, A.; Saucedo-García, M. Plant Defensive Responses Triggered by *Trichoderma* spp. as Tools to Face Stressful Conditions. *Horticulturae* **2022**, *8*, 1181. [CrossRef]

horticulturae

Article

Exogenous Melatonin Application Induced Morpho-Physiological and Biochemical Regulations Conferring Salt Tolerance in *Ranunculus asiaticus* L.

Eman Abdelhakim Eisa [1,2], Péter Honfi [2], Andrea Tilly-Mándy [2] and Iman Mirmazloum [3,*]

1 Department of Floriculture and Dendrology, the Hungarian University of Agriculture and Life Science (MATE), 1118 Budapest, Hungary
2 Botanical Gardens Research Department, Horticulture Research Institute, Agricultural Research Center (ARC), Giza 12619, Egypt
3 Department of Plant Physiology and Plant Ecology, Institute of Agronomy, Hungarian University of Agriculture and Life Sciences, 1118 Budapest, Hungary
* Correspondence: mirmazloum.seyediman@uni-mate.hu

Abstract: This study presents the effects of exogenous melatonin application at different concentrations (50, 100, and 200 μM) on the morphological, physiological, and antioxidant defense systems of the buttercup plant under salinity stress (4.5 and 5.5 $dS \cdot m^{-1}$ EC). Expectedly, the salinity stress negatively affected the plant growth parameters, cell membrane stability, and POX enzyme activity of *R. asiaticus* compared to non-stressed plants. However, in a dose-dependent manner, exogenous melatonin foliar application decreased the salt stress-induced symptoms of retarded vegetative growth, physiological characteristics, and oxidative stress level. The results obtained, revealed the significant effectiveness of exogenous melatonin treatment at 200 μM concentration under salt stress conditions by enhancing the plant growth traits such as chlorophyll and carotenoids content, relative water content, proline content, peroxidase enzyme activity (POD), and by the decreased electrolyte leakage rate, and Na^+ content, as well as delaying the emergence of flower buds under salinity stress. The salt tolerance index percentages (STI%) for all estimated characters are also calculated for all studied parameters. This study uncovered the beneficial effect of melatonin in reducing salt stress symptoms that can be used to reduce the salinity effect in ranunculus plant production.

Keywords: buttercup; melatonin; proline; salt stress; POD activity; STI

Citation: Eisa, E.A.; Honfi, P.; Tilly-Mándy, A.; Mirmazloum, I. Exogenous Melatonin Application Induced Morpho-Physiological and Biochemical Regulations Conferring Salt Tolerance in *Ranunculus asiaticus* L.. *Horticulturae* **2023**, *9*, 228. https://doi.org/10.3390/horticulturae9020228

Academic Editors: Agnieszka Hanaka, Małgorzata Majewska and Barbara Hawrylak-Nowak

Received: 16 January 2023
Revised: 6 February 2023
Accepted: 7 February 2023
Published: 8 February 2023

1. Introduction

Salinity, particularly in arid and semi-arid regions, is the main environmental concern with detrimental effects on soil and consequent agricultural output [1]. The salt-affected land area has increased and will continue to expand due to persistently unsustainable agricultural practices and climate change [2], posing a serious threat to agricultural areas that appears to be worsening [3]. As a result, 30% of farmland will be affected in the next quarter-century, and as much as 50% will be impracticable by 2050 [4]. Therefore, it is now more crucial than ever to learn how plants adapt and cope with salt stress by understanding the potential inhibition and tolerance mechanisms. In many ways, osmotic and ionic stress caused by salt exposure alters the major biological activities of many plants, including sodium ion (Na+) toxicity, nutritional imbalance, physiological water shortage, metabolic disturbances, oxidative damage, and photo-inhibition [5,6]. Commonly, the cut flowers are not suggested as crops for saline soils or water recycling systems because they are considered intolerant to high salt concentration of irrigation water and soil [7]. Salinity, in particular, can affect the quality of the blooms, shorten plants stems, and lower the production in flora crops [7–10]. In light of this, it is crucial to test more floricultural varieties for salinity tolerance from environmental and economic points of view for sustainable development.

Ranunculus asiaticus, also known as the buttercup, a member of the *Ranunculaceae* family, is an indigenous species from Asia and the Mediterranean [11]. The buttercup has delightful and odorless flowers that have a diameter of around 5 cm and come in various colors, including yellow, orange, red, pink, and white. Ranunculus's leaves are light green and grow to a maximum height of about 40 cm [12]. Due to its gorgeous inflorescences on strong and long stems, it is considered as an annual field-grown cut flower and is highly prized as a bedding or potted plant [13]. Depending on the variety, ranunculus can be propagated by tuberous roots or seeds [14,15]. Tuberous roots are more preferred since they often result in plants with earlier flowering time and more blooms per stem than seed-propagated plants [15]. A rosette of leaves with long petioles is produced from the tuberous roots of *R. asiaticus* in the fall after the first rain rehydrates the dried tissue in its natural Mediterranean habitat [16]. Six to eight fully developed leaves signal the onset of flowering, which continues through February and May. After shoot senescence, the tuberous roots are removed, stored over the summer, and sold for the following year's plantings in gardens and flowerbeds [15]. Buttercups like partial exposure to the sun and prefer moderate low temperatures (night/day regime 5–10/12–25 °C; optimum day 16 °C) and medium to high light intensities [16]. According to Rauter et al. [17], ranunculus physiological characteristics and visual growth quality declined when the EC of irrigation water raised from 0.5 to 5.5 $dS{\cdot}m^{-1}$, resulting in a rise in root-zone salinity. Similar observations were reported by Valdez-Aguilar et al. [18] on the sensitivity of ranunculus to saline conditions when irrigation solution EC value was raised from 2.0 to 6.0 $dS{\cdot}m^{-1}$.

One of the most recent strategies that tried to reduce the impact of biotic and abiotic stressors on plants is the application of plant bio-stimulators or growth hormones to promote adaptation [19]. Melatonin (MT) is the common name for N-acetyl-5-methoxytryptamine, a derivative of the essential amino acid, tryptophan, which is also a multi-regulatory biological hormone that is vital for plant and animal lives [20]. MT is reported to be involved in the circadian rhythm [21], immunological stimulation [22], antioxidant systems [23], and the seasonal cycle of reproduction [24]. MT was first detected in plants in 1995 [25]. The involvement of MT in plant growth regulation and development, including chlorophyll production and photosynthesis, callus induction, flowering, rhizogenesis, and the ageing of the leaves, has been documented [26–29]. Additionally, melatonin is also famous for its frequent application to combat biotic and abiotic stresses in plants since its ROS scavenging ability was confirmed [30,31]. Recent studies showed that MT-treated plants were more tolerant to salinity stress, had lower levels of reactive oxygen species, electrolyte leakage, and cell damage, and had higher growth parameters than untreated plants [32,33]. The positive effect of MT application for plants exposed to stress was confirmed in different species such as in watermelon [34], cucumber [28,35], melon [36], basil [37], and maize [29]. However, no such trial (melatonin application) has been conducted on *Ranunculus asiaticus* grown under salinity stress. Therefore, the present work intends to exhibit the efficacy of MT in enhancing the salinity tress tolerance of buttercup plants in terms of morphological and physiological reactions associated with oxidative stress. In this regard, the salt tolerance index (STI%) for all the studied parameters is determined to be utilized as an indicant of salt tolerance.

2. Materials and Methods

2.1. Material, Cultivation Conditions, and Trial Layout

The pot trial was undertaken in a research greenhouse of the Floriculture and Dendrology Department at The Hungarian University of Agriculture and life science (MATE) (Budapest, Hungary). The planting process started on 1 October 2021, and the harvest occurred 150 days later. The average day and night temperature of the glasshouse was 20–15 °C, with 60% relative humidity. Healthy seedlings (30-days old) of *R. asiaticus* with four to five true leaves (size, 5–6 cm) were transplanted into ø (9 × 9 × 10) cm plastic pots; one plant per each pot and 135 plants per trial. The design of the experimental groups was a randomized complete block design (RCBD) in a split-plot arrangement; the main plots consisted of three different salinity levels: (1) the control plants typically watered with 80% of the field's capacity; no MT; no salt treatment; (2) salinity treatment at EC 4.5 dS·m^{-1}; (3) salinity treatment EC 5.5 dS·m^{-1}, and the sub-plots included the different levels of MT treatments associated with salinity levels at EC 4.5 and 5.5 dS·m^{-1} (15 plants/subgroup). In each pot, the growing medium consisted of a uniform mixture of Klassmann TS3 Baltic peat (constituents are shown in Table S1), supplemented with 3 kg/m^3 Osmocote Exact Potassium Dominant (Scotts, NSW, Australia) and 1 kg/m^3 (soluble carbonate).

2.2. *Salinity Treatment and MT Application*

Prior to salinity treatments, the ranunculus seedlings were grown in separate pots for four weeks. The salt solution preparation involved adding NaCl and $CaCl_2$ dihydrate at a molar ratio of 2:1 to attain electrical conductivity (EC) values of 4.5 and 5.5 dS·m^{-1} [17]; the electrical conductivity of the solution was determined using a handheld EC meter (Milwaukee EC 60 Inc, Szeged, Hungary). After six weeks of culture and salt treatments, different concentrations (0, 50, 100, and 200 µM in tap water (ethanol/water (v/v) = 1/10,000)) [38,39] of exogenous melatonin (Thermo Fischer, Geel, Belgium) were sprayed four times in two-week intervals using a handgun sprayer. The leaves were sprayed until complete wetness and dripping. Tap water was used for regular irrigation.

2.3. *Morphological Characteristics*

Plants (10 individual plants per treatment) were sampled two weeks after the fourth foliar application to estimate the growth parameters, such as the plant height (cm), (measured from the medium surface to the shoot apex using a meter rod), the number of leaves, (counted manually), fresh and dry weights (shoots and leaves in g). The total leaf area (Area Meter 350, ADC Bioscietific Ltd., Hoddesdon, UK) and the number of flower buds that appeared were also recorded for analysis.

2.4. *Physiological and Biochemical Assesments*

The five uppermost young fully expanded leaf samples were instantly frozen in liquid nitrogen and kept at −80 °C pending inspection. The photosynthetic pigments were measured for the acetone (80%) extract samples of each treatment and the absorbances (644, 663, and 480 nm) were recorded in a UV-VIS Spectrophotometer (Genesys 10S, Waltham, MA, USA). The carotenoid content was computed following the method of Lichtenthaler et al. [40], and the relative water content (RWC) was determined using the methods described by Turk and Erdal [41] and the formula RWC = (FW-DW)/(TW-DW) × 100.

The leaf's proline content (PC) was determined by the methods of Bates et al. [42] at 520 nm. A standard curve was constructed to calculate the proline content in µmol g^{-1} leaf FW. The extent of electrolyte leakage (EL) was assessed via the methods of Reddy et al. [43]. The peroxidase enzyme activity (POD) was also determined spectrophotometrically (470 nm) using the standard guaiacol method [44].

Sodium content of the leaves was determined with a flame photometer according to Campbell's method [45] and is expressed as grams per 100 g dry weight (g 100 g^{-1} DW). The Salt Tolerance Index (STI) was determined as a percentage (%) for each of the analyzed traits, as proposed by Sbei et al. [46]. STI = (T salt/T cont.) × 100, where T salt represents the characteristic's average value recorded under the saline conditions induced by 5.5 EC, and T cont. represents the same traits' average value under control conditions.

2.5. *Statical Analysis*

The experiment was set up in a completely randomized design. The Two-Way MANOVA followed by UNIANOVA for the variables with Bonferroni's correction was run for all dependent variables, between factors at two levels: (1) treatments (Cont (no salinity and no MT), Salinity EC 4.5 and Salinity EC 5.5), (2) melatonin concentrations (0, 50, 100, 200 µM). Assumptions were as if the normality of the residuals for all dependent variables are accepted by Kolmogorov–Smirnov's ($p > 0.05$) [47]. The homogeneity of variances by

Leven's F test was satisfied with all dependent variables $p > 0.05$ [48]. Tukey's post hoc test was used for factor level comparisons [49,50], while Dunnett's test was used to compare the control treatment with two salinity groups. Pairwise within-subject effects were compared via Bonferroni's method. All statistics were gathered using the software IBM SPSS27 [51].

3. Results

Statistical analysis of the obtained data revealed highly significant differences amongst multivariate factors and for the interaction effect of treatment levels (Wilk's lambda < 0.001) [52]. Thus we compared the two salinity levels with the control for the MT concentration effect, and compared the MT concentration levels with the applied salinity levels, then followed up with Univariate ANOVA for different variables using the Bonferroni's correction to see significant differences in all individual variables $p > 0.05$ [53].

3.1. The Effect of Melatonin on Plant Morphology under Salinity Stress

The applied salinity stress at both levels had an obvious negative impact on the recorded growth parameters of buttercup plants (Table 1 and Figure 1). Comparing the control (non-stressed plants), salt-stressed plants at the EC 4.5 (0 MT) and EC 5.5 (0 MT) exhibited a substantial decrease in shoot length by (31.78 and 35.64%), leaf number by (32.23 and 34.03%), total leaf area by (58.34 and 60.98%), shoot FW by (29.32 and 36.61%) and shoot DW by (42.15 and 46.08%), respectively. However, foliar melatonin treatments (50, 100, and 200 µM) considerably enhanced the plant development and improved all vegetative parameters under salinity stress. Plants treated with 200 µM MT showed the best performance under both level of salinity treatments, manifested in the remarkable increase in shoot length (23.37 and 30.04%), leaf numbers (28.32 and 21.14%), total leaf area (58.01 and 58.79%), shoot FW (29.92 and 42.33%) and shoot DW (45.20 and 41.82%), compared with stressed plants of EC 4.5 MT0 and EC 5.5 MT0, respectively, when compared with the NO-MT treated plants (Table 2).

Table 1. The effects of exogenous melatonin application on growth parameters of *R. asiaticus* under salinity stress.

Treatments	Shoot Length (Cm)	No of Leaves	Leaf Area (Cm^2)	FW (g)	DW (g)
Control (non-stressed)	18.88 ± 0.2 aA	6.67 ± 0.15 aA	47.96 ± 0.2 aA	18.11 ± 0.2 aA	3.06 ± 0.02 aA
			S1 Treatments		
MT (0)	12.88 ± 0.12 eB	4.52 ± 0.12 eB	19.98 ± 0.2 eB	12.80 ± 0.1 eB	1.77 ± 0.02 eB
MT (50)	14.29 ± 0.22 dB	5.07 ± 0.11 dB	22.47 ± 0.2 dB	13.90 ± 0.1 dB	2.07 ± 0.01 dB
MT (100)	15.62 ± 0.02 cB	5.40 ± 0.18 cB	27.99 ± 0.1 cB	15.72 ± 0.2 cB	2.25 ± 0.01 cB
MT (200)	15.89 ± 1.3 bB	5.80 ± 0.02 bB	31.57 ± 0.2 bB	16.63 ± 0.2 bB	2.57 ± 0.02 bB
			S2 Treatments		
MT (0)	12.15 ± 0.2 eC	4.40 ± 0.02 eC	18.71 ± 0.2 eC	11.48 ± 0.2 eC	1.65 ± 0.02 eC
MT (50)	14.20 ± 0.1 dC	4.87 ± 0.01 dC	20.60 ± 0.3 dC	13.64 ± 0.3 dC	1.80 ± 0.01 dC
MT (100)	15.55 ± 0.2 cC	5.27 ± 0.01 cC	26.25 ± 0.12 cC	14.67 ± 0.1 cC	2.24 ± 0.02 cC
MT (200)	15.80 ± 0.1 bC	5.33 ± 0.11 bC	29.71 ± 0.4 bC	16.34 ± 0.2 bC	2.34 ± 0.02 bC

Control (no stress–no melatonin), **S1**: plants under salt stress EC 4.5 dS·m^{-1} were sprayed with MT (0 µM, 50 µM, 100 µM, and 200 µM); **S2**: plants under salt stress EC 5.5 dS·m^{-1} were sprayed with MT (0 µM, 50 µM, 100 µM, and 200 µM); different letters are for significantly different groups (Tukey/Dunnett $p < 0.05$). The lowercase letters are for significant differences amongst melatonin concentrations under fixed salinity treatments, and the uppercase letters for significant differences in the salinity treatments with the control group under fixed melatonin concentration levels ($n \geq 10$).

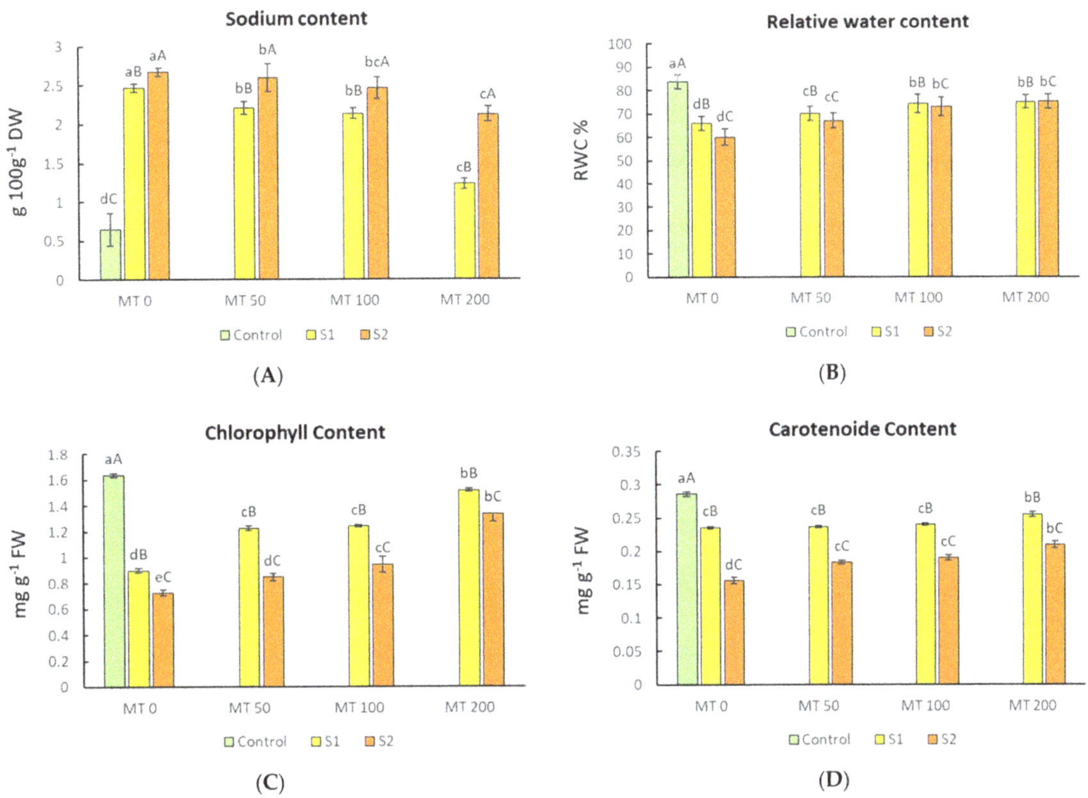

Figure 1. The effects of exogenous melatonin application (0, 50, 100, and 200 μM) on Na^+ level (**A**), RWC (**B**), chlorophyll (**C**), and carotenoids content (**D**) of *R. asiaticus* under saline conditions. Control: no-stress and no melatonin; S1: plants under salt stress EC 4.5 $dS \cdot m^{-1}$; S2: plants under salt stress EC 5.5 $dS \cdot m^{-1}$; different letters are for significantly different groups (Tukey/Dunnett $p < 0.05$). The lowercase letters are for significant differences amongst melatonin concentrations under fixed salinity treatments, and the uppercase letters for significant differences in the salinity treatments with the control group under fixed melatonin concentration levels ($n \geq 5$).

Table 2. Salt tolerance index (STI)% of *R. asiaticus* traits.

Traits	STI %
Shoot length	64.4 ± 0.2
Number of leaves	65.9 ± 0.2
Total leaf area	39.0 ± 0.03
Shoot fresh weight (FW)	63.4 ± 0.0
Shoot dry weight (DW)	53.9 ± 0.3
Emergence of Flower buds	74.3 ± 2.5
Total chlorophyll content	44.4 ± 1.2
Total carotenoids content	54.5 ± 2.1
Leaf Na^+ content	383.1 ± 23.7
Relative water content (RWC)	71.7 ± 0.3
Proline content	202.5 ± 2.1
Electrolyte leakage (El)	309.9 ± 0.3
Peroxidase (POD)	321.3 ± 28.5

The values represent the mean of at least five replicates for the comparison of tolerance index of all estimated traits under salt stress induced by EC 5.5 $dS \cdot m^{-1}$ without melatonin application.

3.2. Changes in Sodium Ion Level and Relative Water Content (RWC) of the Leaves

Our results indicate the sensitivity of ranunculus to salinity with increased Na^+. The Na^+ concentration in dry leaves increased by 280 and 311% upon 4.5 (S1) and 5.5 (S2) $dS{\cdot}m^{-1}$ EC of irrigation water, respectively, when compared to control plants (Figure 1A). The application of exogenous MT at different concentrations (50, 100, and 150 µM) resulted in a marked reduction in Na^+ content in leaf samples of *R. asiaticus* plants under both salinity levels (Figure 1A). The foliar application of 200 MT at both S1 and S2 levels revealed a significant effect and reduced the Na^+ content by 50.4 and 20.5% in comparison to salinity-stressed plants without MT application (S10MT and S20MT, respectively).

The relative water content (RWC) of a plant tissue is one of the most reliable indicators of its water status and capacity for survival under stressful conditions. Our data revealed that saline conditions negatively affected the leaves' RWC, particularly at EC 5.5 $dS{\cdot}m^{-1}$ which had the most severe effect (Figure 1B). Compared with plants under normal conditions (non-stressed plants), stressed plants, S1 0MT and S2 0MT exhibited a significant decrease (21.1 and 28.3%, respectively) in RWC.

On the other hand, the foliar melatonin treatment showed a slight improvement in RWC under the saline conditions. The increase in RWC was noted in the presence of 100 and 200 µM melatonin compared with non-MT-treated stressed plants which was still lower than that of control plants (Figure 1B). These outcomes imply that the foliar melatonin treatment has more favorable impacts on plant biomass and development upon salt stress, and that the 200 µM melatonin could better protect the plants under saline conditions than 100 and 50 µM melatonin.

3.3. Changes in Photosynthetic Pigments of Leaves

Saline conditions negatively affected the leaves' total chlorophyll (Chl) and carotenoid (Car) content of buttercup plants, with the influence being most prominent at the higher NaCl concentration of EC 5.5 treatment (Figure 1C,D). The leaves' Chl and Car content of non-MT-treated plants under both salinity levels (S1 0MT and S2 0MT) were estimated for a decrease of 45.12 and 55.49% in cases of chlorophyll and 17.24, and 44.82% in cases of carotenoids, respectively. However, the exogenous MT application at different doses (50, 100, and 150 µM) on *R. asiaticus* plants resulted in a significant increase in leaves' photosynthetic pigments (total Chl and Car) in comparison to stressed plants without MT under both salinity levels (Figure 1C,D). The foliar application of 200 MT at both S1 and S2 levels revealed a significant effect, with the rise of leaves' Chl and Car estimated at 68.89 and 83.56%, respectively, compared with stressed plants without MT application (S10MT and S20MT, respectively). These data indicate that 200 µM MT was able to inhibit the entrance of detrimental ions into the cells, safeguarding the cellular structure.

3.4. Changes in Electrolyte Leakage and Proline Content

The impacts of salinity stress and exogenous melatonin application on membrane integrity and electrolyte leakage (EL) are illustrated in Figure 2A. Saline conditions induced by NaCl 4.5 and 5.5 $dS{\cdot}m^{-1}$ significantly increased the electrolyte leakage by 64.01 and 67.74%, respectively, when compared to control plants. In contrast, in stressed plants subjected to foliar application of melatonin (50 µM, 100 µM, and 200 µM), there was a significant and concentration-dependent decrease in electrolyte leakage. The application of 200 µM MT resulted in the highest reduction of EL by 28.93 and 20.57% under both S1 and S2 salinity levels, when compared with salt-stressed plants without MT treatment (Figure 2A).

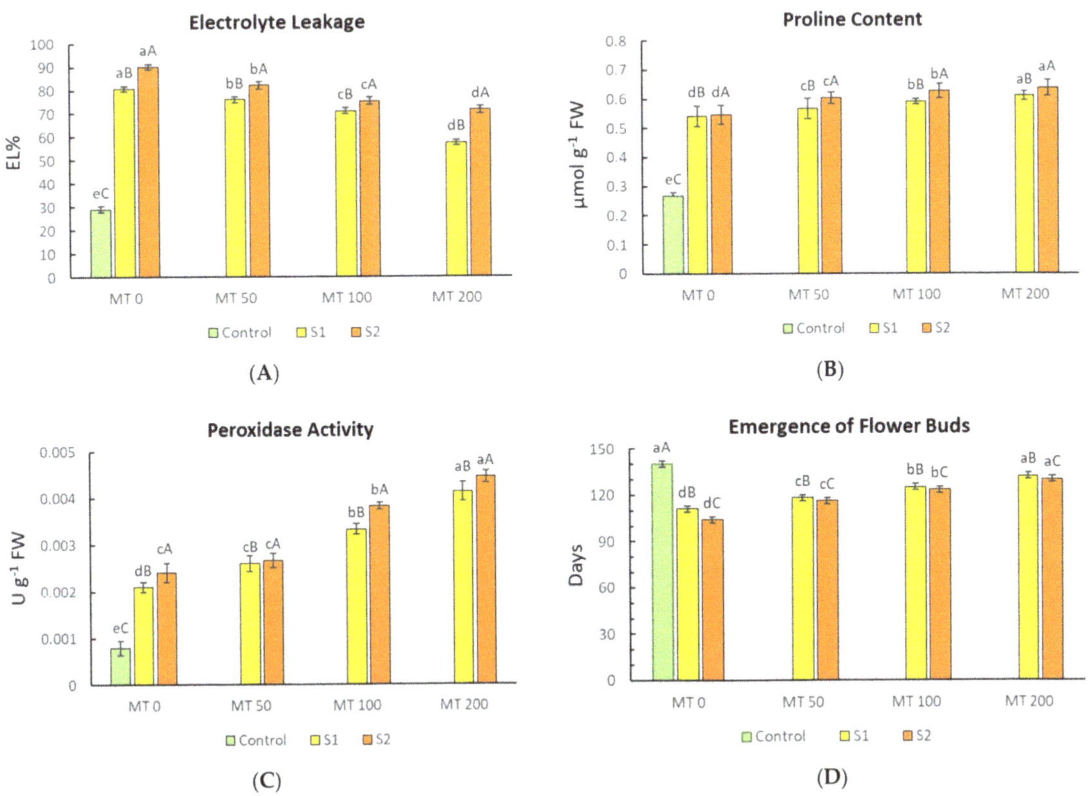

Figure 2. The effects of exogenous melatonin application (0, 50, 100, and 200 μM) on electrolyte leakage (**A**), proline content (**B**), peroxidase enzymes activity (**C**), and flower bud emergence time (**D**) of *R. asiaticus* under saline conditions. Control: no stress and no melatonin; S1: plants under salt stress EC 4.5 dS·m^{-1}; S2: plants under salt stress EC 5.5 dS·m^{-1} different letters are for significantly different groups (Tukey/Dunnett $p < 0.05$). The lowercase letters are for significant differences amongst melatonin concentrations under fixed salinity treatments, while the uppercase letters are for significant differences in the salinity treatments with the control group under fixed melatonin concentration levels ($n \geq 5$).

Proline, a compatible membrane protective solute with its role in maintaining water balance and enhancing cytoplasmic osmotic pressure is known to protect the cells during early dehydration. According to our findings, saline conditions showed a considerable rise in proline content in leaves, with its highest values being recorded at the S2 (EC 5.5) treatment. Stressed plants (S1 0MT and S2 0MT) exhibited a significant increase of about 50% in proline content in comparison with control plants. The foliar melatonin application at 50, 100, and 200 μM resulted in a gradual increase in proline content under the saline conditions when compared to non-MT-stressed plants (Figure 2B).

3.5. Changes in Peroxidase Enzyme Activity (POD)

A significant increase ($p < 0.05$) in POD activity was detected in buttercup plants under both salinity stress conditions (Figure 2C). The POD levels increased by 61.9 and 82.2% when plants were irrigated with saline water with EC of 4.5 and 5.5 dS·m^{-1}, respectively, when compared to non-stressed plants. Under both S1 and S1 salinity conditions, all of the melatonin treatments (50 μM, 100 μM, and 200 μM) significantly increased ($p < 0.05$) the POD activity, with its highest level being detected after 200 μM MT treatment. A 48.8 and 46.7% increase in antioxidant enzyme activity was recorded in plants under S1 and S2 conditions, respectively, when treated with 200 μM MT compared with non-MT-stressed plants under both salinity levels. These findings suggest that the NaCl stress brought the POD activities to a certain level as expected and that the foliar MT application enhanced (in a linear concentration-dependent manner) the POD activity even more to protect the cells from ROS accumulation.

3.6. Changes in Flower Bud Emergence

Our results reveal that both salt stress conditions (S1 and S2) induced the flower buds' appearance (Figure 3). Flowering time of salt stressed plants irrigated with saline water with EC of 4.5 and 5.5 dS·m^{-1} without melatonin treatment occurred 29 and 36 days sooner than in the control plant, respectively. This early flowering time was delayed after the application of melatonin for 22, 24, and 15 days in the case of S1 treatment and 17, 8, and 10 days in the case of S2 salinity treatment when 50, 100, and 200 μM of MT was applied, respectively, in comparison to control plants (unstressed plants without MT).

Figure 3. Morphological changes in *R. asiaticus* plants being treated with different concentrations of exogenous melatonin under salinity stress.

3.7. Salt Tolerance Index (STI)

Table 2 depicts the STI as a percentage of all examined variables between the non-melatonin-treated plants grown under unstressed conditions and stressed plants under S2 salinity level (EC 5.5 dS·m^{-1}). The STI% demonstrated that Na^+ content in leaves was the trait most sensitive to saline conditions, compared to STI% of other variables, which gave 437.9%, followed by POD activity and El value in the second place, which gave 321.3 and 309.9%, respectively. Proline content was in the third place for the salt response, which gave 202.5%. In addition, without significant differences, the remaining traits exhibited STI% values less than 100% and higher than 50%, except total Chl showed the lowest salinity-responsiveness, which gave 44.4%.

4. Discussion

Salinity stress is known to cause plant performance to deteriorate due to osmotic, ionic, or nutritional disorders, represented in the morphological characteristics of subjected plants. As an activator of antioxidant enzymes, melatonin is known to protect plants from oxidative stress [54]. The cut flower species and in particular the ranunculus is reported to be sensitive to salinity stress [17,18]. In the present work, exposing ranunculus plants to salinity stress induced by EC 4.5 and EC 5.5 caused a significant reduction in all the assessed vegetative growth characteristics when compared with unstressed plants (control) where the EC 5.5 was more harmful (Table 1 and Figure 1). Similarly, it has been reported that increasing EC from 2 to 3 dS·m^{-1} significantly stunted plant growth, dry weight, number of leaves, and diameter of flowering stems of white and pink *R. asiaticus* cultivars [18]. Poorter et al. [55] reported that the plant leaf-related parameters were significantly affected by salinity.

Melatonin (MT) is known as an endogenous regulator with an enhancement effect on both plant growth promotion [36,56,57] and also in protecting against abiotic stress [31,58–61] even when applied in small concentrations [62]. Many investigations have demonstrated MT's efficacy as a dose-dependent stimulator of plant growth and development [35,63,64]. We evaluated the effects of melatonin on ranunculus plants under saline conditions and found that foliar spray with 200 µM MT was the most effective concentration to mitigate the adverse effects of stress in most of the traits studied. Our findings were in line with the recent investigation on cotton seedlings treated with 200 µM of melatonin [65]. Exogenous melatonin must be applied to plants at the appropriate concentration, which may be different for distinct plant species [66]. Melatonin. an indoleamine, shares IAA's metabolic precursor, which may explain its influence on plant growth and development [67]. In addition, earlier research reported that environmental stress factors may induce flowering to be able to perpetuate the species' survival [68,69]. Exogenous melatonin was involved in flowering bud growth, as reported by Kolář et al. [70]. Similar to our results (Figure 2D), in Arabidopsis, exogenous MT caused a delay in flowering time [71]. This may occur due to the involvement of exogenous MT in upregulation of flowering locus C (FLC), which suppresses the transcription of flowering locus T (*FT*) [56,72].

Excessive cytoplasmic NaCl levels interrupt ion balance and inhibit plant growth and development [73]. Accordingly, plants must be able to ingest and compartmentalize ions to increase their salt tolerance efficiently. Thus, plants redistribute cytoplasmic salt ions into vacuoles or store them in various tissues to cope with salinity stress [74,75]. Consequently, a more significant Na^+ build-up will be observed in leaves. Likewise, in this experiment, Na^+ content of the leaves dramatically increased at both saline conditions EC 4.5 and 5.5 dS·m^{-1} (Figure 1A). Similar outcomes were also noticed in the cotton leaves [65] and pepper cultivars 'Granada' and 'Nobili' [76]. However, exogenous MT treatment allowed ranunculus plants to retain lower Na^+ in leaves than in non-MT-treated stressed plants (Figure 1A), and higher reduction was observed under S1 conditions by increasing MT levels. These findings agree with Castañares and Bouzo [36] and Wei et al. [77], reporting that MT treatment reduced ion toxicity in melons and rice, respectively when subjected to salt stress. This research also confirmed that 200 µM MT aids in preserving ion equilibrium while exposed to salt stress. Specifically, melatonin's action in the presence of saline stress is to upregulate the transporter genes *NHX1* and *AKT1* [78], which are involved in preserving the ion homeostasis. Due to the extreme sensitivity of plants to salinity, water status is the primary factor in response to stress [79]. Using water relation properties in ranunculus leaves, we found a common reactivity to salinity, consistent with the findings of Stępień and Kłobus [80] on cucumber and on pepper cultivars [76]. Toxic ion accumulation, especially Na^+ and Cl^-, may be to blame for the reduction in RWC by causing leaves to shrink and their stomata to close, thereby lowering the intracellular CO_2 partial pressure [81]. According to Munns' [82] study, higher sodium chloride concentrations led to a greater loss of water from the cells, which manifested as dehydration; thus, the cell growth rate was rapidly reduced in addition to a series of metabolic alterations like those imposed by water stress. MT application in our study, however, increased RWC gradually by increasing MT concentrations and the higher concentration was the most effective under both salinity levels S1 and S2 (Figure 1B) similar to those reported in borage plants [79], likely attributable to enhanced water absorption due to safeguarding the cell membrane [36,80]. It has been proven that melatonin can improve the thickness of a plant's cuticle, which in turn helps to limit water loss. Melatonin treatment has been shown to improve stress tolerance in plants by keeping their turgor and water ratio stable [81].

Photosynthesis relies on chloroplast chlorophyll, which can capture and transmit light energy for plants and is a key indicator of plant physiological properties and performance [82]. In plants subjected to saline conditions, a decrease in total chlorophyll and carotenoids was observed, and a more severe effect was found under S2 conditions (Figure 1C,D). This is correlated with excessive quantities of Na^+ that result in sluggish synthesis or rapid breakdown of photosynthetic pigments, which has an immediate impact on plant growth [83]. An application of melatonin in earlier studies enabled salt-stressed plants to keep their chlorophyll levels closer to normal such as in melon (*Cucumis melo* L.) [36], pistachio [60], maize seedlings [29], and in spearmint plants [84]. Similarly, in the current study, exogenous melatonin enhanced chlorophyll and carotenoid levels in ranunculus plants compared to non-MT-stressed plants. In general, S1 (4.5 EC) showed the higher pigment accumulation with MT application, and 200 μM MT was the most effective concentration applied under both salinity stress levels, supporting the hypothesis that MT pretreatment promotes biosynthesis and reduces chlorophyll breakdown [85]. This effect has been attributed to melatonin's antioxidant properties and its influence on the genes for the chlorophyll-degrading enzymes CLH (chlorophyllase), PAO (pheophorbide an oxygenase), and the red chlorophyll catabolite reductase [56]. Additionally, high melatonin dosages improved chloroplast carotenoid content and lipid-soluble antioxidants [74]. Consequently, melatonin can stimulate carotenoid production and result in lowering the levels of photooxidative degradation [31].

Ion-specific salt degradation mainly occurs in the plasma membrane [86]. Electrolyte leakage can be attributed to a loss of membrane integrity, which reduces plants' ability to hold K^+ as a result of stress [87]. Consequently, plasma membrane electrolyte leakage is regarded as an essential indicator for selecting tolerant plants under salt stress [86]. From our data, both salinity levels increased EL rate, but S2 (EC 5.5) caused the highest EL rate. However, minimizing EL in plants treated with MT afforded a protective effect toward membrane damage under saline conditions (Figure 2A). Similar results were presented by Khan et al. [88] and Zhang et al. [89], who reported that MT treatment reduced EL in sugar beets and tomato under salinity stress, indicating that MT may alleviate the oxidative damage induced by salt conditions. Moreover, melatonin, is present on the hydrophilic side of lipid bilayers and can diffuse through lipid membranes and the cytoplasm [90–92]. Melatonin's ability to organize itself in lipid membranes is concentration-dependent as at lower doses the molecules align themselves parallel to the lipid tail, while at higher doses they arrange themselves parallel to the bilayer [92].

Plants accumulate osmotic regulators as part of their adoptive stress-reduction strategy to preserve intracellular stability and shield their cells from the detrimental consequences of saline stress [64,93]. Ferchichi et al. [94] revealed that proline serves numerous functions, including stability of membranes and proteins, regulating gene expression in response to salinity conditions to maintain the redox balance. Previous investigations have shown that the melatonin-induced buildup of certain osmolytes diminishes cells' osmotic potential, which enhances osmotic adjustment and hence increases plants' water content under stressful environments possibly by upregulating the transcription of genes like *BADH* and *P5CS* [36,89]. Melatonin stimulates root development and aquaporin function, which in turn stimulate water uptake and distribution, and ultimately improves hydration status under salt conditions [95]. This mechanism is supported by the findings of the current study (Figure 2B), when MT enhanced proline accumulation under both salinity levels compared to non-stressed plants. Likewise, Sheikhalipour et al. [64] observed that MT-treated plants showed increased RWC and proline content under salinity conditions when compared with non-MT-treated plants under stress.

Additionally, ROS is a significant indicator of oxidative and salt stress [96]. The antioxidant machinery present in cells is responsible for maintaining ROS homeostasis, which is necessary for cells to avoid irreparable damage and maintain their integrity (antioxidant compounds and enzymes) [77,97]. The primary function of peroxidases as an ROS scavenger is to convert H_2O_2 into water molecules after SOD has converted $O_2^{\bullet-}$ to O_2 and H_2O_2 [98]. Exogenous MT decreased the formation of ROS by scavenging ROS and promoting antioxidant activity [99]. In this investigation, MT treatments, especially those with a concentration of 200 μM MT (Figure 2D), enhanced the POD activity under stressful conditions and the highest activity was noticed under the S2 treatment (EC 5.5). Our results also corroborate the findings of Zhang [28] that MT can improve salt tolerance by upregulating antioxidant enzyme genes and lowering biological macromolecule breakdown to increase antioxidant enzyme activity under high salinity. Similar investigations into rice [77], naked oat [100], and rapeseed [101] showed the same trend.

Depending on the salinity tolerance index (STI%) values under severe stress of EC 5.5 (S1 treatment), the investigated characteristics in this study could be divided into three categories (Table 2). The first category consists of the variables having an STI% greater than 400%, as observed with Na content; the second category is the variables that exhibited STI% of more than 200%, as observed with POD activity, El, and proline. While the final category includes the variables with STI% values below 100%, as observed with flower bud emergence time, RWC, number of leaves, shoot length, shoot fresh weight, carotenoids, shoot dry weight, total chlorophyll, and leaf area, in descending order. Based on the high STI% values for Na content, followed by POD activity, El, and proline, it is conceivable to use these characteristics as obvious markers of ranunculus plants' response to salt stress. Some scientific papers have debated the fractionated STI as a stress-response indicator and considered rather the biomass of the entire plant, for instance, as reported for chickpea [102] and Asian barley [46]. In this work, it can be stated that STI% was computed in a wide variety of ranunculus plant features, similar to an approach by Roshdy et al. in strawberry plants [103], that are more informative for future studies and plant breeding projects.

5. Conclusions

This study investigated the effect of exogenous melatonin treatment on *R. asiaticus* under two salinity levels (EC 4.5 and EC 5.5). The melatonin application enhanced RWC content and photosynthesis pigments, and reduced Na^+ accumulation in leaves, resulting in strengthening the vegetative and growth parameters under saline conditions in a dose-dependent manner. Additionally, our data showed improvement in osmotic regulation capability by increasing osmolyte accumulation (Proline), as well as the protective evidence that exogenous melatonin application in buttercup seedlings enhanced the performance of the antioxidant defense system by diminishing the ROS generation as demonstrated by activation of POD and the decrease in EL. Considering STI% values under the most harmful salinity level (EC 5.5), sodium content in leaves, followed by enzyme activity, EL, and leaf proline content could indicate a Ranunculus plant's salinity stress response. Overall, when applied at the optimal dose of 200 μM in this study, melatonin alleviates salinity stress on morpho-physiological characteristics, improving *R. asiaticus*' tolerance, which can be considered as an effective practice for productions under stress conditions.

Supplementary Materials: The following supporting information can be downloaded at: https://www.mdpi.com/article/10.3390/horticulturae9020228/s1, Table S1: Klassmann TS3 Baltic peat chemical components.

Author Contributions: Conceptualization, A.T.-M. and P.H.; methodology, E.A.E., A.T.-M. and P.H.; validation, E.A.E., A.T.-M. and I.M.; formal analysis, E.A.E.; investigation, E.A.E.; resources, A.T.-M.; data curation, E.A.E. and I.M.; writing—original draft preparation, E.A.E.; writing—review and editing, I.M.; visualization, E.A.E. and I.M.; supervision, A.T.-M. and P.H. All authors have read and agreed to the published version of the manuscript.

Funding: This research received no funding. The APC was paid by the Hungarian University of Agriculture and Life Sciences (MATE).

Institutional Review Board Statement: Not applicable.

Informed Consent Statement: Not applicable.

Data Availability Statement: Not applicable.

Acknowledgments: E.A.E. is thankful to the Tempus Public Foundation of Hungary for the Stipendium Hungaricum scholarship.

Conflicts of Interest: The authors declare no conflict of interest.

References

1. Porcel, R.; Aroca, R.; Ruiz-Lozano, J.M. Salinity Stress Alleviation Using Arbuscular Mycorrhizal Fungi: A Review. *Agron. Sust. Dev.* **2012**, *32*, 181–200. [CrossRef]
2. Qadir, M.; Quillérou, E.; Nangia, V.; Murtaza, G.; Singh, M.; Thomas, R.; Drechsel, P.; Noble, A. Economics of Salt-Induced Land Degradation and Restoration. *Nat. Resour. Forum* **2014**, *38*, 282–295. [CrossRef]
3. Bui, E. Soil Salinity: A Neglected Factor in Plant Ecology and Biogeography. *J. Arid Env.* **2013**, *92*, 14–25. [CrossRef]
4. Chandrasekaran, M.; Boughattas, S.; Hu, S.; Oh, S.; Sa, T. A Meta-Analysis of Arbuscular Mycorrhizal Effects on Plants Grown under Salt Stress. *Mycorrhiza* **2014**, *24*, 611–625. [CrossRef]
5. Shabala, S.; Cuin, T.A. Potassium Transport and Plant Salt Tolerance. *Physiol. Plant* **2007**, *133*, 651–669. [CrossRef]
6. Pandolfi, C.; Mancuso, S.; Shabala, S. Physiology of Acclimation to Salinity Stress in Pea (*Pisum sativum*). *Env. Exp. Bot.* **2012**, *84*, 44–55. [CrossRef]
7. Carter, C.T.; Grieve, C.M. Mineral Nutrition, Growth, and Germination of *Antirrhinum majus* L. (Snapdragon) When Produced Under Increasingly Saline Conditions. *HortScience* **2008**, *43*, 710–718. [CrossRef]
8. Grieve, C.M. Salinity-Induced Enhancement of Horticultural Crop Quality. In *Handbook of Plant and Crop Stress*; Pessarakli, M., Ed.; Taylor and Francis: Boca Raton, FL, USA, 2011; pp. 1173–1194.
9. Grieve, C.M.; Poss, J.A.; Shouse, P.J.; Carter, C.T. Modeling Growth of Matthiola Incana in Response to Saline Wastewaters Differing in Nitrogen Level. *HortScience* **2008**, *43*, 787–1793. [CrossRef]
10. Shillo, R.; Ding, M.; Pasternak, D.; Zaccai, M. Cultivation of Cut Flower and Bulb Species with Saline Water. *Sci. Hort.* **2002**, *92*, 41–54. [CrossRef]
11. De Hertogh, A.A. *Holland Bulb Forcer's Guide*, 5th ed.; Alkemade Printing BV: Lisse, The Netherlands, 1996.
12. Parkin, J. The Glossy Petals of Ranunculus. *Ann. Bot.* **1928**, *4*, 739–755. [CrossRef]
13. Karlsson, M. *Producing Ravishing Ranunculus*; Greenhouse Product News: Sparta, MI, USA, 2003; pp. 44–48. Available online: https://gpnmag.com/article/producing-ravishing-ranunculus/ (accessed on 1 November 2022).
14. Margherita, B.; Giampiero, C.; Pierre, D. Field Performance of Tissue-Cultured Plants of *Ranunculus asiaticus* L. *Sci. Hortic.* **1996**, *66*, 229–239. [CrossRef]
15. Meynet, J. Ranunculus. In *Physiology of Flower Bulbs: A Comprehensive Treatise on the Physiology and Utilization of Ornamental Flowering Bulbous and Tuberous Plants*; De Hertogh, A.A., Le Nard, M., Eds.; Elsevier: Amsterdam, The Netherlands, 1993; pp. 603–610.
16. Horovitz, A. Ranunculus. In *Handbook of Flowering*; Halevy, A., Ed.; CRC Press: Boca Raton, FL, USA, 1985; Volume 4, pp. 155–161.
17. Rauter, S.; Sun, Y.; Stock, M. Visual Quality, Gas Exchange, and Yield of Anemone and Ranunculus Irrigated with Saline Water. *Horttechnology* **2021**, *31*, 763–770. [CrossRef]
18. Valdez-Aguilar, L.A.; Grieve, C.M.; Poss, J.; Mellano, M.A. Hypersensitivity of *Ranunculus asiaticus* to Salinity and Alkaline PH in Irrigation Water in Sand Cultures. *HortScience* **2009**, *44*, 138–144. [CrossRef]
19. Kamran, M.; Wennan, S.; Ahmad, I.; Xiangping, M.; Wenwen, C.; Xudong, Z.; Siwei, M.; Khan, A.; Qingfang, H.; Tiening, L. Application of Paclobutrazol Affect Maize Grain Yield by Regulating Root Morphological and Physiological Characteristics under a Semi-Arid Region. *Sci. Rep.* **2018**, *1*, 15.
20. Qingbo, K.; Jun, Y.; Bomei, W.; Jianhong, R.; Lina, Y.; Xiping, D.; Shiwen, W. Melatonin Mitigates Salt Stress in Wheat Seedlings by Modulating Polyamine Metabolism. *Front. Plant Sci.* **2018**, *9*, 914.
21. Mishima, K. Melatonin as a Regulator of Human Sleep and Circadian Systems. *Nihon. Rinsho. Jpn. J. Clin. Med.* **2012**, *70*, 1139–1144.
22. Hardeland, R.; Madrid, J.; Tan, D.-X.; Reiter, R. Melatonin, the Circadian Multioscillator System and Health: The Need for Detailed Analyses of Peripheral Melatonin Signaling. *J. Pineal Res.* **2012**, *52*, 139–166. [CrossRef]
23. Rodriguez, C.; Mayo, J.C.; Sainz, R.M.; Antolin, I.; Herrera, F.; Martin, V.; Reiter, R.J. Regulation of Antioxidant Enzymes: A Significant Role for Melatonin. *J. Pineal Res.* **2004**, *36*, 1–9. [CrossRef]

24. Barrett, P.; Bolborea, M. Molecular Pathways Involved in Seasonal Body Weight and Reproductive Responses Governed by Melatonin. *J. Pineal Res.* **2012**, *52*, 376–388. [CrossRef]
25. Hattori, A.; Migitaka, H.; Iigo, M.; Itoh, M.; Yamamoto, K.; Ohtani-Kaneko, R.; Hara, M.; Suzuki, T.; Reiter, R.J. Identification of Melatonin in Plants and Its Effects on Plasma Melatonin Levels and Binding to Melatonin Receptors in Vertebrates. *Biochem. Mol. Boil. Int.* **1995**, *35*, 627–634.
26. Tan, D.; Manchester, L.; Korkmaz, A.; Ma, S.; Rosales-Corral, S.; Reiter, R.J. Fundamental Issues Related to the Origin of Melatonin and Melatonin Isomers during Evolution: Relation to Their Biological Functions. *Int. J. Mol. Sci.* **2012**, *15*, 15858–15890. [CrossRef] [PubMed]
27. Wang, P.Y.L.; Liang, D.; Li, C.; Ma, F.; Yue, Z. Delayed Senescence of Apple Leaves by Exogenous Melatonin Treatment: Toward Regulating the Ascorbate–Glutathione Cycle. *J. Pineal Res.* **2012**, *53*, 11–20. [CrossRef] [PubMed]
28. Zhang, H.J.; Yang, R.C.; Wang, L.; Sun, Q.Q.; Li, D.B.; Cao, Y.Y.; Weeda, S.; Zhao, B.; Ren, S.; Guo, Y.D. Melatonin Promotes Seed Germination under High Salinity by Regulating Antioxidant Systems, ABA and GA4 Interaction in Cucumber (*Cucumis sativus* L.). *J. Pineal Res.* **2014**, *57*, 269–279. [CrossRef]
29. Ahmad, S.; Cui, W.; Kamran, M.; Ahmad, I.; Meng, X.; Wu, X.; Su, W.; Javed, T.; El Hamed, A.; Zhikuan, S.; et al. Exogenous Application of Melatonin Induces Tolerance to Salt Stress by Improving the Photosynthetic Efficiency and Antioxidant Defense System of Maize Seedling. *J. Plant Growth Regul.* **2021**, *40*, 1270–1283. [CrossRef]
30. Yin, L.; Wang, P.; Li, M.; Ke, X.; Li, C.; Liang, D.; Wu, S.; Ma, X.; Li, C.; Zou, Y. Exogenous Melatonin Improves Malus Resistance to Marssonina Apple Blotch. *J. Pineal Res.* **2013**, *54*, 426–434. [CrossRef]
31. Szafrańska, K.; Reiter, R.J.; Posmyk, M.M. Melatonin Application to *Pisum sativum* L. Seeds Positively Influences the Function of the Photosynthetic Apparatus in Growing Seedlings during Paraquat-Induced Oxidative Stress. *Front. Plant Sci.* **2016**, *7*, 1663. [CrossRef] [PubMed]
32. Arnao, M.B.; Hernández-Ruiz, J. Functions of Melatonin in Plants: A Review. *J. Pineal Res.* **2015**, *59*, 133–150. [CrossRef]
33. Shi, H.; Jiang, C.; Ye, T.; Tan, D.X.; Reiter, R.J.; Zhang, H.; Liu, R.; Chan, Z. Comparative Physiological, Metabolomic, and Transcriptomic Analyse Sreveal Mechanisms of Improved Abiotic Stress Resistance in Bermudagrass [*Cynodon dactylon* (L). Pers.] by Exogenous Melatonin. *J. Exp. Bot.* **2015**, *66*, 681–694. [CrossRef]
34. Li, H.; Chang, J.; Chen, H.; Wang, Z.; Gu, X.; Wei, C.; Zhang, Y.; Ma, J.; Yang, J.; Zhang, X.; et al. Exogenous Melatonin Confers Salt Stress Tolerance to Watermelon by Improving Photosynthesis and Redox Homeostasis. *Front. Plant Sci.* **2017**, *8*, 295. [CrossRef]
35. Wang, L.Y.; Liu, J.L.; Wang, W.X.; Sun, Y. Exogenous Melatonin Improves Growth and Photosynthetic Capacity of Cucumber under Salinity-Induced Stress. *Photosynthetica* **2016**, *54*, 19–27. [CrossRef]
36. Castañares, J.L.; Bouzo, C.A. Effect of Exogenous Melatonin on Seed Germination and Seedling Growth in Melon (*Cucumis melo* L.) Under Salt Stress. *Hortic. Plant J.* **2019**, *5*, 79–87. [CrossRef]
37. Bahcesular, B.; Yildirim, E.D.; Karaçocuk, M.; Kulak, M.; Karaman, S. Seed Priming with Melatonin Effects on Growth, Essential Oil Compounds and Antioxidant Activity of Basil (*Ocimum basilicum* L.) under Salinity Stress. *Ind. Crops Prod.* **2020**, *146*, 112165. [CrossRef]
38. Zhang, Y.P.; Yang, S.J.; Chen, Y.Y. Effects of Melatonin on Photosynthetic Performance and Antioxidants in Melon during Cold and Recovery. *Biol. Plant.* **2017**, *61*, 571–578. [CrossRef]
39. Bidabadi, S.S.; Vander, W.J.; Sabbatini, P. Exogenous Melatonin Improves Glutathione Content, Redox State and Increases Essential Oil Production in Two *Salvia* Species under Drought Stress. *Sci. Rep.* **2020**, *10*, 6883. [CrossRef]
40. Lichtenthaler, H.K.; Wellburn, A.R. Determinations of Total Carotenoids and Chlorophylls a and b of Leaf Extracts in Different Solvents. *Biochem. Soc. Trans.* **1983**, *11*, 591–592. [CrossRef]
41. Turk, H.; Erdal, S. Melatonin Alleviates Cold-Induced Oxidative Damage in Maize Seedlings by up-Regulating Mineral Elements and Enhancing Antioxidant Activity. *J. Plant Nutr. Soil Sci.* **2015**, *178*, 433–439. [CrossRef]
42. Bates, L.; Waldren, R.; Teare, I. Rapid Determination of Free Proline for Water-Stress Studies. *Plant Soil* **1973**, *39*, 205–207. [CrossRef]
43. Reddy, A.R.; Chaitanya, K.V.; Vivekanandan, M. Drought-Induced Responses of Photosynthesis and Antioxidant Metabolism in Higher Plants. *J. Plant Physiol.* **2004**, *161*, 1189–1202. [CrossRef]
44. Zhang, J.; Kirkham, M.B. Drought-Stress-Induced Changes in Activities of Superoxide Dismutase, Catalase, and Peroxidase in Wheat Species. *Plant Cell Physiol.* **1994**, *35*, 785–791. [CrossRef]
45. Korkmaz, D. Precipitation Titration: Determination of Chloride by the Mohr Method. *Methods* **2017**, *4*, 1–6.
46. Sbei, H.; Shehzad, T.; Harrabi, M.; Okuno, K. Salinity Tolerance Evaluation of Asian Barley Accessions (*Hordeum vulgare* L.) at the Early Vegetative Stage. *J. Arid L. Stud.* **2014**, *24*, 183–186.
47. West, S.G.; Finch, J.F.; Curran, P.J. Structural Equation Models with Nonnormal Variables: Problems and Remedies. In *Structural Equation Modeling: Concepts, Issues, and Applications*; Hoyle, R.H., Ed.; Sage Publications, Inc.: New York, NY, USA, 1995; pp. 56–75.
48. Brown, M.B.; Forsythe, A.B. Robust Tests for the Equality of Variances. *J. Am. Stat. Assoc.* **1974**, *69*, 364–367. [CrossRef]
49. Garson, G.D. *Testing Statistical Assumptions*; Statistical Associates Publishing: Asheboro, NC, USA, 2012.
50. Tabachnick, B.; Fidell, L. IBM SPSS Statistics for Windows. In *Using Multivar. Stat. Pearson, Boston. IBM Corp. Released 2020. IBM SPSS Stat. Wind. 27.0.*; IBM Corp.: Armonk, NY, USA, 2013.

51. IBM Corporation. *Released 2020. IBM SPSS Statistics for Windows, Version 27.0.*; IBM Corp.: Armonk, NY, USA, 2020.
52. Olson, C.L. On Choosing a Test Statistic in Multivariate Analysis of Variance. *Psychol. Bull.* **1976**, *83*, 579. [CrossRef]
53. Barbara, G.T.; Linda, S.F. *Using Multivariate Statistics*, 7th ed.; Pearson: New York, NY, USA, 2013; ISBN-13: 9780135350904; ISBN-10: 0134790545.
54. Manchester, L.C.; Coto-Montes, A.; Boga, J.A.; Andersen, L.P.; Zhou, Z.; Galano, A.; Vriend, J.; Tan, D.X.; Reiter, R.J. Melatonin: An ancient molecule that makes oxygen metabolically tolerable. *J. Pineal Res.* **2015**, *59*, 403–419. [CrossRef] [PubMed]
55. Poorter, H.; Niinemets, U.; Poorter, L.; Wright, I.; Villar, R. Causes and Consequences of Variation in Leaf Mass per Area (LMA): A Meta-Analysis. *New Phytol.* **2010**, *182*, 565–588. [CrossRef] [PubMed]
56. Ruiz, M.B.A.J.H.; Arnao, M.B.; Hernández-Ruiz, J. Melatonin in Flowering, Fruit Set and Fruit Ripening. *Plant Reprod.* **2020**, *33*, 77–87. [CrossRef]
57. Alyammahi, O.; Gururani, M.A. Chlorophyll-a Fluorescence Analysis Reveals Differential Response of Photosynthetic Machinery in Melatonin-Treated Oat Plants Exposed to Osmotic Stress. *Agronomy* **2020**, *10*, 1520. [CrossRef]
58. Sadak, M.S.; Abdalla, A.M.; Abd Elhamid, E.M.; Ezzo, M.I. Role of Melatonin in Improving Growth, Yield Quantity and Quality of *Moringa oleifera* L. Plant under Drought Stress. *Bull. Natl. Res. Cent.* **2020**, *44*, 18. [CrossRef]
59. Han, Q.; Huang, B.; Ding, C.; Zhang, Z.; Chen, Y.; Hu, C.; Zhou, L.; Huang, Y.; Liao, J.; Yuan, S.; et al. Osystem II in Cold-Stressed Rice Seedlings. *Front. Plant Sci.* **2017**, *8*, 785. [CrossRef]
60. Kamiab, F. Exogenous Melatonin Mitigates the Salinity Damages and Improves the Growth of Pistachio under Salinity Stress. *J. Plant Nutr.* **2020**, *43*, 1468–1484. [CrossRef]
61. Imran, M.; Latif Khan, A.; Shahzad, R.; Aaqil Khan, M.; Bilal, S.; Khan, A.; Kang, S.; Lee, I. Exogenous Melatonin Induces Drought Stress Tolerance by Promoting Plant Growth and Antioxidant Defence System of Soybean Plants. *AoB Plants* **2021**, *13*, plab026. [CrossRef]
62. Zhang, N.; Sun, Q.; Zhang, H.; Cao, Y.; Weeda, S.; Ren, S.; Guo, Y.D. Roles of Melatonin in Abiotic Stress Resistance in Plants. *J. Exp. Bot.* **2015**, *66*, 647–656. [CrossRef]
63. Wei, W.; Li, Q.T.; Chu, Y.N.; Reiter, R.J.; Yu, X.M.; Zhu, D.H.; Zhang, W.K.; Ma, B.; Lin, Q.; Zhang, J.S.; et al. Melatonin Enhances Plant Growth and Abiotic Stress Tolerance in Soybean Plants. *J. Exp. Bot.* **2015**, *66*, 695–707. [CrossRef]
64. Sheikhalipour, M.; Mohammadi, S.A.; Esmaielpour, B.; Zareei, E.; Kulak, M.; Ali, S.; Nouraein, M.; Bahrami, M.K.; Gohari, G. Exogenous Melatonin Increases Salt Tolerance in Bitter Melon by Regulating Ionic Balance, Antioxidant System and Secondary Metabolism-Related Genes. *BMC Plant Biol.* **2022**, *33*, 380. [CrossRef]
65. Jiang, D.; Lu, B.; Liu, L.; Duan, W.; Meng, Y.; Li, J.; Zhang, K.; Sun, H.; Zhang, Y.; Dong, H.; et al. Exogenous Melatonin Improves the Salt Tolerance of Cotton by Removing Active Oxygen and Protecting Photosynthetic Organs. *BMC Plant Biol.* **2021**, *21*, 331. [CrossRef]
66. Bajwa, V.S.; Shukla, M.R.; Sherif, S.M.; Murch, S.J.; Saxena, P.K. Role of Melatonin in Alleviating Cold Stress in *Arabidopsis thaliana*. *J. Pineal Res.* **2014**, *56*, 238–245. [CrossRef]
67. Hernández-Ruiz, J.; Cano, A.; Arnao, M.B. Melatonin: A Growth-Stimulating Compound Present in Lupin Tissues. *Planta* **2004**, *220*, 140–144. [CrossRef]
68. Sharp, R.G.; Else, M.A.; Cameron, R.W.; Davies, W.J. Water Deficits Promote Flowering in Rhododendron via Regulation of Pre and Post Initiation Development. *Sci. Hortic.* **2009**, *120*, 511–517. [CrossRef]
69. Wada, K.C.; Takeno, K. Stress-Induced Flowering. *Plant Signal. Behav.* **2010**, *5*, 944–947. [CrossRef]
70. Kolář, J.; Johnson, C.H.; Macháčková, I. Exogenously Applied Melatonin (N -Acetyl-5-Methoxytryptamine) Affects Flowering of the Short-Day Plant *Chenopodium rubrum*. *Physiol. Plant* **2003**, *118*, 605–612. [CrossRef]
71. Shi, H.; Wei, Y.; Wang, Q.; Reiter, R.; He, C. Melatonin Mediates the Stabilization of DELLA Proteins to Repress the Floral Transition in *Arabidopsis*. *J. Pineal Res.* **2016**, *60*, 373–379. [CrossRef] [PubMed]
72. Zhang, Z.; Hu, Q.; Liu, Y.; Cheng, P.; Cheng, H.; Liu, W.; Xing, X.; Guan, Z.; Fang, W.; Chen, S.; et al. Strigolactone Represses the Synthesis of Melatonin, Thereby Inducing Floral Transition in *Arabidopsis thaliana* in an FLC-Dependent Manner. *J. Pineal Res.* **2019**, *67*, e12582. [CrossRef] [PubMed]
73. Yuan, F.; Lyu, M.J.A.; Leng, B.Y.; Zhu, X.G.; Wang, B.S. The Transcriptome of NaCl-Treated Limonium Bicolor, Leaves Reveals the Genes Controlling Salt Secretion of Salt Gland. *Plant. Mol. Biol.* **2016**, *91*, 241–256. [CrossRef] [PubMed]
74. Zhu, J.K. Regulation of Ion Homeostasis under Salt Stress. *Curr. Opin. Plant Boil.* **2003**, *6*, 441–445. [CrossRef] [PubMed]
75. Yuan, F.; Lyu, M.J.; Leng, B.Y.; Zheng, G.Y.; Feng, Z.T.; Li, P.H.; Zhu, X.G.; Wang, B.S. Comparative Transcriptome Analysis of Developmental Stages of the Limonium Bicolor Leaf Generates Insights into Salt Gland Differentiation. *Plant Cell Environ.* **2015**, *38*, 1637–1657. [CrossRef]
76. Hand, M.J.; Taffouo, V.D.; Nouck, A.E.; Nyemene, K.P.J.; Tonfack, B.; Meguekam, T.L.; Youmbi, E. Effects of Salt Stress on Plant Growth, Nutrient Partitioning, Chlorophyll Content, Leaf Relative Water Content, Accumulation of Osmolytes and Antioxidant Compounds in Pepper (*Capsicum annuum* L.) Cultivars. *Not. Bot. Hortic. Agrobot. Cluj-Napoca* **2017**, *45*, 481–490. [CrossRef]
77. Wei, L.; Zhao, H.; Wang, B.; Wu, X.; Lan, R.; Huang, X.; Chen, B.; Chen, G.; Jiang, C.; Wang, J.; et al. Exogenous Melatonin Improves the Growth of Rice Seedlings by Regulating Redox Balance and Ion Homeostasis under Salt Stress. *J. Plant Growth Regul.* **2022**, *41*, 2108–2121. [CrossRef]

78. Li, C.; Wang, P.; Wei, Z.; Liang, D.; Liu, C.; Yin, L.; Jia, D.; Fu, M.; Ma, F. The Mitigation Effects of Exogenous Melatonin on Salinity-Induced Stress in *Malus hupehensis*. *J. Pineal Res.* **2012**, *53*, 298–306. [CrossRef]
79. Farouk, S.; AL-Huqail, A.A. Sustainable Biochar and/or Melatonin Improve Salinity Tolerance in Borage Plants by Modulating Osmotic Adjustment, Antioxidants, and Ion Homeostasis. *Plants* **2022**, *11*, 765. [CrossRef]
80. Hu, W.; Zhang, J.; Wu, Z.; Loka, D.A.; Zhao, W.; Chen, B.; Wang, Y.; Meng, Y.; Zhou, Z.; Gao, L. Effects of Single and Combined Exogenous Application of Abscisic Acid and Melatonin on Cotton Carbohydrate Metabolism and Yield under Drought Stress. *Ind. Crops Prod.* **2022**, *176*, 114302. [CrossRef]
81. Liu, J.; Wang, W.; Wang, L.; Sun, Y. Exogenous Melatonin Improves Seedling Health Index and Drought Tolerance in Tomato. *Plant Growth Regul.* **2015**, *77*, 317–326. [CrossRef]
82. Liang, C.; Zheng, G.; Li, W.; Wang, Y.; Hu, B.; Wang, H.; Wu, H.; Qian, Y.; Zhu, X.-G.; Tan, D.-X.; et al. Melatonin Delays Leaf Senescence and Enhances Salt Stress Tolerance in Rice. *J. Pineal Res.* **2015**, *59*, 91–101. [CrossRef]
83. Muller, O.; Cohu, C.M.; Stewart, J.J.; Protheroe, J.A.; Demmig-Adams, B.; Adams, W.W. Association between Photosynthesis and Contrasting Features of Minor Veins in Leaves of Summer Annuals Loading Phloem via Symplastic versus Apoplastic Routes. *Physiol Plant.* **2014**, *152*, 174–183. [CrossRef]
84. Gohari, G.; Farhadi, H.; Panahirad, S.; Zareei, E.; Labib, P.; Jafari, H.; Mahdavinia, G.; Hassanpouraghdam, M.B.; Ioannou, A.; Kulak, M.; et al. Mitigation of Salinity Impact in Spearmint Plants through the Application of Engineered Chitosan-Melatonin Nanoparticles. *Int. J. Biol. Macromol.* **2023**, *224*, 893–907. [CrossRef]
85. Liang, D.; Ni, Z.; Xia, H.; Xie, Y.; Lv, X.; Wang, J.; Lin, L.; Deng, Q.; Luo, X. Exogenous Melatonin Promotes Biomass Accumulation and Photosynthesis of Kiwifruit Seedlings under Drought Stress. *Sci. Hortic.* **2019**, *246*, 34–43. [CrossRef]
86. Ashraf, M.; Ali, Q. Relative Membrane Permeability and Activities of Some Antioxidant Enzymes as the Key Determinants of Salt Tolerance in Canola (*Brassica napus* L.). *Environ. Exp. Bot.* **2008**, *63*, 266–273. [CrossRef]
87. Parkash, V.; Singh, S. A Review on Potential Plant-Basedwater Stress Indicators for Vegetable Crops. *Sustainability* **2020**, *12*, 3945. [CrossRef]
88. Khan, T.A.; Saleem, M.; Fariduddin, Q. Melatonin Influences Stomatal Behavior, Root Morphology, Cell Viability, Photosynthetic Responses, Fruit Yield, and Fruit Quality of Tomato Plants Exposed to Salt Stress. *J. Plant Growth Regul.* **2022**, 1–25. [CrossRef]
89. Zhang, P.; Liu, L.; Wang, X.; Wang, Z.; Zhang, H.; Chen, J.; Liu, X.; Wang, Y.; Li, C. Beneficial Effects of Exogenous Melatonin on Overcoming Salt Stress in Sugar Beets (*Beta vulgaris* L.). *Plants* **2021**, *10*, 886. [CrossRef]
90. Saneoka, H.; Moghaieb, R.E.; Premachandra, G.S.; Fujita, K. Nitrogen Nutrition and Water Stress Effects on Cell Membrane Stability and Leaf Water Relations in *Agrostis palustris* Huds. *Environ. Exp. Bot.* **2004**, *52*, 131–138. [CrossRef]
91. Catala, A. The Ability of Melatonin to Counteract Lipid Peroxidation in Biological Membranes. *Curr. Mol. Med.* **2007**, *7*, 638–649. [CrossRef] [PubMed]
92. Huang, B.; Chen, Y.-E.; Zhao, Y.-Q.; Ding, C.-B.; Liao, J.-Q.; Hu, C.; Zhou, L.-J.; Zhang, Z.-W.; Yuan, S.; Yuan, M. Exogenous Melatonin Alleviates Oxidative Damages and Protects Photosystem II in Maize Seedlings under Drought Stress. *Front. Plant Sci.* **2019**, *10*, 677. [CrossRef] [PubMed]
93. Zhu, J.K. Plant Salt Tolerance. *Trends Plant Sci.* **2001**, *6*, 66–71. [CrossRef] [PubMed]
94. Ferchichi, S.; Hessini, K.; Dell'Aversana, E.; D'Amelia, L.; Woodrow, P.; Ciarmiello, L.; Fuggi, A.; Carillo, P. Hordeum Vulgare and Hordeum Maritimum Respond to Extended Salinity Stress Displaying Different Temporal Accumulation Pattern of Metabolites. *Funct. Plant Biol.* **2018**, *45*, 1096–1109. [CrossRef]
95. Qiao, Y.; Ren, J.; Yin, L.; Liu, Y.; Deng, X.; Liu, P.; Wang, S. Exogenous Melatonin Alleviates PEG-Induced Short-Term Water Deficiency in Maize by Increasing Hydraulic Conductance. *BMC Plant Biol.* **2020**, *20*, 218. [CrossRef]
96. Ahanger, M.A.; Aziz, U.; Alsahli, A.A.; Alyemeni, M.N.; Ahmad, P. Influence of Exogenous Salicylic Acid and Nitric Oxide on Growth, Photosynthesis, and Ascorbate-Glutathione Cycle in Salt Stressed *Vigna angularis*. *Biomolecules* **2020**, *10*, 42. [CrossRef]
97. Liang, L.; Li, D.; Chen, Y.; Cheng, J.; Zhao, G.; Fahima, T.; Yan, J. Selenium Mitigates Salt-Induced Oxidative Stress in Durum Wheat (*Triticum durum* Desf.) Seedlings by Modulating Chlorophyll Fluorescence, Osmolyte Accumulation, and Antioxidant System. *3 Biotech* **2020**, *10*, 368. [CrossRef]
98. Hu, Z.; Fan, J.; Xie, Y.; Amombo, E.; Liu, A.; Gitau, M.M.; Khaldun, A.B.M.; Chen, L.; Fu, J. Comparative Photosynthetic and Metabolic Analyses Reveal Mechanism of Improved Cold Stress Tolerance in Bermudagrass by Exogenous Melatonin. *Plant Physiol. Biochem.* **2016**, *100*, 94–104. [CrossRef]
99. Shi, H.; Chen, Y.; Tan, D.-X.; Reiter, R.J.; Chan, Z.; He, C. Melatonin Induces Nitric Oxide and the Potential Mechanisms Relate to Innate Immunity against Bacterial Pathogen Infection in *Arabidopsis*. *J. Pineal Res.* **2015**, *59*, 102–108. [CrossRef]
100. Gao, W.; Feng, Z.; Bai, Q.; HeWang, J.Y. Melatonin-Mediated Regulation of Growth and Antioxidant Capacity in Salt-Tolerant Naked Oat under Salt Stress. *Int. J. Mol. Sci.* **2019**, *20*, 1176. [CrossRef]
101. Zeng, L.; CaiJS, L.; Lu, G.; Li, C.; Fu, G.; Zhang, X.; Ma, H.; Liu, Q.; Zou, X.; Cheng, Y. Exogenous Application of a Low Concentration of Melatonin Enhances Salt Tolerance in Rapeseed (*Brassica napus* L.) Seedlings. *J. Integr. Agric.* **2018**, *17*, 328–335. [CrossRef]

102. Mann, A.; Kaur, G.; Kumar, A.; Sanwal, S.K.; Singh, J.; Sharma, P.C. Physiological Response of Chickpea (*Cicer arietinum* L.) at Early Seedling Stage under Salt Stress Conditions. *Legum. Res.* **2019**, *42*, 625–632. [CrossRef]
103. Roshdy, A.E.-D.; Alebidi, A.; Almutairi, K.; Al-Obeed, R.; Elsabagh, A. The Effect of Salicylic Acid on the Performances of Salt Stressed Strawberry Plants, Enzymes Activity, and Salt Tolerance Index. *Agronomy* **2021**, *11*, 775. [CrossRef]

horticulturae

MDPI

Article

The Effective Role of Nano-Silicon Application in Improving the Productivity and Quality of Grafted Tomato Grown under Salinity Stress

Eman G. Sayed [1,*], Abdel Wahab M. Mahmoud [2], Mohamed M. El-Mogy [1], Mahmoud A. A. Ali [3], Mahmoud A. M. Fahmy [4] and Ghada A. Tawfic [1]

1 Department of Vegetable Crops, Faculty of Agriculture, Cairo University, Giza 12613, Egypt; elmogy@agr.cu.edu.eg (M.M.E.-M.); ghadatawfic2@yahoo.com (G.A.T.)
2 Department of Agricultural Botany, Plant Physiology Division, Faculty of Agriculture, Cairo University, Giza 12613, Egypt; mohamed.mahmoud@agr.cu.edu.eg
3 Department of Horticulture, Faculty of Agriculture, Ain Shams University, Cairo 11566, Egypt; mahmoud_adel489@agr.asu.edu.eg
4 Horticulture Department, Faculty of Agriculture, Beni-Suef University, Beni-Suef 62517, Egypt; mahmoud1351985@yahoo.com
* Correspondence: emaan.gamal2020@yahoo.com; Tel.: +20-100-636-2604

Abstract: This study aims to determine the influence of grafting and nano-silicon fertilizer on the growth and production of tomatoes (*Solanumlycopersicum* L.) under salinity conditions. A commercial tomato hybrid (cv. Strain B) was used as a scion and two tomato phenotypes were used as rootstocks: *S. pimpinellifolium* and Edkawy. The rootstock effect was evaluated by growing plants at two NaCl concentrations plus the control (0, 4000, and 8000 ppm NaCl). Nano-silicon foliar application (0.5 ppm) after 20, 28, and 36 days from transplanting was also used to mitigate salinity stress. Antioxidants, hormones, and proline were evaluated for a better understanding of the physiological changes induced by salinity and grafting. The results showed that grafting either on *S. pimpinellifolium* or Edkawy combined with nano-silicon application enhanced shoot and root growth, fruit yield, and fruit quality. The Edkawy rootstock was more effective than the *S. pimpinellifolium* rootstock in terms of counteracting the negative effect of salinity. Higher levels of mineral contents, GA3, ABA, and proline were detected in shoots that were subjected to grafting and nano-silicon application compared to the control treatment. This study indicates that grafting and nano-silicon application hold potential as alternative techniques to mitigate salt stress in commercial tomato cultivars.

Keywords: *Solanum lycopersicum* L.; salinity; quality; nano-silicon; plant hormones

Citation: Sayed, E.G.; Mahmoud, A.W.M.; El-Mogy, M.M.; Ali, M.A.A.; Fahmy, M.A.M.; Tawfic, G.A. The Effective Role of Nano-Silicon Application in Improving the Productivity and Quality of Grafted Tomato Grown under Salinity Stress. *Horticulturae* **2022**, *8*, 293. https://doi.org/10.3390/horticulturae8040293

Academic Editors: Agnieszka Hanaka, Małgorzata Majewska and Barbara Hawrylak-Nowak

Received: 14 February 2022
Accepted: 28 March 2022
Published: 30 March 2022

1. Introduction

The tomato plant (*Solanum lycopersicum* L.) is nowadays a very popular crop as a means of preventing many human diseases [1,2] and its fruit is well known for carotenoids like lycopene that seem to be active in cancer prevention, cardiovascular risk, and cellular aging [3]. Tomato is one of the most common and widely consumed vegetable crops in the world, and in Egypt a high-quality yield is a must for commercial success. Salinity is one of the most important abiotic stress factors limiting crop growth and productivity worldwide [4]. High salinity leads to a decrease in plant growth, biomass, yield, photosynthesis, and water use efficiency [5], as salinity stress negatively impacts the morphological, biochemical, and physiological processes of plants [6,7]. Salt stress reduces the growth and yield of grafted and non-grafted tomato plants, but the appropriate combination of scion and rootstocks can help mitigate the negative effects of salinity. An effective and sustainable method of improving the performance of commercial cultivars that are susceptible to abiotic stresses is to use resistant genotypes as rootstocks [8]. The local Egyptian cultivar Edkawy exhibits greater salinity tolerance by exhibiting greater growth stability as salinity increases [9]. In addition, *S. pimpinellifolium* is identified as an important source of genes that can help mature tomato plants cope with stress [10]. Conversely, it has been demonstrated that salinity eustress may contribute to augmenting organoleptic components of fruit quality, such as soluble carbohydrates and health-promoting phytochemicals [6,11]. Moreover, numerous experiments have indicated the beneficial role of silicon (Si) against different stresses, including salinity in tomatoes [12]. Exogenous Si spraying has been a recent eco-friendly strategy to improving plant salinity stress response [13] as it acts by increasing plant biomass [14] by reducing Na^{+} and Cl^{-} ion uptake into plants [15]. Silicon is found in nature as crystalline, amorphous, or weakly crystalline complex silicate minerals [16,17]. It also aids plant growth in a variety of ways by improving antioxidant activity, mineral absorption, organic acid anion and phenolic compound exudation, photosynthetic rates, the accumulation of suitable solutes, hydration status, and the control of plant growth regulators [18–28], as well as considerably lowering the deleterious impact of salinity on chlorophyll levels [29]. Nanotechnologies and plant biotechnology have attracted a lot of attention in agriculture in recent years [30,31] on account of their potential to increase plant productivity, improve plant tolerance to environmental stress conditions, improve nutrient use efficiency, and decrease hazardous environmental consequences [31–33]. Crop improvement experiments have been carried out with silicon nano-particles (n-Si) and many studies have found that increasing n-Si concentrations improves plant development and tolerance to hydroponic conditions [34]. Tomato plants treated with 1 mM n-Si displayed increased germination rates and seedling dry weight, thereby showing higher salinity tolerance at 50 mM NaCl compared with the controls [35,36]. A similar effect was shown in lentil seeds [37]. Nano-silicon sprays increased the amount of chlorophyll in stressed plants and enhanced physiological parameters, such as transpiration rate, photosynthetic rate, stomatal conductance, and photochemical efficiency, in *Indocalamus barbatus* [38]. Moreover, (n-Si) enhanced the nutritional quality of potatoes that had been exposed to salt [39], and n-Si spray significantly improved plant height, stem diameter, ground cover, canopy spread, and other growth characteristics in safflower [40]. Based on this body of evidence, the purpose of this study is to assess whether grafting and foliar application of n-Si have a positive effect in terms of improving the yield and quality of a tomato hybrid by evaluating plant growth characteristics, yield, and fruit quality traits.

2. Materials and Methods

2.1. Plant Material

In this study, the hybrid cv. Strain B tomato obtained from Ferry-Morse seed CO USA was employed as a scion. The seeds of a tomato rootstock with documented salinity tolerance features, *S. pimpinellifolium* L. (line AusTRCF31212), were obtained from the International Tomato Genetic Resource Center in the United States. In addition, Edkawy (obtained from Haraz company, Cairo, Egypt) was used in this study as a local cultivar with recognized tolerance qualities.

2.2. Grafting of Test Plants

Seeds of the scion and rootstock were sown on 21 March 2020 and 17 March 2021 in seedling trays filled with peat moss and vermiculite (1:1 *v:v*). Scion seedlings were grafted on rootstock manually via tongue grafting on 20 and 15 April in 2020 and 2021, respectively. Both rootstock and scion grafts were sliced obliquely at a 40° angle to the perpendicular axis at a sufficient depth to allow for greater vascular bundle overlap. Clips were applied at the grafting site to secure the graft in place, and the grafted plants were maintained under a clear polyethylene tunnel cover for 5–7 days under 90–95% RH and 50% shading at a temperature of 27.9 °C. The polyethylene cover was gradually opened to acclimatize the grafted plants to the greenhouse environment. Three sets of grafted plants were produced using the commercial tomato hybrid (cv. Strain B) as a scion grafted on one of two tomato rootstocks, namely, Edkawy and *S. pimpinellifolium*, respectively, or on the same hybrid (cv. Strain B) to serve as a basis for the control treatment.

2.3. *Greenhouse Experiment*

2.3.1. Growth Conditions

Transplantation of grafted plants was carried out in the greenhouse at the Eastern Farm of the Faculty of Agriculture, Cairo University, Giza, Egypt on 28 April in 2020 and on 23 April in 2021. The three groups of successfully grafted tomato plants were transplanted in pots 60 cm^2 in diameter, each one filled with a 1:1:1 mixture of peat moss, vermiculite, and perlite and containing two seedlings, and were each subjected to (i) three salinity levels—0 ppm, 4000 ppm, and 8000 ppm of NaCl solution—and (ii) treatment with or without nano-silicon (n-Si) foliar application (0.5 ppm) three times after 20 and 28 and 36 days from transplanting. The reported design resulted in 18 treatment combinations. Physiological and biochemical investigations were performed 60 days after transplantation.

2.3.2. Salinity Treatments

Thirty days after transplantation, the NaCl solution was applied to the levels of saline treatments, which contained 0, 4000, and 8000 ppm NaCl (Technogene chemical company, Dokki, Egypt). The saline treatments were continued until the experiment was completed (190 days after seedling transplantation). A completely randomized design was used for the treatments and each treatment was replicated six times.

2.3.3. Preparation of Nano-Silicon

Nano-silicon in the form of silicon tetrachloride ($SiCl_4$) was purchased from Sigma Chemical Co. (St. Louis, MO, USA). The nano-silicon (Figure 1) was synthesized using the method described by Zhu and Gong [41] and published elsewhere [42]. The morphologies and sizes of the nano-particles (n-Si) were investigated using a JEOL 1010 transmission electron microscope at 80 kV (JEOL, Tokyo, Japan). One drop of the nano-particle solution was spread onto a carbon-coated copper grid for transmission electron microscopy (TEM) analysis. TEM imaging was carried out in the TEM lab of the Faculty of Agriculture, Cairo University Research Park (FA-CURP) to determine the nano-size of the silicon particles.

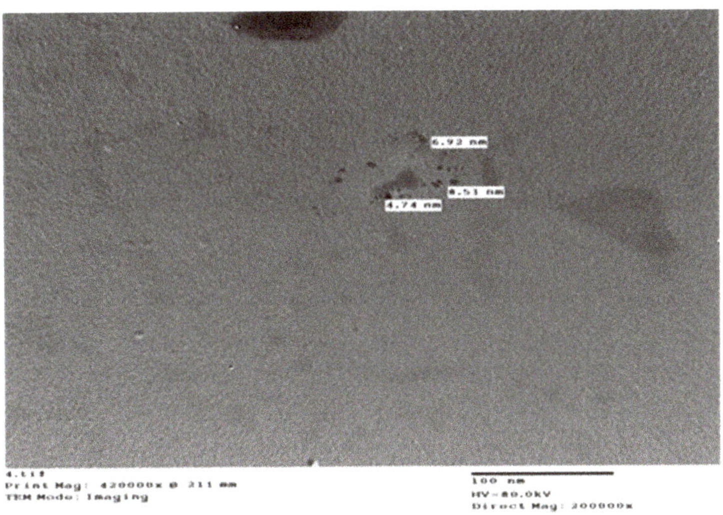

Figure 1. Scanning electron microscopy (SEM) for the prepared silicon nano-particles. Silicon nano-particle size from 4.75 to 6.92 nm.

2.3.4. Plant Growth and Yield

Plant growth was measured as a function of shoot length, plant fresh weight, leaf number, and leaf dry weight. Leaf chlorophyll content was recorded using a SPAD Meter.

The fruits of each plot were picked during harvesting, and the number of fruits per plant, fruit fresh weight, and total yield per plant were recorded.

2.3.5. Characteristics of Fruit Quality

At the maturity stage, ten mature fruits per experimental unit were chosen to measure the following data. Total soluble solids percentage (TSS%) was measured using a Zeiss laboratory refractometer, and the ascorbic acid (AA) content of ripe fruits was evaluated via the titration method using 2,6-dichlorophenol indophenol according to the Society for Analytical Chemistry's Official Method 967.21. Five fruits were selected in each replicate to determine their firmness. The firmness of ripe fruits was determined using a Force Gauge Mode M4-200 (ELECTROMATIC Equipment Co., Inc., Cedarhurst, NY, USA) with a 1 mm diameter flat probe. The firmness readings of tomato fruits were taken at two opposing points of the equatorial region and expressed in newtons.

2.3.6. Physiological Parameters

To evaluate photosynthesis and leaf stomatal conductance, analysis was conducted via an infrared gas analyzer, the LICOR 6400 Portable Photosynthesis System (IRGA, Licor Inc., Lincoln, NE, USA), on the fifth leaves of twenty plants chosen from each treatment with six replications. Measurements were taken from 9 a.m. to 2 p.m., with a light intensity of around 1300 mol m^{-2} s^{-1} and 80% RH. The leaf chamber's temperature ranged from 25.2 to 27.9 °C, and the volume gas flow rate was 400 mL/min. The CO_2 content in the air was 398 μmol mol^{-1}.

2.3.7. Activity of Antioxidant Enzymes, Gibberellic Acid, Abscisic Acid Content, and Leaf Proline

The leaf samples used to measure the activity of peroxidase (POD) were prepared according to Bates et al. [43]. The free proline content was analyzed as described by Bates et al. [43]. After homogenisation, 500 mg of freeze-dried materials in 5 mL of sulphosalicylic acid at 3% (*w/v*) was filtered using Whatman No. 1 filter paper. The filtrate was combined with acetic acid and a ninhydrin acid reagent (2% *v/v*). Then, for 45 min, the mixture was immersed in a boiling water bath at 100 °C. The tubes were then filled with 4 mL of toluene and soaked for 20 s, after which the reaction was quenched by submerging the tubes in an ice-bath. The free proline was measured spectrophotometrically against a blank reagent at 520 nm. The contents of gibberellic acid (GA3) and abscisic acid (ABA) in tomato leaves were analyzed using the method described by Fales et al. [44]. A 15 mL aliquot of a methanol/butylated hydroxytoluene (80% *v/v*) solution was mixed with the homogenized freeze-dried samples, and GA3 and ABA were extracted and quantified according to AOAC guidelines [45].

2.3.8. Mineral Composition in Tomato Plants

The tomato leaf samples were dried for two days at 75 °C in a forced air oven and then coarsely pulverized for the determination of endogenous nutrients. Sulfuric acid (5 mL) and perchloric acid were used to digest a 0.2 g dried sample. The mixture was then heated for ten minutes and 0.5 mL of perchloric acid was added while continuing to heat the mixture until clear [46,47]. The total nitrogen (N) content of the dried leaf samples was analyzed using the AOAC-recommended modified micro-Kjeldahl method described by Singh et al. [46] The phosphorus (P) content was determined colorimetrically using the chlorostannous molybdophosphoric blue color method in sulfuric acid, according to [46]. A flame photometer (CORNINGM410, Halstead, UK) was used to test the content of potassium (K), magnesium (Mg), calcium (Ca), and sodium (Na). An atomic absorption spectrophotometer with air-acetylene fuel (PyeUnicam Company, model SP-1900, Ventura, CA, USA) was used to examine the iron (Fe) and zinc (Zn) concentrations.

2.3.9. Statistical Analysis

A randomized complete block design with two factors was used to analyze all the data obtained from six replicates during two growing seasons using the computer application "MSTATC" [48]. The LSD test was used to assess changes across treatment modalities at a 5% probability level [49].

3. Results

3.1. Plant Growth Parameters

Plant growth parameters were affected by salinity levels and grafting (Table 1). The results of the present experiment indicated that among the tested water salinity concentrations, plant growth parameters decreased with increasing salinity.

Table 1. Effects of water salinity, grafting, foliar application of nano-silicon (n-Si), and their interactions on the vegetative growth of tomato plants (combined 2020 and 2021 seasons).

Salinity Levels	Rootstocks + (n-Si)	Plant Height (cm)	Number of Leaves/Plant	Shoot Fresh Weight (g)	Shoot Dry Weight (g)	Root Fresh Weight (g)	Root Dry Weight (g)
0 ppm	Strain B hybrid	161.7 ^{ij}	24.7 ^{e}	306.8 ^{d}	49.37 ^{e}	63.2 ^{e}	17.6 ^{ef}
	Edkawy	239.0 ^{ab}	35.3 ^{a}	440.7 ^{a}	62.50 ^{c}	81 ^{ab}	21.7 ^{b}
	S. pimpinellifolium	221.0 $^{b–d}$	25.7 ^{de}	320.7 ^{cd}	65.37 ^{b}	75.3 ^{c}	18.8 ^{de}
	Strain B hybrid + (n-Si)	199.3 $^{e–g}$	30.0 ^{c}	343.7 ^{c}	53.87 ^{d}	78.0 ^{bc}	20.3 $^{b–d}$
	Edkawy + (n-Si)	248.0 ^{a}	37.0 ^{a}	463.3 ^{a}	66.43 ^{b}	84.0 ^{a}	25.7 ^{a}
	S. pimpinellifolium + (n-Si)	228.3 ^{bc}	34.0 ^{ab}	368.3 ^{b}	69.77 ^{a}	84.0 ^{a}	21.2 ^{bc}
4000 ppm	Strain B hybrid	117.7 ^{l}	18 ^{gh}	195.8 $^{f–h}$	29.33 ^{i}	43.2 ^{h}	16.8 ^{f}
	Edkawy	194.7 $^{f–h}$	29 ^{cd}	314.3 ^{d}	46.70 ^{f}	56.6 ^{f}	19.3 $^{c–e}$
	S. pimpinellifolium	178.7 ^{hi}	25.3 ^{e}	252.7 ^{e}	39.37 ^{g}	56.0 ^{f}	18.2 ^{ef}
	Strain B hybrid + (n-Si)	182.0 ^{gh}	24. 7 ^{e}	198.3 ^{fg}	36.87 ^{h}	50.67 ^{g}	14.8 ^{gh}
	Edkawy + (n-Si)	213.7 $^{c–e}$	31.3 ^{bc}	320.0 ^{cd}	55.27 ^{d}	70.60 ^{d}	21.3 ^{b}
	S. pimpinellifolium + (n-Si)	196.0 $^{e–h}$	31.0 ^{bc}	256.3 ^{e}	49.43 ^{e}	63.67 ^{e}	17.6 ^{ef}
8000 ppm	Strain B hybrid	82.67 ^{m}	14.3 ^{i}	133.7 ^{i}	20.171 ^{g}	33.90 ^{i}	7.2 ^{i}
	Edkawy	158.0 ^{jk}	22.3 ^{ef}	191.7 ^{gh}	35.53 ^{h}	43.27 ^{h}	14.7 ^{h}
	S. pimpinellifolium	178.7 ^{hi}	17.7 $^{g–i}$	188.3 ^{gh}	22.87 ^{k}	49.33 ^{g}	13.2 ^{h}
	Strain B hybrid + (n-Si)	142.0 ^{k}	15.3 ^{hi}	171.7 ^{h}	26.63 ^{j}	40.67 ^{h}	13.5 ^{h}
	Edkawy + (n-Si)	207.0 $^{d–f}$	24.7 ^{e}	216.3 ^{f}	49.53 ^{e}	53.93 ^{fg}	18.5 $^{d–f}$
	S. pimpinellifolium + (n-Si)	189.3 $^{f–h}$	20.0 ^{fg}	191.7 ^{gh}	36.60 ^{h}	53.33 ^{fg}	16.7 ^{fg}
LSD value at 0.05:		18.5	3.7	24.24	2	4.6	1.97

Values followed by the same letter are not significant according to the LSD test ($p \leq 0.05\%$).

In general, it was observed that plant height was inversely proportional to salinity concentration. In the Edkawy cultivar, the n-Si treatment had a positive influence on plant height under no salinity (248.0 cm with 0 ppm) or moderate stress (213.7 cm with 4000 ppm). The leaf count per plant was slightly affected by moderate salinity and the results were comparable to those obtained in the no-salinity treatment (31.3 with 4000 ppm, 35.3 with 0 ppm) with interaction between grafting and nano-silicon (Table 1). Additionally, in the Edkawy rootstock, it was observed that the fresh weights of shoots and roots were higher when the grafting and n-Si treatments were combined. A slight reduction in fresh root weight was observed (70.60 g) under 4000 ppm salinity and n-Si treatment when compared with the control of the same rootstock (81 g) with unstressed plants without nano-silicon application; a similar trend was observed for the dry weight parameter. As can be seen, plant growth parameters decreased with increasing salinity levels, but the combination of grafting and n-Si foliar application mitigated the effects of salinity (Table 1).

3.2. Tomato Fruit Yield and Its Components

Fruit number and fruit weight as well as total fruit yield per plant were significantly reduced with increasing salt levels. The self-grafted Strain B hybrid was the most negatively affected, particularly at 8000 ppm concentration. Grafted tomato plants subjected to the n-Si treatment registered greater fruit numbers. Similarly, significantly higher fruit number, fruit weight, and total yield per plant values were recorded for the Edkawy rootstock combined with n-Si treatment when compared with the untreated control under all three salinity levels. Notably, the n-Si treated Edkawy at 4000 ppm and 8000 ppm yielded fruit numbers comparable to the untreated Edkawy plants at 0 ppm (Figure 2A).

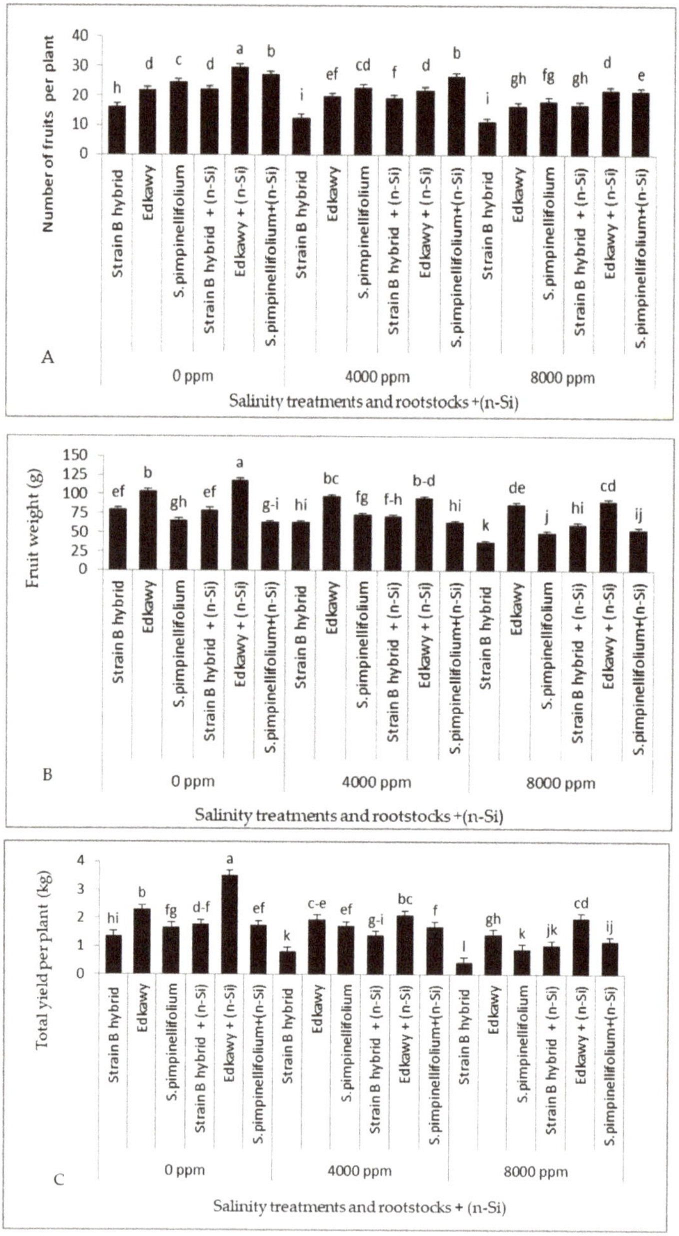

Figure 2. Effects of water salinity levels × rootstocks and foliar spray with (n-Si) interaction on (**A**) average number of fruits, (**B**) average fruit weight (g), and (**C**) average total yield per plant (kg). Vertical bars represent standard errors of the mean; in each bar, values followed by different letters differ significantly at $p = 0.05$ according to the LSD test.

The highest mean fruit weight was recorded using Edkawy combined with nano-silicon under 0 and 4000 ppm of salt concentration (Figure 2B). Plants grafted and combined with n-Si application had higher yields per plant (Figure 2C), although, at 4000 and 8000 ppm, all treatments had a higher value than self-grafted plants. Interaction between Edkawy and n-Si showed higher fruit weight at 0 ppm. Plants grafted on Edkawy and combined with nano-silicon had higher total yields per plant under non-saline water treatment.

3.3. Physiological Traits of Tomato Plants

Transpiration rate, stomatal conductance, photosynthesis, and chlorophyll reading (SPAD) were significantly influenced by salinity levels, grafting, foliar application of n-Si, and the interactions between them (Figure 3). The values obtained for the physiological traits were higher at 0 ppm salinity; however, no statistically significant differences were observed between n-Si and untreated counterparts of the same cultivars. Overall, all grafting combinations, with the exception of the self-grafted Strain B hybrid, showed similar levels of transpiration rates, stomatal conductance, photosynthesis, and SPAD readings, with only minor statistically significant differences. In this case, also, the self-grafted Strain B hybrid had the lowest transpiration rate, stomatal conductance, photosynthesis, and SPAD reading, and, notably, the negative impact of salinity was reduced for its grafted counterpart. Edkawy with nano-silicon and *S. pimpinellifolium* with nano-silicon exhibited better stomatal conductance and photosynthesis under 4000 ppm levels. SPAD showed no significant change between the control and 4000 ppm using Edkawy combined with the nano-silicon application.

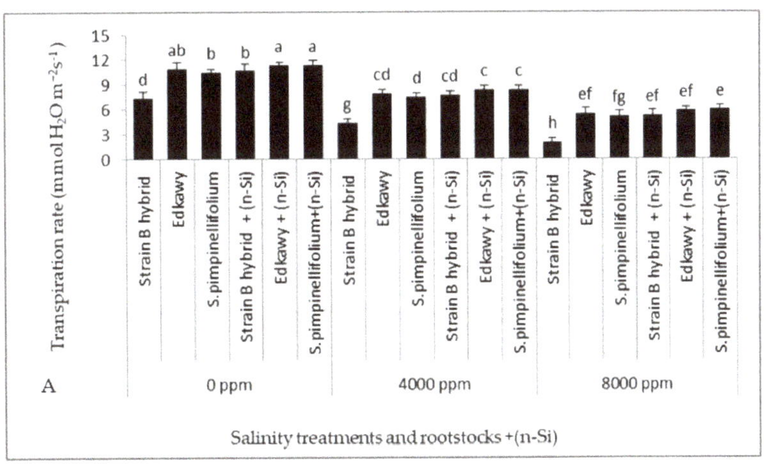

Figure 3. *Cont.*

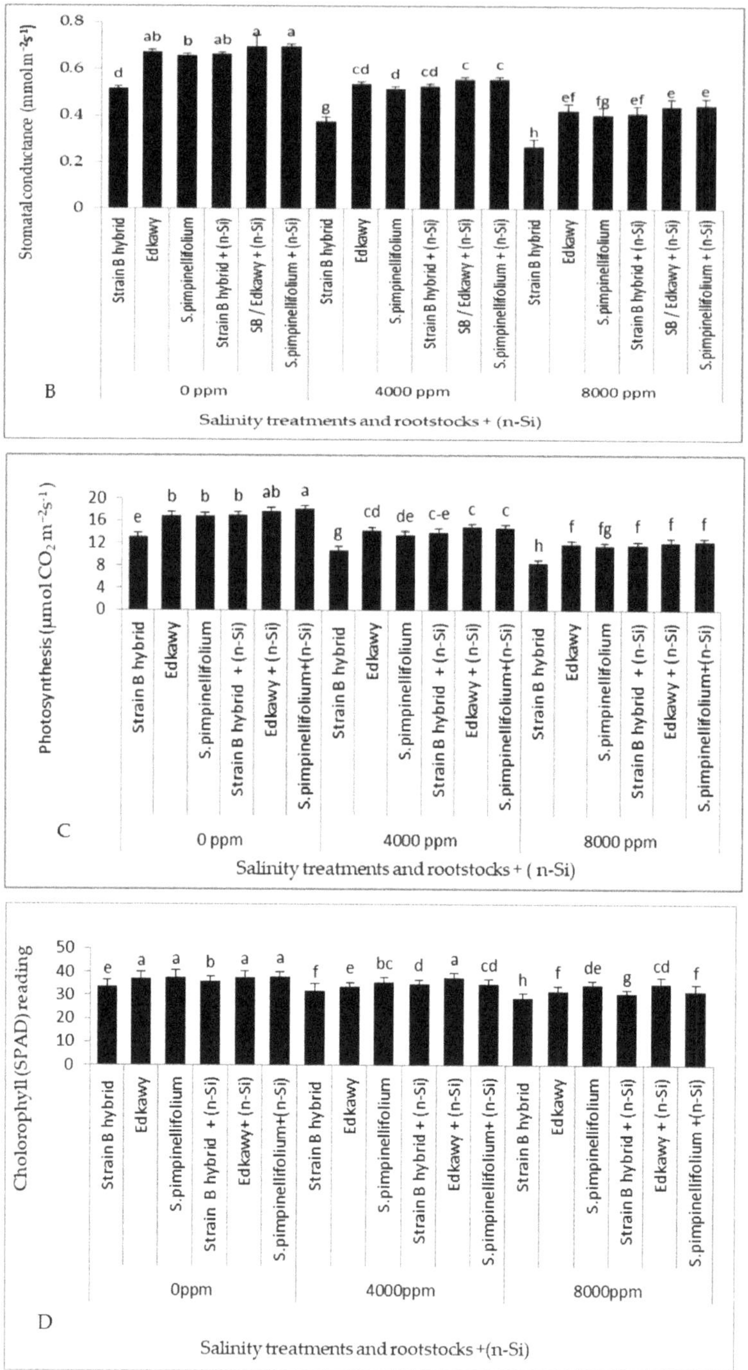

Figure 3. Effects of water salinity levels, rootstocks, foliar spray with (n-Si), and their interactions on (**A**) transpiration rates, (**B**) stomatal conductance, (**C**) photosynthesis and (**D**) SPAD readings. Vertical bars represent standard errors of the mean; in each bar, values followed by different letters differ significantly at $p = 0.05$ according to the LSD test.

3.4. Quality Parameters of Tomato Fruits

Total soluble solids (TSS), firmness, and vitamin C values for tomato fruits were significantly influenced by salinity levels, grafting, foliar application of nano-silicon, and their interactions (Table 2). TSS, firmness, and vitamin C values for tomato fruits were increased considerably as a result of increasing salinity. Tomato fruits from the Edkawy rootstock with nano-silicon and from the *S. pimpinellifolium* rootstock with n-Si showed significantly greater TSS percentages compared to other treatments. Harvested fruits from the self-grafted Strain B hybrid had lower quality parameters than fruits from tomato plants subjected to the 8000 ppm concentration treatment.

Table 2. Effect of water salinity, grafting, foliar application of nano-silicon (n-Si), and their interactions on the quality parameters of tomato fruits (combined 2020 and 2021 seasons).

Salinity Levels	Rootstocks + (n-Si)	TSS% (Brix°)	Firmness (*n*)	Vitamin C (mg/100 g FW)
0 ppm	Strain B hybrid	4.43 h	1.467 j	14.67 j
	Edkawy	5.60 fg	2.150 hi	18.27 hi
	S. pimpinellifolium	6.20 ef	2.08 i	16.33 ij
	Strain B hybrid + (n-Si)	4.33 h	1.67 j	16.33 ij
	Edkawy + (n-Si)	5.67 f	2.43 e–h	19.27 f-h
	S. pimpinellifolium + (n-Si)	6.23 ef	2.37 f–i	18.93 gh
4000 ppm	Strain Bhybrid	5.10 gh	2.22 g–i	17.67 hi
	Edkawy	6.43 ef	3.017 ab	20.60 e–g
	S. pimpinellifolium	6.37 ef	2.62 d-f	20.33 e–g
	Strain B hybrid + (n-Si)	6.07 ef	2.48 e–g	18.67 gh
	Edkawy + (n-Si)	6.53 e	3.017 ab	23.33 b-d
	S.pimpinellifolium + (n-Si)	6.43 ef	3.02 ab	21.33 de
8000 ppm	Strain B hybrid	7.87 d	2.78 b–d	21.07 ef
	Edkawy	9.53 c	3.18 a	23.67 bc
	S. pimpinellifolium	10.43 b	2.98 a–c	23.67 bc
	Strain B hybrid + (n-Si)	8.23 d	2.72 c–e	22.0 de
	Edkawy + (n-Si)	11.30 a	3.12 a	26.67 a
	S. pimpinellifolium + (n-Si)	11.47 a	3.08 a	25.33 ab
	LSD value at 0.05:	0.86	0.28	2.03

Values followed by the same letter are not significant according to the LSD test ($p \leq 0.05\%$).

3.5. Mineral Content in Tomato Shoots

Measured mineral traits in tomato shoots were significantly influenced by salinity, grafting, foliar application of nano-silicon, and their interactions (Table 3). Increasing salinity concentration significantly reduced N, P, K, Ca, and Mg contents in tomato shoots. All treatments accumulated minerals more than the self-grafted strain B hybrid. At 4000 ppm levels, plants grafted and combined with foliar nano-silicon had better shoot mineral concentrations, although the self-grafted strain B hybrid showed the lowest concertation of N, P, K, Ca, and Mg under 8000 ppm levels.

Tested Na, Fe, and Zn in tomato shoots were also considerably affected by salinity, grafting, foliar application of nano-silicon, and their interactions (Table 4). Increasing salinity concentration significantly reduced Fe and Zn contents in tomato shoots, while Na increased with increasing salinity.

Edkawy with nano-silicon exhibited the lowest content of Na and the highest contents of Fe and Zn. On the other hand, the self-grafted strain B hybrid recorded the higher content of Na. Nevertheless, the self-grafted strain B hybrid had the lowest concentrations for Fe and Zn. Salinity, grafting, and foliar application of nano-silicon interaction showed various meaningful influences. The highest Na content appeared with the self-grafted strain B hybrid at 8000 ppm salt levels. Despite this, the lowest Na content appeared using a combination of grafting with nano-silicon under control. *S. pimpinellifolium* combined with nano-silicon had a high Zn content under 4000 ppm levels compared with other treatments.

Table 3. Effect of water salinity, grafting, and foliar application of nano-silicon (n-Si) interactions on the mineral contents of tomato shoots (combined 2020 and 2021 seasons).

Salinity Levels	Rootstocks + (n-Si)	N (%)	P (%)	K (%)	Ca (%)	Mg (%)
0 ppm	Strain B hybrid	3.70 de	0.51 f	4.23 b	0.38 h	0.49 d
	Edkawy	4.29 c	0.64 c	4.72 a	1.11 c	0.58 c
	S. pimpinellifolium	3.83 d	0.57 d	3.93 c	0.99 de	0.48 d
	Strain B hybrid + (n-Si)	4.53 bc	0.66 c	4.49 ab	1.13 c	0.64 b
	Edkawy + (n-Si)	4.87 a	0.89 a	4.71 a	1.45 a	0.67 a
	S. pimpinellifolium + (n-Si)	4.617 ab	0.76 b	4.70 a	1.1 c	0.59 c
4000 ppm	Strain B hybrid	2.58 jk	0.45 g	2.42 g	0.34 h	0.33 h
	Edkawy	2.96 g–i	0.54 ef	3.21 d	1.08 cd	0.37 fg
	S. pimpinellifolium	2.860 h–j	0.55 de	2.85 ef	0.93 e	0.36 f-h
	Strain B hybrid + (n-Si)	3.260 fg	0.56 de	3.11 de	1.06 cd	0.41 e
	Edkawy + (n-Si)	3.52 d–f	0.59 d	3.31 d	1.31 b	0.42 e
	S. pimpinellifolium + (n-Si)	3.45 ef	0.55 de	3.27 d	1.03 cd	0.40 e
8000 ppm	Strain B hybrid	1.987 l	0.41 h	1.90 h	0.6 g	0.19 j
	Edkawy	2.66 i–k	0.51 f	2.40 g	0.69 fg	0.28 i
	S. pimpinellifolium	2.170 l	0.46 g	1.90 h	0.63 g	0.18 j
	Strain B hybrid + (n-Si)	2.52 k	0.44 gh	2.82 f	0.64 g	0.33 gh
	Edkawy + (n-Si)	3.44 ef	0.56 de	3.11 de	0.74 f	0.37 f
	S. pimpinellifolium+ (n-Si)	2.97 gh	0.50 f	2.31 g	0.73 f	0.29 i
LSD value at 0.05:		0.3	0.04	0.3	0.1	0.1

Values followed by the same letter are not significant according to the LSD test ($p \leq 0.05\%$).

Table 4. Effect of water salinity, grafting, and foliar application of (n-Si) interactions on minerals of tomato shoots (combined 2020 and 2021 seasons).

Salinity Levels	Rootstocks + (n-Si)	Na (%)	Fe (ppm)	Zn (ppm)
0 ppm	Strain B hybrid	0.15 f	62.28 cd	44.55 d–g
	Edkawy	0.11 fg	72.01 b	47.77 de
	S. pimpinellifolium	0.13 fg	65.18 c	45.5 d–f
	Strain B hybrid + (n-Si)	0.10 fg	71.28 b	54.62 bc
	Edkawy + (n-Si)	0.09 g	85.30 a	64.15 a
	S. pimpinellifolium + (n-Si)	0.1 fg	82.33 a	58.61 ab
4000 ppm	Strain B hybrid	1.670 b	49.66 jk	26.43 kl
	Edkawy	1.102 de	52.34 g–j	42.31 e–h
	S. pimpinellifolium	1.108 de	50.28 i–k	41.53 f–h
	Strain B hybrid + (n-Si)	1.070 e	53.47 f–i	39.71 g–i
	Edkawy + (n-Si)	1.085 e	59.4 de	48.60 d
	S. pimpinellifolium + (n-Si)	1.140 cd	55.87 fg	49.54 cd
8000 ppm	Strain B hybrid	1.740 a	42.19 m	20.91 i
	Edkawy	1.187 c	51.44 h–k	34.09 ij
	S. pimpinellifolium	1.188 c	48.50 kl	32.18 jk
	Strain B hybrid + (n-Si)	1.17 c	44.94 lm	33.69 j
	Edkawy + (n-Si)	1.11 de	57.06 ef	41.08 f–h
	S. pimpinellifolium + (n-Si)	1.142 cd	54.20 fh	36.65 h–j
LSD value at 0.05:		0.05	3.58	5.7

Values followed by the same letter are not significant according to the LSD test ($p \leq 0.05\%$).

3.6. Plant Hormones, Antioxidant Enzymes, and Proline Content in Tomato Shoots

Salinity, grafting, foliar application of nano-silicon, and their interactions result in variations and changes in the activity of the most important hormones (GA3, ABA) and POD antioxidant enzymes as well as the amino acid proline. Increasing salinity concentrations increase the content of GA3, ABA, and proline (Figure 4A,B,D). Greater activity of POD was documented at 4000 and 8000 ppm compared with non-saline water (Figure 4C). The self-grafted Strain B hybrid recorded the lowest GA3, ABA, and proline content, as well as the lowest POD activity. The higher GA3, ABA, and proline contents were registered for Edkawy combined with nano-silicon under 8000 ppm. Moreover, Edkawy combined with nano-silicon registered the greatest POD activity at 8000 ppm (Figure 4C), while the self-grafted Strain B hybrid exhibited the lowest contents of GA3, ABA, and proline with 0 ppm. Correspondingly, the self-grafted Strain B hybrid had the lowest POD activity with non-saline water. The concentration of proline was significantly increased with increasing salinity levels (Figure 4D). A higher proline content was recorded for Edkawy combined with nano-silicon under 8000 ppm.

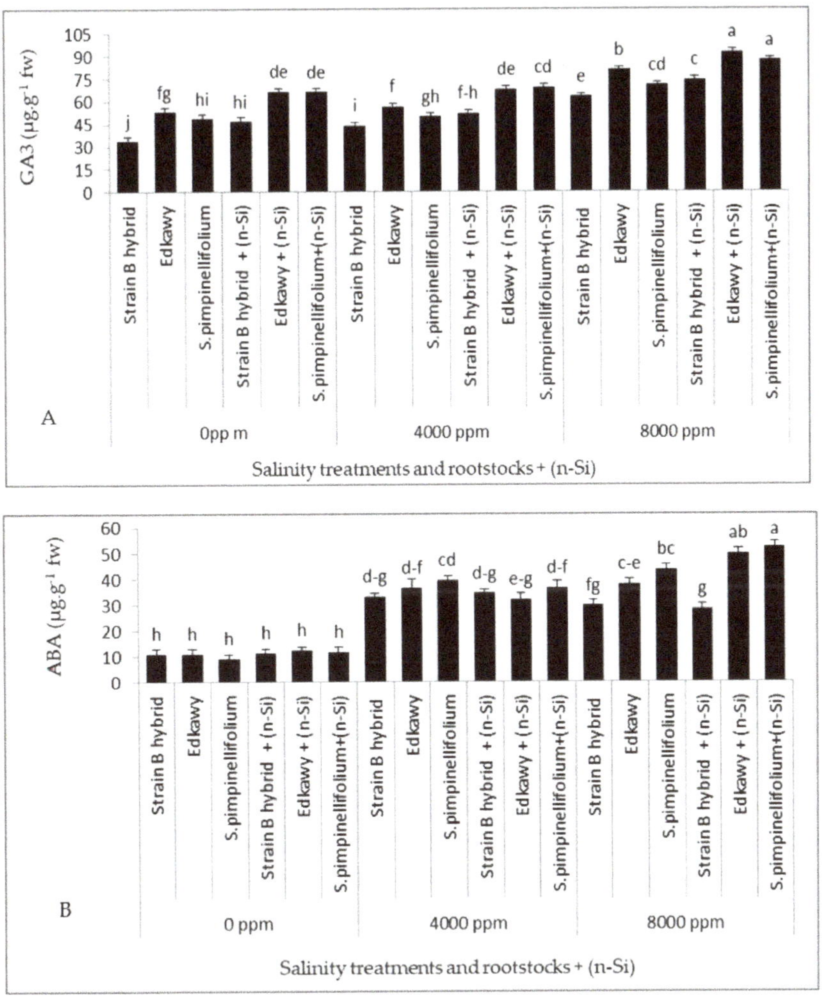

Figure 4. *Cont.*

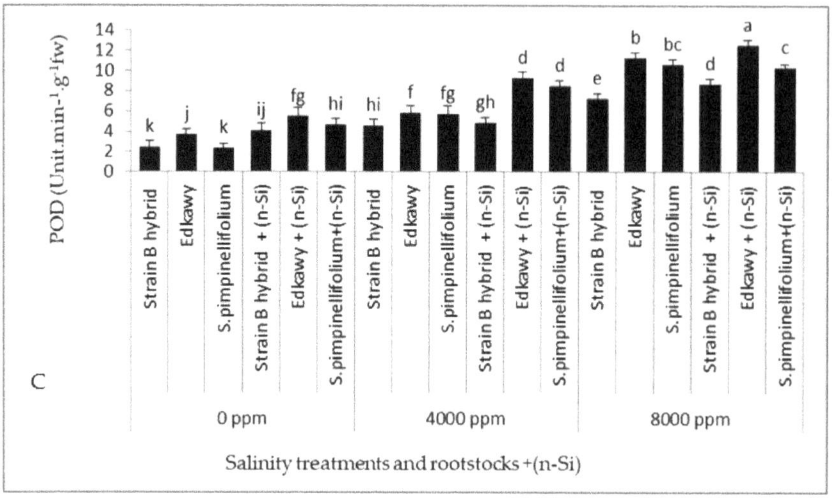

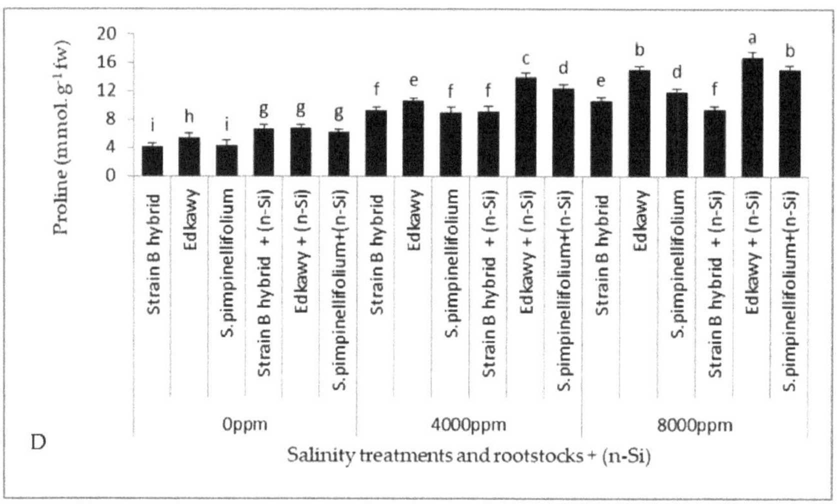

Figure 4. Effect of water salinity levels × rootstocks and foliar spray with (n-Si) and their interactions on (**A**) GA_3, (**B**) ABA, (**C**) POD, and (**D**) proline. Vertical bars represent standard errors of the means; in each bar, values followed by different letters differ significantly at $p = 0.05$ according to the LSD test.

3.7. Clustering Analysis

Cluster analysis included all growth traits (i.e., plant height, the number of leaves per plant, fresh and dry weights of shoots and roots per plant), all yield components (i.e., the number of fruits per plant, average fruit weight, and fruit yield (kg) per plant), all mineral compositions in tomato plants (total nitrogen (N), phosphorus (P), potassium (K), magnesium (Mg), calcium (Ca), sodium (Na), iron (Fe), and zinc (Zn) concentrations), physiological traits (i.e., transpiration rate, stomatal conductance, photosynthesis, and chlorophyll reading), and activity of antioxident enzymes, gibberellic acid, abscisic acid content, and leaf proline. Figure 5 illustrates a heatmap showing the relationships among the salinity levels and grafting combinations based on the tested parameters. Heatmap analysis clearly identified the overall variations among all treatments. The grafting treatment alone increased all studied traits compared with the non-salt-stressed control. However, foliar application of nano-silicon (n-Si) on the stressed grafted tomato exerted a positive role through decreasing oxidative injury by enhancing photosynthetic performance, increasing antioxidant enzyme concentrations, and significantly improving plant growth and yield.

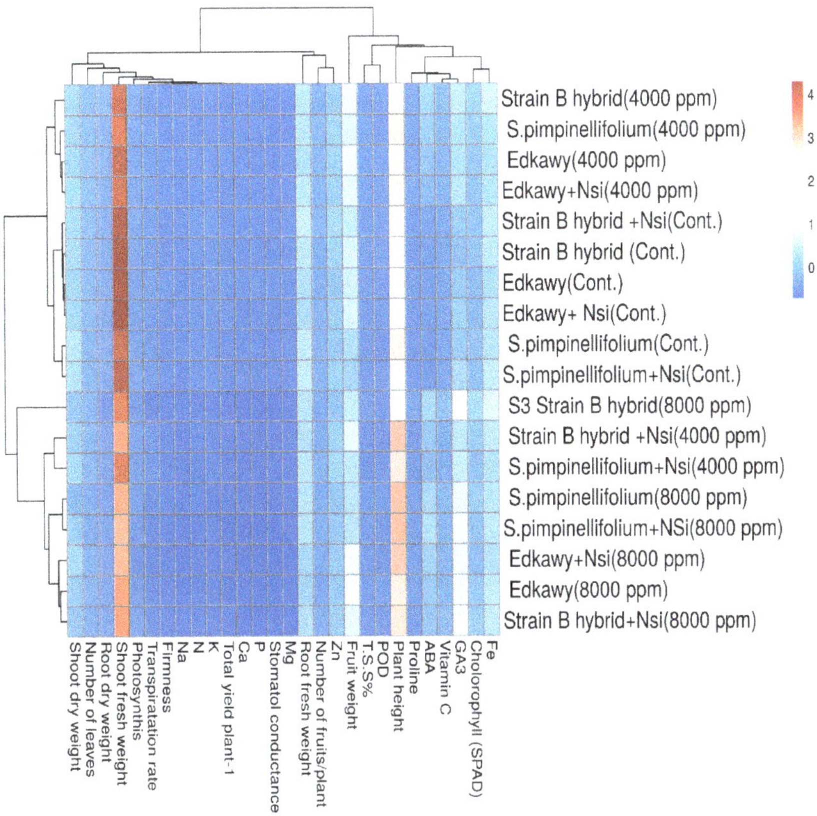

Figure 5. Heatmap of salinity stress, rootstocks, and measured parameters of tomato samples. The differences in the response variables between all studied treatments are visualized in the heatmap diagram. Columns represent the individual response variables, while rows represent the treatments. Lower numerical values are colored blue, whereas higher numerical values are colored red (see the scale at the bottom right corner of the heatmap).

4. Discussion

For the first time, this study introduced an approach to study the effect of grafting combined with foliar application of nano-silicon to improve salt tolerance in tomato. Recently, a few studies have aimed to improve salt tolerance in tomatoes utilizing grafting [50,51]. The tomato plant is a moderately salt-tolerant plant. Plant growth parameters are a perfect indicator for evaluating several abiotic stresses on plants (Table 1). The present data indicated that grafting with foliar application of nano-silicon is an effective approach for enhancing salt tolerance in tomato, as under saline conditions the plant growth parameters of most grafted combinations and nano-silicon combinations were significantly higher than their counterparts without nano-silicon application. Considerable reductions were observed in plant height, number of leaves, and shoot and root weights with higher salinity levels, confirming trends observed in other studies [50–52]. The combination of grafting and nano-silicon application could enhance the salt tolerance of tomato plants. The results also indicated that root growth was negatively affected by salt stress, which also affected shoot and plant height and lowered water and mineral uptake [50,52]. The trends observed were to be expected, since high salinity levels create water stress via osmotic imbalance at the soil–root interface and impair a plant's ability to take up water through the roots [53], thereby negatively impacting fresh and dry weights (Table 1). This could induce plasmolysis, which, in turn, might alter the plant's cellular and macromolecular metabolism and result in a slowing down or halting of plant growth [54]. Additionally, high salt concentrations might be damaging to both shoots and roots due to ionic toxicity brought about by excess Na+ ions, as they generate ionic imbalances within plants [50,54,55]. In this research, a high amount of Na^+ documented in the shoot parts of tomato plants under salinity stress could be the result of Na+ accumulation in cellular vacuoles, whose function is to regulate osmotic balance inside the cell and maintain the photosynthetic rate at nominal levels [53]. It should be stated that the rootstock genotype had a prominent role in the performance of grafted tomato crops under stress conditions [56]. It is also worth mentioning that the data presented here suggest that the scion of the hybrid Strain B exhibited higher growth when grafted onto the rootstock of cvs Edkawy and *S. pimpinellifolium* as compared to self-grafted plants. However, there is a significant contribution made by the shoot genotype to growth [57]. Since cross-grafting involves tissue wounding and reunion, similar to what is observed in wound healing or self-grafting, it may be inferred that additional stress tolerance may be imparted by specific compatibility factors. These may include specific root/shoot signals [58], including RNA transport [59]. The foliar application of nano-silicon combined with grafting had a growth-promoting effect on unstressed tomato plants. The shoot and root fresh weights of grafted plants subjected to nano-silicon treatment increased significantly compared to the untreated grafted plants; the increase in shoot fresh weight might be described by the higher number of leaves (Table 1). The exogenous nano-silicon application may enhance the growth and yield of tomato plants [30–34], as was observed in various instances. In this study, the combination of grafting with nano-silicon application was associated with significantly higher fruit number, fruit weight, and total yield per plant values under non-saline conditions, although the smallest values were obtained at high salt levels (Figure 2). Tomato fruit yield might be described by the fruit weight and the number of fruits. Reduction in fruit weight or fruit number leads to reduced tomato fruit yield [55]. Salinity negatively impacts and decreases water and nutrient availability to the plant, causing a reduced photosynthetic rate. Such water stress is understood to result in a reduction in the number of fruits, total fruit yield, and fruit fresh weight [50,52,54]. In contrast, the data demonstrate that grafting with nano-silicon treatment counteracted salinity stress and resulted in higher fruit number, fruit fresh weight, and ultimately total fruit yield in the tomato plants compared with the controls. The appreciated effect of nano-silicon might be owed to its role in increasing RNA polymerase expression and the activity of ribosomal proteins, which stimulate stress tolerance and lower the transpiration rate and oxidative stress, thereby stabilizing the photosynthetic rate, and ultimately improving fruit yield [29,31,34]. The highest mean fruit weight was recorded for Edkawy combined with nano-silicon under 0 and 4000 ppm salt concentrations (Figure 2). Furthermore, the scion of hybrid Strain B produced a higher fruit crop when grafted onto the rootstock of cvs Edkawy and *S. pimpinellifolium* as compared to ungrafted plants, confirming the results obtained by [60]. It should be mentioned that *S. pimpinellifolium* is reputed to be rich in genes involved in biotic and abiotic stress responses in comparison with other varieties of cultivated tomato [10], which could potentially explain the trends observed. Chlorophyll is the major pigment involved in photosynthesis in plants [42]. The results indicated that under salt stress, chlorophyll readings (SPAD) were decreased, while nano-silicon addition increased chlorophyll readings (SPAD) (Figure 3C). The increase in chlorophyll readings (SPAD) with the application of nano-silicon under salinity stress might be associated with improved antioxidant defense and decreased oxidative damage due to the added nano-silicon, as observed in previous studies [31,35,47,51]. Potato plants treated with nano-silicon under salinity stress showed improved growth in terms of plant height, fresh and dry weights, and leaf chlorophyll contents [31]. In this study, transpiration rate, stomatal conductance, and photosynthesis were significantly decreased by increasing salinity levels, while this effect was significantly alleviated by grafting along with foliar application of nano-silicon alleviates (Figure 3A,B). Transpiration rate and photosynthesis are correlated with stomatal conductance [31,34,35]. According to the results obtained, stomatal conductance showed decreased values under salinity stress, resulting in decreased transpiration rates and photosynthesis, similar to what was observed by [61,62]. The application of nano-silicon in combination with grafting is known to enhance stomatal conductance, transpiration rate, and photosynthesis, in the latter case by improving the activity of enzymes involved in photosynthesis [33,35,36,61,62]. Furthermore, the benefits of silicon nano-particles for plants grown under salinity stress are linked to an increased photosynthetic rate, stomatal conductance, and water use efficiency, all of which improve crop plant tolerance to salinity [31]. Silicon increases the stomatal conductance of plants regarding improved water status in plants under water deficit due to increased water uptake by the roots [62–65], and this increases transpiration rates and photosynthesis. Edkawy with nano-silicon and *S. pimpinellifolium* with nano-silicon recorded better stomatal conductance and photosynthesis under 4000 ppm levels. According to the results (Table 2) and previous reports, measures of tomato fruit quality, such as TSS, vitamin C content, and firmness, are changed due to salinity [55,61]. TSS content in tomato fruits is an essential factor in manipulating tomato quality. The data generated in this study indicated that TSS increased with increasing salt levels. Previous research found that TSS in tomato increased as salinity increased [11,61,66]. Foliar application of nano-silicon combined with increasing salinity increased the TSS of tomato plants, which might be referred to higher accumulations of metabolites and the direct modification of starch into soluble sugar [31,33]. In this experiment we found a significant increase in TSS in tomato fruits with grafting under the 8000 ppm level. Tomato fruit firmness increased (Table 2) with increasing salinity. Similar results have been reported elsewhere [11,67,68], and it has been found that nano-silicon increases firmness and fruit quality [11,26,39]. Increasing fruit firmness could be referred to the powerful bonding of silica to the cellulose structure [39,69]. Vitamin C (ascorbic acid) is an antioxidant that protects the body from free radical damage. According to the data, vitamin C increased with increasing salinity levels and also with nano-silicon treatment. Similar data were obtained for tomatoes [26,31,39] using nano-silicon and salinity to enhance vitamin C. The results for tomato fruits from Edkawy (Tables 3 and 4) showed that increasing salinity concentration significantly reduced N, P, K, Ca, Mg, Fe, and Zn contents in tomato shoots. On the contrary, Na content was increased with rising salt levels in tomato plants. It was reported that N, P, K, Ca, Mg, Fe, and Zn contents were lowered at high salts levels [31,55]. The results showed a decrease in tomato growth parameters and yield due to less water uptake which led to decreased mineral absorption under salinity. The higher Na amount in plants under saline conditions could be referred to the accumulation of Na inside vacuoles. On the other hand, the data for nano-silicon combined with grafting showed increased absorption of minerals under saline conditions, except Na was decreased [26]. The self-grafted Strain B hybrid showed the lowest concentrations of N, P, K, Ca, Mg, Fe, and Zn under 8000 ppm levels. Phytohormones play a major part in plant development under stress and normal conditions. Hormones such as ABA and GA have been described that regulate plant growth and development [70]. In Figure 4, it can be seen from the data that GA_3 concentration reduced under salt stress but increased with the application of silicon combined with grafting. Nano-silicon mitigates the harmful effects of NaCl on plant growth by improving endogenous GA3 in soybean [31,71]. In this study (Figure 4B), ABA increased with increasing salinity levels, in a similar fashion to what has been reported in previous studies [68], and was reduced when no silicon treatment was applied, as was also observed in [72,73]. Furthermore, nano-silicon reduced Na absorbance and decreased ABA concentration in the leaves of salt-stressed plants, resulting in an increase in stomatal aperture and CO_2 supply from the stomatal cavity to the CO_2 fixation site [31]. In this research, POD increased with rising salt stress levels (Figure 4C). On the other hand, nano-silicon application alleviated this effect of salt. Using nano-silicon enhances antioxidant activity, hence it plays a protective function against salt stress [28,31,59]. Compatible solutes such as proline (non-enzymatic) act as ROS scavengers and increase resistance to drought, salinity, and cold stresses [45,47,69]. In this study, the concentration of proline was significantly increased with increasing salinity levels (Figure 4D). A previous study on tomatoes yielded a similar result [74]. Higher contents of GA3, ABA, and proline were registered by Edkawy combined with nano-silicon under 8000 ppm. This could be related to the beneficial effects of nano-silicon on lipid peroxidation, plasma membrane stability, and osmolyte accumulation, all of which result in increasing concentrations of scavenging reactive oxygen species, predominantly hydrogen peroxide and superoxide [31,75].

5. Conclusions

Based on our findings, yield improvement can be achieved for tomato plants with no known tolerance to salt stress by grafting their scions on rootstocks of genotypes possessing a salinity tolerance. However, outstanding crop gain would be obtained from shoot–root exchange between salt-adapted genotypes. According to the obtained results in this study, it may be concluded that grafting combined with foliar application of nano-silicon is a unique technique for improving salt tolerance and reducing salt damage in tomato plants. Plant growth, fruit yield, fruit quality, especially vitamin C content and TSS percentage, mineral contents, and GA3, ABA, and proline levels of grafted tomato combined with foliar application of nano-silicon were significantly higher than the self-grafted Strain B hybrid under saline stress conditions. Based on these findings, it is advisable to grow the cross-grafted hybrid Strain B/Edkawy or hybrid Strain B/*Solanum pimpinellifolium* combined with foliar spray application of nano-silicon when production of tomato under high salinity conditions cannot be avoided.

Author Contributions: Conceptualization, E.G.S. and A.W.M.M.; methodology, E.G.S.; software, E.G.S. and M.M.E.-M.; validation, E.G.S., G.A.T., A.W.M.M. and E.G.S.; formal analysis, E.G.S. and G.A.T.; investigation, M.M.E.-M.; resources E.G.S., A.W.M.M. and M.A.A.A.; data curation, E.G.S.; writing—original draft preparation, E.G.S. and M.A.A.A.; writing—review and editing, E.G.S., A.W.M.M. and M.A.M.F.; visualization, E.G.S.; supervision, E.G.S. and M.M.E.-M.; project administration, E.G.S.; funding acquisition, E.G.S.; G.A.T. and M.A.M.F. All authors have read and agreed to the published version of the manuscript.

Funding: This research was funded by Cairo University, the Faculty of Agriculture, Giza, Egypt.

Institutional Review Board Statement: Not applicable.

Informed Consent Statement: Not applicable.

Data Availability Statement: Not applicable.

Acknowledgments: The authors are grateful to the Department of Vegetable Crops, the Faculty of Agriculture, and Cairo University for supplying some of the equipment and facilities used to accomplish this research.

Conflicts of Interest: The authors declare no conflict of interest.

References

1. Di Cesare, L.F.; Migliori, C.; Ferrari, V.; Parisi, M.; Campanelli, G.; Candido, V.; Perrone, D. Effects of irrigation-fertilization and irrigation-mycorrhization on the alimentary and nutraceutical properties of tomatoes. In *Irrigation Systems and Practices in Challenging Environments*; TECH Press: Rijeka, Croatia, 2012.
2. Abdel-Monaim, M.F. Induced Systemic Resistance in Tomato Plants against Fusarium Wilt Disease. *Int. Res. J. Microbiol.* **2012**, *3*, 14–23.
3. Gerster, H. The potential role of lycopene for human health. *J. Am. Coll. Nutr.* **1997**, *16*, 109–126. [CrossRef] [PubMed]
4. Ahmad, P.; Ahanger, M.A.; Alam, P.; Alyemeni, M.N.; Wijaya, L.; Ali, S.; Ashraf, M. Silicon (Si) Supplementation Alleviates NaCl Toxicity in Mung Bean [*Vigna radiata* (L.) Wilczek] Through the Modifications of Physio-biochemical Attributes and Key Antioxidant Enzymes. *J. Plant Growth Regul.* **2019**, *38*, 70–82. [CrossRef]
5. Shahid, S.A.; Zaman, M.; Heng, L. Soil Salinity: Historical Perspectives and a World Overview of the Problem. In *Guideline for Salinity Assessment, Mitigation and Adaptation Using Nuclear and Related Techniques*; Springer: Cham, Switzerland, 2018; pp. 43–53.

6. Singh, K.; Chatrath, R. *Application of Physiology in Wheat Breeding*; International Maize and Wheat Improvement Center (CIMMYT): Texcoco, Mexico, 2001.
7. Ashraf, M. Some important physiological selection criteria for salt tolerance in plants. *Flora-Morphol. Distrib. Funct. Ecol. Plants* **2004**, *199*, 361–376. [CrossRef]
8. El-Mogy, M.M.; Atia, M.A.M.; Dhawi, F.; Fouad, A.S.; Bendary, E.S.A.; Khojah, E.; Samra, B.N.; Abdelgawad, K.F.; Ibrahim, M.F.M.; Abdeldaym, E.A. Towards Better Grafting: SCoT and CDDP Analyses for Prediction of the Tomato Rootstocks Performance under Drought Stress. *Agronomy* **2022**, *12*, 153. [CrossRef]
9. Santa-Cruz, A.; Martinez-Rodriguez, M.M.; Perez-Alfocea, F.; Romero-Aranda, R.; Bolarin, M.C. The rootstockeffect on the tomato salinity response depends on the shoot genotype. *Plant Sci.* **2002**, *162*, 825–831. [CrossRef]
10. Razali, R.; Bougouffa, S.; Morton, M.J.L.; Lightfoot, D.J.; Alam, I.; Essack, M.; Arold, S.T.; Kamau, A.A.; Schmöckel, S.M.; Pailles, Y.; et al. The Genome Sequence of the Wild Tomato *S. pimpinellifolium* Provides Insights Into SalinityTolerance. *Front. Plant Sci.* **2018**, *9*, 1402. [CrossRef] [PubMed]
11. Rouphael, Y.; Petropoulos, S.A.; Cardarelli, M.; Colla, G. Salinity as eustress or for enhancing the quality of vegetables. *Sci. Hortic.* **2018**, *234*, 361–369. [CrossRef]
12. Hoffmann, J.; Berni, R.; Hausman, J.F.; Guerriero, G. A Review on the Beneficial Role of Silicon against Salinity in Non-Accumulator Crops: Tomato as a Model. *Biomolecules* **2020**, *10*, 1284. [CrossRef]
13. Almeida, D.M.; Margarida Oliveira, M.; Saibo, N.J.M. Regulation of Na+ and K+ homeostasis in plants: Towards improved salt stress tolerance in crop plants. Genet. *Mol. Biol.* **2017**, *40*, 326–345. [CrossRef]
14. Meng, Y.; Yin, Q.; Yan, Z.; Wang, Y.; Niu, J.; Zhang, J.; Fan, K. Exogenous Silicon Enhanced Salt Resistance by Maintaining K+/Na+ Homeostasis and Antioxidant Performance in Alfalfa Leaves. *Front. Plant Sci.* **2020**, *11*, 1183. [CrossRef] [PubMed]
15. Tuna, A.L.; Kaya, C.; Higgs, D.; Murillo-Amador, B.; Aydemir, S.; Girgin, A.R. Silicon improves salinity tolerance in wheat plants. *Environ. Exp. Bot.* **2008**, *62*, 10–16. [CrossRef]
16. Sommer, M.; Kaczorek, D.; Kuzyakov, Y.; Breuer, J. Silicon pools and fluxes in soils and landscapes—A review. *J. Plant Nutr. Soil Sci.* **2006**, *169*, 310–329. [CrossRef]
17. Frew, A.; Weston, L.A.; Reynolds, O.L.; Gurr, G.M. The role of silicon in plant biology: A paradigm shift in research approach. *Ann. Bot.* **2018**, *121*, 1265–1273. [CrossRef] [PubMed]
18. Barcelo, J.; Guevara, P.; Poschenrieder, C. Silicon amelioration of aluminium toxicity in teosinte (*Zea mays* L. ssp. mexicana). *Plant Soil* **2004**, *154*, 249–255. [CrossRef]
19. Cocker, K.M.; Evans, D.E.; Hodson, M.J. The amelioration of aluminium toxicity by silicon in wheat (*Triticumaestivum* L.): Malate exudation as evidence for an in planta mechanism. *Planta* **1998**, *204*, 318–323. [CrossRef]
20. Shen, X.; Xiao, X.; Dong, Z.; Chen, Y. Silicon effects on antioxidative enzymes and lipid peroxidation in leaves and roots of peanut under aluminum stress. *Acta Physiol. Plant.* **2014**, *36*, 3063–3069. [CrossRef]
21. Sivanesan, I.; Jeong, B.R. Silicon promotes adventitious shoot regeneration and enhances salinity tolerance of Ajuga multiflora bunge by altering activity of antioxidant enzyme. *Sci. World J.* **2014**, *2014*, 521703. [CrossRef]
22. Kim, Y.H.; Khan, A.L.; Waqas, M.; Lee, I.J. Silicon regulates antioxidant activities of crop plants under abiotic-induced oxidative stress: A review. *Front. Plant Sci.* **2017**, *8*, 510. [CrossRef]
23. Tripathi, D.; Bashri, G.; Shweta Singh, S.; Ahmad, P.; Singh, V.; Prasad, S.; Dubey, N.; Chauhan, D. Chapter 19 Efficacy of Silicon against Aluminum Toxicity in Plants: An Overview. In *Silicon in Plants: Advances and Future Prospects*; CRC Press: Boca Raton, FL, USA, 2016; pp. 241–258. [CrossRef]
24. Ahanger, M.A.; Alyemeni, M.N.; Wijaya, L.; Alamri, S.A.; Alam, P.; Ashraf, M.; Ahmad, P. Potential of exogenously sourced kinetin in protecting *Solanumlycopersicum* from NaCl-induced oxidative stress through up-regulation of the antioxidant system, ascorbate-glutathione cycle and glyoxalase system. *PLoS ONE* **2018**, *13*, e0202175. [CrossRef]
25. Ahmad, R.; Hussain, S.; Anjum, M.A.; Khalid, M.F.; Saqib, M.; Zakir, I.; Hassan, A.; Fahad, S.; Ahmad, S. Oxidative Stress and Antioxidant Defense Mechanisms in Plants Under Salt Stress. In *Plant Abiotic Stress Tolerance*; Springer: Cham, Switzerland, 2019. [CrossRef]
26. Stamatakis, A.; Papadantonakis, N.; Lydakis-Simantiris, N.; Kefalas, P.; Savvas, D. Effects of silicon and salinity on fruit yield and quality of tomato grown hydroponically. *Acta Hortic.* **2003**, *609*, 141–147. [CrossRef]
27. Al-Aghabary, K.; Zhu, Z.; Shi, Q. Influence of Silicon Supply on Chlorophyll Content, Chlorophyll Fluorescence, and Antioxidative Enzyme Activities in Tomato Plants Under Salt Stress. *J. Plant Nutr.* **2005**, *27*, 2101–2115. [CrossRef]
28. Romero-Aranda, M.R.; Jurado, O.; Cuartero, J. Silicon alleviates the deleterious salt effect on tomato plant growth by improving plant water status. *J. Plant Physiol.* **2006**, *163*, 847–855. [CrossRef] [PubMed]
29. Seal, P.; Das, P.; Biswas, A.K.; Seal, P.; Das, P.; Biswas, A.K. Versatile Potentiality of Silicon in Mitigation of Biotic and Abiotic Stresses in Plants: A Review. *Am. J. Plant Sci.* **2018**, *9*, 1433–1454. [CrossRef]
30. Haghighi, M.; Pessarakli, M. Influence of silicon and nano–silicon on salinity tolerance of cherry tomatoes (*Solanum lycopersicum* L.) at early growth stage. *Sci. Hortic.* **2013**, *161*, 111–117. [CrossRef]
31. Tondey, M.; Kalia, A.; Singh, A.; Dheri, G.S.; Taggar, M.S.; Nepovimova, E.; Krejcar, O.; Kuca, K. Seed Priming and Coating by Nano-Scale Zinc Oxide Particles Improved Vegetative Growth, Yield and Quality of Fodder Maize (*Zea mays*). *Agronomy* **2021**, *11*, 729. [CrossRef]

32. Mahmoud, A.W.M.; Samy, M.M.; Sany, H.; Eid, R.R.; Rashad, H.M.; Abdeldaym, E.A. Nano-potassium, Nano-silicon, and Biochar Applications Improve Potato Salt Tolerance by Modulating Photosynthesis, Water Status, and Biochemical Constituents. *Sustainability* **2022**, *14*, 723. [CrossRef]
33. Nasrallah, A.K.; Kheder, A.A.; Kord, M.A.; Fouad, A.S.; El-Mogy, M.M.; Atia, M.A.M. Mitigation of Salinity Stress Effects on Broad Bean Productivity Using Calcium Phosphate Nanoparticles Application. *Horticulturae* **2022**, *8*, 75. [CrossRef]
34. Yuvakkumar, R.; Elango, V.; Rajendran, V.; Kannan, N.S.; Prabu, P. Influence of nanosilica powder on the growth of maize crop. *Int. J. Green Nanotechnol.* **2011**, *3*, 180–190. [CrossRef]
35. Suriyaprabha, R.; Karunakaran, G.; Yuvakkumar, R.; Rajendran, V.; Kannan, N. Silica nanoparticles for increased silica availability in maize (*Zea mays* L.) seeds under hydroponic conditions. *Curr. Nanosci.* **2012**, *8*, 902–908. [CrossRef]
36. Haghighi, M.; Afifipour, Z.; Mozafarian, M. The effect of N-Si on tomato seed germination under salinity levels. *J. Biol. Environ. Sci.* **2012**, *6*, 87–90.
37. Sabaghnia, N.; Janmohammadi, M. Graphic analysis of nano-silicon by salinity stress interaction on germination properties of lentil using the biplot method. *Agric. For.* **2014**, *60*, 29–40.
38. Xie, Y.; Li, B.; Tao, G.; Zhang, Q.; Zhang, C. Effects of nano–silicon dioxide on photosynthetic fluorescence characteristics of Indocalamus barbatus McClure. *J. Nanjing For. Univ.* **2012**, *36*, 59–63.
39. Mahmoud, A.W.M.; Sweaty, H.M. Comparison between commercial and nano-nPk in presence of nano- ze-polite on sage plant yield and its components under water stress. *Agriculture* **2020**, *66*, 24–39. [CrossRef]
40. Janmohammadi, M.; Amanzadeh, T.; Sabaghnia, N.; Ion, V. Effect of nano-silicon foliar application on safflower growth under organic and inorganic fertilizer regimes. *Bot. Lith.* **2016**, *22*, 53–64. [CrossRef]
41. Zhu, Y.; Gong, H. Beneficial effects of silicon on salt and drought tolerance in plants. *Agron. Sustain. Dev.* **2014**, *34*, 455–472. [CrossRef]
42. Mahmoud, A.W.M.; Abdeldaym, E.A.; Abdelaziz, S.M.; El-Sawy, M.B.I.; Mottaleb, S.A. Synergetic Effects of Zinc, Boron, Silicon, and Zeolite Nano-particles on Confer Tolerance in Potato Plants Subjected to Salinity. *Agronomy* **2019**, *10*, 19. [CrossRef]
43. Polle, A.; Otter, T.; Mehne-Jakobs, B. Effect of magnesium deficiency on antioxidative systems in needles of Norway spruce [Picea abies (L) Karst.] grown with different ratios of nitrate and ammonium as nitrogen souices. *New Phytol.* **1994**, *128*, 621–628. [CrossRef]
44. Bates, L.S.; Waldren, R.P.; Teare, I.D. Rapid determination of free proline for water stress studies. *Plant Soil* **1973**, *39*, 205–207. [CrossRef]
45. Fales, T.M.; Jaouni, J.F.; Babashak, I. Simple device for preparing ethereal diazomethane without resorting to codistillation. *Ann. Chem.* **1973**, *45*, 2302–2303. [CrossRef]
46. AOAC (Association of Official Analytical Chemists). *Official Methods of Analysis*; Association of Official Analytical Chemists: Washington, DC, USA, 1990.
47. Singh, A.K.; Dubey, R.S. Changes in chlorophyll a and b contents and activities of photosystems 1 and 2 in rice seedlings inducedby NaCl. *Photosynthetica* **1995**, *31*, 489–499.
48. Gomez, K.N.; Gomez, A.A. *Statistical Procedures for Agricultural Research*, 2nd ed.; John Wiley and Sons: New York, NY, USA, 1984.
49. Snedecor, G.W.; Cochran, W.G. *Statistical Methods*, 7th ed.; Iowa State University Press: Ames, IA, USA, 1980; p. 507.
50. Abdeldym, E.A.; El-Mogy, M.M.; Abdellateaf, H.R.L.; Atia, M.A.M. Genetic Characterization, Agro-Morphological and Physiological Evaluation of Grafted Tomato under Salinity Stress Conditions. *Agronomy* **2020**, *10*, 1948. [CrossRef]
51. Singh, H.; Kumar, P.; Kumar, A.; Kyriacou, M.C.; Colla, G.; Rouphael, Y. Grafting Tomato as a Tool to Improve Salt Tolerance. *Agronomy* **2020**, *10*, 263. [CrossRef]
52. Cuartero, J.; Fernández-Muñoz, R. Tomato and salinity. *Sci. Hortic.* **1998**, *78*, 83–125. [CrossRef]
53. Munns, R.; Tester, M. Mechanisms of Salinity Tolerance. Annu. *Rev. Plant Biol.* **2008**, *59*, 651–681. [CrossRef]
54. Abdelaal, K.A.A.; Mazrou, Y.S.; Hafez, Y.M. Silicon Foliar Application Mitigates Salt Stress in Sweet Pepper Plants by Enhancing Water Status, Photosynthesis, Antioxidant Enzyme Activity and Fruit Yield. *Plants* **2020**, *9*, 733. [CrossRef]
55. Chaichi, M.R.; Keshavarz-Afshar, R.; Lu, B.; Rostamza, M. Growth and nutrient uptake of tomato in response to application of saline water, biological fertilizer, and surfactant. *J. Plant Nutr.* **2017**, *40*, 457–466. [CrossRef]
56. Ali, E.G.; Mohamed, M.F.; Farghali, M.A.; Abdel-Rahman, M.S. Productivity of Grafted Tomato Grown in the Summer Season Under The New Valley Environmental Conditions. *Assiut J. Agric. Sci.* **2014**, *45*, 2.
57. Mohamed, F.M.; Abd ElKader, M.F.; Shalaby, G.I. New Potential Hybrid 'Assiut-15' for Production of Tomato under Adverse High Temperature Conditions. *Proc. Third Sci. Conf. Agric. Sci.* **2002**, 385–392.
58. Aloni, B.; Cohen, R.; Karni, L.; Aktas, H.; Edelstein, M. Hormonal signaling in rootstock–scion interactions. *Sci. Hortic.* **2010**, *127*, 119–126. [CrossRef]
59. Harada, T. Grafting and RNAtransport via phloem tissue in horticultural plants. *Sci. Hortic.* **2010**, *125*, 545–550. [CrossRef]
60. Effat, A.; Sanaa, A.; Horeya, M.; Ramzy, M.; Yehia, A.; Manal, A. Comparative Performance of Five Genotypes of Tomato to Salt Stress. *Alex. Sci. Exch. J.* **2018**, *39*, 3.
61. Zhang, Y.; Shi, Y.; Gong, H.-J.; Zhao, H.-L.; Li, H.-L.; Hu, Y.-H.; Wang, Y.-C. Beneficial effects of silicon on photosynthesis of tomato seedlings under water stress. *J. Integr. Agric.* **2018**, *17*, 2151–2159. [CrossRef]
62. Shah, S.H.; Houborg, R.; McCabe, M.F. Response of Chlorophyll, Carotenoid and SPAD-502 Measurement to Salinity and Nutrient Stress in Wheat (*Triticumaestivum* L.). *Agronomy* **2017**, *7*, 61. [CrossRef]

63. Kusumi, K.; Hirotsuka, S.; Kumamaru, T.; Iba, K. Increased leaf photosynthesis caused by elevated stomatal conductance in a rice mutant deficient in SLAC1, a guard cell anion channel protein. *J. Exp. Bot.* **2012**, *63*, 5635–5644. [CrossRef]
64. Saglam, A.; Terzi, R.; Demiralay, M. Effect of polyethylene glycol induced drought stress on photosynthesis in two chickpea genotypes with different drought tolerance. *Acta Biol. Hung.* **2014**, *65*, 178–188. [CrossRef]
65. Gong, H.; Chen, K. The regulatory role of silicon on water relations, photosynthetic gas exchange, and carboxylation activities of wheat leaves in field drought conditions. *Acta Physiol. Plant.* **2012**, *34*, 1589–1594. [CrossRef]
66. Al-Ismaily, S.S.; Al-Yahyai, R.A.; Al-Rawahy, S.A. Mixed Fertilizer can Improve Fruit Yield and Quality of Field-Grown Tomatoes Irrigated with Saline Water. *J. Plant Nutr.* **2014**, *37*, 1981–1996. [CrossRef]
67. Abdelgawad, K.; El-Mogy, M.M.; Mohamed, M.I.A.; Garchery, C.; Stevens, R.G. Increasing Ascorbic Acid Content and Salinity Tolerance of Cherry Tomato Plants by Suppressed Expression of the Ascorbate Oxidase Gene. *Agronomy* **2019**, *9*, 51. [CrossRef]
68. El-Mogy, M.M.; Garchery, C.; Stevens, R. Irrigation with salt water affects growth, yield, fruit quality, storability and marker-gene expression in cherry tomato. *Acta Agric. Scand. Sect. B Soil Plant Sci.* **2018**, *68*, 727–737. [CrossRef]
69. Dehghanipoodeh, S.; Ghobadi, C.; Baninasab, B.; Gheysari, M.; Bidabadi, S.S. Effects of potassium silicate and nano-silica on quantitative and qualitative characteristics of a commercial strawberry (Fragaria × ananassa cv. 'camarosa'). *J. Plant Nutr.* **2016**, *39*, 502–507. [CrossRef]
70. Khan, A.; Khan, A.L.; Muneer, S.; Kim, Y.H.; Al-Rawahi, A.; Al-Harrasi, A. Silicon and Salinity: Crosstalk in Crop-Mediated Stress Tolerance Mechanisms. *Front. Plant Sci.* **2019**, *10*, 1429. [CrossRef]
71. Lee, S.K.; Sohn, E.Y.; Hamayun, M.; Yoon, J.Y.; Lee, I.J. Effect of silicon on growth and salinity stress of soybean plant grown under the hydroponic system. *Agrofor. Syst.* **2010**, *80*, 333–340. [CrossRef]
72. Pena-Cortes, H.; Sanchez-Serrano, J.J.; Mertens, R.; Willmitzer, L.; Prat, S. Abscisic acid is involved in the wound-induced expression of the proteinase inhibitor II gene in potato and tomato. *Proc. Natl. Acad. Sci. USA* **1989**, *86*, 9851. [CrossRef] [PubMed]
73. Khan, A.; Kamran, M.; Imran, M.; Al-Harrasi, A.; Al-Rawahi, A.; Al-Amri, I.; Lee, I.J.; Khan, A.L. Silicon and salicylic acid confer high-pH stress tolerance in tomato seedlings. *Sci. Rep.* **2019**, *9*, 19788. [CrossRef] [PubMed]
74. Kahlaoui, B.; Hachicha, M.; Misle, E.; Fidalgo, F.; Teixeira, J. Physiological and biochemical responses to the exogenous application of proline of tomato plants irrigated with saline water. *J. Saudi Soc. Agric. Sci.* **2018**, *17*, 17–23. [CrossRef]
75. Kafi, M.; Nabati, J.; Ahmadi-Lahijani, M.J.; Oskoueian, A. Silicon compounds and potassium sulfate improve salinity tolerance of potato plants through instigating the defense mechanisms, cell membrane stability, and accumulation of osmolytes. *Commun. Soil Sci. Plant Anal.* **2021**, *52*, 843–858. [CrossRef]

horticulturae

MDPI

Article

Iodine Enhances the Nutritional Value but Not the Tolerance of Lettuce to NaCl

Giuseppe Maglione [1], Ermenegilda Vitale [2], Giulia Costanzo [2], Franca Polimeno [1], Carmen Arena [2] and Luca Vitale [3,*]

[1] Institute for the Animal Production System in the Mediterranean Environment (ISPAAM), Department of Biology, Agriculture and Food Sciences (DiSBA), National Research Council (CNR), P.le E. Fermi 1, Loc. Granatello, 80055 Portici, Italy; giuseppe.maglione@cnr.it (G.M.); franca.polimeno@cnr.it (F.P.)

[2] Department of Biology, University of Naples Federico II, Via Cinthia 21, 80126 Naples, Italy; ermenegilda.vitale@unina.it (E.V.); giulia.costanzo2@unina.it (G.C.); c.arena@unina.it (C.A.)

[3] Institute for Agricultural and Forestry Systems in the Mediterranean (ISAFoM), Department of Biology, Agriculture and Food Sciences (DiSBA), National Research Council (CNR), P.le E. Fermi 1, Loc. Granatello, 80055 Portici, Italy

* Correspondence: luca.vitale@cnr.it; Tel.: +39-0649-932-7526

Abstract: Positive stress or essential and nonessential elements can improve nutritive values (biofortification) of edible plants. In the present study, we evaluate (i) the effect of moderate salinity on lettuce biofortification, evaluated as nutritional bioactive compound accumulation, and (ii) the role of iodine in enhancing salt tolerance by increasing photorespiration and the content of antioxidants in lettuce. Physiological (gas exchange and chlorophyll fluorescence emission) and biochemical (photosynthetic pigment and bioactive compound) analyses were performed on lettuce plants grown under moderate salinity (50 mM NaCl alone or 50 mM NaCl in combination with iodine, KIO_3). Our results show that NaCl + iodine treatment improves the nutritional value of lettuce in terms of bioactive compounds acting as antioxidants. More specifically, iodine enhances the accumulation of photosynthetic pigments and polyphenols, such as anthocyanins, under salt but does not improve the salt tolerance. Our findings indicate that iodine application under moderate salinity could be a valid strategy in plant biofortification by improving nutritional bioactive compound accumulation, thus exercising functional effects on human health.

Keywords: nonessential elements; moderate salinity; biofortification; gas exchange; *Lactuca sativa* L.

Citation: Maglione, G.; Vitale, E.; Costanzo, G.; Polimeno, F.; Arena, C.; Vitale, L. Iodine Enhances the Nutritional Value but Not the Tolerance of Lettuce to NaCl. *Horticulturae* **2022**, *8*, 662. https://doi.org/10.3390/horticulturae8070662

Academic Editors: Agnieszka Hanaka, Małgorzata Majewska and Barbara Hawrylak-Nowak

Received: 22 June 2022
Accepted: 19 July 2022
Published: 20 July 2022

1. Introduction

Biofortification consists in developing crops with bioavailable micronutrients in edible parts [1,2], and generally it can be achieved by breeding, agronomic, and transgenic approaches [3]. The application of positive stress (eustress) also can trigger an accumulation of bioactive compounds in edible plants [4].

Salinity, in particular an elevated salt concentration, may affect plant growth in several ways [5–8], negatively affecting the crop yield. However, at low or moderate salt concentration, yields are mildly affected or not affected at all [9], and nutritive and/or bioactive compounds can be accumulated [10,11]. Santander et al. [12] found a greater phenolic concentration and antioxidant activity in lettuce plants grown under moderate salinity (50 mM NaCl) with no effect on photosynthetic activity and plant biomass; however, severe salinity (150–200 mM NaCl) reduced the antioxidant capacity and plant biomass. An improvement of functional compounds was also found in *Perilla frutescens*, a novel food, exposed from mild to moderate salinity conditions [13]. Thus, appropriate irrigation management of horticultural crops through a mild–moderate salinity could be a means to enhance the nutraceutical value of crops because a salt-induced reshuffling of plant metabolism would increase the accumulation of nutritional bioactive compounds, thus enriching the functional quality of fresh vegetables with positive outcomes on human wellbeing.

Biofortification of vegetables to increase the mineral concentration in edible organs can also be achieved by adding essential and nonessential beneficial micronutrients [3], generally via mineral fertilizers (agronomic fortification). Several studies on mineral biofortification have been carried out in the past [14], and recently the attention has been focused on the use of iodine (I). For a long time, this element has been considered a nonessential nutrient for plants; however, now it has gained a rising interest because recent research has shown evidence for a nutritional role of iodine in vegetables [15]. Plants can utilize two iodine forms: iodide (I^-) and iodate (IO_3). Owing to their ability to absorb and accumulate exogenous iodine into the edible organs, horticultural crops are the best candidates to test the outcomes of I biofortification [16]. To date, it is not well understood which form of iodine is more adequate in inducing valuable results on plant biomass or inducing plant biofortification; however, it is ascertained that iodine concentrations ranging from 10^{-6} to 10^{-4} M exert positive outcomes [17].

It has been demonstrated that iodine can improve salt tolerance in plants [18], depending on variables such as the sources of iodine, their concentration, and type of application. Salt tolerance is obtained by an increase in antioxidant content, and strictly depends on the species and the form of the ion that is applied. The last aspect is very interesting because negative effects have been reported in plants fertilized with I^-, whereas positive results were derived from the application of IO_3^- [19]. In particular, the growth with IO_3^- improves nitrogen metabolism and photorespiration in lettuce plants [20].

In the present study, we evaluate, through combined physiological and biochemical analyses: (i) the effect of the moderate salinity on the biofortification of lettuce, a species largely cultivated and consumed for human diet worldwide, assessing the nutritional bioactive compounds accumulation; (ii) the role of IO_3^- in improving the salt tolerance and promoting a higher antioxidant–bioactive compound accumulation. Based on the findings of Blasco et al. [20], we hypothesize that the addition of iodine to soil leads to an increase in photorespiration and antioxidant content in lettuce plants, thereby enhancing their tolerance to moderate salinity.

2. Materials and Methods

2.1. Plant Material and Growth Conditions

Seedlings of lettuce cv Bionda liscia were transplanted 20 days after sowing (DAS) in 0.5 L pots filled with peat soil (90% peat and 10% sand, organic carbon content 31% *w/w*, C/N 25) and grown during autumn–winter 2021–2022. After one week from transplanting, plants were subjected to three treatments: (i) fertilization with nutrient solution (Control, EC: 1.8 dS m^{-1}, pH: 7.5), (ii) fertilization with nutrient solution added with NaCl (3 g L^{-1}, 52 mM, EC: 7.0 dS m^{-1}, pH: 7.5), and (iii) fertilization with nutrient solution added with NaCl (3 g L^{-1}, 52 mM, EC: 7.0 dS m^{-1}, pH: 7.5) plus KIO_3 (0.09 mg L^{-1}, 53 µg I). Plants were grown in a greenhouse equipped with a white LED illumination system (T5-60, 14-Watt LED T5 tube, Driwei) at the following growth conditions: 100 µmol photons m^{-2} s^{-1} at the top of canopy and 16 h/8 h light/dark photoperiod. Plants were fertilized weekly—for a total of 6 weeks—until the harvest by supplying 50 mL nutrient solution in which NaCl or NaCl + KIO_3 were dissolved. The control received only the nutrient solution (2 g L^{-1}) with the following mineral composition: total N 20%, N-NO_3 3.6%, N-NH_4 3.6%, N-CH_4N_2O 12.1%, P_2O_5 40%, K_2O 20%, B 0.05%, Cu 0.02%, Fe 0.4%, Mn 0.2%, Mo 0.02%, and Zn 0.02%. Each treatment consisted of 5 plants in three replicates. The plants were harvested 60 DAS, when leaves were moderately expanded. The harvest time was chosen based on a previous study [21], which demonstrated that a high post harvest quality is expected when fresh-cut lettuce presents moderately expanded leaves. At the harvesting, biometrical, ecophysiological, and biochemical determinations were carried out on leaves of different plants. The fresh vegetable quality was evaluated in terms of nutritional bioactive compounds such as: chlorophylls, carotenoids, polyphenols (flavonoids and anthocyanins), and antioxidant capacity. Soluble proteins content was also determined.

Electrical conductivity (EC) and pH were measured at harvesting in a 1:5 soil:distilled water suspension by means of a conductivity meter (Portlab 203) and pH meter (XS Instruments).

2.2. Leaf Gas Exchange and Chlorophyll a Fluorescence Measurement

A Li6400 portable photosynthesis system (Licor Inc., Lincoln, NE, USA) was used to perform simultaneous gas exchange and chlorophyll *a* (Chl a) fluorescence measurement on mature leaves. Measurements were carried out at 400 µmol photons m^{-2} s^{-1}, 25 °C ± 2, RH 45% ± 5, and CO_2 of 400 ppm. To evaluate the potential of photosynthesis and the occurrence of processes other than CO_2 fixation and photorespiration, gas exchange and Chl *a* fluorescence measurements were also measured at elevated CO_2 concentration (800 ppm). Gas exchange parameters: net photosynthesis (A_N), stomatal conductance (G_s), intercellular to ambient CO_2 ratio (C_i/C_a), instantaneous water use efficiency (ratio between net photosynthesis and transpiration—A_N/T_r), were calculated by software operating in the instrument following the equation of von Caemmerer and Farquhar [22]. The fluorescence parameters in the light quantum yield of noncyclic electron transport (Φ_{PSII}) were calculated by Genty et al. [23], while the quantum yield of regulated (Φ_{NPQ}) and non-regulated energy dissipation (Φ_{NO}) were calculated according to Kramer et al. [24]. The total electron transport rate (J_f) was determined as reported in Kral and Edwards [25]), and the electron transport rate to CO_2 (J_c) and to O_2 (J_o) was calculated according to Epron et al. [26] as:

$$J_c = 1/3\ [J_f + 8(A_N + R_d)] \quad (1)$$

$$J_o = 2/3\ [J_f - 4(A_N + R_d)] \quad (2)$$

Dark respiration (R_d) and maximum photochemical efficiency (F_v/F_m) were measured on 30 min dark-adapted leaves following the measurements in the light. Leaves were darkened through an aluminum sheet placed around the leaf, allowing air to circulate.

2.3. Photosynthetic Pigments

The content of total chlorophylls and carotenoids was evaluated following the procedure reported by Lichtenthaler [27]. Pigments were extracted in 100% ice-cold acetone and quantified by measuring absorbance at the wavelengths of 470, 645, and 662 nm. The concentration was expressed in µg cm^{-2}.

2.4. Polyphenols and Soluble Proteins

For all the following assays, samples were preventively powdered with liquid nitrogen and stored at −80 °C until further analysis.

To estimate the total polyphenol content, samples (0.02 g) were extracted in 2 mL of 80% aqueous methanol following the procedure reported in Vitale et al. [28]. The extracts were kept for 1 h at 4 °C and then centrifuged at 11,000 rpm for 5 min. An aliquot (274 µL) was mixed with 274 µL of the Folin Ciocolteau reagent and 1.452 mL of 700 mM sodium carbonate (Na_2CO_3). Samples were incubated for 2 h in darkness. The absorbance was read at 765 nm and the concentration was calculated and expressed as gallic acid equivalent in mg GAE g^{-1} fresh weight (FW) using a gallic acid standard curve.

Total flavonoid content was quantified according to Moulehi et al. [29] and Sun et al. [30]; 250 µL of a diluted methanol sample was added to 75 µL of 5% $NaNO_2$ (sodium nitrite, 150 µL of 10% $AlCl_3$ (aluminum chloride), and 500 µL NaOH (1 M). Then, the mixture was adjusted with distilled water to a final volume of 1.525 mL. After absorbance determination at 510 nm, the flavonoid content was determined using a catechin standard curve and expressed as mg catechin equivalent per gram of fresh weight (mg CE g^{-1}FW).

The anthocyanin content was analyzed according to Mancinelli et al. [31] and Chung et al. [32]. Samples (0.05 g) were extracted with acidified methanol (1% HCl) for 24 h at 4 °C in the dark. Samples were centrifuged and the supernatants were measured spectrophotometrically at 530 and 657 nm. The extinction coefficient of 31.6 M^{-1} cm^{-1}

was used to convert absorbance values into concentrations of anthocyanins, using the following equation: anthocyanin content (μmol g^{-1}) = [(A_{530} − 0.33 × A_{657})/31.6] × [volume (mL)/weight (g)].

The antioxidant capacity was evaluated by the ferric-reducing antioxidant power (FRAP) assay according to George et al. [33] and Vitale et al. [34]. Samples (0.250 g) were extracted with 60:40 (*v*/*v*) methanol/water solution. After centrifugation, an aliquot of extract (150 μL) was mixed with the FRAP reagents (2.5 mL of 300 mM acetate buffer pH 3.6, 250 μL of 10 mM tripyridyltriazine (TPTZ) and 250 μL of 12 mM $FeCl_3$). After the sample incubation for 1 h in the darkness, the absorbance was read at 593 nm. The total antioxidant capacity was quantified as μmol of Trolox (6-hydroxy-2,5,7,8-tetramethylchroman-2-carboxylic acid) equivalents per gram of fresh weight (μmol TE g^{-1} FW) using the Trolox a standard curve.

Total soluble protein content was determined following Bradford [35] and Im et al. [36]. Briefly, samples (0.200 g) were extracted in 0.2 M potassium phosphate buffer (pH 7.8 + 0.1 mM EDTA) and centrifuged. The supernatant was added to the dye reagent, and the absorbance was read at 595 nm. The total soluble proteins were quantified using a bovine serum albumin (BSA) calibration curve and expressed as mg BSA eq g^{-1} FW.

2.5. Statistical Analysis

The statistical software SigmaPlot 12.0 was used to perform graphics and statistical analysis. Data were analyzed by one-way ANOVA followed by the Holm–Sidak post hoc test. Data shown in tables and figures are means ± standard error (SE).

3. Results

3.1. Plant Growth

Mineral fertilization added with salt (NaCl) did not alter soil pH but increased the electrical conductivity, as compared to the control (Table 1).

Table 1. pH and electrical conductivity (EC) in soil at the harvesting of control plants, and plants fertilized with nutrient solution added with salt (NaCl) or with salt plus iodine (NaCl + KIO_3). Data are means ± SE.

Parameters	Control	NaCl	NaCl + KIO_3
pH_{H2O}	7.28 ± 0.03 ^{a}	7.16 ± 0.03 ^{a}	7.18 ± 0.01 ^{a}
EC (dS m^{-1})	1.80 ± 0.00 ^{a}	6.70 ± 0.05 ^{b}	6.55 ± 0.13 ^{b}

Different letters indicate significant differences among treatments. $p < 0.05$.

Salt did not affect the plant growth (Table 2). The shoot biomass and the number of leaves were comparable among the treatments and no visible difference in plant appearance was observed.

Table 2. Main biometrical characteristics at the harvesting of control plants, and plants fertilized with nutrient solution added with salt (NaCl) or with salt plus iodine (NaCl + KIO_3). Data are means ± SE.

Parameters	Control	NaCl	NaCl + KIO_3
Shoot (g FW p^{-1})	36.60 ± 1.40 ^{a}	36.00 ± 1.05 ^{a}	34.20 ± 1.39 ^{a}
N° leaves	35.80 ± 1.11 ^{a}	34.20 ± 0.58 ^{a}	34.40 ± 0.51 ^{a}

Different letters indicate statistically significant differences among treatments. $p < 0.05$.

3.2. Bioactive Compounds and Soluble Proteins

The chlorophyll and carotenoid content was significantly affected by the treatment with salt plus iodine but not by the salt alone (Table 3). The application of only NaCl caused a 74% reduction in total polyphenols, but increased the anthocyanin content and the FRAP antioxidant capacity by 95% and 21%, compared to control, respectively. (Table 3). The NaCl + KIO_3 treatment improved the total polyphenol content in lettuce leaves compared to NaCl treatment, and produced a higher anthocyanin accumulation in leaves. Conversely, the combination NaCl + KIO_3 induced a reduction in antioxidant capacity by 42% compared to control and by 52% compared to NaCl treatments. No effect on soluble proteins and flavonoids was observed among the treatments.

Table 3. Chemical composition of control plants and plants fertilized with nutrient solution with salt (NaCl) added or with salt plus iodine (NaCl + KIO_3). Data are means ± SE.

Parameters	Control	NaCl	NaCl + KIO_3
Chlorophylls (a + b) (μg cm^{-2})	49.78 ± 0.92 [a]	53.02 ± 2.52 [a]	64.68 ± 2.76 [b]
Carotenoids (x + c) (μg cm^{-2})	10.03 ± 0.13 [a]	10.97 ± 0.46 [a]	12.99 ± 0.40 [b]
Total polyphenols (mg GAE g^{-1} FW)	0.42 ± 0.08 [c]	0.11 ± 0.02 [a]	0.18 ± 0.02 [b]
Anthocyanins (μmol g^{-1} FW)	0.021 ± 0.006 [a]	0.041 ± 0.003 [b]	0.119 ± 0.007 [c]
Flavonoids (mg CE g^{-1} FW)	6.78± 0.42 [a]	6.31± 0.34 [a]	6.16± 0.20 [a]
Antioxidant capacity (μmol TE g^{-1} FW)	0.43 ± 0.01 [a]	0.52 ± 0.02 [b]	0.25 ± 0.01 [c]
Soluble proteins (mg BSA eq g^{-1} FW)	1.61± 0.07 [a]	1.47± 0.05 [a]	1.56± 0.03 [a]

Different letters indicate significant difference among treatments. $p < 0.05$.

3.3. Leaf Gas Exchange and Chl a Fluorescence Measurement

Salt treatments did not affect leaf gas exchange both at ambient and at elevated CO_2 concentration (Figure 1). According to the limitation imposed on photosynthesis by the current CO_2 concentration in atmosphere, the net photosynthesis (A_N) measured under elevated CO_2 was significantly greater than under ambient CO_2 (Figure 1a), promoting a greater instantaneous water use efficiency (Figure 1d). Under ambient CO_2 concentration, salt treatments determined a significant increase in Φ_{PSII} as compared to control (Figure 2a). The stimulation of Φ_{PSII} by salt treatments induced a lower thermal dissipation (Φ_{NPQ}) of absorbed light (Figure 2b) but no change in nonregulated energy dissipation (Φ_{NO}) (Figure 2c). Under elevated CO_2 the photochemical efficiency (Φ_{PSII}) reflected the behavior of photosynthesis, resulting higher under elevated than under ambient CO_2 concentration and was not affected by salt treatments (Figure 2a), as observed for Φ_{NPQ} and Φ_{NO}. Under ambient CO_2, a higher J_f/A_G ratio was measured in response to salt treatments compared to the control (Figure 2d). This result indicates an increase in electron flow toward O_2 reduction (photorespiration), in contrast to CO_2 reduction (photosynthesis) (Figure 2e,f), which significantly declined under both NaCl and NaCl + KIO_3 treatments.

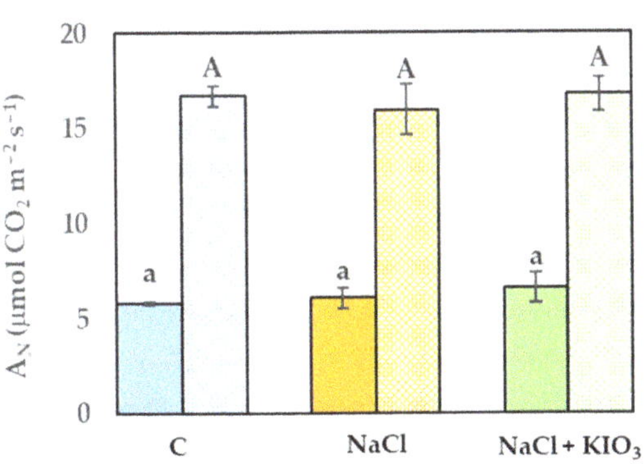

(a)

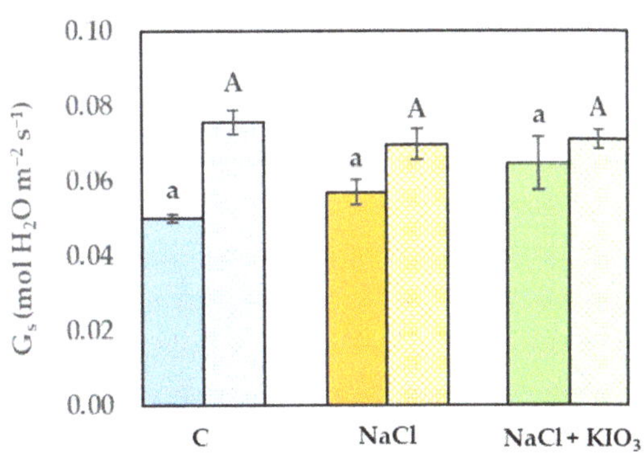

(b)

Figure 1. *Cont.*

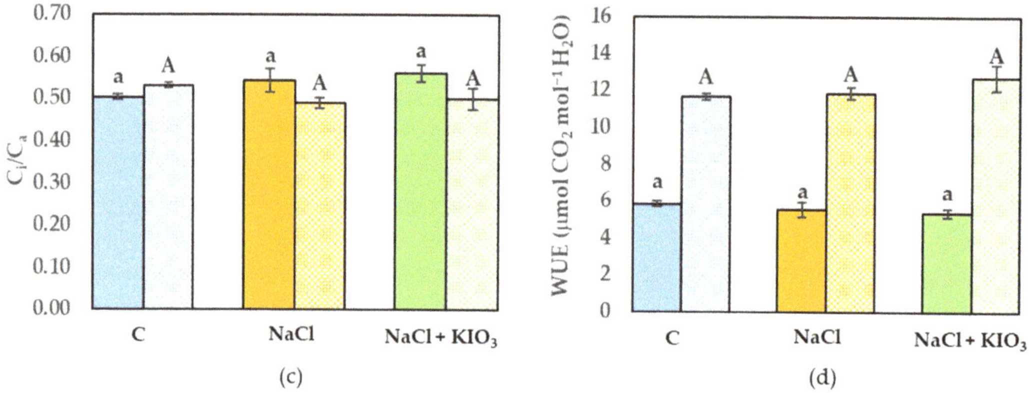

Figure 1. (**a**) Net photosynthesis (A_N); (**b**) stomatal conductance (G_s); (**c**) intercellular to ambient CO_2 concentration ratio (C_i/C_a); (**d**) instantaneous water use efficiency (WUE) measured in plants fertilized with nutrient solution (C), in plants fertilized with nutrient solution *plus* salt (NaCl), and in plants fertilized with nutrient solution *plus* salt and iodine (NaCl + KIO_3). Filled bars: measurements to ambient CO_2; cross-linked bars: measurements to elevated CO_2. Data are means ± SE. Different letters indicate statistical difference among treatments ($p < 0.05$).

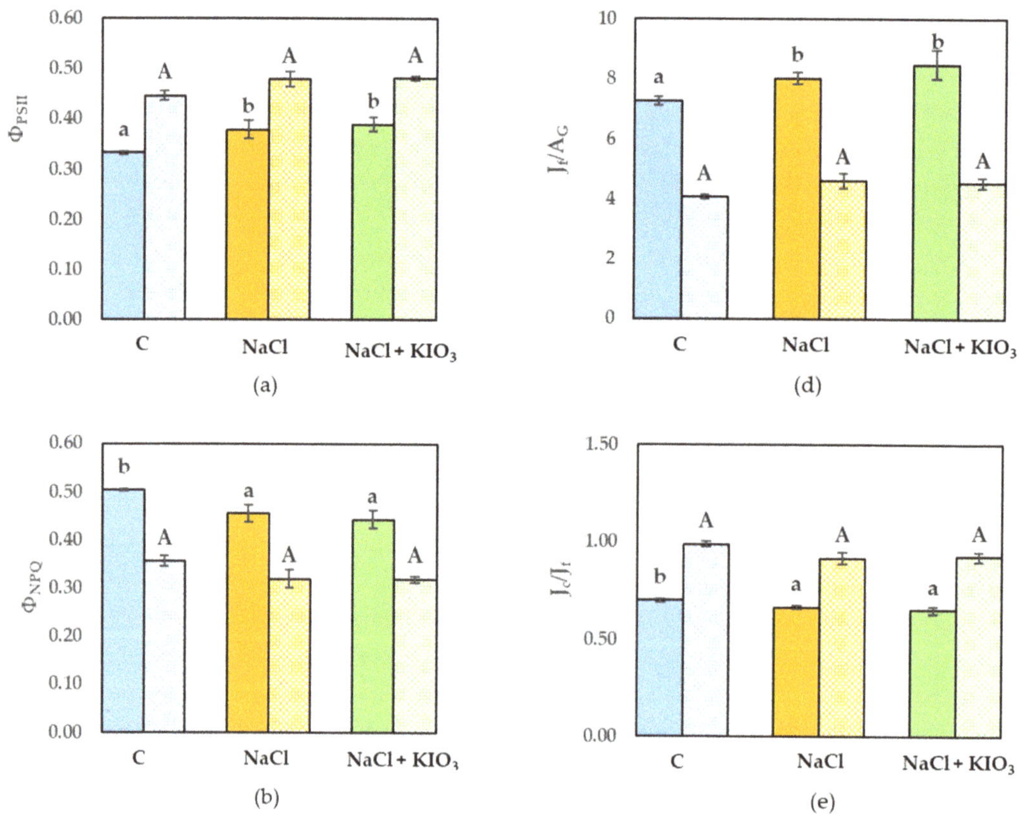

Figure 2. *Cont.*

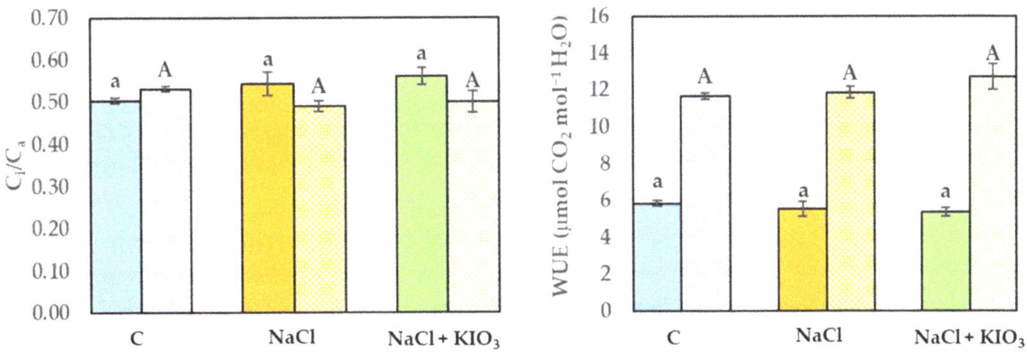

(c) (d)

Figure 2. (**a**) Quantum yield of noncyclic electron transport (Φ_{PSII}); (**b**) quantum yield of regulated energy dissipation (Φ_{NPQ}); (**c**) quantum yield of nonregulated energy dissipation ((Φ_{NO}); (**d**) electron transport rate to gross photosynthesis ratio (J_f/A_G); (**e**) electron transport rate to CO_2 and total electron transport rate ratio (J_c/J_f); (**f**) electron transport rate to O_2 and total electron transport rate ratio (J_o/J_f) measured in plants fertilized with nutrient solution (C), in plants fertilized with nutrient solution *plus* salt (NaCl), and in plants fertilized with nutrient solution *plus* salt and iodine (NaCl + KIO_3). Filled bars: measurements to ambient CO_2; cross-linked bars: measurements to elevated CO_2. Data are means ± SE. Different letters indicate statistical difference among treatments ($p < 0.05$).

4. Discussion

4.1. Nutritional Value of Crops

In agricultural practices, many approaches have been utilized aimed to enhance plant growth and nutritional value of crops, such as mild abiotic stress and the use of essential and nonessential elements [4,37,38].

In the present research, it was found that moderate salt levels do not modify the growth of lettuce plants, but improve their nutraceutical value, depending on the imposed salt treatment. Our findings confirm previous studies which demonstrated that eustress, such as the imposition of a mild to moderate salinity, may enhance the organoleptic components of quality in vegetables [12,39]. In our study, the vegetable quality was evaluated in terms of bioactive substances acting as antioxidants. Iodine increased the nonenzymatic antioxidants under moderate salinity, leading to an accumulation of phenolic compounds and, in particular, anthocyanins, more than under saline treatment alone. It is hypothesized that the combination NaCl + KIO_3 may induce an enhancement of enzymes involved in polyphenol synthesis [14]. Similar results have been previously found in lettuce plants [32] biofortified with KIO_3 at <80 μM, a concentration comparable to that used in our study. These bioactive compounds would act as antioxidants in preventing oxidative damages, and it is likely to suppose that the increase in polyphenols and, in particular, anthocyanins, was associated with plant photoprotection, because a significant increase in chlorophyll and carotenoid content occurred in plants fertilized with iodine under moderate salinity. Our results are consistent with the findings of Medrano Macías et al. [19]. It has been previously demonstrated that anthocyanin synthesis is activated by salt stress [40]; however, our results indicated that it can be potentiated by iodine. Our results also indicated that lettuce plants grown under NaCl treatment supplemented with iodine invested more energy in anthocyanin synthesis compared to the control and NaCl plants. The investment of plants in anthocyanins may have reduced the energy at disposal for the synthesis of other secondary metabolites acting as antioxidants, inducing the significant decline observed for the antioxidant capacity measured in plants fertilized with IO_3^-. On the contrary, the lowest energy engaged in anthocyanin synthesis in plants grown with salt without iodine likely allowed a greater synthesis of antioxidants, enhancing the antioxidant capacity.

Our data also indicated that iodine exerted some effect on lettuce nitrogen metabolism. According to Blasco et al. [41] which reports an enhanced nitrogen assimilation and protein synthesis in plants treated with IO_3^-, the increase in total chlorophyll content in lettuce plants suggests that nitrogen is assimilated in organic compounds that serve as a N donor for their synthesis.

4.2. Stress Tolerance

In our experiment, salt did not influence leaf gas exchange, evidencing that a moderate salt level may trigger plant biofortification without detrimental effects on crop yield and plant shape and architecture. Similarly, no unfavorable effect on net photosynthesis in NaCl + KIO_3 plants was observed, confirming that low iodine amounts (<80 μM) are not toxic for plants [42]. Once absorbed, iodine is translocated to chloroplasts via phloem, where it exerts its effects on photosynthesis [43]. Phenolic compounds can bind iodine through an electrophilic H substitution in the aromatic ring [44], likely protecting the photosynthetic apparatus from the mineral toxicity.

Photosynthesis is compromised by salt, generally provided at high concentration in salt-sensitive plants. On the contrary, salt stress-tolerant plants activated protective responses to overcome this form of abiotic stress [8]. In both salt treatments, salt induces a greater partitioning of electrons of the photosynthetic electron chain toward processes other than CO_2 assimilation, as indicated by the greater J_f/A_G ratio compared to the unstressed treatment. Our data showed that the J_f/A_G ratio in lettuce plants ranges from 4 to 5 under elevated CO_2, a typical value indicating a nonphotorespiratory condition for photosynthetic machinery and, in turn, a negligible electron flow to O_2 as an alternative acceptor to CO_2. Salt, alone or in combination with iodine, induces an increase in the electron transport flow toward photorespiration (J_o/J_f), which becomes the main photochemical process other than CO_2 assimilation in dissipating the absorbed light under stress. The increase in photorespiration determines a reduction in regulated thermal dissipation (Φ_{NPQ}) sustained by the xanthophyll cycle, avoiding the rise of processes linked to photooxidation and photoinhibition of photosystems (which absence is proved by steady values of nonregulated thermal dissipation—Φ_{NO-} among treatments). Blasco et al. [20] reported an increase in photorespiration in lettuce plants treated with IO_3^-. These authors found in plants treated with high IO_3^- concentration, an increase in the hydroxypyruvate reductase activity, a key enzyme of the photorespiration cycle. It is likely to suppose that under treatment with moderate salt levels, as in our experimental conditions, iodine did not enhance salt tolerance through an increase in photorespiration. Nevertheless, in both salt treatments the photorespiration rise could have provided carbonated skeletons for the photosynthetic carbon reduction cycle, thus avoiding the decline of photosynthetic activity.

In conclusion, our data show that moderate salinity improves the nutritional value of lettuce, and that iodine promotes a higher accumulation in bioactive compounds such as chlorophylls and carotenoids, polyphenols, and anthocyanins under conditions of moderate salt levels. It is likely that iodine induces a greater synthesis of anthocyanins at the expense of the synthesis of other secondary metabolites that act as antioxidants. Our data indicate that iodine does not improve salt tolerance through an ameliorating of photorespiration; however, iodine application under moderate salt levels could represent a valid strategy for plant biofortification, as it improves the bioactive compound accumulation with its valuable functional effects on human health.

Author Contributions: Conceptualization, G.M. and L.V.; investigation, G.M., E.V., G.C. and L.V.; data curation, E.V., F.P. and L.V.; writing—original draft preparation, L.V.; writing—review and editing, C.A. and L.V.; supervision, C.A. and L.V. All authors have read and agreed to the published version of the manuscript.

Funding: This research received no external funding.

Institutional Review Board Statement: Not applicable.

Informed Consent Statement: Not applicable.

Data Availability Statement: The data presented in this study are available on request from the corresponding author.

Acknowledgments: The authors thank Fabiana Fusco for her valuable help in biochemical analysis.

Conflicts of Interest: The authors declare no conflict of interest.

References

1. Malorgio, F.; Diaz, K.E.; Ferrante, A.; Mensuali, A.; Pezzarossa, B. Effects of selenium addition on minimally processed leafy vegetables grown in floating system. *J. Sci. Food Agric.* **2009**, *89*, 2243–2251. [CrossRef]
2. Montesano, F.F.; D'Imperio, M.; Parente, A.; Cardinali, A.; Renna, M.; Serio, F. Green bean biofortification for Si through soilless cultivation: Plant response and Si bioaccessibility in pods. *Sci. Rep.* **2016**, *6*, 31662. [CrossRef] [PubMed]
3. Garg, M.; Sharma, N.; Sharma, S.; Kapoor, P.; Kumar, A.; Chunduri, V.; Arora, P. Biofortified crops generated by breeding, agronomy, and transgenic approaches are improving lives of millions of people around the world. *Front. Nutr.* **2018**, *5*, 12. [CrossRef] [PubMed]
4. Rouphael, Y.; Kyriacou, M.C. Enhancing quality of fresh vegetables through salinity eustress and biofortification applications facilitated by soilless cultivation. *Front. Plant Sci.* **2018**, *9*, 1254. [CrossRef]
5. Carillo, P.; Annunziata, M.G.; Pontecorvo, G.; Fuggi, A.; Woodrow, P. Salinity Stress and Salt Tolerance. In *Abiotic Stress in Plants-Mechanisms and Adaptations*; Shanker, A., Venkateswarlu, B., Eds.; Inthec Open: London, UK, 2011; Volume 1, pp. 21–38.
6. Cirillo, C.; De Micco, V.; Arena, C.; Pannico, A.; De Pascale, S.; Rouphael, Y. Biochemical, Physiological and Anatomical Mechanisms of Adaptation of *Callistemon citrinus* and *Viburnum lucidum* to NaCl and $CaCl_2$ Salinization. *Front. Plant. Sci.* **2019**, *10*, 742. [CrossRef]
7. Carillo, P.; Cirillo, C.; De Micco, V.; Arena, C.; De Pascale, S.; Rouphael, Y. Morpho-anatomical, physiological and biochemical adaptive responses to saline water of *Bougainvillea spectabilis* Willd. trained to different canopy shapes. *Agric. Water Manag.* **2019**, *212*, 12–22. [CrossRef]
8. De Micco, V.; Arena, C.; Amitrano, C.; Rouphael, Y.; De Pascale, S.; Cirillo, C. Effects of NaCl and $CaCl_2$ Salinization on Morpho-Anatomical and Physiological Traits of Potted *Callistemon citrinus* Plants. *Forests* **2021**, *12*, 1666. [CrossRef]
9. Maggio, A.; Hasegawa, P.M.; Bressan, R.A.; Consiglio, M.F.; Joly, R.J. Review: Unravelling the functional relationship between root anatomy and stress tolerance. *Aust. J. Plant Physiol.* **2001**, *28*, 999–1004. [CrossRef]
10. Rouphael, Y.; Kyriacou, M.C.; Petropoulos, S.A.; De Pascale, S.; Colla, G. Improving vegetable quality in controlled environments. *Sci. Hort.* **2018**, *234*, 275–289. [CrossRef]
11. Rouphael, Y.; Petropoulos, S.A.; Cardarelli, M.; Colla, G. Salinity as eustressor for enhancing quality of vegetables. *Sci. Hort.* **2018**, *234*, 361–369. [CrossRef]
12. Santander, C.; Vidal, G.; Ruiz, A.; Vidal, C.; Cornejo, P. Salinity Eustress Increases the Biosynthesis and Accumulation of Phenolic Compounds That Improve the Functional and Antioxidant Quality of Red Lettuce. *Agronomy* **2022**, *12*, 598. [CrossRef]
13. Rouphael, Y.; Kyriacou, M.C.; Carillo, P.; Pizzolongo, F.; Romano, R.; Sifola, M.I. Chemical Eustress Elicits Tailored Responses and Enhances the Functional Quality of Novel Food Perilla frutescens. *Molecules* **2019**, *24*, 185. [CrossRef]
14. Buturi, C.V.; Mauro, R.P.; Fogliano, V.; Leonardi, C.; Giuffrida, F. Mineral Biofortification of Vegetables as a Tool to Improve Human Diet. *Foods* **2021**, *10*, 223. [CrossRef] [PubMed]
15. Kiferle, C.; Martinelli, M.; Salzano, A.M.; Gonzali, S.; Beltrami, S.; Salvadori, P.A.; Hora, K.; Holwerda, H.T.; Scaloni, A.; Perata, P. Evidences for a nutritional role of iodine in plants. *Front. Plant Sci.* **2021**, *12*, 616868. [CrossRef] [PubMed]
16. Caffagni, A.; Pecchioni, N.; Meriggi, P.; Bucci, V.; Sabatini, E.; Acciarri, N.; Ciriaci, T.; Pulcini, L.; Felicioni, N.; Beretta, M.; et al. Iodine uptake and distribution in horticultural and fruit tree species. *Ital. J. Agron.* **2012**, *7*, 229–236. [CrossRef]
17. Medrano-Macías, J.; Leija-Martínez, P.; González-Morales, S.; Juárez-Maldonado, A.; Benavides-Mendoza, A. Use of iodine to biofortify and promote growth and stress tolerance in crops. *Front. Plant Sci.* **2016**, *7*, 1146. [CrossRef]
18. Leyva, R.; Sánchez-Rodríguez, E.; Rubio-Wilhelmi, M.M.; Romero, L.; Ruiz, J.M.; Blasco, B. Beneficial effects of exogenous iodine in lettuce plants subjected to salinity stress. *Plant Sci.* **2011**, *181*, 195–202. [CrossRef]
19. Medrano Macías, J.; López Caltzontzit, M.G.; Rivas Martínez, E.N.; Narváez Ortiz, W.A.; Benavides Mendoza, A.; Martínez Lagunes, P. Enhancement to salt stress tolerance in strawberry plants by iodine products application. *Agronomy* **2021**, *11*, 602. [CrossRef]
20. Blasco, B.; Rios, J.J.; Cervilla, L.M.; Sánchez-Rodríguez, E.; Rubio-Wilhelmi, M.M.; Rosales, M.A.; Ruiz, J.M.; Romero, L. Photorespiration process and nitrogen metabolism in lettuce plants (*Lactuca sativa* L.): Induced changes in response to iodine biofortification. *J. Plant Growth Regul.* **2010**, *29*, 477–486. [CrossRef]
21. Chiesa, A.; Frezza, D.; Fraschina, A.; Trinchero, G.; Moccia, S.; León, A. Pre-harvest factors and fresh-cut vegetables quality. *Acta Hortic.* **2003**, *604*, 153–159. [CrossRef]
22. von Caemmerer, S.; Farquhar, G.D. Some relationships between the biochemistry of photosynthesis and the gas exchange of leaves. *Planta* **1981**, *153*, 376–387. [CrossRef] [PubMed]

23. Genty, B.; Briantais, J.M.; Baker, N.R. The relationship between the quantum yield of photosynthetic electron transport and quenching of chlorophyll fluorescence. *Biochim. Biophys. Acta* **1989**, *990*, 87–92. [CrossRef]
24. Kramer, D.M.; Johnson, G.; Kiirats, O.; Edwards, G.E. New fluorescence parameters for the determination of QA redox state and excitation energy fluxes. *Photosynth Res.* **2004**, *79*, 209–218. [CrossRef]
25. Krall, J.P.; Edwards, G.E. Relationship between photosystem II activity and CO_2 fixation in leaves. *Physiol. Plant.* **1992**, *86*, 180–187. [CrossRef]
26. Epron, D.; Godard, D.; Cornic, G.; Genty, B. Limitation of net CO_2 assimilation rate by internal resistances to CO_2 transfer in the leaves of two tree species (*Fagus sylvatica* L. and *Castanea sativa* Mill.). *Plant Cell Environ.* **1995**, *18*, 43–51. [CrossRef]
27. Lichtenthaler, H.K. Chlorophylls and carotenoids: Pigments of photosynthetic biomembranes. In *Methods in Enzymology, PlantCell Membranes*; Academic Press: Cambridge, MA, USA, 1987; pp. 350–382.
28. Vitale, L.; Vitale, E.; Guercia, G.; Turano, M.; Arena, C. Effects of different light quality and biofertilizers on structural and physiological traits of Spinach plants. *Photosynthetica* **2020**, *58*, 932–943. [CrossRef]
29. Moulehi, I.; Bourgou, S.; Ourghemmi, I.; Tounsi, M.S. Variety and ripening impact on phenolic composition and antioxidant activity of mandarin (Citrus reticulate Blanco) and bitter orange (*Citrus aurantium* L.) seeds extracts. *Ind. Crop. Prod.* **2017**, *39*, 74–80. [CrossRef]
30. Sun, B.; da Silva, J.M.R.; Spranger, I. Factors of Vanillin Assay for Catechins and Proanthocyanidins. *J. Agric. Food Chem.* **1998**, *46*, 4267–4274. [CrossRef]
31. Mancinelli, A.L.; Huang Yang, C.P.; Lindquist, P.; Anderson, R.; Rabino, I. Photocontrol of anthocyanin synthesis. *Plant Physiol.* **1975**, *55*, 251–257. [CrossRef]
32. Chung, Y.C.; Chen, S.J.; Hsu, C.K.; Chang, C.T.; Chou, S.T. Studies on the antioxidative activity of *Graptopetalum paraguayense* E. Walther. *Food Chem.* **2005**, *91*, 419–424. [CrossRef]
33. George, B.; Kaur, C.; Khurdiya, D.S.; Kapoor, H.C. Antioxidants in tomato (*Lycopersicum esculentum*) as a function of genotype. *Food Chem.* **2004**, *84*, 45–51. [CrossRef]
34. Vitale, E.; Vitale, L.; Costanzo, G.; Velikova, V.; Tsonev, T.; Simoniello, P.; De Micco, V.; Arena, C. Light spectral composition influences structural and eco-physiological traits of *Solanum lycopersicum* L. cv. 'Microtom' in response to high-LET ionizing radiation. *Plants* **2021**, *10*, 1752. [CrossRef] [PubMed]
35. Bradford, M.M. A rapid and sensitive method for the quantitation of microgram quantities of protein utilizing the principle of protein-dye binding. *Anal. Biochem.* **1976**, *72*, 248–254. [CrossRef]
36. Im, J.; Ko, J.; Kim, H.Y.; Ha, B.K. Biochemical responses of Soybean (*Glycine max* L. Merr.) to proton beam irradiation. *Plant Breed. Biotechnol.* **2017**, *5*, 97–105. [CrossRef]
37. Carillo, P.; Soteriou, G.A.; Kyriacou, M.C.; Giordano, M.; Raimondi, G.; Napolitano, F.; Di Stasio, E.; Mola, I.D.; Mori, M.; Rouphael, Y. Regulated Salinity Eustress in a Floating Hydroponic Module of Sequentially Harvested Lettuce Modulates Phytochemical Constitution, Plant Resilience, and Post-Harvest Nutraceutical Quality. *Agronomy* **2021**, *11*, 1040. [CrossRef]
38. Nunes da Silva, M.; Machado, J.; Osorio, J.; Duarte, R.; Santos, C.S. Non-Essential Elements and Their Role in Sustainable Agriculture. *Agronomy* **2022**, *12*, 888. [CrossRef]
39. Kyriacou, M.C.; Rouphael, Y. Towards a new definition of quality for fresh fruits and vegetables. *Sci. Hortic.* **2018**, *234*, 463–469. [CrossRef]
40. Chutipaijit, S.; Cha-um, S.; Sompornpailin, K. High contents of proline and anthocyanin increase protective response to salinity in Oryza sativa L. spp. indica. *Aust. J. Crop Sci.* **2011**, *5*, 1191–1198.
41. Blasco, B.; Rios, J.J.; Leyva, R.; Cervilla, L.M.; Sánchez-Rodríguez, E.; Rubio-Wilhelmi, M.M.; Rosales, M.Á.; Ruiz, J.M.; Romero, L. Does iodine biofortification affect oxidative metabolism in lettuce plants? *Biol. Trace Elem. Res.* **2010**, *142*, 831–842. [CrossRef]
42. Blasco, B.; Rios, J.J.; Leyva, R.; Melgarejo, R.; Constan-Aguilar, C.; Sanchez-Rodrıguez, E.; Rubio-Wilhelmi, M.M.; Romero, L.; Ruiz, J.M. Photosynthesis and metabolism of sugars from lettuce plants (*Lactuca sativa* L. var. longifolia) subjected to biofortification with iodine. *Plant Growth Regul.* **2011**, *65*, 137–143. [CrossRef]
43. Weng, H.X.; Hong, C.L.; Yan, A.L.; Pan, L.H.; Qin, Y.C.; Bao, L.T.; Xie, L.L. Mechanism of iodine uptake by cabbage: Effects of iodine species and where it is stored. *Biol. Trace Elem. Res.* **2008**, *125*, 59–71. [CrossRef] [PubMed]
44. Incrocci, L.; Carmassi, G.; Maggini, R.; Poli, C.; Saidov, D.; Tamburini, C.; Kiferle, C.; Perata, P.; Pardossi, A. Iodine Accumulation and Tolerance in Sweet Basil (*Ocimum basilicum* L.) With Green or Purple Leaves Grown in Floating System Technique. *Front. Plant Sci.* **2019**, *10*, 1494. [CrossRef] [PubMed]

horticulturae

MDPI

Article

Anatomical and Physiological Performance of Jojoba Treated with Proline under Salinity Stress Condition

M. S. Aboryia [1,*], El-Refaey F. A. El-Dengawy [1], Mostafa F. El-Banna [2], Mervat H. El-Gobba [1], Mahmoud M. Kasem [3], Ahmed A. Hegazy [3], Heba Metwally Hassan [4], Ahmed Abou El-Yazied [5], Hany G. Abd El-Gawad [5,*], Salem Mesfir Al-Qahtani [6], Nadi Awad Al-Harbi [6], Eldessoky S. Dessoky [7], Ismail A. Ismail [7], Mohamed M. El-Mogy [8] and El-Sayed A. EL-Boraie [9]

1 Pomology Department, Faculty of Agriculture, Damietta University, New Damietta 34517, Egypt
2 Agricultural Botany Department, Faculty of Agriculture, Mansoura University, Mansoura 35516, Egypt
3 Vegetable and Floriculture Department, Faculty of Agriculture, Mansoura University, Mansoura 35516, Egypt
4 Botany Department, Faculty of Science, Ain Shams University, Cairo 11566, Egypt
5 Department of Horticulture, Faculty of Agriculture, Ain Shams University, Cairo 11566, Egypt
6 Biology Department, University College of Tayma, Tabuk University, P.O. Box 741, Tabuk 47512, Saudi Arabia
7 Department of Biology, College of Science, Taif University, P.O. Box 11099, Taif 21944, Saudi Arabia
8 Vegetable Crops Department, Faculty of Agriculture, Cairo University, Giza 12613, Egypt
9 Vegetable and Floriculture Department, Faculty of Agriculture, Damietta University, New Damietta 34517, Egypt
* Correspondence: modyaboryia@du.edu.eg (M.S.A.); hany_gamal2005@agr.asu.edu.eg (H.G.A.E.-G.)

Citation: Aboryia, M.S.; El-Dengawy, E.-R.F.A.; El-Banna, M.F.; El-Gobba, M.H.; Kasem, M.M.; Hegazy, A.A.; Hassan, H.M.; El-Yazied, A.A.; El-Gawad, H.G.A.; Al-Qahtani, S.M.; et al. Anatomical and Physiological Performance of Jojoba Treated with Proline under Salinity Stress Condition. *Horticulturae* **2022**, *8*, 716. https://doi.org/10.3390/horticulturae8080716

Academic Editors: Agnieszka Hanaka, Małgorzata Majewska, Barbara Hawrylak-Nowak and Alessandra Francini

Received: 17 June 2022
Accepted: 5 August 2022
Published: 9 August 2022

Abstract: A field trial study was conducted for two consecutive seasons 2020 and 2021 in approximately 8-month-old jojoba plants to evaluate the physiological responses following salt treatment and the role of proline as a foliar application to enhance jojoba tolerance to salinity stress. Jojoba plants were irrigated once a week for four months with diluted seawater in concentrations of 5000, 10,000, and 15,000 ppm and tap water (control). Anti-stress proline was applied four times throughout the experiment, the first at the beginning of the experiment and another three times at 30-day intervals, at concentrations of 0, 300, and 450 ppm. The effect of proline treatments on jojoba plant behavior includes growth vegetative characteristics, namely plant height increase percentage (PHIP), shoot number increase percentage (NSIP), stem diameter increase percentage (SDIP), number of leaves, leaf thickness, leaf area, and fresh and dry weights of leaves, and chemical characteristics, namely chlorophyll a and b, total chlorophyll, carotenoids, leaf mineral contents (N, P, K, Na, and Cl), total phenolic content (TPC), and proline concentration. Moreover, the impacts of proline on hydrogen peroxide (H_2O_2), superoxide anion ($O_2^{\bullet-}$), malondialdehyde (MDA), and ion leakage (IL) under salinity stress were investigated. Briefly, proline at 450 ppm enhanced all studied growth and physiological characteristics and promoted the antioxidant system of jojoba plants compared with the control and other treatments. The anatomical structure of leaves was also examined, and favorable variations in the anatomical structure were detected in the stressed and proline-treated plants. Exogenous application of proline enhanced most of this anatomical characteristic of jojoba leaf under saline stress. In conclusion, proline as a foliar application at 450 ppm under salinity stress of 10,000 ppm enhances jojoba tolerance to salinity stress by modifying the physicochemical and morphological characteristics of jojoba plants.

Keywords: jojoba; proline; malondialdehyde; total phenolic content; climate change; seawater; salt stress; ion leakage

1. Introduction

Environmental pollution is the main reason for the phenomenon of climate change, as it leads to an increase in water and soil salinity due to the rise in seawater, which affects the sustainability of agriculture [1]. Therefore, it is necessary to move towards the selection of plant species that could resist unsuitable environmental conditions, such as salinity [2].

Jojoba (*Simmondsia Chinensis* (Link) Schneider) is a novel plant that is acclimated to unsuitable climates. It is considered an industrial crop called green gold and is grown commercially in hot arid and semi-arid regions. It is dioecious, meaning female and male flowers are carried on separate plants. Jojoba is commonly regarded as among the most powerful strategies for cultivating desert regions. Resistance to extreme temperature, dryness, and salinity, as well as a decreased risk of disease, a lesser demand for chemical fertilizers, and a high level of financial gain, are all compelling reasons to grow jojoba in desert environments [3,4]. In the early phases of its development, the jojoba necessitates the use of only a few well-established techniques and a long-term commitment to combat desert encroachment [5]. Indeed, shrubs of jojoba have deep root systems and grow in both dry and salty conditions, and their growth is satisfactory in brackish water with a salt concentration of 2000 ppm as in the Pacific Ocean and in California. More than a quarter of the world's cultivated fields have saltwater levels high enough to produce salinity stress in growing plants [6]. Because of the advantageous commercial applications of its seed oil in cosmetics and fuel for aircraft and missiles, jojoba has become an interesting alternative crop. Seeds have a high "oil" content, which collectively accounts for about half of the dry weight of the seeds; this "oil" is made up of lipids that contain straight chain liquid wax esters of uniform length.

Salt is becoming a serious agricultural problem, primarily in irrigated arid and semi-arid areas where rainfall is insufficient to wash away the excess soluble salts from the root area, but it can also occur in irrigated agricultural areas, particularly when the irrigation water is poor quality. Biochemical change that occurs in plants exposed to environmental stressors is the production of reactive oxygen species that can damage nucleic acid essential membrane lipid proteins [7]. High salinity levels in the soil decrease the production of various plants in two ways: through osmotic influences and specific ion toxicities [8,9]. Boughalleb et al. [10] found that NaCl (100–300 mM) increased the anatomical characteristics of leaves in *Medicago arborea* such as the increased thickness of lamina, upper and lower epidermis, palisade, and spongy tissues. He claims that the necessity to save water causes the leaves to become succulent, resulting in increased leaf thickness.

Proline is a protein-genic amino acid with special conformational hardness and is necessary for initial metabolism; it has been found in plant tissues under a variety of circumstances, including drought [11] and salinity stress [12]. The suggested function of proline is the regulation of osmosis, which has a system of adaptation to adverse environmental conditions such as salinity. Another proposed function is the maintenance of membranes and protein stability; growth; and the provision of carbon, nitrogen, and energy storage [13]. Proline has been shown to stimulate growth, physiochemical, and anatomical features, as well as enhance the antioxidant mechanism defenses of plants under salt stress [12,14]. Proline is an amino acid that plays an essential role in plant metabolism and growth. It protects plants from diverse stresses and supports plants in faster recuperation from stress. When the proline treatment is applied as a foliar application to plants that are susceptible to stress, proline improves the growth and other physiological properties of the plants [15].

The exogenous application of proline has raised the internal level of proline in beans [16]. In addition, foliar application of proline regulates the expression of several genes related to antioxidant enzymes under salt stress. Among such genes, one gene of 1-pyrroline-5-carboxylate synthetase is responsible for upregulating the salinity stress-induced proline accumulation [17]. Exogenous application of proline enhances plant response to abiotic stressors, notably salt by shielding them from the damaging effects of reactive oxygen species (ROS) by enhancing their endogenous amount and intermediate enzymes [18]. Foliar proline treatment enhanced leaf N, Ca^{2+}, and K^+ concentration in *Cucumis melo* treated with 150 mM salt stress [19].

The present experiment was conducted to study the favorable effect of proline as a foliar application with two concentrations on growth parameters, leaf pigments, proline content, and anatomical characteristics of jojoba plants under salinity stress.

2. Materials and Methods

2.1. Growth Experiment and Salt–Proline Treatments

Jojoba plants were obtained from a private nursery of the Egyptian Gulf Company for Desert Land Reclamation, located in Alexandria, Egypt. They were propagated by placing the tip cuttings in plastic cups containing peat moss soil; after six months, they were transplanted to polyethylene bags (40 × 20 cm) containing 10 kg of mixed, sieved, acid-washed sand, compost, and clay (2:1:1 by volume) (Figure 1A–C). Each polyethylene bag contained one plant, and there were nine bags for each treatment. Plants were irrigated after the transplanting process with tap water (tap water was analyzed according to El-Harouny et al. [20] and presented Ca, 10.89 mg^{-1}; Mg, 9.98 mg^{-1}; Na, 8.15 mg^{-1}; Zn, 9.08 μg^{-1}; Se, 1.57 μg^{-1};) and fertilized for two months by drenching with 1 g L^{-1} NPK + TE(20:20:20) + 7 g L^{-1} humic acid. Seawater was collected from the Mediterranean Sea located at New Damietta, Egypt (31°27′33.5″ N and 31°40′02.1″ E), with electrical conductivity (EC) of 44.64 dS m^{-1}(35,710 ppm) and pH of 8.07. The ion composition of New Damietta seawater was 11.23 g L^{-1} Na^+, 18.21 g L^{-1} Cl^-, 0.83 g L^{-1} Ca^{2+}, 0.92 g L^{-1} Mg^{2+}, 0.56 g L^{-1} K^+, 3.82 g L^{-1} $SO_4{}^{2-}$, and 0.12 g L^{-1} $HCO_3{}^-$. Three dilutions from seawater were prepared by using tap water (5000, 10000, and 15000 ppm) and, along with tap water as control, were used for the irrigation once a week for four months with the foliar application of proline solution once monthly at three concentrations (0, 300, and 450 ppm); the two control treatments that were sprayed with proline at concentrations of 300 and 450 were removed, and the experiment was analyzed using one-way ANOVA. The salt levels were gradually increased to avoid osmotic shock, and the experiment included 10 treatments as follows: T1 = tap water (control) without proline application; T2 = 5000 ppm salinity without proline application; T3 = 10,000 ppm salinity without proline application; T4 = 15,000 ppm salinity without proline application; T5 = 5000 ppm salinity with 300 ppm proline application; T6 = 10,000 ppm salinity with 300 ppm proline application; T7 = 15,000 ppm salinity with 300 ppm proline application; T8 = 5000 ppm salinity with 450 ppm proline application; T9 = 10,000 ppm salinity with 450 ppm proline application; T10 = 15,000 ppm salinity with 450 ppm proline application.

2.2. Measurement of Vegetative Growth Characteristics of Jojoba Plant

2.2.1. Shoot Number Increase Percentage (NSIP)

At the beginning (BS) and end (ES) of the experiment, the shoot number increase percentage was counted. The following equation was used to compute the NSIP: NSIP = [(ES − BS)/BS] × 100.

2.2.2. Plant Height Increase Percentage (PHIP)

The measurement of the plant height (cm) was calculated at the beginning (bH) and the end (eH) of the investigation by measuring the length of the main stem from the soil surface to the end of the plant. The PHIP was calculated using the following equation: PHIP = [(eH − bH)/bH] × 100.

2.2.3. Stem Diameter Increase Percentage (SDIP)

The plant stem diameter at the beginning of the experiment (bD) and at the end of the experiment (eD) on the same height starting from the soil surface was measured, where the measurement position was marked with red paint to confirm this on the plant stem. The SDIP was calculated using the following equation: SDIP = [(eD − bD)/bD] × 100.

Figure 1. (**A**) The shape of the used tip cutting in propagation; (**B**) the tip cutting shape after rooting; (**C**) the jojoba plant propagation in a greenhouse located in Alexandria, Egypt (30°53′08.7″ N and 28°45′0.7.5″ E); (**D**) the shape of the plant at the beginning of the experiment; (**E**,**F**) the shape of the plant at the end of the experiment (after four months of the experiment).

2.2.4. Number of Leaves, Leaf Thickness, and Leaf Area (LA)

The number of all mature leaves plant^{-1} was counted as a number for each replicate (in 9 seedlings per treatment). Leaf thickness was determined using a digital caliper (Digital Caliper Model 500, China). The length and width of the medial leaves were measured. Then, LA was calculated according to an equation given previously [21] as follows: leaf area (cm^2) = 0.70 × (leaf length × leaf width) − 1.06.

2.2.5. Visible Quality

Leaf injury severity was recorded according to Sun et al. [22], using a scale of 0–5 (visible score) at the end of the experiment, where 0 = dead, 1 = severe visible damage (more than 90% of leaves with necrosis and injury), 2 = moderate visible damage (50% to 90%), 3 = slight visible damage (20% to 50%), 4 = minimal visible damage (less than 20%), and 5 = no visible damage.

2.2.6. Leaf Fresh and Dry Weight

Twenty new mature leaves of each plant were weighed to calculate the average fresh weight, the same leaves were dried at 70 °C for 72 h until the weight was completely stable in two successive weights, and then the average dry weight was estimated.

2.3. Chemical Characteristics of Jojoba Plant

2.3.1. Pigment Measurements

Fresh leaf samples (0.5 g) were added to 5 mL DMF (dimethylformamide). To allow the DMF (dimethylformamide) to leach the pigments from the sample, the suspension was sonicated for 15 min at 4 °C and then held at 4 °C for 16 h. The extracting solution was centrifuged for 5 min at 16,000 rpm to remove any suspended material. Then the optical density of the clarified supernatant (1 mL) was measured using a spectrophotometer at two wavelengths (662 nm (E 662) and 650 nm (E 650)) according to [23]. In addition, carotenoids were measured using wavelength 480 nm according to Wellburn [24].

Chlorophyll a content = (12 × (E 662)) − (3.11 × (E 650))
Chlorophyll b content = (20.78 × (E 650)) − (4.88 × (E 662))
Total chlorophyll content = (17.67 × (E 650)) + (7.12 × (E 662))
Carotenoids (μg/mL) = (1000 × A480 − 0.89 × chla − 52.02 × chlb)/245

2.3.2. Proline Determination

A weight of fresh leaves (0.5 g) was homogenized in a 5 mL solution of sulfosalicylic acid (3% w/v) and centrifuged at 10,000 rpm for 10 min. A 2.0 mL volume of the supernatant was used with ninhydrin reagent and toluene solution to measure the proline content by spectrophotometer at 515 nm according to the method of Bates et al. [25]. The proline amount (mg/g DW) was determined against a standard curve of L-proline.

2.3.3. Soluble Carbohydrate Content (SCC)

SCC was extracted according to Kerepsi et al. [26]. Dry leaf powder (0.1 g) was boiled for 50 min in 10 mL distilled water with agitation and then filtered through qualitative filter paper. An aliquot (0.5 mL) of this filtrate was used to determine the SCC according to Dubois et al. [27] using D (+)-glucose as standard. The obtained results were recorded as mg/g DW.

2.3.4. Total Phenolic Content (TPC)

TPC was measured in the leaf samples using Folin–Ciocalteu reagent as described by Ainsworth and Gillespie [28]. Gallic acid was used as a standard solution in the aqueous form in the concentration range of 100 to 600 ppm. The absorbance was measured at 760 nm. TPC was represented as mg gallic acid (GAE) per 1 g fresh weight (FW).

2.3.5. Ion Leakage (IL%) and Malondialdehyde (MDA) Accumulation

To measure the percentage of ion leakage (IL%), a 5 g sample (fresh jojoba leaves) was placed in 20 mL 0.4 M mannitol for 3 h at 24 °C, and then the electrical conductivity sample was first detected (R1). Thereafter, all samples were heated in a H_2O bath at 100 °C for 30 min to quantify the final leakage after the sample reached room temperature (R2). The percentage of ion leakage (IL%) was expressed as IL (%) = (R1/R2) × 100 [29]. About 2.5 g (jojoba leaves) was used to determine the MDA content; the ground sample was mixed well with thiobarbituric acid, 500 μL of butylated hydroxytoluene ($C_{15}H_{24}O$; 2%, w/v), and 25 mL of HPO_3 in ethyl alcohol (5%, w/v). As a result of establishing the 1,1,3,3-tetraethyoxypropane concentrations of TBARS ranging from 0 to 2 mM that were equivalent to MDA in the limit from 0–1 mM, the calibration curves were constructed. Stoichiometrically, tetraethyoxypropane is converted into malondialdehyde via the acid-heating stage of the testing [30].

2.3.6. $O_2^{\bullet-}$ and H_2O_2 Production Rate

A fresh plant sample of jojoba leaves was mixed with 3 mL of a 50 mM KH_2PO_4 buffer (pH 7.8) under refrigeration at 4 °C. The reagent was mixed with polyvinylpyrrolidone (PVP 1% w/v) and centrifuged at 10,000 rpm for 15 min at 4 °C, and the degree of $O_2^{\bullet-}$ production was evaluated by the creation of nitrite from NH_2OH in the presence of $O_2^{\bullet-}$, as described by Yang et al. [31]. The optical density was measured at 530 nm. To measure the creation level of $O_2^{\bullet-}$ from the reaction equation of NH_2OH with $O_2^{\bullet-}$, a standard curve with NO_2 was utilized. The creation level of $O_2^{\bullet-}$ was recognized as mmol min^{-1} g^{-1} FW. The H_2O_2 content was established following the process described by Xu et al. [32]. One gram of jojoba leaves was mixed with 5 mL acetone. After centrifugation at 6000 rpm for 15 min at 4 °C, the obvious extraction was filled. One milliliter of the latter clear mining was added to 0.2 mL ammonia and 0.1 mL titanium sulfate (5%) and then centrifuged at

6000 rpm for 10 min at 4 °C. The pellets (titanium–peroxide complex) that were created were dispersed in 3 mL of sulfuric acid 10% (v/v) and centrifuged at 5000 rpm for 10 min at 4 °C. The optical density of the subsequent supernatant was measured at 410 nm. Using H_2O_2 as a standard curve, the H_2O_2 content was expressed and then identified as mmol min^{-1} g^{-1} FW.

2.3.7. Leaf Mineral Content Determination

Total nitrogen (N), phosphorus (P), and potassium (K) contents were determined by taking 0.3 g of samples from dried leaves out of each replicate and wet digesting them with a mixture of concentrated sulfuric and perchloric acids and calculating N, P, and K as indicated previously [33]. Na^+ and Cl^- were analyzed by using the method of [34] followed by inductively coupled plasma atomic emission spectroscopy (plasma View Duo iCAP7400) according to [35]. Na^+ and Cl^- concentrations were expressed as percentages.

2.4. Anatomical Study

In conjunction with the anatomical investigation (135 days from the beginning of applying the treatments), leaf specimens (5 × 5 mm) were taken from the midrib of the middle part of the 4th leaf including the main midvein were taken in the 2nd season. Specimens were fixed for 48 h in FAA solution (formaldehyde–acetic acid–alcohol), washed gently with sterile water, dehydrated in a series of ethanol, cleared in ethanol:xylene (3:1–1:1–1:3% and 100% xylene), and embedded in paraffin wax (52–54 °C melting points). Sectionswere made at 10–15 µm thickness using a rotary microtome, double stained with safranin–light green, cleared in clove oil, and mounted in Canada balsam according to Ruzin [36]. A light microscope (Olympus CX41, Davao City, Philippines) connected with a digital camera (TUCSEN, USB2, H Series, Fuzhou, China) was used to examine the chosen sections (five sections from each treatment) to visualize the following microscopic characteristics: thicknesses of lamina (µm), upper and lower epidermis, palisade and spongy tissue, as well as main vascular bundle dimensions (length and width in µm) of leaf mesophyll.

2.5. Statistical Analysis

The average data for two growth seasons (2020–2021) for the present study were examined statistically. In a complete randomized block design (CRBD), data were subjected to analysis of variance (ANOVA) using a one-way analysis by using the statistical program SPSS, with three replications. The means of all examined treatments results have been contrasted utilizing Duncan's multiple range test at $p \leq 0.05$.

3. Results

3.1. Vegetative Growth Characteristics of Jojoba Plant

Results as an average of the two tested seasons are presented in Figure 2. The data of the plant height increase percentage (PHIP), shoot number increase percentage (NSIP), and stem diameter increase percentage (SDIP) indicated that PHIP, NSIP, and SDIP gradually decreased with different seawater salinity levels after four months from stress. Utilizing a high concentration of seawater (15,000 ppm) resulted in the highest value of reduction which reached 6.44, 24.68, and 27.87%, respectively, as compared to the control. The highest increase in PHIP, NSIP, and SDIP was achieved with the proline at 450 ppm at the level of salinity 10,000 ppm and reached 59.50, 98.00, and 90.45%, respectively, followed by the 300 ppm proline at the same level of salinity. It was observed that using the foliar application of proline at 450 ppm under 10,000 ppm level salinity (Figure 3) resulted in a significant increase in the number of leaves (154.66), leaf thickness (0.91 mm), and leaf area (16.63 cm^2) as compared to the control that recorded 117.5, 0.59, and 11.68, respectively. Jojoba plants showed minor leaf injury after four months when irrigated with saline water at the level of 10,000 ppm and treated with proline at 450 ppm with a visual quality of 4.66 compared with other treatments, while the lowest value of visual quality was 2.66 under the effect of salinity stress at 15,000 ppm. The value of injured leaves increased slightly in the jojoba plant when the salinity level of irrigation water was increased from 5000 ppm to 15,000 ppm. The results indicated that there were no injured leaves under irrigation with tap water (control). It is clear from Figures 1 and 2 that the concentration of proline at 450 ppm positively increased all vegetative growth characteristics, modified the undesirable effects, and had the ability to overcome the deleterious impact of salt on the growth characteristics of the jojoba plant that was grown under 10,000 ppm salt compared to control and other treatments.

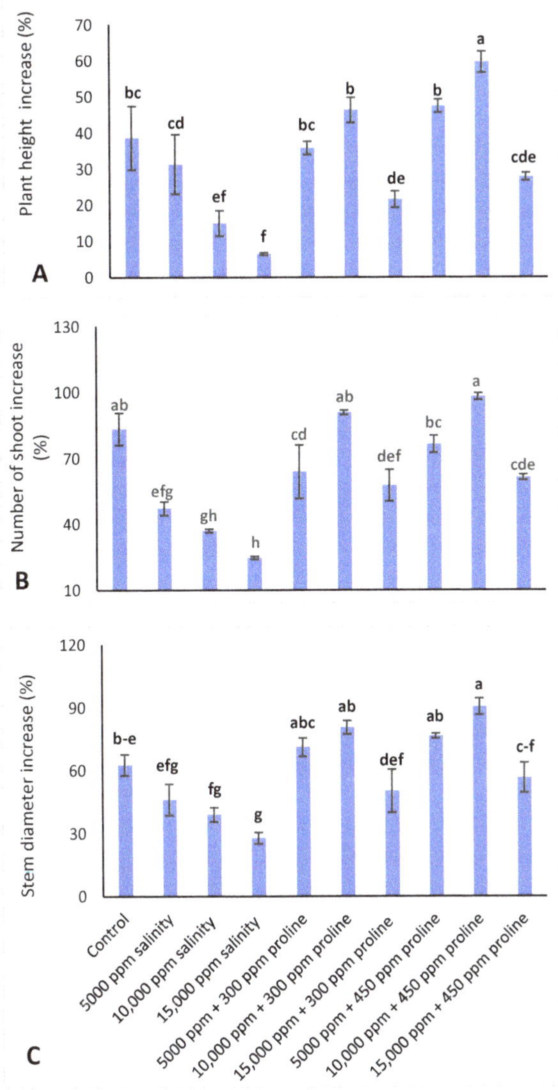

Figure 2. The effect of different levels of seawater salinity and exogenous proline at three concentrations, 0 ppm, 300 ppm, and 450 ppm on plant height increase percentage (**A**), number of shoot increase percentage (**B**), and stem diameter increase percentage (**C**). Data are means of two seasons (2020 and 2021) and three replicates (n = 3) for each season. The mean values ± SE of each parameter followed by different alphabetical letters are significantly different according to Duncan's multiple range test at $p \leq 0.05$.

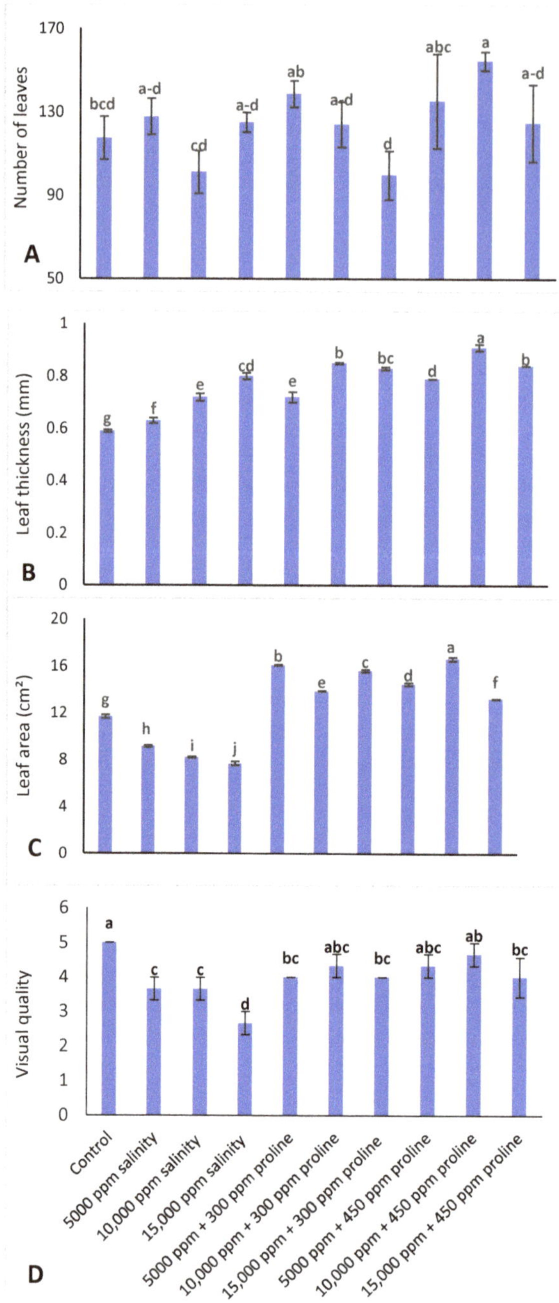

Figure 3. Effect of different levels of seawater salinity and exogenous proline at three concentrations, 0 ppm, 300 ppm, and 450 ppm, on number of leaves (**A**), leaf thickness (mm) (**B**), leaf area (cm^2) (**C**), and visual quality (**D**). Data are means of two seasons (2020 and 2021) and three replicates (n = 3) for each season. The mean values ± SE of each parameter followed by different alphabetical letters are significantly different according to Duncan's multiple range test at $p \leq 0.05$.

3.2. Leaf Fresh Weight and Dry Weight

As shown in Figure 4, leaves fresh and dry weights of jojoba plants significantly increased with the exogenous application of both proline concentrations (i.e., 300 and 450 ppm) compared to the control (tap water) and untreated plants (no proline). Maximum values of leaf fresh weight and leaf dry weight were obtained with the proline level of 450 ppm under 10,000 ppm salinity.

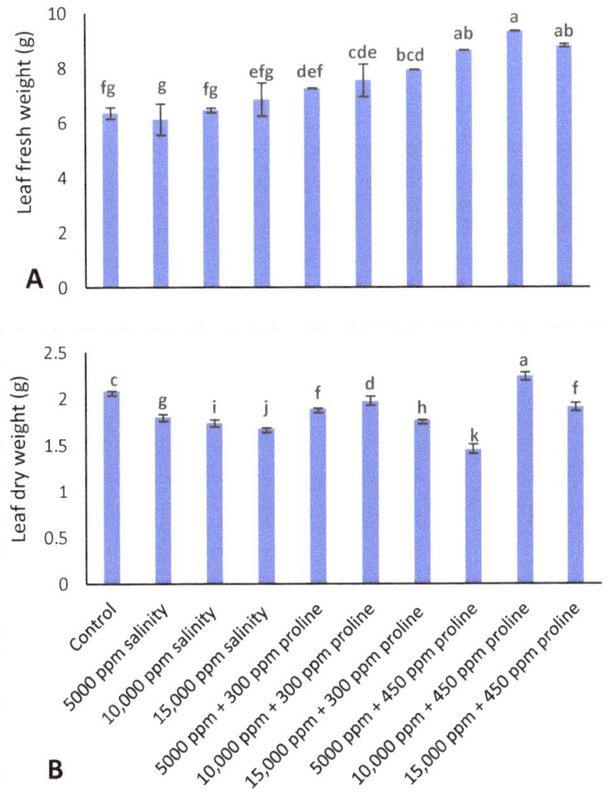

Figure 4. Effect of different levels of seawater salinity and exogenous proline at three concentrations, 0 ppm, 300 ppm, and 450 ppm, on leaf fresh weight (**A**) and leaf dry weight (**B**). Data are means of two seasons (2020 and 2021) and three replicates (n = 3) each season. The mean values ± SE of each parameter followed by different alphabetical letters are significantly different according to Duncan's multiple range test at $p \leq 0.05$.

3.3. Chemical Characters of Jojoba Plant

3.3.1. Leaf Pigments

The findings show how varying amounts of seawater salinity irrigation and foliar application of proline (300 and 450 ppm) affect the content of chlorophyll a, chlorophyll b, total chlorophyll, and carotene in the leaves of jojoba plants (Figure 5). The results showed that as salinity levels increased, total chlorophyll levels decreased. The highest levels of chlorophyll a, chlorophyll b, total chlorophyll, and carotene were 84.20, 26.03, 110.24, and 24.5 µg cm^{-2} in T9, while the lowest levels were 63.73, 16.9, 80.55, and 18.9 µg cm^{-2} for the irrigation with 15,000 ppm seawater for jojoba plants, respectively. The data clearly confirmed that the foliar application of jojoba plants by proline treatment has a highly efficient effect in enhancing the leaf content of pigments compared to non-sprayed plants under salinity stress. Remediation jojoba plants with proline (300 and 450 ppm) maintained leaf pigment contents under different levels of salinity, especially using the foliar application of proline at 450 ppm under the 10,000 ppm level of salinity.

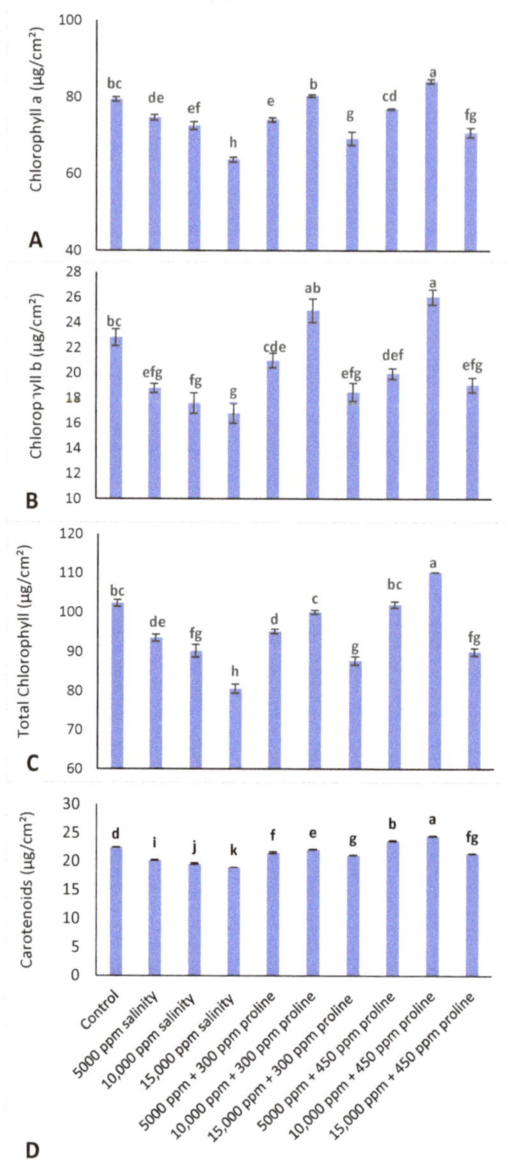

Figure 5. Effect of different levels of seawater salinity and exogenous proline at three concentrations, 0 ppm, 300 ppm, and 450 ppm, on (**A**) chlorophyll a ($\mu g/cm^2$), (**B**) chlorophyll b ($\mu g/cm^2$), (**C**) total chlorophyll ($\mu g/cm^2$), (**D**) and carotenoids ($\mu g/cm^2$). Data are means of two seasons (2020 and 2021) and three replicates ($n = 3$) each season. The mean values ± SE of each parameter followed by different alphabetical letters are significantly different according to Duncan's multiple range test at $p \leq 0.05$.

3.3.2. Ion Leakage (IL%) and Malondialdehyde (MDA)

Ion leakage (IL%) and malondialdehyde (MDA) slightly increased under different salinity levels and reached 27.83% and 29.66 μM g $^{-1}$ FW, respectively, under the salinity of 15,000 ppm as compared to those of the control plant (tap water), which reached 10.5% and 20.5 μM g $^{-1}$ FW, respectively. The frequency of accumulation was affected by salt levels up to 15,000 ppm as well as proline treatments. Foliar application of proline at 450 ppm decreased the accumulation of ion leakage (IL%) to 13% and malondialdehyde (MDA) to 21 μM g^{-1} FW in the salinized jojoba plants, especially for those plants irrigated with 10,000 ppm diluted seawater (Figure 6). When jojoba was sprayed with proline at a concentration of 450 ppm under a salt seawater irrigation level of 10,000 ppm, it was shown to be more tolerant to salt stress.

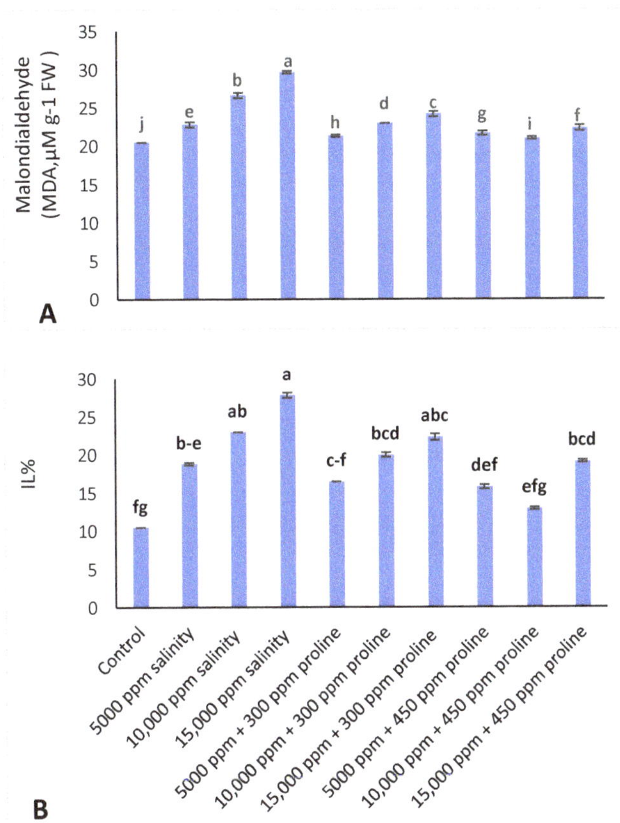

Figure 6. Effect of different levels of seawater salinity and exogenous proline at three concentrations, 0 ppm, 300 ppm, and 450 ppm, on MDA (μM g $^{-1}$ FW) (**A**) and IL (%) (**B**). Data are means of two seasons (2020 and 2021) and three replicates (n = 3) each season. The mean values ± SE of each parameter followed by different alphabetical letters are significantly different according to Duncan's multiple range test at $p \leq 0.05$.

3.3.3. $O_2^{\bullet-}$ and H_2O_2 Accumulation Rate

Results indicated that jojoba plants irrigated with saline water significantly affected the $O_2^{\bullet-}$ and H_2O_2 accumulation rate. After four months of salt stress, $O_2^{\bullet-}$ and H_2O_2 accumulation was markedly increased in all different seawater treatments as compared to the control, except in the jojoba plants treated with proline 450 ppm under the salinity level of 10,000 ppm, in which $O_2^{\bullet-}$ and H_2O_2 accumulation increased slightly and reached

0.28 mmol min^{-1} g^{-1} FW and 0.09 mmol min^{-1} g^{-1} FW, respectively. The use of proline at 450 ppm decreased the accumulation rate of $O_2^{\bullet-}$ and H_2O_2 under the salinity level of 10,000 ppm (Figure 7).

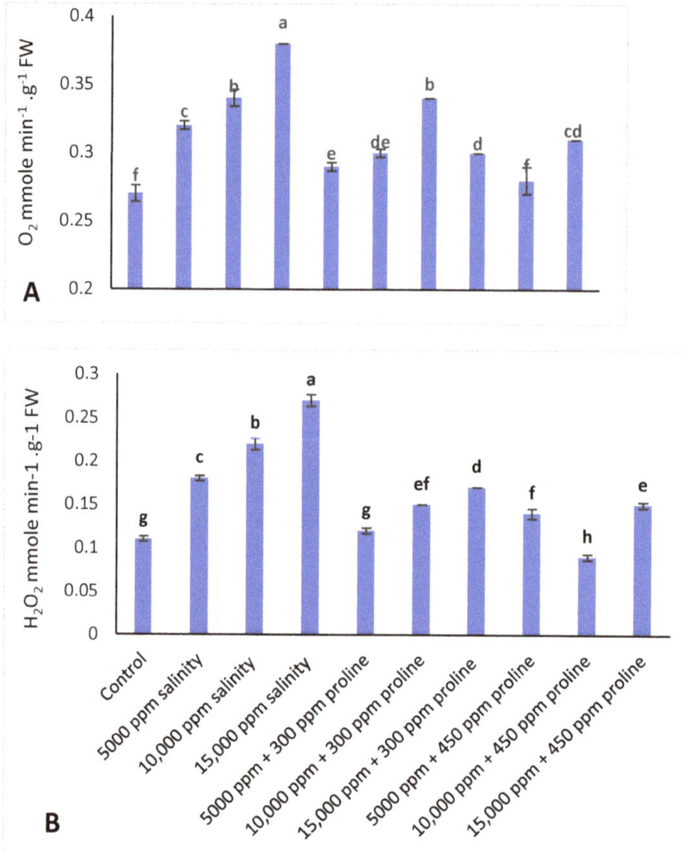

Figure 7. Effect of different levels of seawater salinity and exogenous proline at three concentrations, 0 ppm, 300 ppm, and 450 ppm, on $O_2^{\bullet-}$ (mmol min^{-1} g^{-1} FW) (**A**) and H_2O_2 (mmol min^{-1} g^{-1} FW) (**B**). Data are means of two seasons (2020 and 2021) and three replicates (n = 3) each season. The mean values ± SE of each parameter followed by different alphabetical letters are significantly different according to Duncan's multiple range test at $p \leq 0.05$.

3.3.4. Proline Content, Soluble Carbohydrate Content (SCC), and Total Phenols

The results shown in Figure 8 evidence a significant increase in the proline content, soluble carbohydrate content (SCC), and total phenol content of jojoba leaves under the impact of irrigation with seawater at 5000 ppm compared to the control. In addition, irrigation with a higher salinity level (10,000 and 15,000 ppm) caused a significant increase in the proline content of jojoba leaves compared to non-salty water (control). The changes in SCC and phenol contents were not significant. The increase in salinity level from 5000 ppm to 15,000 ppm led to a significant decrease in the soluble carbohydrate content (SCC) and total phenol content. Treating jojoba plants with proline at both concentrations (300 and 450 ppm) affected the proline content of jojoba plants under the studied levels of seawater irrigation compared to non-treated plants. The use of proline at 450 ppm led to a significant decrease in the proline content at the salinity level of 10,000 ppm and a significant increase in soluble carbohydrate content (SCC) and total phenol content. From the presented results, we concluded that jojoba plants treated with proline at 450 ppm were more efficient at the 10,000 ppm salt level.

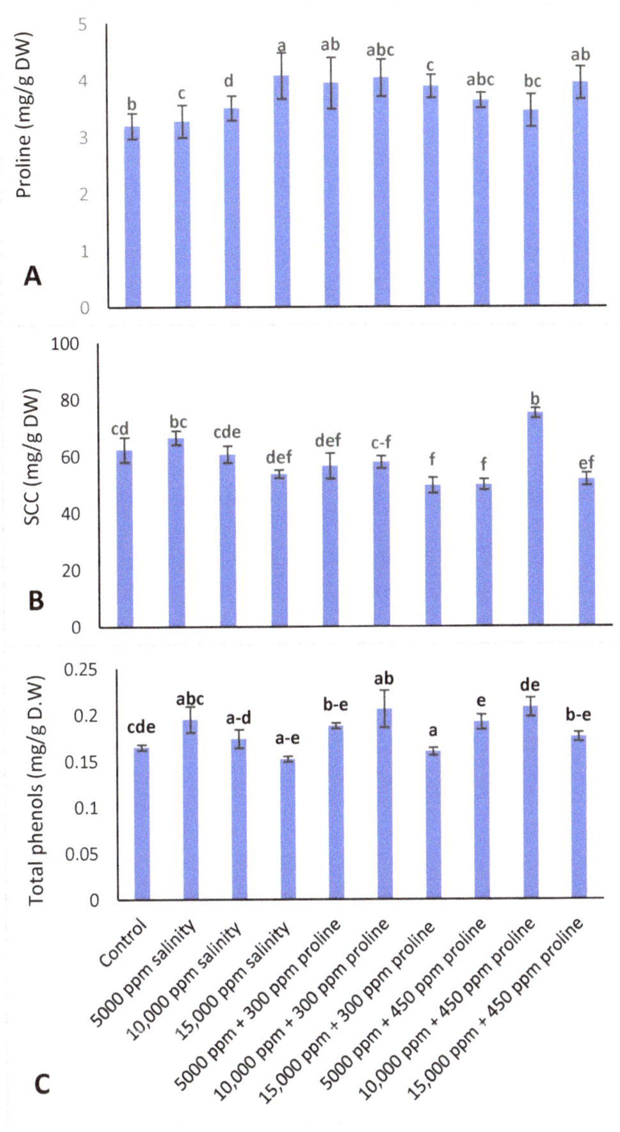

Figure 8. Effect of different levels of seawater salinity and exogenous proline at three concentrations, 0 ppm, 300 ppm, and 450 ppm, on proline content (mg/g DW) (**A**), soluble carbohydrate content (SSC) (mg/g DW), (**B**) and total phenol content (mg/g DW) (**C**). Data are means of two seasons (2020 and 2021) and three replicates (n = 3) each season. The mean values ± SE of each parameter followed by different alphabetical letters are significantly different according to Duncan's multiple range test at $p \leq 0.05$.

3.3.5. Leaf Mineral Concentration

It is observable in Figure 9 that salinity at 5000 ppm increased the concentrations of N, P, and K in the leaves of jojoba plants. Salinity concentrations at 10,000 and 15,000 decreased these concentrations significantly. The highest concentration of N, P, and K was registered in plants irrigated with 10,000 ppm of salinity and sprayed with 450 ppm of proline; they had concentrations of 52, 41, and 55 mg/100 g DW, respectively, compared to the control plants. When salt levels in seawater irrigation were raised, the levels of Na^+ and Cl^- gradually increased. Plants that were irrigated with 10,000 ppm salinity and received a foliar treatment of 450 ppm proline had the lowest levels of Na^+ and Cl^- (0.4 and 1.1 ppm, respectively), when compared to the other salinity concentrations (Figure 10).

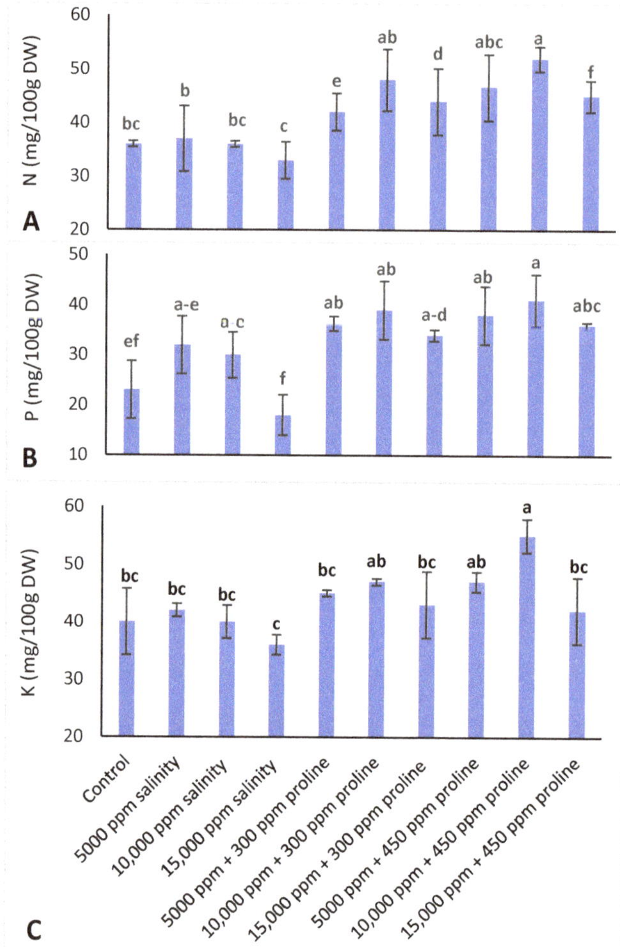

Figure 9. Effect of different levels of seawater salinity and exogenous proline at three concentrations, 0 ppm, 300 ppm, and 450 ppm, on N content (mg/100 g DW) (**A**), P content (mg/100 g DW) (**B**), and K content (mg/100 g DW) (**C**). Data are means of two seasons (2020 and 2021) and three replicates (n = 3) each season. The mean values ± SE of each parameter followed by different alphabetical letters are significantly different according to Duncan's multiple range test at $p \leq 0.05$.

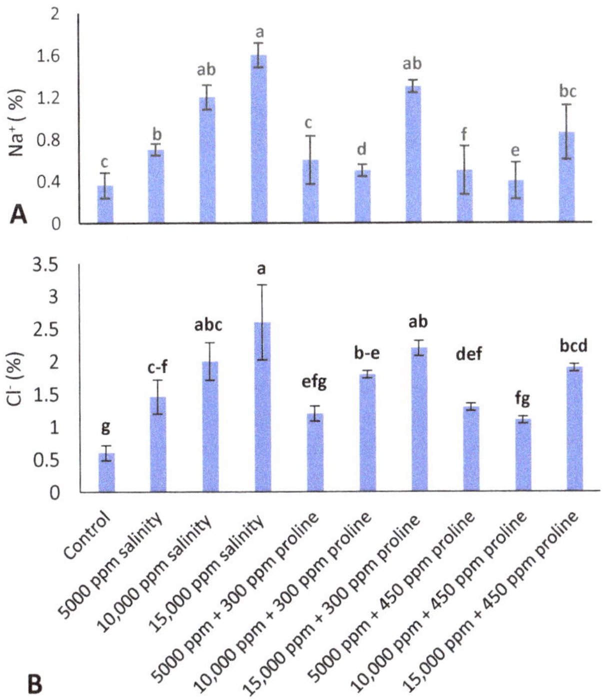

Figure 10. Effect of different levels of seawater salinity and exogenous proline at three concentrations, 0 ppm, 300 ppm, and 450 ppm, on Na^+ content (**A**) and Cl^- content (%) (**B**). Data are means of two seasons (2020 and 2021) and three replicates (n = 3) each season. The mean values ± SE of each parameter followed by different alphabetical letters are significantly different according to Duncan's multiple range test at $p \leq 0.05$.

3.3.6. Anatomical Characterization of the Leaf Structure

Data in Table 1 and Figures 11–13 showed that salinity adversely affected the leaf anatomical structure of jojoba plants. A prominent increase was noted in lamina thicknesses "succulence" with increasing salinity levels (by 19.81, 21.78, and 26.08% at 5000, 10,000, and 15,000 ppm, respectively) compared to control plants. The increment in the thickness of lamina could be attributed to the increase in the thickness of both upper and lower epidermis by 12.82% and 28.21%, 33.33% and 32.43%, and 21.62% and 24.32%, respectively. In addition, the thickness of palisade tissue increased by 12.29 and 38.57% and that of spongy tissue increased by 20.74 and 17.39, respectively, under low and moderate salinity levels. Under high salinity levels, the thickness of palisade tissue was decreased by 27.14%, and the spongy tissue increased in thickness by 38.13% in comparison with the control plants. Additionally, under all salinity levels, the dimensions of the main vascular bundle (thickness and width) were decreased compared with the control.

Table 1. Anatomical modification of jojoba leaves as affected by seawater salinity and exogenous proline. Data are means of the second season (2021) and five replicates (n = 5). The mean values ± SE of each parameter in the same column followed by different alphabetical letters are significantly different according to Duncan's multiple range test at $p \leq 0.05$.

Treatments	Lamina Thickness		Upper Epidermis Thickness		Palisade Tissue Thickness		Spongy Tissue Thickness		Lower Epidermis Thickness		Dimensions of Main Vascular Bundle Thickness		Width	
	µm	± % to S_0 *	µm	± % to S_0	µm	± % to S_0	µm	± % to S_0	µm	± % to S_0	µm	± % to S_0	µm	± % to S_0
T1	144.20 ± 2.61e	0.00	7.72 ± 0.37e	0.00	24.50 ± 0.55cd	0.00	104.65 ± 3.15d	0.00	7.33 ± 0.24c	0.00	98.00 ± 1.75b	0.00	98.70 ± 0.42b	0.00
T2	172.76 ± 2.36c	+19.81	8.71 ± 0.37de	+12.82	28.00 ± 0.78b	+14.29	126.35 ± 1.28b	+20.74	9.70 ± 0.48a	+32.43	84.35 ± 0.65e	−13.93	92.05 ± 1.30c	−6.74
T3	175.61 ± 1.45bc	+21.78	9.90 ± 0.31bc	+28.21	33.95 ± 0.89a	+38.57	122.85 ± 1.16b	+17.39	8.91 ± 0.76a-c	+21.62	81.20 ± 0.89f	−17.14	77.35 ± 0.65de	−21.63
T4	181.80 ± 1.70ab	+26.08	10.30 ± 0.24ab	+33.33	17.85 ± 1.28e	−27.14	144.55 ± 1.88a	+38.13	9.11 ± 0.20ab	+24.32	78.75 ± 0.55f	−19.64	74.55 ± 1.18e	−24.47
T5	159.89 ± 1.40d	+10.88	8.32 ± 0.24de	+7.69	27.65 ± 1.02bc	+12.86	116.20 ± 0.89c	+11.04	7.72 ± 0.48bc	+5.41	94.15 ± 0.65c	−3.93	98.70 ± 0.78c	0.00
T6	130.10 ± 1.37f	−9.78	10.30 ± 0.39ab	+33.33	18.55 ± 1.42e	−24.29	91.35 ± 2.02e	−12.71	9.90 ± 0.54a	+35.14	90.65 ± 0.65d	−7.50	89.25 ± 0.95a	−9.57
T7	124.83 ± 4.29f	−13.43	9.31 ± 0.39b-d	+20.51	23.80 ± 1.62d	−2.86	84.00 ± 3.79f	−19.73	7.72 ± 0.37bc	+5.41	80.50 ± 1.66f	−17.86	78.05 ± 5.35de	−20.92
T8	183.32 ± 0.96a	+ 27.13	9.11 ± 0.19cd	+17.95	21.70 ± 1.80d	−11.43	144.20 ± 2.85a	+37.79	8.32 ± 0.24a–c	+13.51	107.80 ± 1.18a	+10.00	105.00 ± 2.18b	+6.38
T9	140.82 ± 1.25e	−2.34	8.91 ± 0.54cd	+15.38	24.50 ± 0.55cd	0.00	98.70 ± 1.18d	−5.69	8.71 ± 0.48a–c	+18.92	94.15 ± 1.02c	−3.93	102.20 ± 0.89ab	+3.55
T10	145.65 ± 1.79e	+ 1.01	11.09 ± 0.37a	+43.59	22.05 ± 0.70d	−10.00	103.60 ± 2.17d	−1.00	8.91 ± 0.70a–c	+21.62	86.80 ± 0.42e	−11.43	82.95 ± 1.18d	−15.96

* S_0 = control; T1 = control "tap water" without proline application; T2 = 5000 ppm salinity without proline application; T3 = 10,000 ppm salinity without proline application; T4 = 15,000 ppm salinity without proline application; T5 = 5000 ppm salinity with 300 ppm proline application; T6 = 10,000 ppm salinity with 300 ppm proline application; T7 = 15,000 ppm salinity with 300 ppm proline application; T8 = 5000 ppm salinity with 450 ppm proline application; T9 = 10,000 ppm salinity with 450 ppm proline application; T10 = 15,000 ppm salinity with 450 ppm proline application.

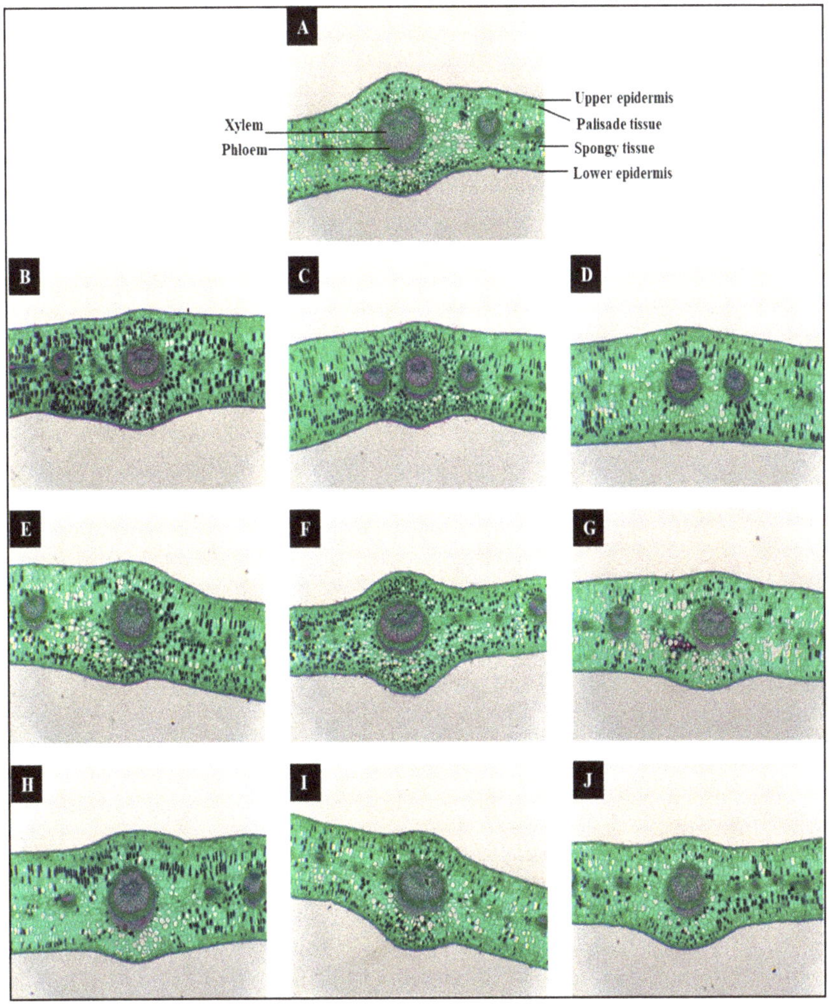

Figure 11. Microphotographs of transverse sections through the blade of leaves on the median portion of jojoba plants (135 days from the beginning of applying treatments) as affected by salinity and proline. (**A**) Control "tap water" without proline application; (**B**) 5000 ppm salinity without proline application; (**C**) 10,000 ppm salinity without proline application; (**D**) 15,000 ppm salinity without proline application; (**E**) 5000 ppm salinity with 300 ppm proline application; (**F**) 10,000 ppm salinity with 300 ppm proline application; (**G**) 15,000 ppm salinity with 300 ppm proline application; (**H**) 5000 ppm salinity with 450 ppm proline application; (**I**) 10,000 ppm salinity with 450 ppm proline application; (**J**) 15,000 ppm salinity with 450 ppm proline application. (Obj. × 4 and Oc. × 10.).

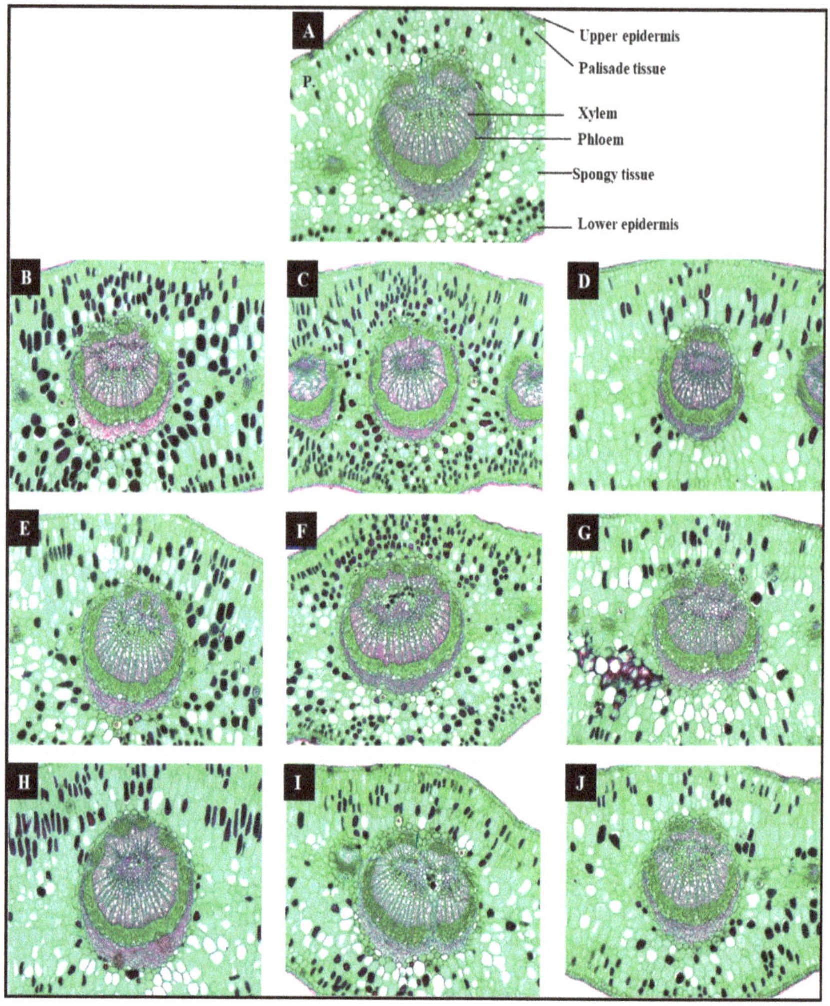

Figure 12. Microphotographs of transverse sections through the blade of leaves on the median portion of jojoba plants (135 days from the beginning of applying treatments) as affected by salinity and proline. (**A**), Control "tap water" without proline application; (**B**), 5000 ppm salinity without proline application; (**C**), 10,000 ppm salinity without proline application; (**D**), 15,000 ppm salinity without proline application; (**E**), 5000 ppm salinity with 300 ppm proline application; (**F**), 10,000 ppm salinity with 300 ppm proline application; (**G**), 15,000 ppm salinity with 300 ppm proline application; (**H**), 5000 ppm salinity with 450 ppm proline application; (**I**), 10,000 ppm salinity with 450 ppm proline application; (**J**), 15,000 ppm salinity with 450 ppm proline application. (Obj. × 10 and Oc. × 10.).

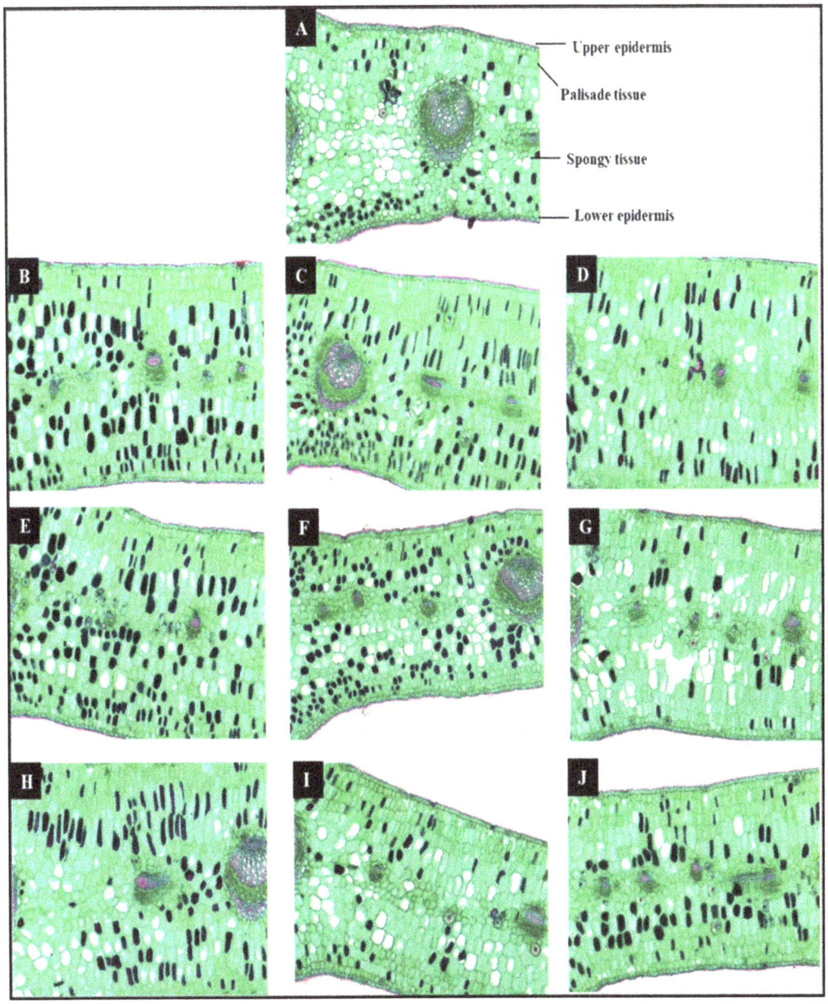

Figure 13. Microphotographs of transverse sections through the blade of leaves on the median portion of jojoba plants (135 days from the beginning of applying treatments) as affected by salinity and proline. (**A**) Control "tap water" without proline application; (**B**) 5000 ppm salinity without proline application; (**C**) 10,000 ppm salinity without proline application; (**D**) 15,000 ppm salinity without proline application; (**E**) 5000 ppm salinity with 300 ppm proline application; (**F**) 10,000 ppm salinity with 300 ppm proline application; (**G**) 15,000 ppm salinity with 300 ppm proline application; (**H**) 5000 ppm salinity with 450 ppm proline application; (**I**) 10,000 ppm salinity with 450 ppm proline application; (**J**) 15,000 ppm salinity with 450 ppm proline application. (Obj. × 10 and Oc. × 10.)

Generally, notably, improvements in leaf anatomy for salt-stressed plants with proline at 300 or 450 ppm induced a reduction in the deleterious effects of salinity on anatomical characteristics. Proline reduced the sharp decrease in the dimensions of the main vascular bundle compared with control; the most efficient treatment was 450 ppm proline. Under low salinity levels, the application of 450 ppm proline increases the thickness and width of the main vascular bundle by 10.00 and 6.38%, respectively. However, a moderate level led to an increase in the width of the main vascular bundle by 3.55%, while the thickness was decreased by 3.93% as compared to control.

4. Discussion

Jojoba (*Simmondsia chinensis* (Link) Schneider) is a unique plant that has adapted to harsh conditions such as heat, drought, and salinity. Salinity stress has been seen as a massive problem and a vital aspect of the world's long-term agricultural sustainability. It has an impact on plant growth and causes significant production and quality decreases in stressed plants. Proline improves plant tolerance to various abiotic stressors by increasing their endogenous level and intermediate enzymes. Vegetative parameters have been measured to evaluate the effects of salinity stress and proline as an anti-stress treatment on jojoba plants. Our data indicated that the jojoba plant's vegetative characteristics were negatively correlated with increasing seawater salinity levels. Values of plant height increase percentage (PHIP), shoot number increase percentage (NSIP), stem diameter increase percentage (SDIP) (Figure 2), number of leaves, leaf thickness, leaf area, and visual quality (Figure 3) were reduced with rising seawater irrigation salinity levels, especially at higher salinity levels (15,000 ppm). These results are in agreement with previous works [37,38] on pomegranates. In contrast, foliar application of proline in different concentrations minimizes the damaging impacts of salt stress on vegetative characteristics of jojoba plants. Amino acids are well-known biostimulants that have a beneficial effect on plant growth and significantly alleviate the damage caused by abiotic stress [39,40]. Proline is an amino acid that plays an essential role in plant metabolism and growth. It protects plants from diverse stresses and supports plants in recuperating faster from stress. When proline treatment was applied as a foliar application to plants that are susceptible to stress, proline improved the growth and other physiological properties of the plants [17]. Moreover, El-Sherbeny et al. [41] found that a foliar application of 100 mg L^{-1} proline increased the plant height, number of branches, and fresh and dry weights of leaves of *Beta vulgaris* L. plants. Dawood et al. [14] indicated that the foliar application of 25 mM proline caused significant increases in growth parameters of *Vicia faba* compared with the control. The idea that certain amino acids might alter the growth and development of plants via their impact on gibberellin biosynthesis could explain the regulatory effect of amino acids on growth [42]. Furthermore, according to Lea and Fowden [43], amino acids, which are the building blocks of proteins, can perform a variety of other tasks within the regulation of nitrogen metabolism, transport, and storage of nitrogen. Çiçek and Çakirlar [44] reported that salinity stress determines the capacity of plant cells to absorb water and some nutrients dissolved in the soil and reduces plant growth. In contrast, the foliar application of proline, especially the level of 6 mM, significantly improved internal proline levels in lupine plants, increasing their tolerance to salt stress and therefore enhancing lupine vegetative growth properties.

Using seawater irrigation under a high level of salinity up to 15,000 ppm causes inhibited growth resulting in the reduction in fresh and dry weights of the jojoba plant. Spraying proline in both concentrations of 300 and 450 ppm under different salinity levels presents beneficial effects on growth, especially under the salt level of 10,000 ppm. In a previous work on *Medicago sativa* callus cells, exogenously introducing proline into the culture medium under salt stress led to an increase in dry weight and also an increased content of free proline in the callus cells [45]. The increase in plant biomass due to the foliar application of consistent osmolytes can be attributed to an active role of these osmolytes in the osmotic transformation of plants, which in turn improved water uptake and enhanced plant growth. Proline's favorable benefits on plant vegetative development in a variety of plant species maintained under stress conditions might be linked to its dual role as a nutrient and an osmoprotectant [46].

Leaf chlorophyll, one of several biochemical attributes, is an essential property reflecting plant health status and is related to plant water availability and nutrient content [14]. In our study, detrimental effects of salt stress on leaf chlorophyll content were reported several crops [12,47]. The decline in chlorophyll levels in most stressed plants may be due to disorganization of the thylakoid membranes, with more degradation than synthesis of chlorophyll via the formation of proteolytic enzymes such as chlorophyllase, which are responsible for breaking down chlorophyll and damaging the photosynthetic apparatus [48].

This reduces the rate of photosynthesis in plants [49] and inhibits ion accumulation [50]. In contrast, the application of 450 ppm proline caused a significant increase in photosynthetic pigments in salt-stressed jojoba plants. These improved chlorophyll concentrations can be attributed to this treatment stimulating chlorophyll biosynthesis and/or inhibiting its breakdown. Furthermore, these increases in chlorophyll concentrations could be attributed to more efficient scavenging of ROS by proline and other antioxidant compounds [12]. It is possible that proline's beneficial influence on chlorophyll concentrations during salt stress is related to the stability of photosynthetic processes. Abdelhamid et al. [12] also reported that carotenoids help in the production of vital nutrients related to photosynthesis. These pigments give fruits red, yellow, or orange colors. The beneficial effects of additional antioxidants on plant survival under salt stress associated with partial inhibition of ROS formation have been described in several papers [51–53]. Proline application significantly increased carotenoid concentrations in the leaves of jojoba plants. Carotenoids have been recognized as protecting the photosynthetic apparatus against photoinhibition damage caused by singlet oxygen ($^1O_2$) created by chlorophyll's triplet excited state. Carotenoids directly deactivate and also quench singlet oxygen in the triplet excited state of chlorophyll, thereby indirectly reducing the formation of singlet oxygen species [54].

The accumulation of MDA and IL% increased significantly at $p \leq 0.05$. Malondialdehyde (MDA) is a natural product of lipid peroxidation and has traditionally been used as an indicator of the degree of cell damage from stress [55]. Plant exposure to salt stress causes negative effects, the most important of which is an imbalance of cellular ions, leading to ion toxicity and osmotic stress, while high levels of salinity induce the generation of reactive oxygen species (ROS), which are considered highly toxic oxygen derivatives, especially superoxide ($O_2^{\bullet-}$) and hydrogen peroxide (H_2O_2), and as a result of the Fenton reaction in plants [56,57], the most hazardous hydroxyl radicals ($OH^{\bullet}$) develop, which can interact with a variety of important macromolecules and metabolites, causing damage to cellular structures [58,59]. Reactive oxygen species (ROS) detoxification pathways play a defensive role in the salt stress response by removing toxic radicals generated from the mitochondrial and chloroplast electron transport chains. Proline protects plants from stress by maintaining osmoregulation and detoxifying ROS, thus conserving membrane integrity and stabilizing enzymes and other proteins [60,61].

ROS are produced in plant cells under normal physiological conditions, either in a radical or non-radical form [62]. Excessive ROS generation, on the other hand, causes oxidative damage to the cell's proteins, lipids, nucleic acids, and plasma membrane. To protect cells from oxidative damage, plants must produce low-molecular-weight non-enzymatic antioxidants such as proline, glutathione, and ascorbate, as well as enzymatic antioxidants such as peroxidase (POX), superoxide dismutase (SOD), ascorbate peroxidase (APX), and catalase (CAT) to ward off oxidative stress [57,59]. One of the effective protective mechanisms of the plant against hyperstress is the increase in the content of internal compatible solutes such as proline, glycine betaine, and sorbitol [57]. Free proline is generated in many plants in response to the application of a wide range of biotic and abiotic stresses. The ability of proline to regulate osmotic adaptation, settle subcellular systems, and scavenge harmful oxygen species has been the target of most research. High levels of proline are synthesized under stress conditions and also maintain the NAD(P)+/NAD(P)H ratio [63]. Proline is a multifunctional amino acid and also a signaling molecule acting as a plant growth regulator by triggering cascade signaling processes [18]. Proline is preferred as a common osmolyte in plants and is upregulated in response to different stresses [18,59]. Its accumulation in plants protects them from drought and salt stress. Exogenous application of proline enhances plant response to abiotic stressors, notably salt, by shielding them from the damaging effects of reactive oxygen species (ROS) [18]. Plants tend to enhance their endogenous level of proline under continuously increasing levels of salinity. This study focused on the adverse impact of salt stress on plants, and how plants survive under irrigation with salt water by increasing their endogenous level of proline.

Our results show that salt stress or the exogenous implementation of proline under salt stress increased the content of proline in the jojoba plants (Figure 8). The increased levels of proline under salt stress were also documented in wheat and common bean [64,65]. It has been suggested that the accumulation of proline may be generated by the increase in proteolysis or by the reduction in protein synthesis. Because proline is involved in the osmotic potential of leaves and consequently in osmotic adjustment, plants benefit from a greater proline content under salt stress.

Phenolic compounds protect plants from a range of biotic and abiotic stresses. The oxidation of phenols generates many defenses that alter plant physiology and metabolism, which in turn allow the plant to resist diverse stresses either directly or through the mediation of various plant signaling pathways. From our results (Figure 8) we deduced that the use of foliar proline at a level of 450 ppm was more efficient at the salinity of 10,000 ppm, which resulted in a significant increase in the content of SCC and polyphenols in the leaves of the jojoba plant. Such findings confirmed that proline application mitigates adverse salinity and increases polyphenols and SCC [66,67].

Salt stress not only increases sodium (Na^+) and chloride (Cl^-) in plants but also causes reductions in calcium (Ca^{2+}), potassium (K^+), magnesium (Mg^{2+}), nitrate (NO_3^-), sulfur (S), and other vital nutrients, leading to general nutrient insufficiency [68]. The favorable impacts of foliar proline on plant toleration to salinity stress have been related to raised intake of nutrients in many investigations. It was documented that using foliar proline application increased P, K, NO_3^-, and NO_2^- contents in *Phaseolus vulgaris* under different levels of salinity [50]. Similarly, foliar proline treatment enhanced leaf N, Ca^{2+}, and K^+ concentration in *Cucumis melo* treated with 150 mM salt stress [19]. In addition, foliar application of proline increased Ca^{2+} and K^+ in *Sorghum bicolor* under salt stress [68]. It was reported that the use of foliar proline can raise the uptake of N, P, K^+, and S in *Zea mays* under salt conditions [69]. Besides nutrient absorption, the activities of some enzymes involved in nutrient absorption are catalyzed by foliar proline under salinity stress. Nitrate reductase is one of the key enzymes involved in nitrogen absorption, and foliar proline promotes its activity in *Helianthus annuus* [70] and *Cucurbita melo* [46] exposed to salt stress. Some authors have suggested that proline may provide a good way to accumulate and recycle nitrogen under salt stress [71,72]. Exogenous proline was also utilized as a source of nitrogen (N) by *Vigna radiata* L. seedlings under stress conditions [73]. Elevated salt levels increase sodium (Na^+) and chloride (Cl^-) contents in plants and reduce the levels of other cations such as potassium ($K^{+)}$ and calcium (Ca^{2+}), causing mineral nutrient imbalance [74]. Clearly, maintaining ion homeostasis is one of the adaptive systems that resistant plants use to manage salt stress. These mechanisms might help the plant reduce the harmful effects of ions such as Na^+ and Cl^-, which can cause a variety of injuries to lipids, proteins, and nucleic acids [75]. Recently, Gholami Zali et al. [66] found that salt-stressed *Zea mays* subjected to foliar treatment with proline exhibited a decrease in both Na^+ and Cl^- concentrations and an increase in the K^+ concentration and the K^+/Na^+ ratio. Identical effects have been registered for *Sorghum bicolor* [68]. Previous work established that external proline relieved the adverse influence of 120 mM salt, increased K^+ level, and decreased Na^+ level in *Helianthus annuus* [70]. Under salinity, foliar proline enhanced salt resistance of *Olea europaea* by preserving low Na^+, high K^+, and decreased Na^+/K^+ and Na^+/Ca^{2+} ratios in plant leaves [76]. Salt tolerance is not usually induced by proline in this way. Exogenous proline has little influence on Na^+ and K^+ levels in the leaves of cucumber, although it does increase leaf water content below 100 mM NaCl [77]. This increased water content is due to the external proline can dilute the salt and therefore restrict salt toxicity, leading to better plant growth.

In accordance with this study, the prominent increase in leaf lamina thickness with increasing levels of salinity and succulence for jojoba plants was recorded under different salinity levels, indicating a corresponding increase in mesophyll tissue which could be attributed to the necessity to conserve water rendering the leaves succulent. In general, the increased leaf thickness in this investigation is compatible with the results of Roussos et al. [78] and Gonzalez et al. [79] on *jojoba Simmondsia chinensis* Link, Akcin et al. [80] on *Spergularia marina*, and Debez et al. [81] on Cakile. Additionally, [82] showed a significant increase in leaf thickness due to salt treatment with 250 mM NaCl (spongy mesophyll thickness increased by 38% while palisade mesophyll cell length was increased by 50% over controls). Moreover, Boughalleb et al. [10] found that NaCl (100–300 mM) increased the anatomical characteristics of leaves in *Medicago arborea*, including the thickness of lamina, upper and lower epidermis, palisade, and spongy tissues. The authors claim that the necessity to save water causes the leaves to become succulent, resulting in increased leaf thickness. In this regard, Waisel et al. [83] suggested that the succulence exercises a dilution effect upon the toxic ions from the cells and upon the salts accumulated in plant organs, thus permitting the plant to cope with higher salt amounts. Additionally, [84] reported that succulence is one of the mechanisms that halophytes utilize to deal with high internal ion concentrations.

Additionally, previous works reported that salinity in halophyte plants catalyzes vacuolization in the parenchyma cells; these anatomical changes may help in the storage of ions inside the plant organs and protect the cytoplasm from toxic ion levels [10,85]. In addition, [86] found that salt-tolerant plants have a tendency to increase the leaf thickness, consequently inducing the succulence of leaves. Roussos et al. [78] reported that leaves tended to be thicker with increasing salinity, which is probably the result of high water accumulation within their tissues. In this regard, Debez et al. [84] pointed out that increased succulence possibly aids in storing additional water by increasing vacuolar volume at higher salt concentrations.

Under all salinity levels, the dimensions of the main vascular bundle (thickness and width) were decreased. In this regard, previous studies found that salinity decreased the dimensions of the main vascular bundle of the mature leaf of *Fragaria x ananassa* Duch. [87], *Vigna unguiculata* [88], and *Vitis vinifera* [89]. This reduction may be due to the restriction of cell division and expansion as well as hampering procambial activity [90]. In addition, Kozlowski [91] stated that under saline conditions, the xylem tissue was much more abundant and narrower than that in normal conditions. Since large vessels cannot provide a good flow of water, plants have developed these adaptation mechanisms [92]. Additionally, Abd Elbar et al. [93] explained that the decrement in the dimensions of the main vascular bundle caused by abiotic stress could decrease water translocation on one hand but on the other hand could help protect the water column from embolism.

It is evident from the current studies that were notable improvements in leaf anatomy for salt-stressed plants receiving proline treatment, which induced a reduction in the deleterious effects of salinity on anatomical characteristics. These results are generally in agreement with the results obtained by Abdelaal [94], Hussein et al. [95], and Dawood et al. [14] for mung bean, jatropha, and faba bean, respectively. Foliar application of proline enhanced most of these anatomical characteristics in the leaves of stressed plants, and these results suggested that the treatment with 450 ppm proline had the ability to reduce the deleterious effect of salinity stress on the histological structure of jojoba leaves. On the contrary, treatment with proline reduced succulence (lamina thickness) which resulted from exposure to salinity conditions in comparison with the control and salinity treatment as well, except for the treatment with proline under low and high salinity levels, where the thickness of the blade increased by 27.13 and 1.01%, respectively.

5. Conclusions

The obtained results from the present investigation showed that the foliar spraying of proline at a concentration of 450 ppm under a salinity level of 10,000 ppm relieves the adverse impact of salty stress on the vegetative growth and physiological and anatomical characteristics of the jojoba plant (*Simmondsia chinensis* (Link) Schneider). The jojoba is a salinity-tolerant plant that produces proline in suitable and unsuitable conditions, but in concentrations that vary according to the severity of the stress. In the current study, the jojoba plant was exposed to salinity at 5000 ppm, which is considered harmless and used as a nutrition method, meaning that the plant is in an appropriate condition and does not need an external addition of proline. The exposure of the jojoba plant to salinity at 10,000 ppm is considered a stress concentration that is not severe, so the plant's internal defense systems push the plant to produce proline, but in concentrations that are insufficient to help the plant withstand stress; in this case, foliar spraying the plant with proline may help it tolerate salinity stress, adapt to unsuitable conditions, and improve its physiological and morphological properties. Exposure to a 15,000 ppm concentration of salinity leads to severe stress and failure of the internal defense systems to produce sufficient proline. In this case, the plant may need outside intervention by foliar spraying of proline at a high rate of over 450 ppm to help it adapt and withstand harsh conditions. This research needs further studies to clarify the physiological role of proline in the jojoba plant under the influence of diluted seawater salinity and under normal conditions.

Author Contributions: Conceptualization, M.S.A. and E.-S.A.E.-B.; methodology, M.S.A., E.-S.A.E.-B., M.H.E.-G., M.F.E.-B. and M.M.K.; software, M.S.A., A.A.H., H.M.H. and M.M.E.-M.; validation, M.S.A., A.A.E.-Y. and N.A.A.-H.; formal analysis, M.S.A., H.G.A.E.-G. and S.M.A.-Q.; investigation, M.S.A., E.-R.F.A.E.-D., M.M.K., M.F.E.-B. and E.S.D.; resources, M.S.A., I.A.I. and E.-S.A.E.-B.; data curation, M.S.A., M.H.E.-G., M.F.E.-B. and H.M.H.; writing—original draft preparation, M.S.A., E.-R.F.A.E.-D., M.F.E.-B., M.M.K. and E.-S.A.E.-B.; writing—review and editing, M.S.A., M.H.E.-G., A.A.H., E.-S.A.E.-B., M.F.E.-B., M.M.K. and M.M.E.-M.; visualization, M.S.A., H.M.H., A.A.E.-Y., M.F.E.-B. and E.S.D.; supervision, M.S.A., E.-R.F.A.E.-D., M.F.E.-B., A.A.H. and E.-S.A.E.-B.; project administration, N.A.A.-H. and I.A.I.; funding acquisition, H.G.A.E.-G. and S.M.A.-Q. All authors have read and agreed to the published version of the manuscript.

Funding: The current work was funded by Ain Shams University, Faculty of Agriculture, Egypt.

Institutional Review Board Statement: Not applicable.

Informed Consent Statement: Not applicable.

Data Availability Statement: Data are contained within the article.

Acknowledgments: The authors extend their appreciation to Taif University for funding current work under Taif University Researchers Supporting Project (number TURSP-2020/120), Taif University, Taif, Saudi Arabia.

Conflicts of Interest: Authors declare no conflict of interest.

References

1. Bannari, A.; Al-Ali, Z.M. Assessing Climate Change Impact on Soil Salinity Dynamics between 1987–2017 in Arid Landscape Using Landsat TM, ETM+ and OLI Data. *Remote Sens.* **2020**, *12*, 2794. [CrossRef]
2. Benzioni, A.; Nerd, A.; Rosenärtner, Y.; Mills, D. Effect of NaCl Salinity on Growth and Development of Jojoba Clones: I. Young Plants. *J. Plant Physiol.* **1992**, *139*, 731–736. [CrossRef]
3. Bafeel, S.O.; Galal, H.K.; Basha, A.Z. Effect of Seawater Irrigation on Growth and Some Metabolites of Jojoba Plants. *Am. J. Agric. Environ. Sci.* **2016**, *16*, 49–59. [CrossRef]
4. Botti, C.; Prat, L.; Palzkill, D.; Cánaves, L. Evaluation of jojoba clones grown under water and salinity stresses in Chile. *Ind. Crops Prod.* **1998**, *9*, 39–45. [CrossRef]
5. Ahmad, P.; Azooz, M.M.; Prasad, M.N.V. Ecophysiology and responses of plants under salt stress. *Ecophysiol. Responses Plants Salt Stress* **2013**, *9781461447474*, 1–510. [CrossRef]
6. Cho, U.H.; Park, J.O. Mercury-induced oxidative stress in tomato seedlings. *Plant Sci.* **2000**, *156*, 1–9. [CrossRef]
7. Cabrera-Bosquet, L.; Molero, G.; Bort, J.; Nogués, S.; Araus, J.L. The combined effect of constant water deficit and nitrogen supply on WUE, NUE and Δ13C in durum wheat potted plants. *Ann. Appl. Biol.* **2007**, *151*, 277–289. [CrossRef]
8. Ma, S.; Guo, S.; Chen, J.; Sun, J.; Wang, Y.; Shu, S. Enhancement of salt-stressed cucumber tolerance by application of glucose for regulating antioxidant capacity and nitrogen metabolism. *Can. J. Plant Sci.* **2020**, *100*, 253–263. [CrossRef]
9. Slabbert, M.M.; Krüger, G.H.J. Antioxidant enzyme activity, proline accumulation, leaf area and cell membrane stability in water stressed *Amaranthus* leaves. *S. Afr. J. Bot.* **2014**, *95*, 123–128. [CrossRef]
10. Boughalleb, F.; Denden, M.; Tiba, B. Ben Anatomical changes induced by increasing NaCl salinity in three fodder shrubs, *Nitraria retusa*, *Atriplex halimus* and *Medicago arborea*. *Acta Physiol. Plant.* **2009**, *31*, 947–960. [CrossRef]
11. Tabot, P.T.; Adams, J.B. Salt secretion, proline accumulation and increased branching confer tolerance to drought and salinity in the endemic halophyte *Limonium linifolium*. *S. Afr. J. Bot.* **2014**, *94*, 64–73. [CrossRef]

12. Abdelhamid, M.T.; Rady, M.M.; Osman, A.S.; Abdalla, M.A. Exogenous application of proline alleviates salt-induced oxidative stress in *Phaseolus vulgaris* L. *Plants* **2015**, *88*, 439–446. [CrossRef]
13. Mansour, M.M.F. Nitrogen Containing Compounds and Adaptation of Plants to Salinity Stress. *Biol. Plant.* **2000**, *43*, 491–500. [CrossRef]
14. Dawood, M.G.; Taie, H.A.A.; Nassar, R.M.A.; Abdelhamid, M.T.; Schmidhalter, U. The changes induced in the physiological, biochemical and anatomical characteristics of *Vicia faba* by the exogenous application of proline under seawater stress. *S. Afr. J. Bot.* **2014**, *93*, 54–63. [CrossRef]
15. Hayat, S.; Hayat, Q.; Alyemeni, M.N.; Wani, A.S.; Pichtel, J.; Ahmad, A. Role of proline under changing environments: A review. *Plant Signal. Behav.* **2012**, *7*, 1456–1466. [CrossRef]
16. Aggarwal, M.; Sharma, S.; Kaur, N.; Pathania, D.; Bhandhari, K.; Kaushal, N.; Kaur, R.; Singh, K.; Srivastava, A.; Nayyar, H. Exogenous proline application reduces phytotoxic effects of selenium by minimising oxidative stress and improves growth in bean (*Phaseolus vulgaris* L.) seedlings. *Biol. Trace Elem. Res.* **2011**, *140*, 354–367. [CrossRef]
17. Kim, G.B.; Nam, Y.W. A novel Δ1-pyrroline-5-carboxylate synthetase gene of *Medicago truncatula* plays a predominant role in stress-induced proline accumulation during symbiotic nitrogen fixation. *J. Plant Physiol.* **2013**, *170*, 291–302. [CrossRef]
18. Yang, S.L.; Lan, S.S.; Gong, M. Hydrogen peroxide-induced proline and metabolic pathway of its accumulation in maize seedlings. *J. Plant Physiol.* **2009**, *166*, 1694–1699. [CrossRef]
19. Kaya, C.; Tuna, A.L.; Ashraf, M.; Altunlu, H. Improved salt tolerance of melon (*Cucumis melo* L.) by the addition of proline and potassium nitrate. *Environ. Exp. Bot.* **2007**, *60*, 397–403. [CrossRef]
20. Mona, A.E.-H.; Sahar, A.E.-D.; Sohayla, M.A.; Hasan, N.A.; Ragia, H. MJFMCT_Volume 16_Issue 2_Pages 1–15.pdf. *Mansoura J. Forensic Med. Clin. Toxicol.* **2008**, *16*, 1–15.
21. Ahmed, F.; Morsy, M.H. A new method for measuring leaf area in different fruit species. *Minia J. Agric. Res. Dev.* **1999**, *19*, 97–105.
22. Sun, Y.; Niu, G.; Perez, C. Relative Salt Tolerance of Seven Texas Superstar® Perennials. *HortScience* **2015**, *50*, 1562–1566. [CrossRef]
23. Porra, R.J. The chequered history of the development and use of simultaneous equations for the accurate determination of chlorophylls a and b. *Photosynth. Res.* **2002**, *73*, 149–156. [CrossRef] [PubMed]
24. Wellburn, A.R. The Spectral Determination of Chlorophylls a and b, as well as Total Carotenoids, Using Various Solvents with Spectrophotometers of Different Resolution. *J. Plant Physiol.* **1994**, *144*, 307–313. [CrossRef]
25. Bates, L.S.; Waldren, R.P.; Teare, I.D. Rapid determination of free proline for water-stress studies. *Plant Soil* **1973**, *39*, 205–207. [CrossRef]
26. Kerepesi, I.; Tóth, M.; Boross, L. Water-Soluble Carbohydrates in Dried Plant. *J. Agric. Food Chem.* **1996**, *44*, 3235–3239. [CrossRef]
27. Dubois, M.; Gilles, K.A.; Hamilton, J.K.; Rebers, P.A.; Smith, F. Colorimetric Method for Determination of Sugars and Related Substances. *Anal. Chem.* **2002**, *28*, 350–356. [CrossRef]
28. Ainsworth, E.A.; Gillespie, K.M. Estimation of total phenolic content and other oxidation substrates in plant tissues using Folin–Ciocalteu reagent. *Nat. Protoc.* **2007**, *2*, 875–877. [CrossRef]
29. Hakim, A.; Purvis, A.C.; Mullinix, B.G. Differences in chilling sensitivity of cucumber varieties depends on storage temperature and the physiological dysfunction evaluated. *Postharvest Biol. Technol.* **1999**, *17*, 97–104. [CrossRef]
30. Iturbe-Ormaetxe, I.; Escuredo, P.R.; Arrese-Igor, C.; Becana, M. Oxidative Damage in Pea Plants Exposed to Water Deficit or Paraquat. *Plant Physiol.* **1998**, *116*, 173–181. [CrossRef]
31. Yang, H.; Wu, F.; Cheng, J. Reduced chilling injury in cucumber by nitric oxide and the antioxidant response. *Food Chem.* **2011**, *127*, 1237–1242. [CrossRef] [PubMed]
32. Xu, M.; Dong, J.; Zhang, M.; Xu, X.; Sun, L. Cold-induced endogenous nitric oxide generation plays a role in chilling tolerance of loquat fruit during postharvest storage. *Postharvest Biol. Technol.* **2012**, *65*, 5–12. [CrossRef]
33. Shehata, S.A.; El-Mogy, M.M.; Mohamed, H.F.Y. Postharvest quality and nutrient contents of long sweet pepper enhanced by supplementary potassium foliar application. *Int. J. Veg. Sci.* **2019**, *25*, 196–209. [CrossRef]
34. Anderson, D.L.; Henderson, L.J. Comparing Sealed Chamber Digestion with Other Digestion Methods Used for Plant Tissue Analysis. *Agron. J.* **1988**, *80*, 549–552. [CrossRef]
35. Munter, R.C.; Halverson, T.L.; Anderson, R.D. Quality assurance for plant tissue analysis by ICP-AES. *Commun. Soil Sci. Plant Anal.* **2008**, *15*, 1285–1322. [CrossRef]
36. Ruzin, S.E. Plant Micro Technique and Microscopy. Oxford University Press, Inc., Oxford.-References-Scientific Research Publishing. 1999. *REVIEWS New Phytol.* **2000**, *148*, 57–58.
37. Qaderi, M.M.; Kurepin, L.V.; Reid, D.M. Growth and physiological responses of canola (*Brassica napus*) to three components of global climate change: Temperature, carbon dioxide and drought. *Physiol. Plant.* **2006**, *128*, 710–721. [CrossRef]
38. Khayyat, M.; Tehranifar, A.; Davarynejad, G.H.; Sayyari-Zahan, M.H. Vegetative growth, compatible solute accumulation, ion partitioning and chlorophyll fluorescence of "Malas-e-Saveh" and "Shishe-Kab" pomegranates in response to salinity stress. *Photosynthetica* **2014**, *52*, 301–312. [CrossRef]
39. Sadak, M.S.; Abdelhamid, M.T.; Schmidhalter, U. Effect of foliar application of aminoacids on plant yield and some physiological parameters in bean plants irrigated with seawater. *Acta Biológica Colomb.* **2015**, *20*, 141–152. [CrossRef]
40. Ali, Q.; Athar, H.; Zulqurnain, M. Abiotic Stress Tolerance to Plants. *Acta Physiol. Plant.* **2019**, *41*, 1.
41. El-Sherbeny, M.R.; Teixeira Da Silva, J.A. Foliar treatment with proline and tyrosine affect the growth and yield of beetroot and some pigments in beetroot leaves. *J. Hortic. Res.* **2014**, *21*, 95–99. [CrossRef]

42. Waller, G.R.; Nowacki, E.K. Alkaloid Biology and Metabolism in Plants. *Alkaloid Biol. Metab. Plants* **1978**, 157–189. [CrossRef]
43. Lea, P.J.; Fowden, L. Asparagine Metabolism in Higher Plants. *Biochem. Und Physiol. Der Pflanz.* **1975**, *168*, 3–14. [CrossRef]
44. Çiçek, N.; Çakirlar, H. The effect of salinity on some physiological parameters in two maize cultivars. *Bulg. J. Plant Physiol.* **2002**, *28*, 66–74.
45. Ehsanpour, A.A.; Fatahian, N. Effects of salt and proline on *Medicago sativa* callus. *Plant Cell Tissue Organ Cult.* **2003**, *73*, 53–56. [CrossRef]
46. Yan, Z.; Guo, S.; Shu, S.; Sun, J.; Tezuka, T. Effects of proline on photosynthesis, root reactive oxygen species (ROS) metabolism in two melon cultivars (*Cucumis melo* L.) under NaCl stress. *Afr. J. Biotechnol.* **2013**, *10*, 18381–18390. [CrossRef]
47. Abdeldym, E.A.; El-Mogy, M.M.; Abdellateaf, H.R.L.; Atia, M.A.M. Genetic Characterization, Agro-Morphological and Physiological Evaluation of Grafted Tomato under Salinity Stress Conditions. *Agronomy* **2020**, *10*. [CrossRef]
48. LI, R.-H.; Guo, P.-G.; Michael, B.; Stefania, G.; Salvatore, C. Evaluation of Chlorophyll Content and Fluorescence Parameters as Indicators of Drought Tolerance in Barley. *Agric. Sci. China* **2006**, *5*, 751–757. [CrossRef]
49. Mafakheri, A.; Siosemardeh, A.; Bahramnejad, B.; Struik, P.C.; Sohrabi, E. Effect of drought stress on yield, proline and chlorophyll contents in three chickpea cultivars. *Aust. J. Crop Sci.* **2010**, *4*, 580–585.
50. Jaleel, C.A.; Sankar, B.; Sridharan, R.; Panneerselvam, R. Soil salinity alters growth, chlorophyll content, and secondary metabolite accumulation in Catharanthus roseus. *Turk. J. Biol.* **2008**, *32*, 79–83.
51. Banu, M.N.A.; Hoque, M.A.; Watanabe-Sugimoto, M.; Matsuoka, K.; Nakamura, Y.; Shimoishi, Y.; Murata, Y. Proline and glycinebetaine induce antioxidant defense gene expression and suppress cell death in cultured tobacco cells under salt stress. *J. Plant Physiol.* **2009**, *166*, 146–156. [CrossRef] [PubMed]
52. Ben Ahmed, C.; Ben Rouina, B.; Sensoy, S.; Boukhriss, M.; Ben Abdullah, F. Exogenous Proline Effects on Photosynthetic Performance and Antioxidant Defense System of Young Olive Tree. *J. Agric. Food Chem.* **2010**, *58*, 4216–4222. [CrossRef] [PubMed]
53. Egbichi, I.; Keyster, M.; Ludidi, N. Effect of exogenous application of nitric oxide on salt stress responses of soybean. *S. Afr. J. Bot.* **2014**, *90*, 131–136. [CrossRef]
54. Foyer, C.H.; Harbinson, J. Oxygen Metabolism and the Regulation of Photosynthetic Electron Transport. In *Causes Photooxidative Stress Amelior. Def. Syst. Plants*, 1st ed.; CRC Press: Boca Raton, FL, USA, 2019; p. 42. ISBN 9781351070454. [CrossRef]
55. Ma, J.; Du, G.; Li, X.; Zhang, C.; Guo, J. A major locus controlling malondialdehyde content under water stress is associated with *Fusarium* crown rot resistance in wheat. *Mol. Genet. Genom.* **2015**, *290*, 1955–1962. [CrossRef]
56. Reddy, A.R.; Chaitanya, K.V.; Vivekanandan, M. Drought-induced responses of photosynthesis and antioxidant metabolism in higher plants. *J. Plant Physiol.* **2004**, *161*, 1189–1202. [CrossRef]
57. Nounjan, N.; Nghia, P.T.; Theerakulpisut, P. Exogenous proline and trehalose promote recovery of rice seedlings from salt-stress and differentially modulate antioxidant enzymes and expression of related genes. *J. Plant Physiol.* **2012**, *169*, 596–604. [CrossRef]
58. Verma, K.K.; Song, X.P.; Tian, D.D.; Singh, M.; Verma, C.L.; Rajput, V.D.; Singh, R.K.; Sharma, A.; Singh, P.; Malviya, M.K.; et al. Investigation of Defensive Role of Silicon during Drought Stress Induced by Irrigation Capacity in Sugarcane: Physiological and Biochemical Characteristics. *ACS Omega* **2021**, *6*, 19811–19821. [CrossRef]
59. Bil Der, T.; Yildiz, M.; Terzï, H.; BİLGİSİ Araştırma Makalesi-Bitkisel Üretim Sorumlu Yazar, E. Effect of NaCl Stress on Chlorophyll Biosynthesis, Proline, Lipid Peroxidation and Antioxidative Enzymes in Leaves of Salt-Tolerant and Salt-Sensitive Barley Cultivars. *J. Agric. Sci.* **2013**, *19*, 79–88. [CrossRef]
60. Ashraf, M.; Foolad, M.R. Roles of glycine betaine and proline in improving plant abiotic stress resistance. *Environ. Exp. Bot.* **2007**, *59*, 206–216. [CrossRef]
61. Szabados, L.; Savouré, A. Proline: A multifunctional amino acid. *Trends Plant Sci.* **2010**, *15*, 89–97. [CrossRef]
62. Munns, R. Comparative physiology of salt and water stress. *Plant. Cell Environ.* **2002**, *25*, 239–250. [CrossRef]
63. Hare, P.D.; Cress, W.A. Metabolic implications of stress-induced proline accumulation in plants. *Plant Growth Regul.* **1997**, *21*, 79–102. [CrossRef]
64. Khatkar, D.; Kuhad, M.S. Short-Term Salinity Induced Changes in Two Wheat Cultivars at Different Growth Stages. *Biol. Plant.* **2000**, *43*, 629–632. [CrossRef]
65. Semida, W.M.; Taha, R.S.; Abdelhamid, M.T.; Rady, M.M. Foliar-applied α-tocopherol enhances salt-tolerance in *Vicia faba* L. plants grown under saline conditions. *S. Afr. J. Bot.* **2014**, *95*, 24–31. [CrossRef]
66. Gholami Zali, A.; Ehsanzadeh, P. Exogenous proline improves osmoregulation, physiological functions, essential oil, and seed yield of fennel. *Ind. Crops Prod.* **2018**, *111*, 133–140. [CrossRef]
67. Farissi, M.; Aziz, F.; Bouizgaren, A.; Ghoulam, C. La symbiose Légumineuses-rhizobia sous conditions de salinité: Aspect Agro-physiologique et biochimique de la tolérance [Legume-rhizobia symbiosis under saline conditions: Agro-physiological and biochemical aspects of tolerance]. *Int. J. Innov. Sci. Res.* **2014**, *11*, 96–104.
68. Freitas, P.A.F.; Carvalho, H.H.; Costa, J.H.; Miranda, R.S.; Saraiva, K.D.C.; Oliveira, F.D.B.; Coelho, D.G.; Prisco, J.T.; Gomes-Filho, E. Salt acclimation in sorghum plants by exogenous proline: Physiological and biochemical changes and regulation of proline metabolism. *Plant Cell Rep.* **2019**, *38*, 403–416. [CrossRef]
69. Alam, R.; Das, D.K.; Islam, M.R.; Murata, Y.; Hoque, M.A. Exogenous proline enhances nutrient uptake and confers tolerance to salt stress in maize (*Zea mays* L.). *Progress. Agric.* **2016**, *27*, 409–417. [CrossRef]
70. Khan, A.; Iqbal, I.; Ahmad, I.; Nawaz, H.; Nawaz, M. Role of proline to induce salinity tolerance in sunflower (*Helianthus annus* L.). *Sci. Technol. Dev.* **2014**, *33*, 88–93.

71. Heuer, B. Role of Proline in Plant 9 Role Response to Drought and Salinity. In *Handbook of Plant and Crop Stress*; CRC Press: Boca Raton, FL, USA, 2010; pp. 213–238.
72. Ben Rejeb, K.; Abdelly, C.; Savouré, A. How reactive oxygen species and proline face stress together. *Plant Physiol. Biochem.* **2014**, *80*, 278–284. [CrossRef]
73. Posmyk, M.M.; Janas, K.M. Effects of seed hydropriming in presence of exogenous proline on chilling injury limitation in *Vigna radiata* L. seedlings. *Acta Physiol. Plant.* **2007**, *29*, 509–517. [CrossRef]
74. Zhu, J.K. Regulation of ion homeostasis under salt stress. *Curr. Opin. Plant Biol.* **2003**, *6*, 441–445. [CrossRef]
75. Rizwan, M.; Ali, S.; Ibrahim, M.; Farid, M.; Adrees, M.; Bharwana, S.A.; Zia-ur-Rehman, M.; Qayyum, M.F.; Abbas, F. Mechanisms of silicon-mediated alleviation of drought and salt stress in plants: A review. *Environ. Sci. Pollut. Res.* **2015**, *22*, 15416–15431. [CrossRef] [PubMed]
76. Ben Ahmed, C.; Magdich, S.; Ben Rouina, B.; Sensoy, S.; Boukhris, M.; Ben Abdullah, F. Exogenous proline effects on water relations and ions contents in leaves and roots of young olive. *Amino Acids* **2011**, *40*, 565–573. [CrossRef] [PubMed]
77. Huang, Y.; Bie, Z.; Liu, Z.; Zhen, A.; Wang, W. Protective role of proline against salt stress is partially related to the improvement of water status and peroxidase enzyme activity in cucumber. *Soil Sci. Plant Nutr.* **2010**, *55*, 698–704. [CrossRef]
78. Roussos, P.A.; Tsantili, E.; Pontikis, C.A. Responses of jojoba explants to different salinity levels during the proliferation stage in vitro. *Ind. Crops Prod.* **2006**, *23*, 65–72. [CrossRef]
79. Gonzalez, A.J.; Larraburu, E.E.; Llorente, B.E. *Azospirillum brasilense* mitigates anatomical alterations produced by salt stress in jojoba in vitro plants. *Vegetos* **2021**, *34*, 725–737. [CrossRef]
80. Akcin, T.A.; Akcin, A.; Yalcin, E. Anatomical Adaptations to Salinity in *Spergularia marina* (Caryophyllaceae) from Turkey. *Proc. Natl. Acad. Sci. India Sect. B-Biol. Sci.* **2015**, *85*, 625–634. [CrossRef]
81. Debez, A.; Saadaoui, D.; Ramani, B.; Ouerghi, Z.; Koyro, H.W.; Huchzermeyer, B.; Abdelly, C. Leaf H+-ATPase activity and photosynthetic capacity of *Cakile maritima* under increasing salinity. *Environ. Exp. Bot.* **2006**, *57*, 285–295. [CrossRef]
82. Bongi, G.; Loreto, F.; Nazionale, C.; Alta, V.M. Gas-Exchange Properties of Salt-Stressed Olive (*Olea europea* L.) Leaves. *Plant Physiol.* **1989**, *90*, 1408–1416. [CrossRef]
83. Waisel, Y.; Eshel, A.; Agami, M. Salt balance of leaves of the mangrove *Avicennia Marina*. *Physiol. Plant.* **1986**, *67*, 67–72. [CrossRef]
84. Debez, A.; Ben Hamed, K.; Grignon, C.; Abdelly, C. Salinity effects on germination, growth, and seed production of the halophyte *Cakile maritima*. *Plant Soil* **2004**, *262*, 179–189. [CrossRef]
85. Shannon, M.C.; Grieve, C.M.; Francois, L.E. Whole-Plant Response to Salinity. In *Plant-Environment Interactions*; Wilkinson, R.E., Ed.; Marcel Dekker: New York, NY, USA, 1994; pp. 199–244. Available online: https://www.scirp.org/%28S%28vtj3fa45qm1ean45vvffcz55%29%29/ (accessed on 1 February 2021).
86. Botti, C.; Palzkill, D.; Muñoz, D.; Prat, L. Morphological and anatomical characterization of six jojoba clones at saline and non-saline sites. *Ind. Crops Prod.* **1998**, *9*, 53–62. [CrossRef]
87. El-Banna, M.F.; Abdelaal, K.A.A. Response of Strawberry Plants Grown in the Hydroponic System to Pretreatment with H_2O_2 before Exposure to Salinity Stress. *J. Plant Prod.* **2018**, *9*, 989–1001. [CrossRef]
88. El-Taher, A.M.; Abd El-Raouf, H.S.; Osman, N.A.; Azoz, S.N.; Omar, M.A.; Elkelish, A.; Abd El-Hady, M.A.M. Effect of Salt Stress and Foliar Application of Salicylic Acid on Morphological, Biochemical, Anatomical, and Productivity Characteristics of Cowpea (*Vigna unguiculata* L.) Plants. *Plants* **2021**, *11*, 115. [CrossRef]
89. El-Banna, M.F.; AL-Huqail, A.A.; Farouk, S.; Belal, B.E.A.; El-Kenawy, M.A.; El-Khalek, A.F.A. Morpho-Physiological and Anatomical Alterations of Salt-Affected Thompson Seedless Grapevine (*Vitis vinifera* L.) to Brassinolide Spraying. *Horticulturae* **2022**, *8*, 568. [CrossRef]
90. Kanwal, S.; Ashraf, M.; Shahbaz, M.; Iqbal, M.Y. Influence of saline stress on growth, gas exchange, mineral nutrients and non-enzymatic antioxidatns in mungbean [(*Vigna Radiata* (L.) Wilczek]. *Pak. J. Bot.* **2013**, *45*, 763–771.
91. Kozlowski, T.T. Responses of woody plants to flooding and salinity. *Tree Physiol.* **1997**, *17*, 490. [CrossRef]
92. Akcin, T.A.; Akcin, A.; Yalcın, E. Anatomical changes induced by salinity stress in *Salicornia freitagii* (Amaranthaceae). *Rev. Bras. Bot.* **2017**, *40*, 1013–1018. [CrossRef]
93. Abd Elbar, O.H.; Farag, R.E.; Shehata, S.A. Effect of putrescine application on some growth, biochemical and anatomical characteristics of *Thymus vulgaris* L. under drought stress. *Ann. Agric. Sci.* **2019**, *64*, 129–137. [CrossRef]
94. Abdelaal, K. Effect of salicylic acid and abscisic acid on morpho-physiological and anatomical characters of faba bean plants (*Vicia faba* L.) under drought stress. *J. Plant Prod.* **2015**, *6*, 1771–1788. [CrossRef]
95. Hussein, M.M.; Abo-Leila, B.H.; Metwally, S.A.; Leithy, S.Z. Anatomical structure of jatropha leaves affected by proline and salinity conditions. *J. Appl. Sci. Res.* **2012**, *8*, 491–496.

horticulturae

MDPI

Article

Synergistic Impact of Melatonin and Putrescine Interaction in Mitigating Salinity Stress in Snap Bean Seedlings: Reduction of Oxidative Damage and Inhibition of Polyamine Catabolism

Hossam S. El-Beltagi [1,2,*], Ahmed Abou El-Yazied [3], Hany G. Abd El-Gawad [3], Mahmoud Kandeel [4,5], Tarek A. Shalaby [6,7], Abdallah Tageldein Mansour [8,9], Nadi Awad Al-Harbi [10], Salem Mesfir Al-Qahtani [10], Abdulmalik A. Alkhateeb [1] and Mohamed F. M. Ibrahim [11,*]

1 Agricultural Biotechnology Department, College of Agriculture and Food Sciences, King Faisal University, Al-Ahsa 31982, Saudi Arabia
2 Biochemistry Department, Faculty of Agriculture, Cairo University, Giza 12613, Egypt
3 Department of Horticulture, Faculty of Agriculture, Ain Shams University, Cairo 11566, Egypt
4 Department of Biomedical Sciences, College of Veterinary Medicine, King Faisal University, Al-Ahsa 31982, Saudi Arabia
5 Department of Pharmacology, Faculty of Veterinary Medicine, Kafrelsheikh University, Kafrelsheikh 33516, Egypt
6 Department of Arid Land Agriculture, College of Agricultural and Food Science, King Faisal University, Al-Ahsa 31982, Saudi Arabia
7 Horticulture Department, Faculty of Agriculture, Kafrelsheikh University, Kafr El-Sheikh 33516, Egypt
8 Department of Aquaculture and Animal Production, College of Agriculture and Food Sciences, King Faisal University, Al-Ahsa 31982, Saudi Arabia
9 Department of Fish and Animal Production, Faculty of Agriculture (Saba Basha), Alexandria University, Alexandria 21526, Egypt
10 Biology Department, University College of Tayma, University of Tabuk, Tabuk 47512, Saudi Arabia
11 Department of Agricultural Botany, Faculty of Agriculture, Ain Shams University, Cairo 11566, Egypt
* Correspondence: helbeltagi@kfu.edu.sa (H.S.E.-B.); ibrahim_mfm@agr.asu.edu.eg (M.F.M.I.)

Citation: El-Beltagi, H.S.; El-Yazied, A.A.; El-Gawad, H.G.A.; Kandeel, M.; Shalaby, T.A.; Mansour, A.T.; Al-Harbi, N.A.; Al-Qahtani, S.M.; Alkhateeb, A.A.; Ibrahim, M.F.M. Synergistic Impact of Melatonin and Putrescine Interaction in Mitigating Salinity Stress in Snap Bean Seedlings: Reduction of Oxidative Damage and Inhibition of Polyamine Catabolism. *Horticulturae* **2023**, *9*, 285. https://doi.org/10.3390/horticulturae9020285

Academic Editors: Agnieszka Hanaka, Małgorzata Majewska and Barbara Hawrylak-Nowak

Received: 20 January 2023
Revised: 16 February 2023
Accepted: 17 February 2023
Published: 20 February 2023

Abstract: While the individual influences of melatonin (MT) and polyamines (PAs) have been widely studied under various abiotic stresses, little is known about their interaction under salinity stress. In the present study, salt stress applied by 50 mM of sodium chloride (NaCl) on snap bean seedlings has been supplemented with 20 μM of MT and/or 100 μM of putrescine (Put) (individually and in combination). The results indicated that under salinity stress, the combination of MT + Put achieved the highest significant increase in shoot fresh and dry weight, chlorophyll (Chl *a*), Chl *a* + *b*, carotenoids, total soluble sugars, proline, K, Ca, and cell membrane stability index (CMSI), as well as catalase (CAT) and peroxidase (POX) activities. This improvement was associated with an obvious decrease in Na, Na/K ratio, and oxidative damage as indicated by reducing leaf contents of methylglyoxal (MG), hydrogen peroxide (H_2O_2), and the rate of lipid peroxidation (malondialdehyde; MDA). Moreover, the combination of MT + Put demonstrated a significant decrease in the activities of diamine oxidase (DAO) and polyamine oxidase (PAO) leading to the reduction of the rate of polyamine oxidation. Meanwhile, MT applied individually gave the highest significant increase in leaf relative water content (RWC), Chl *b*, superoxide dismutase (SOD), and ascorbate peroxidase (APX). Conclusively, the combination treatment of MT + Put could decrease the degradation of polyamines and enhance tolerance to salinity stress in snap bean seedlings.

Keywords: antioxidant enzymes; nutrient homeostasis; osmolytes; *Phaseolus vulgaris* L; polyamine oxidation; putrescine and saline conditions

1. Introduction

Salinity stress is one of the biggest impediments to growth and crop production around the world. Each year, there is a significant increase in salt affected soil, which is estimated at 2 million hectares (about 1%) of the world's agricultural lands [1]. Furthermore, soil salinization is predicted to increase in the future due to frequent climate change scenarios, i.e., rising of sea level, temperature, and global warming, leading to increased evaporation and further salinization [2]. This issue represents a major serious threat to sustainable agriculture and food security worldwide [3]. The deleterious effects of salinity stress on plants consist of: (1) the osmotic stress, which is related to the lower water potential of soil, as well as preventing a plant's water uptake from soils with high salt concentration; (2) the toxic effect or ionic stress, which is caused by particular ionic species such as Na^+ and Cl^-; (3) the oxidative stress that is a result of the excessive release of reactive oxygen species (ROS) i.e., superoxide anion, hydrogen peroxide, singlet oxygen, and hydroxyl radical; and (4) the combined influence of these stresses [4–7].

The snap bean (*Phaseolus vulgaris* L.) is considered a rich source of protein, fibers, vitamins, and micronutrients [8,9]. Its consumption as a human food source is estimated to be around 50% of the total legumes consumed worldwide [10]. Several previous studies have demonstrated that consumption of snap beans can reduce the risk of heart diseases [11,12] and cancers [13,14], leading to improved human health and longevity [15]. As a glycophyte, the snap bean is very salt-sensitive plant with threshold of 1 dS m^{-1} [16,17]. Therefore, there is an urgent need for research to develop plants that are tolerant under saline conditions.

Melatonin (N-acetyl-5-methoxytryptamine) is an auxin-like biomolecule due to the presence of an indole ring in its structure [18]. It is ubiquitous in all kingdoms including animals, algae, plants, and microorganisms [19,20]. In recent years, melatonin has been suggested as a new plant growth regulator with a wide spectrum of effects [21]. It plays an important role in seed germination [22], rhizogenesis [23], delaying leaf senescence [24], and alleviating the effects of biotic [25] and abiotic stresses [26]. In this context, melatonin has been found to protect plants against heat stress [27], chilling [28], heavy metals [29,30], drought [31,32], and salinity stress [1,33,34]. It can reduce chlorophyll degradation, protect photosynthetic machinery, and regulate the redox status of the salt-stressed plants [35,36]. Additionally, melatonin has the ability to adjust the osmotic potential and increase the antioxidant capacity of plants under saline conditions [37,38]. It has been shown that melatonin can play several roles in various signaling processes which are related to the evocation of systemic salt tolerance i.e., nitric oxide [39], calcium/calmodulin (Ca^{2+}/CaM) [40], and ROS [41].

Polyamines (PAs) are multifunctional polycationic plant growth regulators that affect several developmental, physiological, and biochemical aspects [42]. They can regulate DNA synthesis, cell division, seed germination, fruit set, and development [43,44]. In higher plants, putrescine (Put; diamine with two positive charges), spermidine (Spd; triamine with three positive charges), and spermine (Spm; tetraamine with four positive charges) are most abundant polyamines [45]. However, Put is considered the most common PA, and it is directly produced from the non-proteinogenic amino acid, ornithine, through the activity of ornithine decarboxylase (ODC), or indirectly produced from arginine through arginine decarboxylase (ADC) [46]. Under environmental stresses, the positive charges of PAs can serve in cell membranes to stabilize through binding to the membranes negatively charged macromolecules i.e., phospholipids and proteins [44]. Moreover, these positive charges can protect photosystem II (PSII) in the isolated thylakoid membranes under photoinhibition conditions [47]. PAs play a key role in signaling and ion homeostasis through affecting various cell membrane transporters [48]. Several lines of evidence suggest that PAs are involved in plant stress responses to various adverse environmental conditions i.e., heat stress ([49], heavy metals [50], drought [51], and salinity [52,53]. Additionally, PAs are responsible for the scavenging of free radicals [54], enhancement of antioxidant systems [55], accumulation of osmolytes [56], and regulation of gene expression [57].

Despite the fact that the influences of melatonin and PAs on plants have been widely and individually studied under various abiotic stresses, little is known about the possible link between melatonin and PAs under salinity stress in snap bean plants. Therefore, this study was conducted to investigate the effect of melatonin and Put, either individually or in combination, on salt-stressed snap bean seedlings.

2. Materials and Methods

2.1. Plant Material and Growth Conditions

Phaseolus vulgaris L. (Colter HMX 2117 cv.) served as the study's plant material. Using 0.5% sodium hypochlorite (NaOCl; *w*/*v*) for 4 min, seeds were sterilized before being rinsed five times with distilled water. The seedlings were cultivated in black plastic pots with a diameter of 13 cm and a capacity of 700 cm^3, which contained an equivalent amount of pre-washed sand. The seedlings were consistently watered twice a week with 250 mL of Hoagland's solution at 1/2 strength [58]. After two weeks, pots were irrigated daily with a modified nutrient solution containing 20 µM melatonin (MT; Bio Basic, Markham, ON, Canada), and/or 100 µM putrescine (Put; Sisco Research Laboratories Pvt. Ltd., Mumbai, India). These concentrations of melatonin and putrescine were determined by doing an initial experiment to measure the concentration of malondialdehyde (MDA), which is used as a measure of lipid oxidation after treatment with several concentrations of melatonin and putrescine. The best results were (with a decrease in MDA concentration) at 20 µM and/or 100 µM putrescine Figure S1.

Additionally, salinity was applied using sodium chloride (NaCl) at 50 mM through the nutrient solution. The experimental design (Complete Randomized Design) included eight different treatments: (1) 1/2 strength Hoagland's solution as a control, (2) 1/2 strength Hoagland's solution + 100 µM Put, (3) 1/2 strength Hoagland's solution + 20 µM MT, (4) 1/2 strength Hoagland's solution + 100 µM Put + 20 µM MT, (5) 1/2 strength Hoagland's solution + 50 mM NaCl, (6) 1/2 strength Hoagland's solution + 50 mM NaCl + 100 µM Put, (7) 1/2 strength Hoagland's solution + 50 mM NaCl + 20 µM MT, and (8) 1/2 strength Hoagland's solution + 50 mM NaCl + 100 µM Put + 20 µM MT. The concentration of MT and Put was determined by a preliminary study (Supplementary 1). All pots were kept under greenhouse conditions (24.3 ± 5.3 day/night temperature and 73.4 ± 2.6% relative humidity recorded by a digital thermo-hygrometer placed in the middle of greenhouse, Art. No.30.5000/30.5002, TFA, Wertheim, Baden-Württemberg, Germany). One month after sowing, samples were taken to determine the different growth and biochemical parameters.

2.2. Plant Growth and Leaf Pigments

A digital balance was used to calculate the fresh weights of the shoot and the root. In accordance with Lichtenthaler and Wellburn [59], the concentration of chlorophyll (Chl) a and b, as well as carotenoids in fresh leaves were assessed spectrophotometrically. The fresh weight (0.2 g) of fully inflated leaves was ground in 80% acetone. The absorbance of the extract was measured versus a blank of pure 80% acetone at 663, 644, and 452.5 nm for Chl a, Chl b, and carotenoid contents, respectively. The results are expressed as mg/g FW.

2.3. Leaf Relative Water Content and Osmolytes

In accordance with Abd El-Gawad et al. [60], the relative water content (RWC) of the leaves was estimated. Six completely expanded leaf discs were weighed (FW), immediately submerged in distilled water for two hours at 25 °C, and their turgid weight (TW) was noted. After that, discs were properly dried for 24 h at 110 °C in an oven (DW). The following formula was used to determine relative water content (RWC):

$$RWC\ (\%) = \frac{FW - DW}{TW - DW} \times 100 \quad (1)$$

The ninhydrin assay was used, according Bates et al. [61], to spectrophotometrically measure proline at 520 nm. Using a method published by Chow and Landhäusser [62], total soluble sugars were measured at 490 nm using phenol and sulfuric acid.

2.4. Determination of Na, K, and Ca

Crushed samples weighing 10 g were weighed in porcelain crucibles. The samples were then dried for 5 h in an oven, charred on a hot plate, and then ash-dried for 3 h at an initial temperature of 100 °C, which automatically rose to the ultimate temperature of 500 °C. The results of the destruction were allowed to cool in the desiccator. A few drops of demineralized water were then dropped on the wall of the crucible's porcelain until wet, dissolved in 5 mL of nitric acid, put into a 100-mL volumetric flask, the crucible's porcelain was rinsed three times with 10 mL of demineralized water, put into the same volumetric flask, and diluted with demineralized water until the marking line. The atomic absorption spectrometer (AAS-Hitachi, Tokyo, Japan) was used to determine Na, K, and Ca using hollow cathode lamp at a wavelength 589.0, 766.5, and 422.7 nm respectively [62]. Results are expressed as mg/g DW.

2.5. Determination of Cell Membrane Stability Index and Oxidative Damage

With few adjustments, the cell membrane stability index (CMSI) was calculated as described by Abd Elbar et al. [63]. Eight leaf discs measuring 1.8 cm in diameter were shaken for 24 h in 10 mL of deionized water. Then, using an electrical conductivity meter (EC; DOH-SD1, TC-OMEGA, USA/Canada), the contents' electrical conductivity (EC1) values were determined. Then, samples were autoclaved at 120 °C for 20 min to determine the values of EC_2. Cell membrane stability index (CMSI) was calculated using the following equation:

$$CMSI = \left[1 - \left(\frac{EC1}{EC2}\right)\right] \times 100 \quad (2)$$

With certain changes, the amount of methylglyoxal (MG) was calculated in accordance with Hossain et al. [64]. Fresh leaves were homogenized in 3 mL of 0.5 M perchloric acid before being incubated on ice for 15 min. The mixture was centrifuged for 10 min at 10,000 rpm at 4 °C. After adding charcoal to the supernatant, it was de-colored, left at ambient temperature for 15 min, and then centrifuged at 10,000 rpm for 10 min. Before utilizing this supernatant for the MG assay, it was neutralized by allowing it to stand in a saturated potassium carbonate solution at room temperature for 15 min before centrifuging it once more for 10 min at 10,000 rpm. The MG was estimated using the neutralized supernatant. One milliliter of the reaction mixture, which included 250 µL of 7.2 mM 1, 2-diaminobenzene, 100 µL of 5 M perchloric acid, and 650 µL of the neutralized supernatant, was added first. A UV spectrophotometer was used to measure the absorbance at 335 nm after 25 min. The methodology outlined by Velikova et al. [54] was used to measure the hydrogen peroxide (H_2O_2) content. A total of 0.5 g of leaf tissues were thoroughly powdered in a 5 mL ice bath containing 0.1% (*w*/*v*) TCA. A total of 0.5 mL of a 10 mM potassium phosphate buffer (PH 7.0) containing 1 M KI was added to the extract. The amount of H_2O_2 was determined spectrophotometrically at 390 nm by reference to a standard curve prepared with H_2O_2 solution. Malondialdehyde (MDA) was used to measure the level of lipid peroxidation in the leaf tissues by reactions with thiobarbituric acid (TBA) [65].

2.6. Measurement of the Activities of Antioxidant Enzymes

Fresh leaves (0.5 g) were homogenized in 4 mL of 0.1 M sodium phosphate buffer (pH 7.0), containing 1% (*w*/*v*) polyvinylpyrrolidone (PVP), and 0.1 mM ethylenediaminetetraacetic acid (EDTA). The supernatant was then utilized for tests after centrifugation at 10,000× *g* for 20 min at 4 °C. The Bradford method was used to assess soluble proteins [66]. Using a spectrophotometer (UV-1601PC; Shimadzu, Tokyo, Japan), all investigated enzyme activity and protein content in the crude enzyme extract were determined as follows:

The superoxide dismutase (SOD) assay was developed using the Beyer and Fridovich [67] approach. A total of 100 L of crude enzyme, 50 mM phosphate buffer (pH 7.8), 75 mM nitroblue tetrazolium (NBT), 13 mM L-methionine, 0.1 mM EDTA, and 0.5 mM riboflavin were all included in the reaction mixture, which had a total volume of 3 mL. Riboflavin was added to start the reaction, and then a 20 W fluorescent bulb was used to illuminate the mixture for 20 min. The quantity of enzyme needed to provide a 50% inhibition in the rate of NBT degradation at 560 nm was determined to be one enzyme activity unit.

In accordance with Cakmak et al. [68], the decrease in absorbance at 240 nm was used to evaluate the catalase (CAT) activity. A total of 15 mM H_2O_2 was present in the reaction mixture, which had a total volume of 3 mL containing 50 mM phosphate buffer (pH = 7). The addition of 50 μL of crude enzyme started the reaction. The activity was determined using the extinction coefficient for H_2O_2 (ε = 40 mM^{-1} cm^{-1}). The breakdown of one mole of H_2O_2 per minute was used to define one unit of enzyme activity.

In accordance with Nakano and Asada [69], ascorbate peroxidase (APX) activity was assessed. For three minutes, the decrease in absorbance at 290 nm was observed. A total of 100 μL of crude enzyme, 50 mM of phosphate buffer (pH 7), 0.1 mM of EDTA, 0.5 mM of ascorbic acid, and 0.1 mM of H_2O_2 made up the reaction mixture, which had a total volume of 3 mL. The reaction was started by the addition of H_2O_2. A unit of enzyme activity was established as the quantity of enzyme needed to oxidize 1 μmol of ascorbate per minute. The extinction coefficient (ε = 2.8 mm^{-1} cm^{-1}) was used to determine the rate of ascorbate oxidation.

With a few modest adjustments, the Dias and Costa [70] method was used to measure the peroxidase (POX) activity. The assay mixture (100 mL) contains 80 mL of 50 mM phosphate buffer (pH = 6.6), 10 mL of 0.3% H_2O_2, and 10 mL of 1% (v/v) guaiacol. To begin the reaction, 2.9 mL of the assay mixture was mixed with 100 μL of the crude enzyme. The absorbance at 470 nm was measured every 30 s for 3 min.

2.7. Determination the Activities of Polyamines Catabolism Enzymes

The method of Su et al. [71] was used to determine diamine oxidase (DAO) and polyamine oxidase (PAO). Briefly, leaf tissues were well ground in an ice bath with 0.1 mM potassium phosphate buffer (pH 6.5). The extract was centrifuged at 8000× g for 20 min at 4 °C. From the supernatant, 0.2 mL was combined with 2.5 mL of potassium phosphate buffer (100 mM, pH 6.5), 0.2 mL of 4-aminoantipyrine/N, N-dimethylaniline, and 0.1 mL horseradish peroxidase (250 U mL^{-1}). The addition of 15 mL of 20 mM putrescine as a substrate was used to measure the activity of DAO, and the addition of 15 mL of 20 mM spermidine as a substrate was used to measure the activity of PAO. One activity unit was defined as a change in absorbance at 555 nm of 0.01 in value.

2.8. Statistics

SAS [72] software was used to carry out a one-way ANOVA procedure. Three replicates' worth of means and standard deviations were generated, and a Duncan's multiple range test ($p \leq 0.05$) was used to identify any variations in means that were statistically significant.

3. Results

Data in Figure 1A,B indicate that all plant growth parameters were decreased by salinity stress. Under non-saline conditions, all treatments had increased shoot fresh weight and root fresh weight compared to the control. The best treatment was the dual application of MT and Put. Additionally, under saline stress, MT treatment alone or with Put also increased shoot fresh weight and root fresh weight compared to the control. However, the individual application of Put did not enhance shoot fresh weight and root fresh weight compared to the control. Our results in Figure 1C–E indicate that all leave pigments including chlorophyll-a, chlorophyll b, total chlorophyll, and carotenoids were decreased by salinity stress. Additionally, all leaf pigments were increased by all treatments compared to the control under non-saline conditions. In addition, all treatments increased leaf pigments compared to the control treatment under saline conditions. The best treatments were the application of MT alone and when combined with Put.

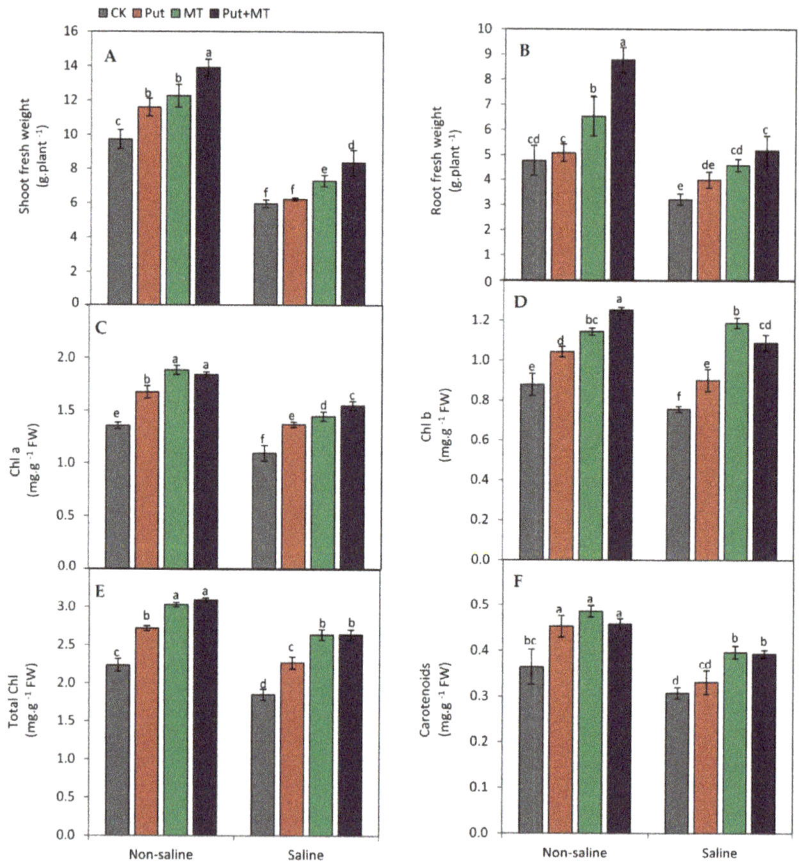

Figure 1. Effect of melatonin (MT; 20 µM), putrescine (Put; 100 µM), and their combination on the growth and leaf pigments of snap bean seedlings grown under control and salt stress (50 mM NaCl). (**A**) Shoot fresh weight, (**B**) shoot dry weight, (**C**) chlorophyll *a*, (**D**) chlorophyll *b*, (**E**) total chlorophyll, and (**F**) carotenoids. The upper error bar denotes standard deviation, while various lowercase letters above the bars denote significant differences according to Duncan's multiple range tests ($p \leq 0.05$).

Our results in Figure 2A indicate that RWC was decreased by salinity treatment. Under non-saline conditions, the RWC was not affected by any treatment compared with the control. However, under salinity stress all treatment significantly increased RWC compared to the control. Additionally, both the MT and MT + Put treatments showed higher RWC than the Put treatment. Figure 2B shows that the total soluble sugars were increased by salinity stress. Under both saline and non-saline conditions, all treatments (without significant difference between them) significantly increased total soluble sugars compared to the control treatment. Proline content increased under salinity stress compared to non-saline conditions (Figure 2C). However, under non-saline conditions, no changes in proline content were observed in all treatments. However, under a salinity stress condition, all treatment significantly increased the content of proline compared to the control treatment. Additionally, the dual application of MT and Put had a higher content of proline compared with the Put and control treatments.

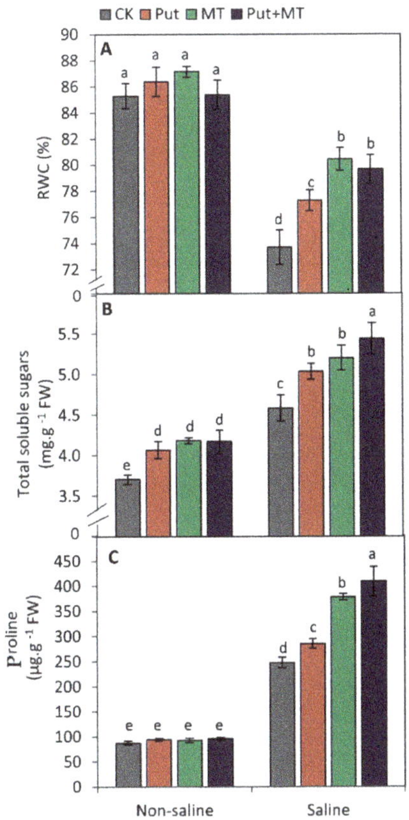

Figure 2. Effect of melatonin (MT; 20 μM), putrescine (Put; 100 μM), and their combination on the leaf relative water content (RWC) and osmolytes of snap bean seedlings grown under control and salt stress (50 mM NaCl). (**A**) RWC, (**B**) total soluble sugars, and (**C**) proline. The upper error bar denotes standard deviation, while various lowercase letters above the bars denote significant differences according to Duncan's multiple range tests ($p \leq 0.05$).

The effect of salinity and treatments on Na, K, Na/K, and Ca contents is presented in Figure 3A–D. As expected, Na content and Na/K ratio increased under saline conditions compared with normal conditions (Figure 3A,C). All treatments had no effect on Na content and Na/K ratio under non-saline conditions. However, under saline conditions, all treatments had significantly decreased Na content and Na/K ratio compared to the control treatment. Our results in Figure 3B show that K content was decreased by saline treatment compared with non-saline conditions. Under non-saline conditions, all treatments had significantly increased K content compared to the control treatment. However, under saline conditions, both MT and MT + Put treatments enhanced the content of K compared with other treatments. Additionally, the individual application of Put did not affect the content of K.

The results in Figure 3D show that Ca content was decreased by saline treatment compared with non-saline conditions. Under non-saline conditions, all treatments had significantly increased Ca content compared to the control treatment. Additionally, under saline conditions, all treatments (without significant difference between them) enhanced the content of Ca compared with other treatments.

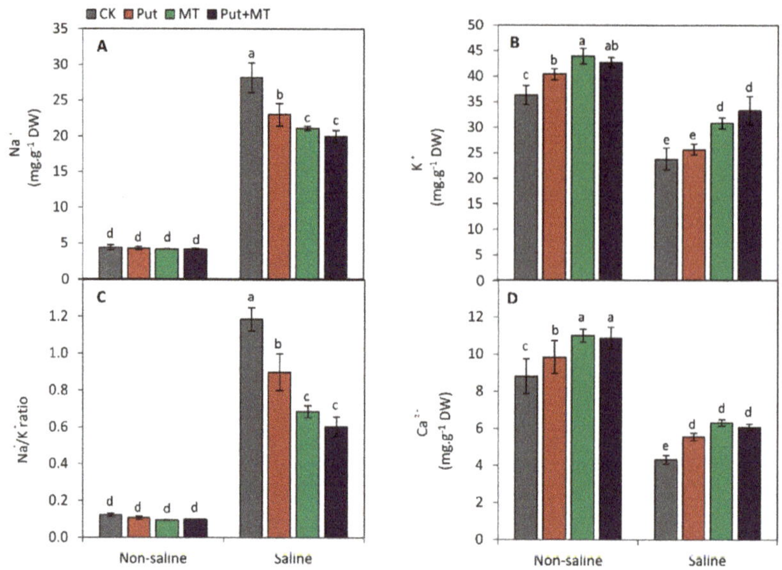

Figure 3. Effect of melatonin (MT; 20 µM), putrescine (Put; 100 µM), and their combination on the nutrient homeostasis of snap bean seedlings grown under control and salt stress (50 mM NaCl). (**A**) Na, (**B**) K, (**C**) Na/K ratio, and (**D**) Ca. The upper error bar denotes standard deviation, while various lowercase letters above the bars denote significant differences according to Duncan's multiple range tests ($p \leq 0.05$).

The results in Figure 4A show that CMSI was decreased by saline treatment. Under normal conditions, all treatments (without significant difference between them) had increased CMSI compared to the control treatment. The same trend of results was observed under saline conditions. Figure 4B shows that MG was increased by salinity stress. No significant difference in MG was observed among all treatments (except the MT treatment) compared to the control under non-saline conditions. However, all treatments had a significantly decreased content of MG compared to the control. Moreover, the dual application was the most effective treatment.

The content of H_2O_2 was increased by salinity treatment (Figure 4C). MT treatment showed a lower H_2O_2 content compared to all treatments under non-saline conditions. However, under saline conditions, all treatments significantly decreased the H_2O_2 content compared to the control, with the superior treatment being the combination of MT + Put. The content of MDA was increased by saline stress (Figure 4D). Moreover, there were no significant differences among all treatments with regards to MDA content under non-saline conditions. However, under saline stress, all treatments significantly decreased the content of MDA compared to the control treatment.

As expected, the activity of all tested antioxidant enzymes (SOD, CAT, POX, and APX) was increased by salinity stress (Figure 5A–D). Generally, under non-saline conditions, there were no significant differences between all of the treatments. However, all treatments showed a higher activity of all antioxidant enzymes under saline conditions compared to the control. The applications of MT + Put and MT were more effective than the Put treatment.

The results in Figure 6A,B show that DAO and PAO contents were increased by salinity stress. There was no significant difference between all treatments in DAO and PAO contents under non-saline conditions. However, under saline stress, all treatments significantly decreased the contents of DAO and PAO compared to the control treatment.

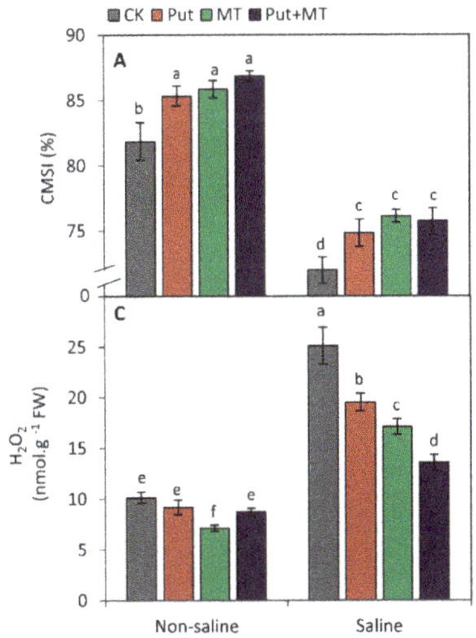

Figure 4. Effect of melatonin (MT; 20 µM), putrescine (Put; 100 µM), and their combination on the membrane stability and oxidative damage of snap bean seedlings grown under control and salt stress (50 mM NaCl). (**A**) Cell membrane stability index (CMSI), (**B**) methylglyoxal (MG), (**C**) H_2O_2, and (**D**) malondialdehyde (MDA). The upper error bar denotes standard deviation, while various lowercase letters above the bars denote significant differences according to Duncan's multiple range tests ($p \leq 0.05$).

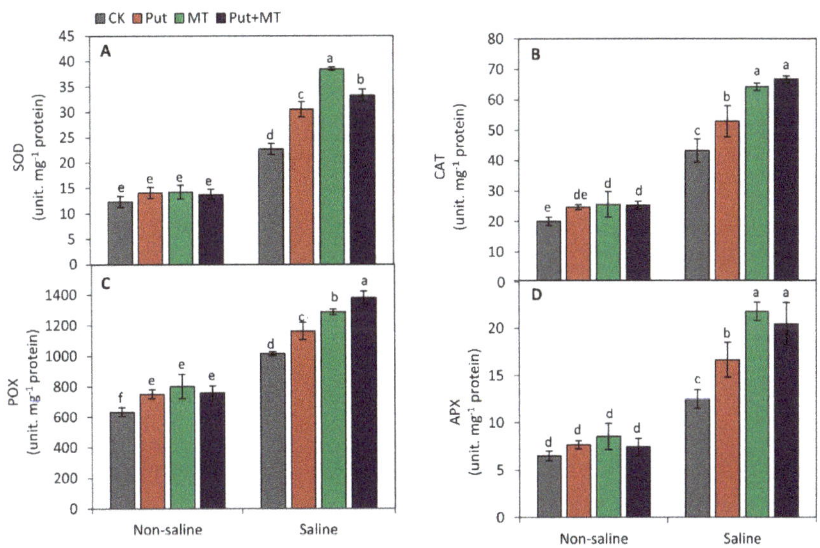

Figure 5. Effect of melatonin (MT; 20 µM), putrescine (Put; 100 µM), and their combination on the activities of antioxidant enzymes of snap bean seedlings grown under control and salt stress (50 mM NaCl). (**A**) Superoxide dismutase (SOD), (**B**) catalase (CAT), (**C**) peroxidase (POX), and (**D**) ascorbate peroxidase (APX). The upper error bar denotes standard deviation, while various lowercase letters above the bars denote significant differences according to Duncan's multiple range tests ($p \leq 0.05$).

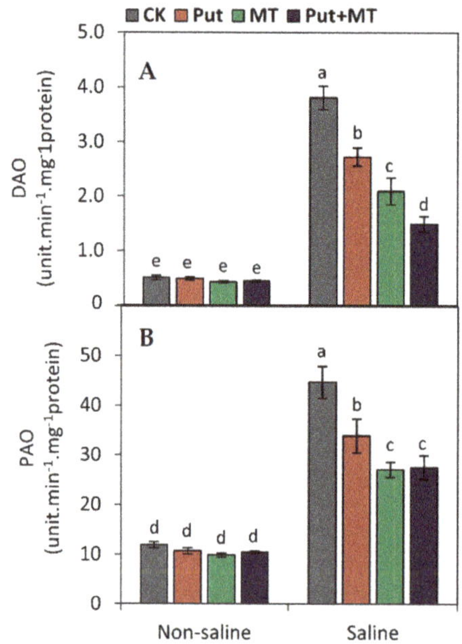

Figure 6. Effect of melatonin (MT; 20 μM), putrescine (Put; 100 μM), and their combination on the oxidation of polyamines of snap bean seedlings grown under control and salt stress (50 mM NaCl). (**A**) Diamine oxidase (DAO) and (**B**) polyamine oxidase (PAO). The upper error bar denotes standard deviation, while various lowercase letters above the bars denote significant differences according to Duncan's multiple range tests ($p \leq 0.05$).

4. Discussion

Green bean plants are considered one of the most saline-sensitive crops [73]. By limiting water intake and ion toxicity, salinity stress has an impact on the water status of the tissue, metabolic processes, and plant growth [74]. Salinity stress dramatically reduces growth by lowering the amount of chlorophyll and the rate of photosynthesis [26]. Likewise, our results showed that growth parameters such as the fresh weight of shoots and roots were decreased by salinity stress. In the current study, salinity stress decreased the leaf pigments, cell division, water uptake, and nutrient homeostasis [7,75,76]. In this study, MT enhanced growth and leaf pigments under saline conditions. Melatonin can regulate plant growth, photosynthetic machinery, and antioxidant capacity on the one hand, while delaying leaf senescence and suppressing ABFs-mediated abscisic acid (ABA) biosynthesis and chlorophyll degradation on the other hand [35,77,78]. Under salinity stress, MT has been shown to have positive effects on growth, chlorophyll content, photosynthesis process, and stomatal conductance in common bean crops [37] and other crops i.e., naked oat [77] and rice [79].

The low molecular weight polycations known as polyamines (PAs) are present in all living organisms and are involved in and/or regulate a wide range of physiological processes, including cell division, plant growth, gene expression control, cell proliferation, modulation of cell signaling, and membrane stabilization [80]. Some previous works have shown that treating different crops with polyamines reverses the damaging effects of salt stress and improves their tolerance [53,81,82].

In our study, the individual application of Put was not effective for enhancing growth under saline conditions (Figure 1A,B). In contrast with our results, Zhao and Qin [83] found that exogenous application of Put enhanced the root growth in barley seedlings under salt stress. This difference might be due to the differences between plant types and application method. The dual application of MT and Put had superior effect. Thus, the enhancement might be mainly related to MT application.

Preserving chlorophyll activity is critical for preventing leaf senescence, which reduces yields. Photosynthetic activity depends on the photosynthetic pigments such as chlorophylls and carotenoids. In this study, the Put treatment increased leaf pigments under normal and saline conditions (Figure 1C–F), which was similar to the results of previous studies conducted on pumpkins [84]. This enhancement of leaf pigments by Put application may be due to the role of Put in preventing the synthesis of ethylene leads to inhibit the degradation of chlorophyll, resulting in an increase in photosynthetic pigments [85]. These effects could also be attributed to the positive effects of Put on chlorophyll levels via thylakoid membrane protection at the site of the chlorophyll–protein complex [86].

Our results indicated that MT enhanced leaf pigments (chlorophyll *a*, chlorophyll *b*, total chlorophyll, and carotenoids) under normal and saline conditions. Similarly, Erdal [87] found that chlorophyll contents in maize seedlings were increased by MT application compared with the control. In accordance with our results, Arnao and Hernández-Ruiz [88] found that exogenous MT treatment protected chlorophyll from degradation in barley. Melatonin's role in chlorophyll preservation was later demonstrated in other crops such as tomatoes [89]. These results could be due to the fact that MT enhanced the gene expression encoding two photosystem I subunits, two photosystem II elements, and ferredoxin PetF, while decreasing the expression of genes encoding the chlorophyll-degrading enzyme chlorophyllase [90]. In agreement with our results (Figure 1F), it has been previously reported that MT application also preserves the carotenoid content [91].

It has been well known that abiotic stresses, such as salinity, affect the metabolism of some compounds including carbohydrates and amino acids [34]. The proline mostly consists of carbohydrates [37]. In this study, and previous studies [37,92], we found that salinity increased the content of proline and total soluble sugars in bean plants (Figure 2B,C). A previous study has shown the role of proline in the resistance of plants to salinity stress by controlling the activity of enzymes and controlling the permeability of the cell membrane [93]. A previous study [37] and our study found that MT treatment increased the levels of proline and total soluble sugars in salinity-stressed plants. This shows the role of MT in combatting salinity by increasing total soluble sugars and proline. In the current study, Put supplementation improved proline synthesis compared to the control. Similarly, previous studies observed enhanced proline content when treated with Put under salt stress in some crops [94,95]. Additionally, Put application under abiotic stresses, such as salinity, increases endogenous Put levels in the plant, increasing plant resistance to saline stress [48].

Relative water content (RWC) is a vital physiological parameter for water status that supports the ability of plants to survive in stressed conditions. In this study, RWC was decreased by saline stress (Figure 2A) which could be explained by the negative effect of salinity on water absorption and availability [96]. Our results also indicated that MT treatment increased RWC under salinity stress. Some previous works reported that Put application enhanced RWC under salt stress [97,98].

All abiotic stresses negatively affect the cell membrane and enzymatic systems by raising ROS levels, including H_2O_2, and change their balance [95]. The exposure of plants to salinity leads to damage to the cell membrane, which causes an increase in the amount of MDA content [99]. In this study, we recorded the same previous results that H_2O_2, MDA, and MG were increased under salinity stress. In this study, MT treatment resisted the negative effects of salinity by decreasing H_2O_2 and MDA content by raising transcription levels [100] and stimulating the activity of antioxidant enzymes (Figure 5A–D). The same results were recorded previously [37,99].

It has been well known that PAs has a role in scavenging ROS under abiotic stresses, in addition to its function as an antioxidant [101]. In this study, the application of Put caused a significant reduction in H_2O_2, MDA, and MG. In accordance with our results, previous works indicated that Put applications reduced MDA and MG toxicity in mung bean plants [102,103].

Antioxidant enzymes work to resist the harmful effect caused by ROS inside the plant cells. For example, SOD converts O_2^- to H_2O_2, while CAT converts H_2O_2 to H_2O and molecular oxygen inside plant cells. Additionally, APX converts H_2O_2 to monodehydroascorbate [104]. The results in this study showed that MT treatment alleviated the harmful effects of salinity by increasing the activity of antioxidant enzymes. Our results are in agreement with Li et al. [105] who found that MT treatment mitigates the negative effect of salinity by increasing the activity of SOD and CAT enzymes. In this study, MT treatment enhanced the activity of all antioxidant enzymes. Previous works indicated the positive role of MT in ROS-scavenging by increasing the activity of antioxidant enzymes [37,104–107].

The data in Figure 5A–D indicated that Put application increased the activities of the antioxidant enzymes. Put application has been shown to induce POX activity under saline conditions, and this activation can be interpreted as a conformational change caused by polyamine binding [94]. Additionally, exogenous application of Put improved the activity of CAT in pepper seedlings grown under salt stress conditions, according to Ekinci et al. [108].

Our results in Figure 3A–D show that MT treatment decreased the Na content and Na/K ratio while there were increased K and Ca contents under saline stress. Our results are in agreement with Li et al. [105] who found that MT treatment decreased the Na content and Na/K ratio in rice plants under salinity conditions. Additionally, Qu et al. [109] found that MT treatment increased the content of K, which might be due to the improvement of stomatal conductance by MT treatment under salinity stress [37].

The results in Figure 3A–D show that Put treatment reduced Na uptakes. Similarly, previous studies indicated that Put applications inhibit or reduce the uptake of Na from the soil in some crops such as rice and mung beans [52,103] due to the role of Put for controlling the balance of cation and anion [84]. Additionally, when exogenous Put was added, there was a decrease in Na and an increase in K, which was associated with root growth promotion [110]. Ca content was clearly reduced in the salt-stressed seedlings (Figure 3D). In this work, Put application alleviated the harmful effect of saline stress and enhanced Ca content in plant. This result is in agreement with Nahar et al. [103] who found that Put application increased Ca content in mung bean shoots and roots.

Our findings showed salinity-induced increases in DAO and PAO activity. By regulating H_2O_2 signalling, which is created by stress-induced PAO activity leading to spermidine, spermine, and spermine oxidation, plant polyamine oxidases (DAO and PAO) play key roles in various stress tolerance and programmed cell death (PCD) processes [111]. In contrast, the treatments of MT and MT + Put diminished the activities of DAO and PAO. Melatonin can accelerate the biosynthesis of PAs through increasing their metabolic flow from the precursor amino acids arginine and methionine [1]. Moreover, this response has been found to be correlated with the suppression of polyamine oxidase (PAO) and diamine oxidase (DAO) activities [1].

5. Conclusions

The present study revealed that the combination treatment of MT + Put mitigated the salinity-induced oxidative damage of snap bean seedlings. This effect was correlated with suppressing the activties of polyamine oxidases (DAO and PAO) and enhancing several features of tolerance to salinity stress. These aspects included the improvement of growth, leaf pigments, CMSI, RWC, osmolytes, and antioxidant enzymes. Conversely, there was an obvious decrease in the oxidative damage as indicated by the reduction in the leaf content of methyglycoxal, H_2O_2, and MDA. Further future studies at the molecular level are required to understand the crosstalk between melatonin and polyamines in regulating plant tolerance to salinity stress.

Supplementary Materials: The following supporting information can be downloaded at: https://www.mdpi.com/article/10.3390/horticulturae9020285/s1. Figure S1. Effect of different concentrations of melatonin and putrescine on the membrane lipid oxidation of snap bean seedlings grown under control and salt stress (50 mM NaCl).

Author Contributions: Conceptualization, H.S.E.-B., A.A.E.-Y., H.G.A.E.-G., T.A.S., A.T.M., M.K. and M.F.M.I.; methodology, H.S.E.-B., A.T.M., A.A.A., N.A.A.-H., S.M.A.-Q. and M.F.M.I.; software, H.S.E.-B. and M.F.M.I.; validation, H.S.E.-B., A.A.E.-Y., H.G.A.E.-G., M.K., T.A.S., A.T.M., A.A.A. and M.F.M.I.; formal analysis, H.S.E.-B., A.A.E.-Y., H.G.A.E.-G., M.K., T.A.S., A.T.M. and M.F.M.I.; investigation, H.S.E.-B. and M.F.M.I.; resources, A.A.A., N.A.A.-H. and S.M.A.-Q.; data curation, H.S.E.-B. and M.F.M.I.; writing—original draft preparation, H.S.E.-B., M.K., A.A.E.-Y., H.G.A.E.-G., T.A.S. and M.F.M.I.; writing—review and editing, H.S.E.-B., A.A.E.-Y., H.G.A.E.-G., M.K., T.A.S., A.T.M. and M.F.M.I.; visualization, H.S.E.-B., A.A.E.-Y., H.G.A.E.-G., M.K., T.A.S., A.T.M. and M.F.M.I.; supervision, H.S.E.-B. and M.F.M.I.; project administration, H.S.E.-B., H.G.A.E.-G. and M.F.M.I.; funding acquisition, H.S.E.-B., M.K., T.A.S., and A.T.M. All authors have read and agreed to the published version of the manuscript.

Funding: This research work was supported and funded by the Deputyship for Research and Innovation, Ministry of Education in Saudi Arabia (Project number INST076).

Institutional Review Board Statement: Not applicable.

Informed Consent Statement: Not applicable.

Data Availability Statement: All data are available within the manuscript.

Acknowledgments: The authors extend their appreciation to the Deputyship for Research and Innovation, Ministry of Education in Saudi Arabia for funding this research work.

Conflicts of Interest: The authors declare no conflict of interest.

References

1. Ke, Q.; Ye, J.; Wang, B.; Ren, J.; Yin, L.; Deng, X.; Wang, S. Melatonin mitigates salt stress in wheat seedlings by modulating polyamine metabolism. *Front. Plant Sci.* **2018**, *9*, 914. [CrossRef]
2. Shahid, S.A.; Zaman, M.; Heng, L. Soil Salinity: Historical Perspectives and a World Overview of the Problem. In *Guideline for Salinity Assessment, Mitigation and Adaptation Using Nuclear and Related Techniques*; Zaman, M., Shahid, S.A., Heng, L., Eds.; Springer International Publishing: Cham, Switzerland, 2018; pp. 43–53. [CrossRef]
3. Mukhopadhyay, R.; Sarkar, B.; Jat, H.S.; Sharma, P.C.; Bolan, N.S. Soil salinity under climate change: Challenges for sustainable agriculture and food security. *J. Environ. Manag.* **2021**, *280*, 111736. [CrossRef]
4. El-Beltagi, H.S.; Ahmad, I.; Basit, A.; Shehata, W.F.; Hassan, U.; Shah, S.T.; Haleema, B.; Jalal, A.; Amin, R.; Khalid, M.A. Ascorbic acid enhances growth and yield of sweet peppers (*Capsicum annum*) by mitigating salinity stress. *Gesunde Pflanz.* **2022**, *74*, 423–433. [CrossRef]
5. Mansour, M.M.F. The plasma membrane transport systems and adaptation to salinity. *J. Plant Physiol.* **2014**, *171*, 1787–1800. [CrossRef]
6. Ramadan, K.M.A.; Alharbi, M.M.; Alenzi, A.M.; El-Beltagi, H.S.; Darwish, D.B.; Aldaej, M.I.; Shalaby, T.A.; Mansour, A.T.; El-Gabry, Y.A.; Ibrahim, M.F.M. Alpha Lipoic Acid as a Protective Mediator for Regulating the Defensive Responses of Wheat Plants against Sodic Alkaline Stress: Physiological, Biochemical and Molecular Aspects. *Plants* **2022**, *11*, 787. [CrossRef]
7. Youssef, M.H.M.; Raafat, A.; El-Yazied, A.A.; Selim, S.; Azab, E.; Khojah, E.; El Nahhas, N.; Ibrahim, M.F.M. Exogenous Application of Alpha-Lipoic Acid Mitigates Salt-Induced Oxidative Damage in Sorghum Plants through Regulation Growth, Leaf Pigments, Ionic Homeostasis, Antioxidant Enzymes, and Expression of Salt Stress Responsive Genes. *Plants* **2021**, *10*, 2519. [CrossRef]
8. Celmeli, T.; Sari, H.; Canci, H.; Sari, D.; Adak, A.; Eker, T.; Toker, C. The nutritional content of common bean (*Phaseolus vulgaris* L.) landraces in comparison to modern varieties. *Agronomy* **2018**, *8*, 166. [CrossRef]
9. Ibrahim, M.; Ibrahim, H.A.; Abd El-Gawad, H. Folic acid as a protective agent in snap bean plants under water deficit conditions. *J. Hortic. Sci. Biotechnol.* **2021**, *96*, 94–109. [CrossRef]
10. Câmara, C.R.; Urrea, C.A.; Schlegel, V. Pinto beans (*Phaseolus vulgaris* L.) as a functional food: Implications on human health. *Agriculture* **2013**, *3*, 90–111. [CrossRef]
11. Bazzano, L.A.; He, J.; Ogden, L.G.; Loria, C.; Vupputuri, S.; Myers, L.; Whelton, P.K. Legume consumption and risk of coronary heart disease in US men and women: NHANES I Epidemiologic Follow-up Study. *Arch. Intern. Med.* **2001**, *161*, 2573–2578. [CrossRef]
12. Hertog, M.G.; Feskens, E.J.; Kromhout, D.; Hollman, P.; Katan, M. Dietary antioxidant flavonoids and risk of coronary heart disease: The Zutphen Elderly Study. *Lancet* **1993**, *342*, 1007–1011. [CrossRef]

13. Didinger, C.; Foster, M.T.; Bunning, M.; Thompson, H.J. Nutrition and Human Health Benefits of Dry Beans and Other Pulses. In *Dry Beans and Pulses: Production, Processing, and Nutrition*; John Wiley & Sons, Inc.: Hoboken, NJ, USA, 2022; pp. 481–504. [CrossRef]
14. Sangaramoorthy, M.; Koo, J.; John, E.M. Intake of bean fiber, beans, and grains and reduced risk of hormone receptor-negative breast cancer: The San Francisco Bay Area Breast Cancer Study. *Cancer Med.* **2018**, *7*, 2131–2144. [CrossRef] [PubMed]
15. Darmadi-Blackberry, I.; Wahlqvist, M.L.; Kouris-Blazos, A.; Steen, B.; Lukito, W.; Horie, Y.; Horie, K. Legumes: The most important dietary predictor of survival in older people of different ethnicities. *Asia Pac. J. Clin. Nutr.* **2004**, *13*, 217–220.
16. Alsamadany, H.; Mansour, H.; Elkelish, A.; Ibrahim, M.F. Folic Acid Confers Tolerance against Salt Stress-Induced Oxidative Damages in Snap Beans through Regulation Growth, Metabolites, Antioxidant Machinery and Gene Expression. *Plants* **2022**, *11*, 1459. [CrossRef]
17. Farag, H.A.; Ibrahim, M.F.; El-Yazied, A.A.; El-Beltagi, H.S.; El-Gawad, H.G.A.; Alqurashi, M.; Shalaby, T.A.; Mansour, A.T.; Alkhateeb, A.A.; Farag, R. Applied Selenium as a Powerful Antioxidant to Mitigate the Harmful Effects of Salinity Stress in Snap Bean Seedlings. *Agronomy* **2022**, *12*, 3215. [CrossRef]
18. Zhang, M.; Gao, C.; Xu, L.; Niu, H.; Liu, Q.; Huang, Y.; Lv, G.; Yang, H.; Li, M. Melatonin and indole-3-acetic acid synergistically regulate plant growth and stress resistance. *Cells* **2022**, *11*, 3250. [CrossRef]
19. Arnao, M.B.; Giraldo-Acosta, M.; Castejón-Castillejo, A.; Losada-Lorán, M.; Sánchez-Herrerías, P.; El Mihyaoui, A.; Cano, A.; Hernández-Ruiz, J. Melatonin from Microorganisms, Algae, and Plants as Possible Alternatives to Synthetic Melatonin. *Metabolites* **2023**, *13*, 72. [CrossRef]
20. Zhao, D.; Yu, Y.; Shen, Y.; Liu, Q.; Zhao, Z.; Sharma, R.; Reiter, R.J. Melatonin synthesis and function: Evolutionary history in animals and plants. *Front. Endocrinol.* **2019**, *10*, 249. [CrossRef] [PubMed]
21. Arnao, M.B.; Hernández-Ruiz, J. Melatonin: A New Plant Hormone and/or a Plant Master Regulator? *Trends Plant Sci.* **2019**, *24*, 38–48. [CrossRef]
22. Xiao, S.; Liu, L.; Wang, H.; Li, D.; Bai, Z.; Zhang, Y.; Sun, H.; Zhang, K.; Li, C. Exogenous melatonin accelerates seed germination in cotton (*Gossypium hirsutum* L.). *PLoS ONE* **2019**, *14*, e0216575. [CrossRef]
23. Arnao, M.B.; Hernández-Ruiz, J. Functions of melatonin in plants: A review. *J. Pineal Res.* **2015**, *59*, 133–150. [CrossRef]
24. Zhao, Y.-Q.; Zhang, Z.-W.; Chen, Y.-E.; Ding, C.-B.; Yuan, S.; Reiter, R.J.; Yuan, M. Melatonin: A potential agent in delaying leaf senescence. *Crit. Rev. Plant Sci.* **2021**, *40*, 1–22. [CrossRef]
25. Moustafa-Farag, M.; Almoneafy, A.; Mahmoud, A.; Elkelish, A.; Arnao, M.B.; Li, L.; Ai, S. Melatonin and its protective role against biotic stress impacts on plants. *Biomolecules* **2019**, *10*, 54. [CrossRef]
26. Zhang, N.; Sun, Q.; Zhang, H.; Cao, Y.; Weeda, S.; Ren, S.; Guo, Y.-D. Roles of melatonin in abiotic stress resistance in plants. *J. Exp. Bot.* **2015**, *66*, 647–656. [CrossRef]
27. Buttar, Z.A.; Wu, S.N.; Arnao, M.B.; Wang, C.; Ullah, I.; Wang, C. Melatonin Suppressed the Heat Stress-Induced Damage in Wheat Seedlings by Modulating the Antioxidant Machinery. *Plants* **2020**, *9*, 809. [CrossRef]
28. Wang, M.; Zhang, S.; Ding, F. Melatonin Mitigates Chilling-Induced Oxidative Stress and Photosynthesis Inhibition in Tomato Plants. *Antioxidants* **2020**, *9*, 218. [CrossRef] [PubMed]
29. Debnath, B.; Islam, W.; Li, M.; Sun, Y.; Lu, X.; Mitra, S.; Hussain, M.; Liu, S.; Qiu, D. Melatonin mediates enhancement of stress tolerance in plants. *Int. J. Mol. Sci.* **2019**, *20*, 1040. [CrossRef]
30. Jahan, M.S.; Guo, S.; Baloch, A.R.; Sun, J.; Shu, S.; Wang, Y.; Ahammed, G.J.; Kabir, K.; Roy, R. Melatonin alleviates nickel phytotoxicity by improving photosynthesis, secondary metabolism and oxidative stress tolerance in tomato seedlings. *Ecotoxicol. Environ. Saf.* **2020**, *197*, 110593. [CrossRef]
31. El-Yazied, A.A.; Ibrahim, M.F.; Ibrahim, M.A.; Nasef, I.N.; Al-Qahtani, S.M.; Al-Harbi, N.A.; Alzuaibr, F.M.; Alaklabi, A.; Dessoky, E.S.; Alabdallah, N.M. Melatonin Mitigates Drought Induced Oxidative Stress in Potato Plants through Modulation of Osmolytes, Sugar Metabolism, ABA Homeostasis and Antioxidant Enzymes. *Plants* **2022**, *11*, 1151. [CrossRef] [PubMed]
32. Ibrahim, M.F.; Elbar, O.H.A.; Farag, R.; Hikal, M.; El-Kelish, A.; El-Yazied, A.A.; Alkahtani, J.; El-Gawad, H.G.A. Melatonin counteracts drought induced oxidative damage and stimulates growth, productivity and fruit quality properties of tomato plants. *Plants* **2020**, *9*, 1276. [CrossRef] [PubMed]
33. Gohari, G.; Farhadi, H.; Panahirad, S.; Zareei, E.; Labib, P.; Jafari, H.; Mahdavinia, G.; Hassanpouraghdam, M.B.; Ioannou, A.; Kulak, M. Mitigation of salinity impact in spearmint plants through the application of engineered chitosan-melatonin nanoparticles. *Int. J. Biol. Macromol.* **2023**, *224*, 893–907. [CrossRef] [PubMed]
34. Siddiqui, M.H.; Alamri, S.; Al-Khaishany, M.Y.; Khan, M.N.; Al-Amri, A.; Ali, H.M.; Alaraidh, I.A.; Alsahli, A.A. Exogenous melatonin counteracts NaCl-induced damage by regulating the antioxidant system, proline and carbohydrates metabolism in tomato seedlings. *Int. J. Mol. Sci.* **2019**, *20*, 353. [CrossRef]
35. Varghese, N.; Alyammahi, O.; Nasreddine, S.; Alhassani, A.; Gururani, M.A. Melatonin positively influences the photosynthetic machinery and antioxidant system of Avena sativa during salinity stress. *Plants* **2019**, *8*, 610. [CrossRef]
36. Zhou, X.; Zhao, H.; Cao, K.; Hu, L.; Du, T.; Baluška, F.; Zou, Z. Beneficial roles of melatonin on redox regulation of photosynthetic electron transport and synthesis of D1 protein in tomato seedlings under salt stress. *Front. Plant Sci.* **2016**, *7*, 1823. [CrossRef]
37. Azizi, F.; Amiri, H.; Ismaili, A. Melatonin improves salinity stress tolerance of *Phaseolus vulgaris* L. cv. Pak by changing antioxidant enzymes and photosynthetic parameters. *Acta Physiol. Plant.* **2022**, *44*, 40. [CrossRef]

38. Chen, L.; Liu, L.; Lu, B.; Ma, T.; Jiang, D.; Li, J.; Zhang, K.; Sun, H.; Zhang, Y.; Bai, Z. Exogenous melatonin promotes seed germination and osmotic regulation under salt stress in cotton (*Gossypium hirsutum* L.). *PLoS ONE* **2020**, *15*, e0228241. [CrossRef]
39. Zhao, G.; Zhao, Y.; Yu, X.; Kiprotich, F.; Han, H.; Guan, R.; Wang, R.; Shen, W. Nitric oxide is required for melatonin-enhanced tolerance against salinity stress in rapeseed (*Brassica napus* L.) seedlings. *Int. J. Mol. Sci.* **2018**, *19*, 1912. [CrossRef]
40. Vafadar, F.; Amooaghaie, R.; Ehsanzadeh, P.; Ghanati, F.; Sajedi, R.H. Crosstalk between melatonin and Ca^{2+}/CaM evokes systemic salt tolerance in Dracocephalum kotschyi. *J. Plant Physiol.* **2020**, *252*, 153237. [CrossRef]
41. Chen, Z.; Xie, Y.; Gu, Q.; Zhao, G.; Zhang, Y.; Cui, W.; Xu, S.; Wang, R.; Shen, W. The AtrbohF-dependent regulation of ROS signaling is required for melatonin-induced salinity tolerance in Arabidopsis. *Free Radic. Biol. Med.* **2017**, *108*, 465–477. [CrossRef] [PubMed]
42. Chen, D.; Shao, Q.; Yin, L.; Younis, A.; Zheng, B. Polyamine function in plants: Metabolism, regulation on development, and roles in abiotic stress responses. *Front. Plant Sci.* **2019**, *9*, 1945. [CrossRef] [PubMed]
43. Egea-Cortines, M.; Mizrahi, Y. Polyamines in cell division, fruit set and development, and seed germination. In *Slocum RD and Flores HE (eds) Biochemistry and Physiology of Polyamines in Plants*; CRC Press: Boca Raton, FL, USA, 1991; pp. 143–158.
44. Kuznetsov, V.V.; Shevyakova, N.I. Polyamines and stress tolerance of plants. *Plant Stress* **2007**, *1*, 50–71. [CrossRef]
45. Kaur-Sawhney, R.; Tiburcio, A.F.; Altabella, T.; Galston, A.W. Polyamines in plants: An overview. *J. Mol. Cell Biol.* **2003**, *2*, 1–12.
46. Alcázar, R.; Altabella, T.; Marco, F.; Bortolotti, C.; Reymond, M.; Koncz, C.; Carrasco, P.; Tiburcio, A.F. Polyamines: Molecules with regulatory functions in plant abiotic stress tolerance. *Planta* **2010**, *231*, 1237–1249. [CrossRef]
47. Hamdani, S.; Gauthier, A.; Msilini, N.; Carpentier, R. Positive charges of polyamines protect PSII in isolated thylakoid membranes during photoinhibitory conditions. *Plant Cell Physiol.* **2011**, *52*, 866–873. [CrossRef]
48. Pottosin, I.; Shabala, S. Polyamines control of cation transport across plant membranes: Implications for ion homeostasis and abiotic stress signaling. *Front. Plant Sci.* **2014**, *5*, 154. [CrossRef]
49. Todorova, D.; Katerova, Z.; Shopova, E.; Jodinskienė, M.; Jurkoniene, S.; Sergiev, I. Responses of pea plants to heat stress and spermine treatment. *Zemdirb.-Agric.* **2016**, *103*, 99–106. [CrossRef]
50. Groppa, M.D.; Tomaro, M.L.; Benavides, M.P. Polyamines and heavy metal stress: The antioxidant behavior of spermine in cadmium-and copper-treated wheat leaves. *Biometals* **2007**, *20*, 185–195. [CrossRef] [PubMed]
51. Farooq, M.; Wahid, A.; Lee, D.-J. Exogenously applied polyamines increase drought tolerance of rice by improving leaf water status, photosynthesis and membrane properties. *Acta Physiol. Plant.* **2009**, *31*, 937–945. [CrossRef]
52. Chattopadhayay, M.K.; Tiwari, B.S.; Chattopadhyay, G.; Bose, A.; Sengupta, D.N.; Ghosh, B. Protective role of exogenous polyamines on salinity-stressed rice (*Oryza sativa*) plants. *Physiol. Plant.* **2002**, *116*, 192–199. [CrossRef]
53. Duan, J.; Li, J.; Guo, S.; Kang, Y. Exogenous spermidine affects polyamine metabolism in salinity-stressed Cucumis sativus roots and enhances short-term salinity tolerance. *J. Plant Physiol.* **2008**, *165*, 1620–1635. [CrossRef] [PubMed]
54. Velikova, V.; Yordanov, I.; Edreva, A. Oxidative stress and some antioxidant systems in acid rain-treated bean plants: Protective role of exogenous polyamines. *Plant Sci.* **2000**, *151*, 59–66. [CrossRef]
55. Matkovics, B.; Kecskemeti, V.; Varga, S.I.; Novak, Z.; Kertesz, Z. Antioxidant properties of di-and polyamines. *Comp. Biochem. Physiol.. B Comp. Biochem.* **1993**, *104*, 475–479. [CrossRef] [PubMed]
56. Sengupta, A.; Chakraborty, M.; Saha, J.; Gupta, B.; Gupta, K. Polyamines: Osmoprotectants in plant abiotic stress adaptation. In *Osmolytes and Plants Acclimation to Changing Environment: Emerging Omics Technologies*; Springer: New Delhi, India, 2016; pp. 97–127.
57. Sakamoto, A.; Terui, Y.; Uemura, T.; Igarashi, K.; Kashiwagi, K. Polyamines regulate gene expression by stimulating translation of histone acetyltransferase mRNAs. *J. Biol. Chem.* **2020**, *295*, 8736–8745. [CrossRef]
58. Hoagland, D.R.; Arnon, D.I. The water-culture method for growing plants without soil. *Circular. Calif. Agric. Exp. Stn.* **1950**, *347*, 32.
59. Lichtenthaler, H.K.; Wellburn, A.R. Determinations of total carotenoids and chlorophylls a and b of leaf extracts in different solvents. *Biochem. Soc. Trans.* **1983**, *11*, 591–592. [CrossRef]
60. Abd El-Gawad, H.G.; Mukherjee, S.; Farag, R.; Abd Elbar, O.H.; Hikal, M.; Abou El-Yazied, A.; Abd Elhady, S.A.; Helal, N.; ElKelish, A.; El Nahhas, N. Exogenous γ-aminobutyric acid (GABA)-induced signaling events and field performance associated with mitigation of drought stress in *Phaseolus vulgaris* L. *Plant Signal. Behav.* **2021**, *16*, 1853384. [CrossRef]
61. Bates, L.; Waldren, R.; Teare, I. Rapid determination of free proline for water-stress studies. *Plant Soil* **1973**, *39*, 205–207. [CrossRef]
62. Ieggli, C.; Bohrer, D.; Do Nascimento, P.; De Carvalho, L.; Garcia, S. Determination of sodium, potassium, calcium, magnesium, zinc, and iron in emulsified egg samples by flame atomic absorption spectrometry. *Talanta* **2010**, *80*, 1282–1286. [CrossRef]
63. Abd Elbar, O.H.; Elkelish, A.; Niedbała, G.; Farag, R.; Wojciechowski, T.; Mukherjee, S.; Abou-Hadid, A.F.; El-Hennawy, H.M.; Abou El-Yazied, A.; Abd El-Gawad, H.G. Protective Effect of γ-Aminobutyric Acid Against Chilling Stress during Reproductive Stage in Tomato Plants Through Modulation of Sugar Metabolism, Chloroplast Integrity, and Antioxidative Defense Systems. *Front. Plant Sci.* **2021**, *12*, 663750. [CrossRef] [PubMed]
64. Hossain, M.A.; Hossain, M.Z.; Fujita, M. Stress-induced changes of methylglyoxal level and glyoxalase I activity in pumpkin seedlings and cDNA cloning of glyoxalase I gene. *Aust. J. Crop Sci.* **2009**, 3, 53.
65. Heath, R.L.; Packer, L. Photoperoxidation in isolated chloroplasts: I. Kinetics and stoichiometry of fatty acid peroxidation. *Arch. Biochem. Biophys.* **1968**, *125*, 189–198. [CrossRef] [PubMed]

66. Bradford, M.M. A rapid and sensitive method for the quantitation of microgram quantities of protein utilizing the principle of protein-dye binding. *Anal. Biochem.* **1976**, *72*, 248–254. [CrossRef]
67. Beyer, W.F., Jr.; Fridovich, I. Assaying for superoxide dismutase activity: Some large consequences of minor changes in conditions. *Anal. Biochem.* **1987**, *161*, 559–566. [CrossRef]
68. Cakmak, I.; Strbac, D.; Marschner, H. Activities of hydrogen peroxide-scavenging enzymes in germinating wheat seeds. *J. Exp. Bot.* **1993**, *44*, 127–132. [CrossRef]
69. Nakano, Y.; Asada, K. Hydrogen peroxide is scavenged by ascorbate-specific peroxidase in spinach chloroplasts. *Plant Cell Physiol.* **1981**, *22*, 867–880. [CrossRef]
70. Dias, M.A.; Costa, M.M. Effect of low salt concentrations on nitrate reductase and peroxidase of sugar beet leaves. *J. Exp. Bot.* **1983**, *34*, 537–543. [CrossRef]
71. Su, G.X.; Zhang, W.H.; Liu, Y.L. Involvement of hydrogen peroxide generated by polyamine oxidative degradation in the development of lateral roots in soybean. *J. Integr. Plant Biol.* **2006**, *48*, 426–432. [CrossRef]
72. SAS. *SAS/STAT User's Guide, Release 6.03 ed.*; SAS Inst. Inc.: Cary, NC, USA, 1988.
73. Bayuelo-Jimenez, J.; Jasso-Plata, N.; Ochoa, I. Growth and physiological responses of Phaseolus species to salinity stress. *Int. J. Agron.* **2012**, *2012*, 527673. [CrossRef]
74. Rady, M.M.; Mohamed, G.F. Modulation of salt stress effects on the growth, physio-chemical attributes and yields of Phaseolus vulgaris L. plants by the combined application of salicylic acid and *Moringa oleifera* leaf extract. *Sci. Hortic.* **2015**, *193*, 105–113. [CrossRef]
75. Abdelaal, K.; Alsubeie, M.S.; Hafez, Y.; Emeran, A.; Moghanm, F.; Okasha, S.; Omara, R.; Basahi, M.A.; Darwish, D.B.E.; Ibrahim, M. Physiological and Biochemical Changes in Vegetable and Field Crops under Drought, Salinity and Weeds Stresses: Control Strategies and Management. *Agriculture* **2022**, *12*, 2084. [CrossRef]
76. El-Beltagi, H.S.; Al-Otaibi, H.H.; Parmar, A.; Ramadan, K.; Lobato, A.K.d.S.; El Mogy, M.M. Application of Potassium Humate and Salicylic Acid to Mitigate Salinity Stress of Common Bean. *Life* **2023**, *13*, 448. [CrossRef]
77. Gao, W.; Feng, Z.; Bai, Q.; He, J.; Wang, Y. Melatonin-mediated regulation of growth and antioxidant capacity in salt-tolerant naked oat under salt stress. *Int. J. Mol. Sci.* **2019**, *20*, 1176. [CrossRef]
78. Tan, X.L.; Fan, Z.q.; Kuang, J.f.; Lu, W.j.; Reiter, R.J.; Lakshmanan, P.; Su, X.g.; Zhou, J.; Chen, J.Y.; Shan, W. Melatonin delays leaf senescence of Chinese flowering cabbage by suppressing ABFs-mediated abscisic acid biosynthesis and chlorophyll degradation. *J. Pineal Res.* **2019**, *67*, e12570. [CrossRef]
79. Liang, C.; Zheng, G.; Li, W.; Wang, Y.; Hu, B.; Wang, H.; Wu, H.; Qian, Y.; Zhu, X.G.; Tan, D.X. Melatonin delays leaf senescence and enhances salt stress tolerance in rice. *J. Pineal Res.* **2015**, *59*, 91–101. [CrossRef] [PubMed]
80. Igarashi, K.; Kashiwagi, K. Polyamines: Mysterious modulators of cellular functions. *Biochem. Biophys. Res. Commun.* **2000**, *271*, 559–564. [CrossRef]
81. Sigmaz, B.; Agar, G.; Arslan, E.; Aydin, M.; Taspinar, M.S. The role of putrescine against the long terminal repeat (LTR) retrotransposon polymorphisms induced by salinity stress in Triticum aestivum. *Acta Physiol. Plant.* **2015**, *37*, 251. [CrossRef]
82. Zhang, W.; Jiang, B.; Li, W.; Song, H.; Yu, Y.; Chen, J. Polyamines enhance chilling tolerance of cucumber (*Cucumis sativus* L.) through modulating antioxidative system. *Sci. Hortic.* **2009**, *122*, 200–208. [CrossRef]
83. Zhao, F.; Qin, P. Protective effect of exogenous polyamines on root tonoplast function against salt stress in barley seedlings. *Plant Growth Regul.* **2004**, *42*, 97–103. [CrossRef]
84. Farsaraei, S.; Mehdizadeh, L.; Moghaddam, M. Seed priming with putrescine alleviated salinity stress during germination and seedling growth of medicinal pumpkin. *J. Soil Sci. Plant Nutr.* **2021**, *21*, 1782–1792. [CrossRef]
85. Hamdy, A.E. Effect of GA_3 and NAA on growth, yield and fruit quality of Washington navel orange. *Egypt. J. Hortic.* **2017**, *44*, 33–43. [CrossRef]
86. Abd Elbar, O.H.; Farag, R.E.; Shehata, S.A. Effect of putrescine application on some growth, biochemical and anatomical characteristics of *Thymus vulgaris* L. under drought stress. *Ann. Agric. Sci.* **2019**, *64*, 129–137. [CrossRef]
87. Erdal, S. Melatonin promotes plant growth by maintaining integration and coordination between carbon and nitrogen metabolisms. *Plant Cell Rep.* **2019**, *38*, 1001–1012. [CrossRef]
88. Arnao, M.; Hernández-Ruiz, J. Protective effect of melatonin against chlorophyll degradation during the senescence of barley leaves. *J. Pineal Res.* **2009**, *46*, 58–63. [CrossRef]
89. Li, M.Q.; Hasan, M.K.; Li, C.X.; Ahammed, G.J.; Xia, X.J.; Shi, K.; Zhou, Y.H.; Reiter, R.J.; Yu, J.Q.; Xu, M.X. Melatonin mediates selenium-induced tolerance to cadmium stress in tomato plants. *J. Pineal Res.* **2016**, *61*, 291–302. [CrossRef]
90. Wei, J.; Li, D.X.; Zhang, J.R.; Shan, C.; Rengel, Z.; Song, Z.B.; Chen, Q. Phytomelatonin receptor PMTR 1-mediated signaling regulates stomatal closure in *Arabidopsis thaliana*. *J. Pineal Res.* **2018**, *65*, e12500. [CrossRef]
91. Tan, D.-X.; Hardeland, R.; Manchester, L.C.; Korkmaz, A.; Ma, S.; Rosales-Corral, S.; Reiter, R.J. Functional roles of melatonin in plants, and perspectives in nutritional and agricultural science. *J. Exp. Bot.* **2012**, *63*, 577–597. [CrossRef]
92. Palma, F.; Lluch, C.; Iribarne, C.; García-Garrido, J.M.; Tejera García, N.A. Combined effect of salicylic acid and salinity on some antioxidant activities, oxidative stress and metabolite accumulation in *Phaseolus vulgaris*. *Plant Growth Regul.* **2009**, *58*, 307–316. [CrossRef]
93. Ashraf, M.; Foolad, M.R. Roles of glycine betaine and proline in improving plant abiotic stress resistance. *Environ. Exp. Bot.* **2007**, *59*, 206–216. [CrossRef]

94. Öztürk, L.; Demir, Y. Effects of putrescine and ethephon on some oxidative stress enzyme activities and proline content in salt stressed spinach leaves. *Plant Growth Regul.* **2003**, *40*, 89–95. [CrossRef]
95. Sharma, A.; Slathia, S.; Choudhary, S.P.; Sharma, Y.P.; Langer, A. Role of 24-epibrassinolide, putrescine and spermine in salinity stressed Adiantum capillus-veneris leaves. *Proc. Natl. Acad. Sci. India Sect. B Biol. Sci.* **2014**, *84*, 183–192. [CrossRef]
96. EL-Bauome, H.A.; Abdeldaym, E.A.; Abd El-Hady, M.A.; Darwish, D.B.E.; Alsubeie, M.S.; El-Mogy, M.M.; Basahi, M.A.; Al-Qahtani, S.M.; Al-Harbi, N.A.; Alzuaibr, F.M. Exogenous Proline, Methionine, and Melatonin Stimulate Growth, Quality, and Drought Tolerance in Cauliflower Plants. *Agriculture* **2022**, *12*, 1301. [CrossRef]
97. Sheokand, S.; Kumari, A.; Sawhney, V. Effect of nitric oxide and putrescine on antioxidative responses under NaCl stress in chickpea plants. *Physiol. Mol. Biol. Plants* **2008**, *14*, 355–362. [CrossRef]
98. Vakharia, D.; Kukadia, A.; Parameswaran, M. Polyamines in response to artificial water stress in groundnut seedlings. *Indian J. Plant Physiol.* **2003**, *8*, 383–387.
99. Posmyk, M.M.; Kuran, H.; Marciniak, K.; Janas, K.M. Presowing seed treatment with melatonin protects red cabbage seedlings against toxic copper ion concentrations. *J. Pineal Res.* **2008**, *45*, 24–31. [CrossRef]
100. Zhang, N.; Zhao, B.; Zhang, H.J.; Weeda, S.; Yang, C.; Yang, Z.C.; Ren, S.; Guo, Y.D. Melatonin promotes water-stress tolerance, lateral root formation, and seed germination in cucumber (*Cucumis sativus* L.). *J. Pineal Res.* **2013**, *54*, 15–23. [CrossRef]
101. Kuznetsov, V.V.; Radyukina, N.; Shevyakova, N. Polyamines and stress: Biological role, metabolism, and regulation. *Russ. J. Plant Physiol.* **2006**, *53*, 583–604. [CrossRef]
102. Nahar, K.; Hasanuzzaman, M.; Alam, M.; Fujita, M. Roles of exogenous glutathione in antioxidant defense system and methylglyoxal detoxification during salt stress in mung bean. *Biol. Plant.* **2015**, *59*, 745–756. [CrossRef]
103. Nahar, K.; Hasanuzzaman, M.; Rahman, A.; Alam, M.M.; Mahmud, J.-A.; Suzuki, T.; Fujita, M. Polyamines confer salt tolerance in mung bean (*Vigna radiata* L.) by reducing sodium uptake, improving nutrient homeostasis, antioxidant defense, and methylglyoxal detoxification systems. *Front. Plant Sci.* **2016**, *7*, 1104. [CrossRef]
104. Cao, S.; Shao, J.; Shi, L.; Xu, L.; Shen, Z.; Chen, W.; Yang, Z. Melatonin increases chilling tolerance in postharvest peach fruit by alleviating oxidative damage. *Sci. Rep.* **2018**, *8*, 806. [CrossRef]
105. Li, X.; Yu, B.; Cui, Y.; Yin, Y. Melatonin application confers enhanced salt tolerance by regulating Na^+ and Cl^- accumulation in rice. *Plant Growth Regul.* **2017**, *83*, 441–454. [CrossRef]
106. Zamani, Z.; Amiri, H.; Ismaili, A. Improving drought stress tolerance in fenugreek (Trigonella foenum-graecum) by exogenous melatonin. *Plant Biosyst.-Int. J. Deal. All Asp. Plant Biol.* **2020**, *154*, 643–655. [CrossRef]
107. Zhang, H.J.; Zhang, N.; Yang, R.C.; Wang, L.; Sun, Q.Q.; Li, D.B.; Cao, Y.Y.; Weeda, S.; Zhao, B.; Ren, S. Melatonin promotes seed germination under high salinity by regulating antioxidant systems, ABA and GA_4 interaction in cucumber (*C ucumis sativus* L.). *J. Pineal Res.* **2014**, *57*, 269–279. [CrossRef] [PubMed]
108. Ekinci, M.; Yildirim, E.; Dursun, A.; Mohamedsrajaden, N. Putrescine, spermine and spermidine mitigated the salt stress damage on Pepper (*Capsicum annum* L.) seedling. *Yüzüncü Yıl Üniversitesi Tarım Bilim. Derg.* **2019**, *29*, 290–299. [CrossRef]
109. Qu, M.; Zheng, G.; Hamdani, S.; Essemine, J.; Song, Q.; Wang, H.; Chu, C.; Sirault, X.; Zhu, X.-G. Leaf photosynthetic parameters related to biomass accumulation in a global rice diversity survey. *Plant Physiol.* **2017**, *175*, 248–258. [CrossRef]
110. Shi, K.; Huang, Y.; Xia, X.; Zhang, Y.; Zhou, Y.; Yu, J. Protective role of putrescine against salt stress is partially related to the improvement of water relation and nutritional imbalance in cucumber. *J. Plant Nutr.* **2008**, *31*, 1820–1831. [CrossRef]
111. Yu, Z.; Jia, D.; Liu, T. Polyamine oxidases play various roles in plant development and abiotic stress tolerance. *Plants* **2019**, *8*, 184. [CrossRef]

horticulturae

MDPI

Article

Morpho-Physiological and Anatomical Alterations of Salt-Affected Thompson Seedless Grapevine (*Vitis vinifera* L.) to Brassinolide Spraying

Mostafa F. El-Banna [1,†], Arwa Abdulkreem AL-Huqail [2,*,†], Saad Farouk [1], Bassam E. A. Belal [3], Mosaad A. El-Kenawy [3] and Ahmed F. Abd El-Khalek [4]

1 Agricultural Botany Department, Faculty of Agriculture, Mansoura University, Mansoura 35516, Egypt; el-banna@mans.edu.eg (M.F.E.-B.); gadalla@mans.edu.eg (S.F.)
2 Department of Biology, College of Science, Princess Nourah bint Abdulrahman University, P.O. Box 84428, Riyadh 11671, Saudi Arabia
3 Viticulture Department, Horticulture Research Institute, Agricultural Research Center, Giza 12619, Egypt; belal201039@yahoo.com (B.E.A.B.); emosaad76@yahoo.com (M.A.E.-K.)
4 Department of Horticulture, Faculty of Agriculture, Tanta University, Tanta 31527, Egypt; ahmed.gameal@agr.tanta.edu.eg
* Correspondence: aaalhuqail@pnu.edu.sa
† These authors contributed equally to this work.

Citation: El-Banna, M.F.; AL-Huqail, A.A.; Farouk, S.; Belal, B.E.A.; El-Kenawy, M.A.; Abd El-Khalek, A.F. Morpho-Physiological and Anatomical Alterations of Salt-Affected Thompson Seedless Grapevine (*Vitis vinifera* L.) to Brassinolide Spraying. *Horticulturae* **2022**, *8*, 568. https://doi.org/10.3390/horticulturae8070568

Academic Editors: Agnieszka Hanaka, Małgorzata Majewska and Barbara Hawrylak-Nowak

Received: 22 May 2022
Accepted: 19 June 2022
Published: 22 June 2022

Publisher's Note: MDPI stays neutral with regard to jurisdictional claims in published maps and institutional affiliations.

Abstract: Salinity is one of the most critical crises worldwide that ultimately compromises future food security. Brassinosteroids including brassinolide (BL) are a class of polyhydroxy steroids phytohormones, that play a crucial role in several plant metabolic pathways and boost plants' stress tolerance, but less data is accessible on its function in salt-affected grapevine. The experiment was conducted throughout the 2019 and 2020 experimental seasons at EL-Baramon experimental farm, Horticulture Research Institute, Mansoura, Egypt, to recognize the remediation potential of BL (1 and 2 mg L^{-1}) in lightening salinity (NaCl at 1000, 2000, and 3000 mg L^{-1}) injury on Thompson seedless grapevine seedlings (H4 strain) growth and physio-anatomical attributes. Data advocated that while salinity reduced growth attributes, BL applications substantially improved the overall salt-affected plant performance. Salinity stress significantly decreased photosynthetic pigment, relative water content, and ions percentage (nitrogen, phosphorus, potassium, potassium/sodium ratio). Alternatively, BL spraying significantly ($p \leq 0.05$) increased the photosynthetic pigment, maintaining a favorable potassium/sodium ratio and increasing the ions percentage. Additionally, increasing salinity levels significantly boost plant sodium percentage and induce a membrane malfunction associated with increased membrane permeability; conversely, the application of BL decreased the sodium percentage associated with decreasing membrane permeability relative to non-treated salinized plants. Moreover, salinity and/or BL significantly improved the antioxidant capacity associated with rising proline accumulation and antioxidant enzyme activities. Anatomically, salinity stress considerably modified leaf structure; meanwhile, the spraying with BL drastically mitigates the harmful effects of salinity on leaf anatomy. Additionally, salt-affected plant cells explained various obvious organelles ultrastructural modifications and cellular damage; meanwhile, BL spraying to salt-affected plants repealed the ultrastructural modifications of cell organelles. Taken together, BL, especially 2 mg L^{-1}, has a great potential to boost the salt tolerance of Thompson seedless grapevine seedlings (H4 strain). It improves salt tolerance by sustaining higher photosynthetic pigment concentrations, maintaining ion homeostasis, regulating water status, and stimulating antioxidant capacity as well as maintaining leaf anatomical attributes.

Keywords: antioxidant systems; brassinolide; grapevine; ion accumulation; leaf anatomy; salt stress; ultrastructure

1. Introduction

Grapevine (*Vitis vinifera* L.), which has both monetary significance and positive effect on human health, is considered one of the most tasty, stimulating, and healthful fruits worldwide. The berries are an excellent supply of sugars, minerals, and vitamins [1]. Owing to its wealthy phenolic compounds, the grapevine is extensively consumed in diverse shapes, i.e., fresh, raisins, wine, vinegar, molasses, grapevine juice, etc.; additionally, it is utilized in food additives, pharmaceutical production, and natural cosmetic products [2]. Customer attentiveness to the connection between foods and health, alongside environmental concerns, has improved the requirement for foods with elevated nutritional qualities [3]. Thompson seedless grapevine is the most imperative table grapevine cultivar in Egypt, particularly in the Delta region for local consumption and exportation. Recently, H4 is a promising strain of Thompson seedless grapevine introduced to Egypt in 2012, which has been cultivated extensively owing to its high vigor and fertility, superior yield and high cluster weight [4]. Yet, a huge acreage is situated at the newly reclaimed soils that have several troubles such as salinity. Grapevines are considered moderately sensitive to salinity, and the injury is primarily originating from chloride ions [5].

Salinity is considered one of the prime exigent environmental threats against sustainable food production [6–8]. About 33% of irrigated croplands are classified as salt-affected soil, which may exceed 50% by 2050 [9]. The undesirable impacts of excess salinity on crop development are possibly attributed to osmotic stress, cytotoxicity provoked by excess sodium (Na^+) and chloride (Cl^-), nutritional imbalances, decreased turgor, and leaf anatomical modifications [6,8,10,11]. Likewise, excess Na^+ evoked the excess generation of toxic reactive oxygen species (ROS) that may interrupt cellular functions and negatively affect metabolic processes. ROS generation usually impedes the redox homeostasis, resulting in loosening photosynthetic effectiveness [12], modifying nitrogen and osmolytes assimilation, and decreasing nutrient absorption, changing phytohormones profile and genes expression [13]. The studies by Farouk et al. [6], Farouk and Al-Huqail [8], and Kaur et al. [11] showed that excess salinity activates the antioxidant enzymes in plant tissues. In this regard, superoxide dismutase (SOD) accelerates the conversion of superoxide radicals (O^{-1}) to hydrogen peroxide (H_2O_2), while peroxidase (POD) and catalase (CAT) decompose H_2O_2 into water (H_2O) and O_2 [14,15]. Additionally, salinity induces necrosis of adult leaves, and increasing Na^+ influx and potassium (K^+) leakage leads to a superior Na^+/K^+ ratio in plant tissues [6,8,16]. In this regard, salinity normally disturbs the growth and yield of grapevine as well as induces physiological processes [5]. Additionally, Hatami and Pourakbar [17] found that irrigation grapevines with saline water (50 and 100 mM NaCl) significantly decreased shoot length, shoot fresh and dry weight, chlorophyll concentration, and potassium%, while increasing Na^+ and Cl^-. Crops possess multiple strategies to mitigate salinity injuries, including up-regulation of the antioxidant capacity, osmotic adjustment, and anatomical alteration [17–19].

There are various methods to minimize the destructive impacts of salinity on plants, i.e., scraping, flushing, and leaching to draw off the extra salt from the plant's rhizosphere [20], exploitation of different irrigation practices [21], and enhancement of plant salt tolerance [22]. Nevertheless, owing to their elevated cost and employment prerequisites, these approaches can be ineffective in alleviating the salinization threats. Consequently, developing novel techniques associated with the modulation of plants' own physiological and metabolic adaptive routes for combating the destructive effects of salinity could be decisive for cultivating salt-affected soil or utilizing saline water for irrigation. In this regard, eco-friendly phytohormones occupy energetic functions in regulating numerous biochemical pathways and enhancing plants' stress tolerance [23,24]. Amongst phytohormones, brassinosteroids (BRs) are ubiquitous steroid plant growth substances that occupy prominent functions in various biochemical pathways leading to accelerating plant stress responses [25,26]. BRs regulated stress response owing to a complex progression of biochemical reactions such as activation or repression of key enzymatic reactions, stimulation of protein assimilation, and the assembly of diverse chemical defense materials [23,26–28].

Additionally, BRs application regulates the ROS metabolism and the rise in the antioxidant enzyme activity, plus a superior concentration of ascorbic acid, carotenoids, etc. [26,29]. Moreover, Ali et al. [30] established that BRs also modified the plasma membrane, improved ion absorption, and facilitated the translocation of photosynthesis to the sink, in addition to enhancing metabolic activities within stress environment. Additionally, exogenous application of BRs under salinity conditions, maintained cell organs ultrastructure including nucleus and chloroplast [31]. There are some reports designating that BRs application mitigates the harmful effects of salinity on several crops [32–34]. All of these outcomes designated the magnitude of BRs in defense within stress-induced injury without bad effects on human health [35]. BRs are commonly classified into three groups depending on the number of carbon atoms in their structure, i.e., C27, C28, and C29 [36]. Vardhini et al. [37] stated that brassinolide (BL), 28-homobrassinolide (28-HomoBL) and 24-epibrassinolide (24-EpiBL) are the three bioactive BRs being extensively utilized in agriculture.

Although recent reports have shown that BL will possibly lessen salinity toxicity [26,28], the mechanisms of BL on inducing grapevine salt tolerance are still far from being implicit. Therefore, the experiment aimed to evaluate the role of BL spraying on the growth, several physio-anatomical trials of grapevine seedlings under salinity. We anticipate that the data acquired from the current study will present a reliable hypothetical basis for the expansion of Thompson seedless grapevine (H4 strain) production in the regions that irrigated with salinized water up-to 3000 mg L^{-1}.

2. Materials and Methods

The current experiments were conducted throughout 2019 and 2020 seasons at EL-Baramon experimental farm (31.1195° N, 31.4487° E), Horticulture Research Institute, Mansoura, Egypt, to evaluate the nullifying effect of BL on salt-affected Thompson seedless grapevines (H4 strain) seedlings. The experimental site is distinguished as the arid environment with cool and low rainfall winter and hot dry summer. Average monthly temperature and relative humidity within experimental periods are available in Figure 1.

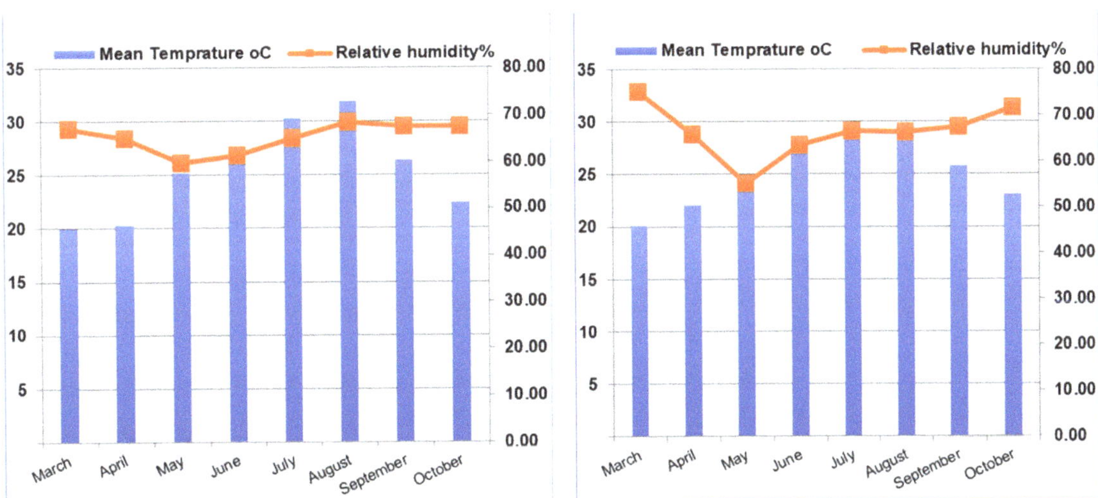

Figure 1. Average temperature and relative humidity of experimental site throughout 2019 and 2020 seasons.

2.1. Experimental Treatments and Design

An open field experiment was conducted in a completely randomized design with three replicates (each replicate included ten plants; in total, the experiment contained 300 plastic bags). In total, there were 10 treatments as indicated in Table 1.

Table 1. Experimental treatments and their abbreviation.

Code	Treatment
T1	Control, irrigated with tap water, 0 salinity (NaCl) without BL application
T2	Irrigated with saline water (1000 mg L^{-1} NaCl) without BL application
T3	Irrigated with saline water (1000 mg L^{-1} NaCl) plus 1 mg L^{-1} BL foliar application
T4	Irrigated with saline water (1000 mg L^{-1} NaCl) plus 2 mg L^{-1} BL foliar application
T5	Irrigated with saline water (2000 mg L^{-1} NaCl) without BL application
T6	Irrigated with saline water (2000 mg L^{-1} NaCl) plus 1 mg L^{-1} BL foliar application
T7	Irrigated with saline water (2000 mg L^{-1} NaCl) plus 2 mg L^{-1} BL foliar application
T8	Irrigated with saline water (3000 mg L^{-1} NaCl) without BL application
T9	Irrigated with saline water (3000 mg L^{-1} NaCl) plus 1 mg L^{-1} BL foliar application
T10	Irrigated with saline water (3000 mg L^{-1} NaCl) plus 2 mg L^{-1} BL foliar application

The concentration used was selected upon the pilot study utilizing 1000, 2000, 3000, 4000, 5000 mg L^{-1} irrigation water for 20 days, and wilting was observed at 4000 and 5000 mg L^{-1}; conversely, there was no visible wilting under 1000, 2000, and 3000 mg L^{-1}. The proper BL concentrations were designated based on earlier investigation [33].

The uniform and healthy cuttings of Thompson seedless grapevine (H4 strain) were taken from one-year-old matured canes (5 years old, grown in the vineyard at EL-Baramon experimental farm, Horticulture Research Institute, Mansoura, Dakahlia Governorate, Egypt). The cuttings were planted on 1st March in bottom holes polyethylene bags (17 × 30 cm) containing 5 kg clay soil (sand 27.16%, silt 24.69%, clay 48.15%, cation exchange capacity 36.5 Cmolc kg^{-1}, pH, 7.8, electric conductivity 0.62 mmose cm^{-1}, organic matter 2.1%, nitrogen 38 mg kg^{-1} soil, phosphorous 11 mg kg^{-1} soil, potassium 282 mg kg^{-1}, calcium 1.88%). The cuttings were irrigated with tap water two times each week for two months in both seasons. After that, the successive seedlings were irrigated using tap water and/or NaCl saline solution (1000, 2000, and 3000 mg L^{-1}) from 1 May till the end of September during the two growing seasons (twice a week with 2 L, in the morning for each irrigation). BL (as a commercial product named © Blank, European group for agricultural development, Alexandria, Egypt, active ingredient, BL 1%, phosphor 20%, and nitrogen 8%) with Tween 20 (0.05%) was sprayed at a rate (1 and 2 mg L^{-1}) on Thompson seedless grape seedlings three times (60, 90 and 120 days from planting). All plastic bags were irrigated with tap water monthly to prevent salinity accumulation. Each plastic bag was given nitrogen (N) in 3 g of ammonium sulfate (20.5% N), phosphorus (P) in 2 g of calcium superphosphate (15.4% P_2O_5), and potassium (K) in 1 g of potassium sulfate (48% K_2O) each month.

2.2. Analyses of Plant Samples

The plant samples were collected after 15 days from the last BL spraying (135 days from planting) for morpho-anatomical and biochemical determinations.

2.3. Ion Determination

For ion estimation, oven-dried plant samples (0.1 g) were entirely digested with H_2SO_4 (98%, 5 mL), at 200 °C, supplemented with a few drops of H_2O_2 (30%, v/v). Once digestion was completed, the sample was brought up to 25 mL with distilled-deionized water. P, N, K, Na^+ were measured as described in Cooper [38], and Motsara and Roy [39], by micro-Kjeldahl technique (N), ammonium molybdate and ascorbic acid protocol (P), and flame photometer (K^+ and Na^+), and then the K^+/Na^+ ratio was calculated.

2.4. The Photosynthetic Pigments

The concentration of chlorophylls and carotenoids was determined following Lichtenthaler and Wellburn [40] protocol, using methanol, and expressed as mg g^{-1} fresh weight.

2.5. Leaf Relative Water Content (LRWC)

The LRWC was estimated by Shams et al. [22] protocol. Leaf pieces (10 mm) were directly weighed for fresh mass (FM) assessment. Afterward, pieces were floated in bi-distilled water at lab. temperature for 24 h to assess the turgor mass (TM). Lastly, leaf pieces were oven-dried at 70 °C for 48 h then recorded as the dry mass (DM). LRWC (%) was designed by the subsequent equation:

$$\text{LRWC (\%)} = \frac{\text{FM} - \text{DM}}{\text{TM} - \text{DM}} \times 100.$$

2.6. Membrane Permeability (MP)

The leaf pieces were rinsed in bi-distilled water to eliminate surface-adhered electrolytes, and afterward they were put in Petri dishes containing deionized water (25 mL) at the lab temperature for 3 h. Electrical conductivity (EC1) in the bath solution was recorded. Subsequently, the leaf pieces were killed by boiling for 60 min, and the conductivity of the bath solution was recorded again (EC2), then calculating MP% following this equation [41],

$$\text{MP\%} = \frac{\text{EC1}}{\text{EC2}} \times 100$$

2.7. Proline Estimation

Proline concentrations (mg g FW^{-1}) were assessed spectrophotometrically following the procedure of Bates et al. [42] using ninhydrin reagent and standard curve.

2.8. Antioxidant Enzymes and Phenols Concentration

Fresh leaf samples were homogenized with 50 mM sodium phosphate buffer (pH 7.8) including 0.2 mM EDTA and 2% insoluble polyvinylpyrrolidone in a cooled mortar and pestle, then centrifuged at 12,000× *g* for 20 min, the supernatant was utilized in enzymatic activities assessment. Peroxidase (POD, EC 1.11.1.7) activity was measured by the increase in absorbance at 470 nm owing to guaiacol oxidation [43]. Polyphenol oxidase (PPO, EC 1.10.3.1) activity was determined according to Augustin et al. [44]. Catalase (CAT, EC 1.11.1.6) activity was deliberate as the decrease in absorbance at 240 nm following the technique of Tian et al. [45].

For phenols (mg equivalents of gallic acid g^{-1} dry weight) determination, 0.5 g oven-dried leaf samples were extracted with 80% ethanol. An aliquot of plant extract was mixed with 1 N Folin–Ciocalteu reagent and Na_2CO_3 and then incubated for 60 min at the lab. temperature, subsequently the absorbance was recorded at 765 nm [46].

2.9. Anatomical Study

Specimens (5 × 5 mm) from the 5th upper leaf including the main midvein were taken in the 2nd year. The specimens were put in formalin aceto alcohol for 48 h, afterward washed and dehydrated in ethanol succession, and embedded in paraffin wax (52–54 °C melting points). Cross-sections were prepared at 12–15 µm by a rotary microtome, stained in toluidine blue, cleared in toluene, and then mounted in Canada balsam. The randomly selected slides were examined with a light microscope (Olympus CX41, Philippines, Tokyo, Japan) outfitted with a digital camera (TUCSEN, USB2, H serial) to visualize the microscopic images.

2.10. Transmission Electron Microscopy (TEM)

Selected leaf blade samples (5 mm^2) (control 'T1', severe salinity 'T8', severe salinity with 2 mg L^{-1} BL 'T10') were double fixed immediately in cold glutaraldehyde (2.5%) and postfixed in osmium tetroxide (1%) for 3 hr. The samples were then dehydrated in a graded alcohol series and embedded in Spurr's resin. The ultrathin sections (50–100 µM) were performed by a Reichert ultramicrotome (Germany). Ultrathin sections were mounted on copper grids (400 mishes), and double-stained for 10 min., with uranyl acetate and Reynolds' lead citrate for 15 min each. Ten stained sections were examined and photographed by using a JEOL 100s transmission electron microscope (Electron Microscope Unit, Mansoura University, Mansoura, Egypt).

2.11. Growth Parameters

Seedling survival percentage, plant height (cm), stem thickness (mm), leaves number plant^{-1}, and mean of leaf surface area (cm^2) of the growing shoot were deliberated using Leaf Area Meter, AM 300 (ADC Bioscientific Ltd., Hoddesdon, UK). Shoot and root dry weights were recorded in g. The coefficient of wood ripening (CWR) was deliberate according to Rizk and Rizk [47]:

$$\text{CWR} = \frac{\text{length of the ripened part of the shoot}}{\text{total length of the shoot}}$$

2.12. Statistical Analysis

Homogeneity of error variance for all variables was determined before the analysis of variance (ANOVA). The outputs displayed that all data fulfilled the homogeneity required to achieve additional ANOVA tests. The data were statistically analyzed using COSTATC statistical package (CoHort software, 2006; Cary, NC, USA). A one-way ANOVA was achieved to examine the impacts of salinity and BL on grapevine plant growth and physiological parameters. Means were separated using Tukey's honestly significant difference (HSD) test at the $p < 0.05$ level of significance, and significant differences were indicated by different letters. Data existed as means ± standard error (SE) of five independent biological samples.

3. Results

3.1. Mineral Nutrient Concentration

Table 2 shows that irrigation with saline water from 0 to 3000 mg L^{-1} provoked a depressing impact on ion percentage except sodium, which was increased with salinity. Specifically, N% was decreased from 2.53% and 2.51% to 1.14% and 1.11%, P% was decreased from 0.364% and 0.362% to 0.209% and 0.201%, and K^+% was also decreased from 0.94% and 0.96% to 0.34% and 0.33% in the 1st and 2nd season, respectively, while Na^+% was increased from 0.29% and 0.31% to 0.94% and 0.96%, respectively, relative to untreated non-salinized plants. Under salinity, BL spraying displayed a greater impact on improving nutrient accumulation over non-treated salt-affected seedlings (Table 2). The Table also indicates that 2 mg L^{-1} BL was more effective than 1 mg L^{-1} BL in increasing ion percentage (N, P, and K) and decreasing Na^+%. The K^+/Na^+ ratio significantly decreased with salinity (Table 2). Conversely, BL spraying improved the K^+/Na^+ ratio in leaves, especially at 2 mg L^{-1} above non-treated plants under such salinity levels.

3.2. Photosynthetic Pigments

Relative to control, the concentration of chlorophyll a declined by 32.11% and 22.91%, 44.95% and 41.98%, and 50.45% and 48.47% under T1, T5, and T8 alone, in the first and second seasons, respectively (Table 3). Likewise, concentrations of chlorophyll b declined by 33.33% and 29.17%, 46.25% and 43.45%, and 51.70% and 51.19%, respectively (Table 3). Accordingly, the total chlorophyll concentration decreased by 32.87% and 25.52%, 45.47% and 42.69%, and 50.95 and 49.41%, while carotenoid concentration decreased by 27.02% and 20.93%, 34.05% and 35.81%, and 23.78% and 40.46% (Table 3) under T1, T5, and T8, respectively. Spraying salt-affected plants with BL concentrations drastically ($p < 0.05$) enhanced the leaves' chlorophyll a, chlorophyll b, total chlorophyll, and carotenoids relative to untreated plants under such salinity levels. The concentration of 2 mg L^{-1} BL was more effective than 1 mg L^{-1} on increasing the concentrations of photosynthetic pigments under salinity (Table 3).

Table 2. Ions percentage and potassium/sodium ratio of grapevine seedlings as affected by salinity (NaCl) and brassinolide (BL) during both growing seasons.

Treatments	Nitrogen%		Phosphorous%		Potassium%		Sodium%		Potassium/Sodium Ratio	
	2019	2020	2019	2020	2019	2020	2019	2020	2019	2020
T1	2.53 ± 0.03 a	2.51 ± 0.02 a	0.364 ± 0.002 a	0.362 ± 0.002 a	0.94 ± 0.01 a	0.96 ± 0.01 a	0.29 ± 0.01 f	0.31 ± 0.01 e	3.25 ± 0.11 a	3.11 ± 0.21 a
T2	1.61 ± 0.03 d,e	1.60 ± 0.02 d,e	0.285 ± 0.003 d	0.280 ± 0.002 d	0.63 ± 0.01 c	0.62 ± 0.01 c,d	0.64 ± 0.01 d	0.66 ± 0.01 c	0.98 ± 0.03 c	0.93 ± 0.01 d
T3	1.90 ± 0.03 c	1.92 ± 0.02 c	0.301 ± 0.002 c	0.305 ± 0.001 c	0.88 ± 0.01 a	0.89 ± 0.01 a	0.57 ± 0.01 e	0.65 ± 0.02 c	1.54 ± 0.02 b	1.37 ± 0.08 b,c
T4	2.14 ± 0.03 b	2.15 ± 0.01 b	0.324 ± 0.002 b	0.331 ± 0.001 b	0.89 ± 0.02 a	0.89 ± 0.01 a	0.54 ± 0.01 e	0.55 ± 0.01 d	1.65 ± 0.06 b	1.61 ± 0.03 b
T5	1.49 ± 0.02 e	1.52 ± 0.03 e	0.227 ± 0.001 g	0.225 ± 0.001 g	0.49 ± 0.01 d	0.47 ± 0.02 e	0.87 ± 0.01 b	0.89 ± 0.01 a	0.56 ± 0.02 d	0.52 ± 0.01 e,f
T6	1.80 ± 0.03 c	1.82 ± 0.02 c	0.243 ± 0.002 f	0.249 ± 0.001 e,f	0.67 ± 0.01 b,c	0.65 ± 0.02 b,c	0.78 ± 0.01 c	0.75 ± 0.02 b	0.85 ± 0.01 c	0.86 ± 0.02 d,e
T7	1.89 ± 0.01 c	1.91 ± 0.02 c	0.277 ± 0.002 d	0.281 ± 0.003 d	0.71 ± 0.01 b	0.72 ± 0.01 b	0.66 ± 0.01 d	0.67 ± 0.01 c	1.07 ± 0.01 c	1.07 ± 0.02 c,d
T8	1.14 ± 0.01 f	1.11 ± 0.01 f	0.209 ± 0.001 h	0.201 ± 0.002 h	0.34 ± 0.01 e	0.33 ± 0.01 f	0.94 ± 0.01 a	0.96 ± 0.01 a	0.36 ± 0.01 d	0.34 ± 0.01 f
T9	1.64 ± 0.02 d	1.70 ± 0.01 d	0.239 ± 0.001 f	0.241 ± 0.001 f	0.48 ± 0.01 d	0.56 ± 0.01 d	0.81 ± 0.01 b	0.76 ± 0.01 b	0.59 ± 0.02 d	0.73 ± 0.01 d,e
T10	1.84 ± 0.02 c	1.85 ± 0.02 c	0.259 ± 0.001 e	0.255 ± 0.002 e	0.69 ± 0.01 b,c	0.70 ± 0.02 b	0.79 ± 0.01 b,c	0.71 ± 0.01 b,c	0.87 ± 0.03 c	0.98 ± 0.02 d
p value	***	***	***	***	***	***	***	***	***	***

Data represent the average of five replicates ± standard error. Different letters in each column indicate significant ($p \leq 0.05$) differences at $p \leq 0.05$ according to Tukey's HSD range at $p \leq 0.05$. Levels of significance are represented *** $p < 0.001$. (T1, 0 NaCl without BL application; T2, 1000 mg L^{-1} NaCl without BL application; T3, 1000 mg L^{-1} NaCl with 1 mg L^{-1} BL application; T4, 1000 mg L^{-1} NaCl with 2 mg L^{-1} BL application; T5, 2000 mg L^{-1} NaCl without BL application; T6, 2000 mg L^{-1} NaCl with 1 mg L^{-1} BL application; T7, 2000 mg L^{-1} NaCl with 2 mg L^{-1} BL application; T8, 3000 mg L^{-1} NaCl without BL application; T9, 3000 mg L^{-1} NaCl with 1 mg L^{-1} BL application; T10, 3000 mg L^{-1} NaCl with 2 mg L^{-1} BL application).

3.3. Physiological Parameters

Physiological trials, i.e., leaf relative water content (LRWC), membrane permeability (MP), and proline concentration were statistically affected by salinity and BL application Figure 2a–c. Salinity levels (1000, 2000, and 3000 mg L^{-1}) caused a significant reduction of 6%, 8%, and 13% in LRWC in the 1st season and 5%, 8%, and 13% in the 2nd season, respectively, compared to the non-saline conditions (Figure 2a). BL application successfully alleviated this impact and improved the LRWC in stressed plants. Under salinity, BL application (1 and 2 mg L^{-1}) proficiently boosted the LRWC over the salt-affected plants with no application of BL. Relative to non-treated salt-affected plants, application of 2 mg L^{-1} BL significantly increased LRWC by 3% in both seasons under low salinity level, and by 4% under moderate salinity level in the second season, as well as by 4% and 11% under high salinity level (Figure 2a) in the 1st and 2nd seasons, respectively. The data presented in Figure 2b indicate that increasing salinity levels significantly increased MP% in plants relative to control seedlings. The maximum increase was recorded under high salinity level, reaching 138% and 150% above control plants in the 1st and 2nd seasons, respectively. BL spraying mitigates the harmful effect of salinity on MP%, relative to untreated plants under such salinity level, and the most effective concentration in this regard was 2 mg L^{-1} BL. The level of proline exhibited an increase of 80%, 143%, and 194% in the 1st season and 84%, 154%, and 199% in the 2nd season, respectively, in response to the salinity levels (1000, 2000, and 3000 mg L^{-1}) of grapevine seedlings, compared to control plants (Figure 2c). Conversely, the influence was more definite under salinity, where BL professionally decreased the liberate of proline by mitigating salinity. Under high salinity level, the application of 2 mg L^{-1} BL caused a 14% and 12% decrease in the 1st and 2nd season, respectively, whereas the application of 1 mg L^{-1} resulted in a 10% and 8% reduction in proline concentration in the second season, relative to untreated salt-affected plants.

Table 3. Photosynthetic pigment concentrations (mg g FW^{-1}) of grapevine leave as affected by salinity (NaCl) and brassinolide (BL) during both growing seasons.

Treatments	**Chlorophyll a**		**Chlorophyll b**		**Total Chlorophyll**		**Carotenoids**	
	2019	**2020**	**2019**	**2020**	**2019**	**2020**	**2019**	**2020**
T1	2.18 ± 0.18 a	2.62 ± 0.12 a	1.47 ± 0.19 a	1.68 ± 0.10 a	3.65 ± 0.37 a	4.31 ± 0.22 a	0.185 ± 0.01	0.215 ± 0.01 a
T2	1.47 ± 0.13 a–c	2.02 ± 0.18 a–c	0.98 ± 0.09 a,b	1.19 ± 0.13 b–d	2.45 ± 0.23 a,b	3.21 ± 0.30 a–d	0.135 ± 0.01	0.170 ± 0.01 a–c
T3	1.76 ± 0.19 a–c	2.36 ± 0.14 a,b	1.12 ± 0.10 a,b	1.42 ± 0.17 a–c	2.88 ± 0.29 a,b	3.78 ± 0.31 a,b	0.154 ± 0.01	0.192 ± 0.01 a,b
T4	1.88 ± 0.24 a,b	2.44 ± 0.09 a,b	1.24 ± 0.16 a,b	1.53 ± 0.09 a,b	3.12 ± 0.40 a,b	3.97 ± 0.19 a,b	0.164 ± 0.01	0.198 ± 0.01 a,b
T5	1.20 ± 0.16 b,c	1.52 ± 0.12 c,d	0.79 ± 0.09 b	0.95 ± 0.06 c,d	1.99 ± 0.26 a,b	2.47 ± 0.18 c,d	0.122 ± 0.01	0.138 ± 0.01 c
T6	1.40 ± 0.12 a–c	1.88 ± 0.11 b–d	0.94 ± 0.09 a,b	1.16 ± 0.07 b–d	2.34 ± 0.22 a,b	3.04 ± 0.18 b–d	0.134 ± 0.01	0.163 ± 0.01 b,c
T7	1.52 ± 0.10 a–c	2.07 ± 0.16 a–c	0.99 ± 0.07 a,b	1.23 ± 0.09 a–d	2.51 ± 0.18 a,b	3.31 ± 0.25 a,c	0.141 ± 0.01	0.176 ± 0.01 a–c
T8	1.08 ± 0.10 c	1.35 ± 0.11 d	0.71 ± 0.08 b	0.82 ± 0.09 d	1.79 ± 0.19 a,b	2.18 ± 0.21 d	0.141 ± 0.02	0.128 ± 0.01 c
T9	1.18 ± 0.10 b,c	1.47 ± 0.07 c,d	0.78 ± 0.07 b	0.91 ± 0.03 d	1.96 ± 0.17 a,b	2.39 ± 0.10 c,d	0.121 ± 0.01	0.140 ± 0.01 c
T10	1.26 ± 0.13 b,c	1.57 ± 0.06 c,d	0.84 ± 0.08 b	0.98 ± 0.03 c,d	2.11 ± 0.21 a,b	2.55 ± 0.10 c,d	0.127 ± 0.01	0.147 ± 0.01 b,c
p value	**	***	**	***	**	***	NS	***

Data represent the average of five replicates ± standard error. Different letters in each column indicate significant ($p \leq 0.05$) differences at $p \leq 0.05$ according to Tukey's HSD range at $p \leq 0.05$. Levels of significance are represented by ** $p < 0.01$ and *** $p < 0.001$; NS, nonsignificant. (T1, 0 NaCl without BL application; T2, 1000 mg L^{-1} NaCl without BL application; T3, 1000 mg L^{-1} NaCl with 1 mg L^{-1} BL application; T4, 1000 mg L^{-1} NaCl with 2 mg L^{-1} BL application; T5, 2000 mg L^{-1} NaCl without BL application; T6, 2000 mg L^{-1} NaCl with 1 mg L^{-1} BL application; T7, 2000 mg L^{-1} NaCl with 2 mg L^{-1} BL application; T8, 3000 mg L^{-1} NaCl without BL application; T9, 3000 mg L^{-1} NaCl with 1 mg L^{-1} BL application; T10, 3000 mg L^{-1} NaCl with 2 mg L^{-1} BL application).

3.4. Antioxidant Enzyme Activities and Phenol Concentration

Salinity was established to have a considerable impact on POD, PPO, and CAT activities, and phenol concentration (Figure 3a–d). The present data recognize that BL spraying lessened the depressing effects of salinity on antioxidant enzyme activities and phenol concentration in grapevine seedlings (Figure 3a–d). A considerable enhancement in the activity of POD was recorded in salt-affected plants (Figure 3a). The POD activity was considerably superior (66%, 58%, 137% in the first season and 66%, 89%, 114% in the second season) in salt treatment (1000, 2000, 3000 mg L^{-1}), respectively, relative to the control. The application of BL significantly lessened POD activity in salt-affected plants without BL application. Under high salinity level, BL applications resulted in a decline by 10% and 13% in the 1st season as well as 5% and 7% in the 2nd season in POD activity at 1 mg L^{-1} BL and 2 mg L^{-1} BL, respectively, compared to salt-affected seedlings only. The PPO activity was established to be drastically influenced by salinity and BL (Figure 3b). Salinity stress increased the PPO activity in both seasons compared to the control. The maximum activity was recorded under a high salinity level (3000 mg L^{-1}) that amplified the activity by 296% and 281% in the 1st and 2nd seasons, respectively, relative to non-salinized control seedlings. BL (1 and 2 mg L^{-1}) application drastically reduced PPO activity relative to untreated stressed seedlings. Figure 3c revealed that CAT activity was drastically influenced by salinity and BL. High salinity level (3000 mg L^{-1}) resulted in a 77% and 52% enhancement compared with the control in the 1st and 2nd seasons, respectively. Application of BL at both rates was established to be efficient in decreasing the CAT activity under salinity. Under severe salinity, 1 mg L^{-1} BL lowered the CAT activity by 5% and 5%, while a 10% and 8% reduction was noted with 2 mg L^{-1} BL application in the 1st and 2nd seasons, respectively, above untreated severe salinity stressed seedlings.

The concentration of phenols in grapevine leaves was significantly ($p < 0.05$) affected by salinity (Figure 3d). Relative to control, increasing salinity level increased the concentration of phenol by 66% and 61%, 107% and 97%, 137% and 122% in T2, T5, and T8 treatments in the 1st and 2nd season, respectively (Figure 3d). Under different saline conditions, the application of BL mitigated the adverse effect of saline stress on the concentration of phenol.

3.5. Leaf Anatomy

Data in Table 4 and Figure 4a–j indicate the anatomical modification of grapevine leaf under salinity and/or BL application. The data clearly show that low salinity level have a stimulation influence on leaf structure that increased the thickness of midrib (TM), width of the midrib (WM), thickness of leaf blade (LB), thickness of the upper epidermis (UE), spongy parenchyma thickness (SP), lower epidermis thickness (LE), thickness of compound vascular bundle (TVB) and width of the compound vascular bundle (WVB) by 8%, 15%, 11%, 43%, 24%, 36%, 34%, and 29%, respectively, while palisade parenchyma thickness (PP) decreased by 19% as compared with non-salinized control plants. On the other hand, severe salinity levels decreased TM, WM, LB, UE, PP, SP, TVB, and WVB, by 14%, 5%, 9%, 6%, 27%, 2%, 6%, and 23%, but they increased LE by 21% relative to non-salinized control plants.

Table 4. Anatomical modification of grapevine leaves as affected by salinity (NaCl) and brassinolide (BL) during the 2nd season.

Treatments	Dimension of Midrib (μm)		Thickness of Leaf Blade (μm)	Upper Epidermis Thickness (μm)	Palisade Tissue Thickness (μm)	Spongy Tissue Thickness (μm)	Lower Epidermis Thickness (μm)	Dimension of Compound Vascular Bundle (μm)	
	Thickness	Width						Thickness	Width
T1	75.83 ± 1.11 d	61.57 ± 0.57 d	15.84 ± 0.20 e,f	1.62 ± 0.10 b	5.87 ± 0.09 d,e	6.53 ± 0.09 d	1.82 ± 0.09 d	45.04 ± 0.22 d	42.76 ± 0.57 d
T2	82.41 ± 0.90 b,c	70.88 ± 1.19 b,c	17.71 ± 0.21 e	2.32 ± 0.09 a	4.75 ± 0.21 e,f	8.14 ± 0.24 b,c	2.48 ± 0.09 a,b	60.58 ± 1.00 a,b	55.24 ± 1.22 b,c
T3	85.88 ± 0.65 a,b	75.48 ± 0.79 b	30.75 ± 0.35 b	1.62 ± 0.1 b	19.42 ± 0.32 a	7.48 ± 0.12 b–d	2.22 ± 0.09 b–d	60.88 ± 0.91 a,b	58.60 ± 0.14 a,b
T4	89.34 ± 1.63 a	76.47 ± 0.79 b	33.76 ± 1.01 a	2.214 ± 0.25 a	19.35 ± 0.52 a	9.91 ± 0.32 a	2.27 ± 0.11 b,c	66.82 ± 1.01 a	57.81 ± 0.91 a,b
T5	75.48 ± 0.79 d	62.12 ± 0.85 c,d	21.64 ± 0.40 d	2.43 ± 0.06 a	6.45 ± 0.05 d	9.93 ± 0.32 a	2.83 ± 0.09 a	53.70 ± 2.45 c	47.27 ± 0.85 c,d
T6	68.80 ± 0.79 e	79.59 ± 0.91 a,b	27.21 ± 0.79 c	1.42 ± 0.06 b	16.13 ± 0.37 b	7.28 ± 0.38 c,d	2.38 ± 0.06 b	46.03 ± 0.91 d	55.24 ± 0.65 b,c
T7	79.00 ± 1.51 c,d	71.87 ± 1.94 b	28.18 ± 0.91 b,c	1.62 ± 0.1 b	16.64 ± 0.50 b	7.64 ± 0.38 b–d	2.27 ± 0.08 b,c	56.03 ± 1.84 b,c	53.06 ± 1.23 b,c
T8	65.09 ± 0.48 e,f	58.41 ± 0.70 d	14.37 ± 0.16 f	1.52 ± 0.11 b	4.25 ± 0.20 f	6.37 ± 0.21 d	2.22 ± 0.09 b–d	41.97 ± 0.91 d	32.86 ± 0.57 e
T9	61.62 ± 0.65 f	56.18 ± 1.18 d	23.63 ± 0.12 d	1.45 ± 0.04 b	14.04 ± 0.12 c	6.26 ± 0.09 d	1.87 ± 0.06 c,d	43.56 ± 0.70 d	41.33 ± 0.95 d,e
T10	82.66 ± 2.47 b,c	87.36 ± 5.26 a	20.98 ± 0.63 d	2.53 ± 0.11 a	6.76 ± 0.26 d	8.85 ± 0.52 a,b	2.83 ± 0.09 a	58.16 ± 2.56 b,c	67.56 ± 5.89 a
p value	***	***	***	***	***	***	***	***	***

Data represent the average of five replicates ± standard error. Different letters in each column indicate significant ($p \leq 0.05$) differences at $p \leq 0.05$ according to Tukey's HSD range at $p \leq 0.05$. Levels of significance are represented *** $p < 0.001$. (T1, 0 NaCl without BL application; T2, 1000 mg L^{-1} NaCl without BL application; T3, 1000 mg L^{-1} NaCl with 1 mg L^{-1} BL application; T4, 1000 mg L^{-1} NaCl with 2 mg L^{-1} BL application; T5, 2000 mg L^{-1} NaCl without BL application; T6, 2000 mg L^{-1} NaCl with 1 mg L^{-1} BL application; T7, 2000 mg L^{-1} NaCl with 2 mg L^{-1} BL application; T8, 3000 mg L^{-1} NaCl without BL application; T9, 3000 mg L^{-1} NaCl with 1 mg L^{-1} BL application; T10, 3000 mg L^{-1} NaCl with 2 mg L^{-1} BL application).

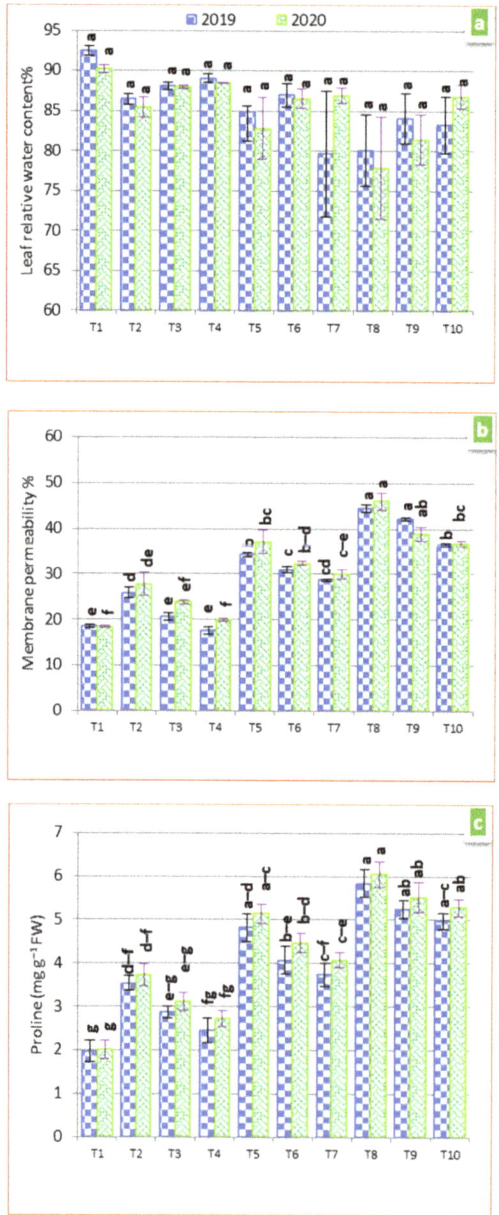

Figure 2. Leaf relative water content (**a**), membrane permeability percentage (**b**), and proline concentration (**c**) of grapevine seedling as affected by salinity (NaCl) and brassinolide (BL) during both growing seasons. Data represent the average of five replicates ± standard error. Different letters in each column indicate significant ($p \leq 0.05$) differences at $p \leq 0.05$ according to Tukey's HSD range at $p \leq 0.05$. (T1, 0 NaCl without BL application; T2, 1000 mg L^{-1} NaCl without BL application; T3, 1000 mg L^{-1} NaCl with 1 mg L^{-1} BL application; T4, 1000 mg L^{-1} NaCl with 2 mg L^{-1} BL application; T5, 2000 mg L^{-1} NaCl without BL application; T6, 2000 mg L^{-1} NaCl with 1 mg L^{-1} BL application; T7, 2000 mg L^{-1} NaCl with 2 mg L^{-1} BL application; T8, 3000 mg L^{-1} NaCl without BL application; T9, 3000 mg L^{-1} NaCl with 1 mg L^{-1} BL application; T10, 3000 mg L^{-1} NaCl with 2 mg L^{-1} BL application).

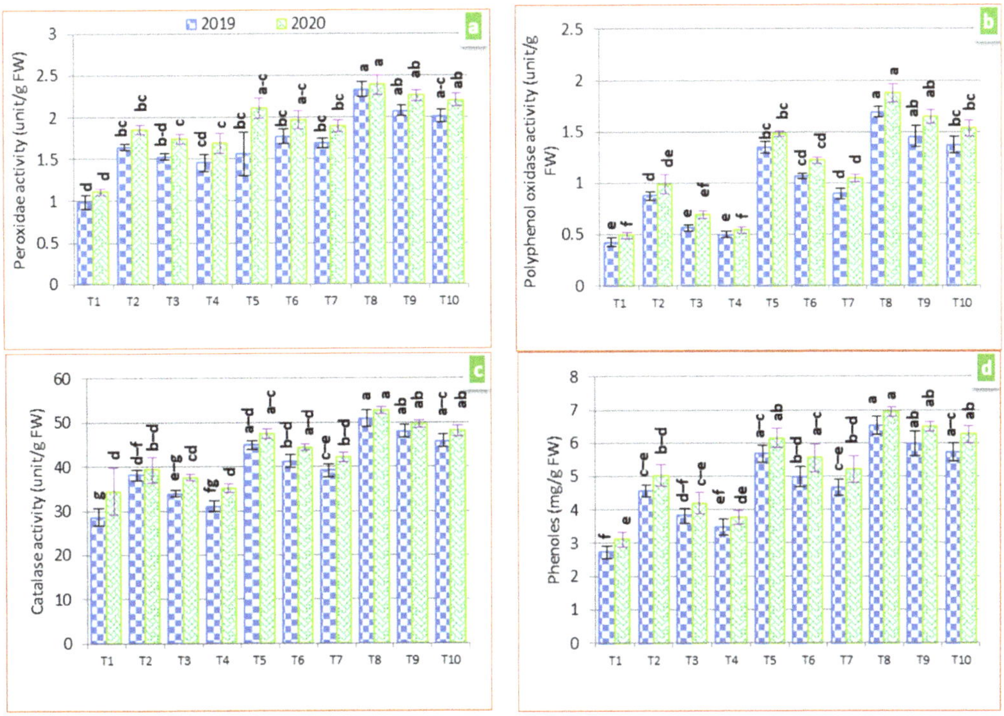

Figure 3. Antioxidant enzyme activities (**a–c**) and phenol concentration (**d**) of grapevine leave as affected by salinity (NaCl) and brassinolide (BL) during both growing seasons. Data represent the average of five replicates ± standard error. Different letters in each column indicate significant ($p \leq 0.05$) differences at $p \leq 0.05$ according to Tukey's HSD range at $p \leq 0.05$. (FW, Fresh weight; mg, milligram; T1, 0 NaCl without BL application; T2, 1000 mg L^{-1} NaCl without BL application; T3, 1000 mg L^{-1} NaCl with 1 mg L^{-1} BL application; T4, 1000 mg L^{-1} NaCl with 2 mg L^{-1} BL application; T5, 2000 mg L^{-1} NaCl without BL application; T6, 2000 mg L^{-1} NaCl with 1 mg L^{-1} BL application; T7, 2000 mg L^{-1} NaCl with 2 mg L^{-1} BL application; T8, 3000 mg L^{-1} NaCl without BL application; T9, 3000 mg L^{-1} NaCl with 1 mg L^{-1} BL application; T10, 3000 mg L^{-1} NaCl with 2 mg L^{-1} BL application).

Foliar application of salt-affected grapevine seedlings by BL levels in special 2 mg L^{-1} enhanced the anatomical feature of the leaves over the control plants. Under low salinity level, the application of 1 mg L^{-1} BL increased TM, WM, LB, PP, SP, LE, TVB, and WVB by 13%, 22%, 94%, 230%, 14%, 21%, 48%, and 37%, respectively, above non-salinized control plants, while spraying 2 mg L^{-1} BL led to an increase of 17%, 24%, 113%, 36%, 229%, 51%, 24%, 48%, and 35%, respectively, relative to non-salinized control plants (Table 4). The same direction was recorded under moderate and severe salinity levels. In this regard, application of 2 mg L^{-1} BL under severe salinity significantly increased TM, MW, LB, UE, PP, SP, LE, TVB, and WVB, by 9%, 41%, 32%, 56%, 15%, 35%, 55%, 29%, and 57%, respectively, above control (Table 4).

3.6. Ultrastructural Characterization of Leaf Mesophyll Cells by TEM

The ultrastructural study demonstrated grapevine leaf mesophyll cells with a bordered cell wall, and unbroken cell membranes, having a granular cytoplasm with many organelles (Figure 5a–o). Salt-affected plant cells illustrated various visible ultrastructural modifications of the organelles and cellular injuries (Figure 5a–o), i.e., nucleus condensation, protoplasm deterioration, and lesser organelles.

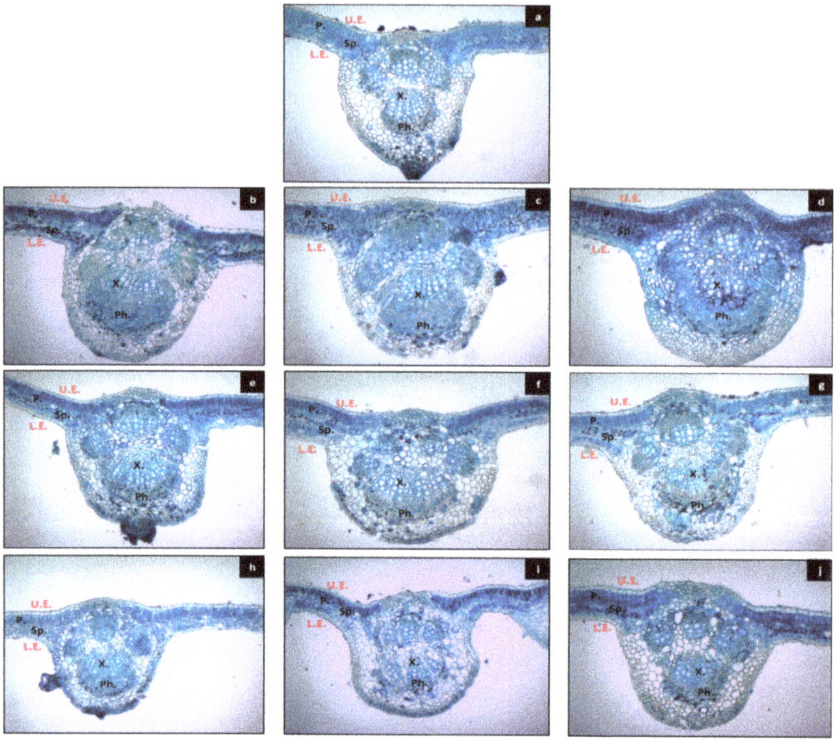

Figure 4. Microphotographs of cross-sections through the blade of leaves on the median portion of grapevine aged 135 days as affected by salinity (NaCl) and brassioloide (BL) (LE, Lower epidermis; P, Palisade parenchyma; SP, Spongy parenchyma; UE, Upper epidermis; X, xylem; Ph, phloem; (**a**), 0 NaCl without BL application; (**b**), 1000 mg L^{-1} NaCl without BL application; (**c**), 1000 mg L^{-1} NaCl with 1 mg L^{-1} BL application; (**d**), 1000 mg L^{-1} NaCl with2 mg L^{-1} BL application; (**e**), 2000 mg L^{-1} NaCl without BL application; (**f**), 2000 mg L^{-1} NaCl with 1 mg L^{-1} BL application; (**g**), 2000 mg L^{-1} NaCl with 2 mg L^{-1} BL application; (**h**), 3000 mg L^{-1} NaCl without BL application; (**i**), 3000 mg L^{-1} NaCl with 1 mg L^{-1} BL application; (**j**), 3000 mg L^{-1} NaCl with 2 mg L^{-1} BL application).

In T1, Figure 5d,e showed that in the control plant the cells having the distinctive chloroplast structure, an ellipsoidal form with well-arranged granum, compactly arranged thylakoid membranes, jointly with several starch grains (1–2 $chloroplast^{-1}$), without plastoglobules. Grana lamellae are completed thylakoids generally oriented parallel to the chloroplast's long axis. Under severe salinity, Figure 5h,i showed that salinity stress-induced clear alternations in chloroplasts, i.e., a decrease in the size of chloroplasts per cell, with the chloroplast becoming rounded and swelling of thylakoids,; Figure 5h,i also showed that the internal membranes were a disoriented lamellar system, a wavy configuration with starch grains was observed from 1–2 in T1 to 2–4 in T8, and the shape of starch grains were converted from the ellipsoidal shape in T1 to the rounded shape in T8; moreover, the number of plastoglobuli was increased. The chloroplasts became misshapen, grana stacking were less regular, and consequently, the thylakoids were loosened and imprecise (Figure 5i). Plants treated with BL demonstrated a distinctive chloroplast ultrastructure with no considerable alterations; the chloroplast was less than control (Figure 5n). BL application maintains the internal structure and grana staking, the number and size of starch grains, and fewer plastoglobule as compared to T8 (Figure 5n).

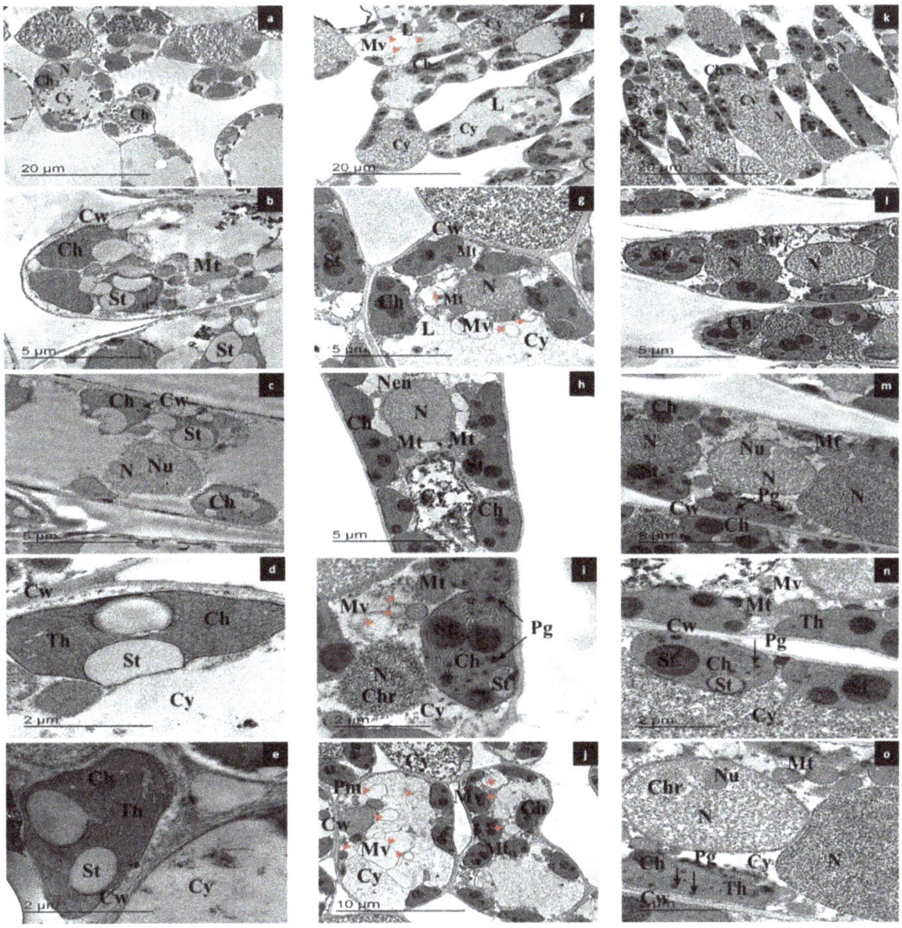

Figure 5. TEM micrograph of mesophyll cells of grapevine leaf, showing alternation in the ultrastructure of cell organelles including chloroplast, nucleus, mitochondria, and cell wall as well as the plasma membrane. (**a**–**e**) (control, T1); (**f**–**j**) (severe salinity, T8), (**k**–**o**) (severe salinity plus 2 mg L^{-1} BL foliar spraying, T10). An overview and details of control cells: (**d**,**e**) showing well-organized chloroplasts with smooth cell walls; (**c**) dispersion of nucleolus and chromatin in nucleus matrix and continuous nuclear envelope; (**b**) soft and thin cell wall with numerous mitochondria; (**f**,**j**) an overview of salt-affected cells indicating devastation of cell organelles, and a decrease in the size of chloroplast and accumulation of lipid droplets in the cytoplasm (white arrows) as well as an increase in the membrane vesicles (red arrows); (**g**) showing membrane vesicles (red arrows), and nucleus appeared without nucleolus; (**h**,**i**) swelling chloroplast with dilations of the thylakoid granum, and an increase in starch grains and plastoglubouline; (**k**) an overview of mesophyll cells in T10; (**i**,**m**,**o**) mesophyll cell having a well-organized nucleus, smooth nuclear envelop, mitochondria; (**n**), well-organized chloroplasts. (TEM, transmission electron microscopy; Ch, chloroplast; N, nucleus; Cy, cytoplasm; Cw, Cell wall; Mt, mitochondria; St, starch grain; Nu, nucleolus; Th, thylakoid; L, lipid droplets; Mv, membrane vesicles; Chr, chromatin; Pg, plastoglubouline; Nen, nuclear envelope; Pm, plasma membrane).

Under normal conditions, the nucleus appeared regular, with a distinctive nuclear envelope, nuclear chromatin, and nucleolus (Figure 5c). On the contrary, severe salinity induced a clear change in the nucleus; i.e., there was a decrease in the size and irregularity in shape, the nuclear envelope was unclear in some cells, the nuclear chromatin was aggregated or condensed as well as the nucleolus vanishing or being absent (Figure 5g–i). However, in T10, spraying salt-affected plants with 2 mg L^{-1} BL nullifies the harmful effect of salinity on nucleus structure, and accordingly, the nucleus appears normal with nucleolus clear (Figure 5l,m,o).

In non-salinized control plants, mitochondria showed regularly with apparent double membranes with a typical distribution of cristae (Figure 5b). In T8 compared to T1, we observed a variation in the number, size, and shape of mitochondria which improved the number of mitochondria, and its size was reduced and the distribution of cristae was indistinct or abnormal (Figure 5g,h). In T10, BL spraying boosted the size and number of mitochondria as compared to T1 and T8 (Figure 5m,o) and appeared normal with a normal distribution of cristae (Figure 5n).

In T1, the cell wall was slim (Figure 5b), while in T8 it was thick (Figure 5g), while the plasma membrane appeared partially separated from the cell wall and not adjacent to it in some cells (Figure 5j). In addition, augmented the plasmolysis of plasma membranes that will increase the number of membrane vesicles (cytoplasmic vesiculation) and disintegration of tonoplast (Figure 5g,j). Additionally, there was an increase in the accumulation of lipid droplets in the cytoplasm (Figure 5f). In contrast, in T10, the treatment with BL led to maintaining the cell wall and plasma membrane structure, and the number of membrane vesicles was decreased as compared to salinity treatment alone (Figure 5n).

3.7. Plant Growth

Salinity levels (1000, 2000, and 3000 mg L^{-1} NaCl) drastically ($p \leq 0.05$) repressed all morphological attributes of grapevine seedlings. Conversely, when BL was sprayed, the undesirable impacts of salinity on morphological attributes were decreased (Figure 6a–h). The undesirable impacts of salinity increased gradually with increasing salinity levels. The severe salinity level (3000 mg L^{-1}) without BL application significantly ($p \leq 0.05$) reduced the survival percentage (by 39% and 40%), plant height (by 36% and 37%), stem thickness (by 36% and 36%), the number of leaves plant^{-1} (by 40% and 38%), leaf area plant^{-1} (by 39% and 38%), shoot DW (by 47% and 48%), root DW (by 39% and 40%), and coefficient of wood ripping (by 38% and 36%) in the 1st and 2nd season, respectively, when relative to the control (Figure 6a–h).

BL spraying at both rates (1 and 2 mg L^{-1}) with salinity resulted in improvement in all growth parameters compared to the untreated samples. The high level of BL (2 mg L^{-1}) was more effective than the low level (1 mg L^{-1}) for increasing morphological trials. Under high salinity level (3000 mg L^{-1}), spraying with 1 mg L^{-1} BL and 2 mg L^{-1} BL significantly increased the survival% (by 19% and 26%), plant height (by 7% and 9%), stem thickness (by 9 and 13%), number of leaves plant^{-1} (by 11% and 18%), leaf area plant^{-1} (by 10% and 17%), shoot DW (by 10% and 14%), root DW (by 10% and 15%), and coefficient of wood ripping (by 15% and 18%) in the first season, in addition to increase the survival% (by 22% and 28%); plant height (by 12% and 17%); stem thickness (by 12% and 12%); number of leaves plant^{-1} (by 13% and 17%); leaf surface area plant^{-1} (by 12% and 14%); shoot DW (by 13% and 19%); root DW (by 13% and 19%) and coefficient of wood ripping (by 12% and 14%), respectively, in the second season, when compared to the untreated treatments (Figure 6). Relative to control, all the treatments with BL provoked a considerable enhancement in plant growth and demonstrated the ability of BL to alleviate saline-related stress on plant growth.

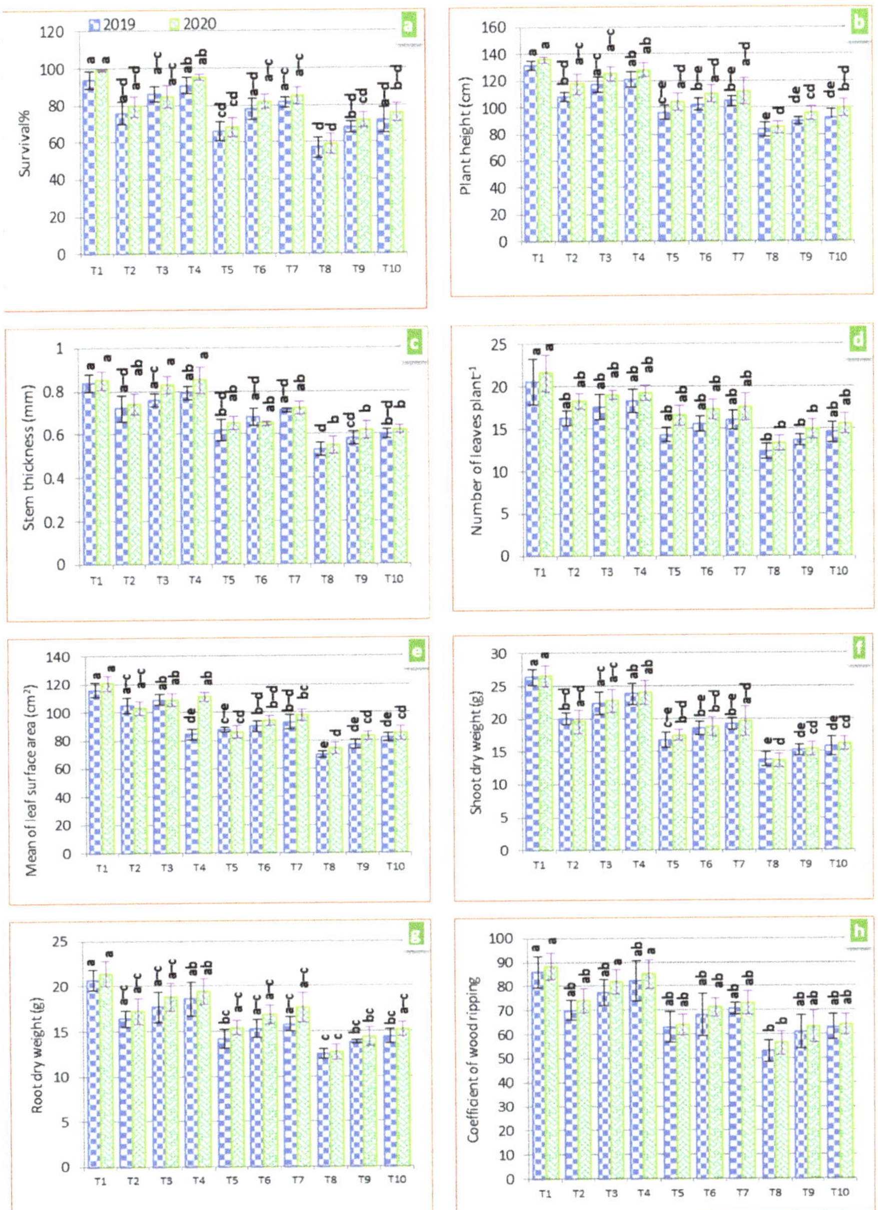

Figure 6. Plant growth trials (**a–h**) of grapevine seedling as affected by salinity (NaCl) and brassinolide (BL) during both growing seasons. Data represent the average of five replicates ± standard error. Different letters in each column indicate significant ($p \leq 0.05$) differences at $p \leq 0.05$ according to Tukey's HSD range at $p \leq 0.05$. (T1, 0 NaCl without BL application; T2, 1000 mg L^{-1} NaCl without BL application; T3, 1000 mg L^{-1} NaCl with 1 mg L^{-1} BL application; T4, 1000 mg L^{-1} NaCl with 2 mg L^{-1} BL application; T5, 2000 mg L^{-1} NaCl without BL application; T6, 2000 mg L^{-1} NaCl with 1 mg L^{-1} BL application; T7, 2000 mg L^{-1} NaCl with 2 mg L^{-1} BL application; T8, 3000 mg L^{-1} NaCl without BL application; T9, 3000 mg L^{-1} NaCl with 1 mg L^{-1} BL application; T10, 3000 mg L^{-1} NaCl with 2 mg L^{-1} BL application).

4. Discussion

Plants undergo salinity critically once they are grown in saline conditions. The drastic impact of salinity as recorded in the current investigation on plant growth was provided with earlier findings for numerous plants [6–8,11,16,17]. The overall decline under salinity might be caused by the negative impact of salinity on different metabolic processes and molecular alterations, i.e., photosynthesis, nutrient homeostasis, stomatal resistance, and ROS production in different plants [6,8,16]. In this concern, the decline in water absorption may result from decreasing hydraulic conductivity and the expression of aquaporins such as plasma membrane intrinsic protein (PIP) and tonoplast intrinsic protein (TIP) [48]. Salinity induced the buildup of salts inside the leaves that caused irreversible injury to the chloroplasts as recorded in the present experiment, resulting in metabolic restrictions of photosynthesis [49]. Additionally, the nutritional imbalance evoked the production of ROS that would induce oxidative stress and decreased cell division, proliferation, and elongation, and finally declined plant growth [14]. Moreover, growth reduction with salinity might partly be owing to the lack of energy, since processes occupied in salt uptake are energy-consuming [45]. The decline in growth-related parameters is possibly attributable to damaged cell development resulting from growth hormone efficiency, leading to a lessening in cell turgor, cell volume, and eventually cell growth and it may also be owing to the blocking up of conductive tissue vessels, thus blocking all translocation that passes throughout these tissues [50]. In the current study, the application of BL in special 2 mg L^{-1} considerably moderates the injury of salinity on plant growth. Comparable trends were recorded in different plant species [26,51]. However, the actual signaling mechanisms are largely unknown. The work with BL biosynthetic mutants in *Arabidopsis thaliana* [52] and *Pisum sativum* [53] have offered strong confirmation that BL signaling plays a vital for plant growth. Friedrichsen et al. [54] also stated that three redundant BL genes encode transcription factors that are necessary for typical growth, demonstrating the requirement of BRs for typical growth. BL controls the transcription of CycD3 (a D-type cyclin gene) throughout which cytokinin activates cell division, and BL mediated CycD3 induction has been recorded to control the de novo synthesis of proteins [55]. The motivating impact of BL on plant growth may result from its effect on physiological processes, including enhancing photoprotection and improving photosynthetic efficiency, improving antioxidant capacity and reducing ROS production, and improving mineral assimilation [56,57]. Accordingly, Anwar et al. [58,59] reported that BL application improved CAT, POD, and SOD activities in tomato and cucumber, respectively. Additionally, BL application improved photosynthetic efficiency by enhancing photochemical quenching coefficient, Rubisco enzyme activity, and over-expression of the large and small subunit genes with increasing CO_2 assimilation rate [27,60]. BL up-regulates water uptake and preservation of plant water potential, which leads to improving RWC as reported in the present study and earlier report [61], and/or reduced Na^+ accumulation and improved K^+ uptake resulted in the avoidance of osmotic and ionic upset to the plant [62]. Additionally, BL accelerates cell division and expansion in the apical meristem, which leads to improving leaf expansion [63].

Nutrient concentrations of grapevine seedlings except Na^+ drastically decreased under salinity; however, BL application mitigated salinity injuries via dropping Na^+ and increased N, P and K (Table 2). The outcomes were compatible with El-Taher et al. [7], Sarwar et al. [10], Hatami and Pourakbar [17] for salinity, Kolomeichuk et al. [27], and Karlidag et al. [64] for BL. Accordingly, Miao et al. [65] proved that BRs application improved root nodulation capacity and nitrogenase activity, resulting in increasing N% in plant tissues. It has been shown that severe salinity could confuse nutrient-ion activities, resulting in ionic imbalance, nutrients shortage, and specific ion toxicity [6,8,66]. Up-regulation of K^+ uptake with evading of Na^+ absorption, efflux of Na^+, and development for osmotic adjustment is an approach typically possessed by the plant for preserving an optimal K^+/Na^+ ratio that is an imperative decisive factor describing plant salinity tolerance. The competition between K^+ and Na^+ resulted in the aggressive uptake as the K^+ transporter lacks discrimination between K^+ and Na^+ ions [67,68]. Currently, salinity enlarged the accumulation of Na^+, which is connected with lessened K^+ concentration, leading to a decline in K^+/Na^+ ratio. Earlier research has stated that under salinity, Na^+ accumulation is associated with the declined in K^+ [6,8,16,17,69–71]. This lessening is connected to the antagonistic routes, since Na^+ uptake by root cells takes place throughout non-selective K^+ channels and high-affinity K^+ transporters caused by physicochemical similarity among Na^+ and K^+ [72]. The preservation of ionic homeostasis under salinity is the requirement to defend the plants alongside the accumulation of lethal ions, with K^+ buildup and Na^+ realization the lowest concentration in grapevine seedlings. As a result, the organization of Na^+ accumulation and therefore a superior K^+/Na^+ ratio might maintain salinity tolerance [73]. Under salinity, BL application can revise the plasma membrane function and boost ion uptake [24]. The ability of BL to maintain plasma membrane structure may have been associated with the considerable reduction in Na^+ and the enhancement of K^+ ions [74]. Salinity tolerance accomplished by BRs amendment is possibly ascribed to its capability to enhance K^+ uptake and restrict Na^+ concentration into xylem, while sustaining an elevated K^+/Na^+ ratio in plant tissues [27,75]. This is possibly owing to the overexpression of salt overly sensitive 1 (SOS1, Na^+/H^+ antiporter), which shifts extra Na^+ outside the cytosol and assists preserve small cytosolic Na^+ levels in tissues, particularly in root epidermal cells and root tips [76]. SOS1 retrieves salinity tolerance mostly by facilitating Na^+ efflux from the cytosol to the rhizosphere [77] through (i) rising Na^+ storage time in vacuoles and dropping Na^+ accumulation in the cytoplasm, and (ii) controlling long-distance Na^+ transport throughout Na^+ repossession between roots and shoots. The elevated shoot K^+/Na^+ ratio might have been implicated in enhancing the plant development with BL application under salinized circumstances.

The current data indicated that grapevine photosynthetic pigments declined with salinity (Table 3). Alternatively, the application of BL causes a significant increase in the concentration of photosynthetic pigments. Similarly, Farouk et al. [6], El-Taher et al. [7], Farouk and Al-Huqail [8], Sarwar et al. [10], and Hatami and Pourakbar [17] indicated that salinity induced a considerable lessening in the chlorophyll level. The decline in chlorophyll under salinity was linked to the activation of chlorophyll degrading enzyme chlorophyllase and ROS production [45,78,79], restricted N absorption [80], and amplified susceptibility of pigment–protein complexes to deprivation [81], plus chloroplasts' ultrastructure [49]. Additionally, there is a reduction in chlorophyll biosynthesis intermediation levels [82] and the expression of ChlD, Chl H, and Chl I-1 genes encoding subunits of Mg-chelatase [83]. Under salinity, the over-production of ROS in cells induces oxidation and, therefore, the deprivation of photosynthetic pigments with the breakdown of the thylakoid membranes and changes in chlorophyll protein complexes [81,84]. However, the application of BL under salt stress restores imprecise chlorophyll accumulation caused by salinity (Table 3). These outcomes were compatible with former research [26,85]. This attenuating effect of BL can be reasoned from the possibility of BRs-induced impact on transcription and/or translation in the synthesis of pigments [26,86]. Additionally, BRs maintain thylakoid membrane, stability and regulate chlorophyll molecules by upregulating chlorophyllase activity. BRs regulate the protection scheme by controlling transcription of defense-related genes and alleviating the difficulty of diverse stresses and by regulating activated Rubisco genes [87]. Consistent with Deng et al. [88], BRs boost the activity of alternative oxidases (AOX) in a respiratory burst oxidase homolog (RBOH)-dependent way. So, a superior activity of AOX controls chloroplast and mitochondria's electron flow through dissipating the extra energy, thus lessening the ROS accretion and increasing the defense of photosynthetic apparatus. Additionally, as recorded in the current research, BL enhanced K^+ absorption, enlarged chloroplast number $cell^{-1}$, preserved chloroplast ultrastructure, or sustained chlorophyll stability by hastened ROS-mitigating activity. Furthermore, carotenoid assimilation was enhanced in grapevine under salinity upon BL spraying, probably by acting as an antioxidant, thus decreasing salinity-accelerated oxidative stress.

Sustaining crop water status-associated trials at an elevated level improves the metabolic pathways that are sustained by osmotic adjustment. Leaf RWC has been considered as a substitute evaluation of plant water status, reflecting the plant's metabolic activity [45]. A reduction in LRWC under salinity has been previously recorded [6,10,71]. This decline can be ascribed to less water accessibility, or to defeat of the plant roots' aptitude to catch up on the water throughout a lessening of the absorbing surface [89]. On the other hand, the LRWC of BL-treated seedlings under salinity were preserved at high altitudes equivalent to salt-affected seedlings that did not receive BL, which consents with the outcomes of earlier [30,90]. This increase may result from the over-accumulation of osmolytes as proline [91] that will preservative tissue LRWC and rapid eradication of ROS [92]. Furthermore, BL application enhanced root development, reinforced water uptake, and controlled the expression of aquaporin-synthesizing genes [93]. This designates that BL application sustained cell membrane stability and sustained water status in salt-affected grapevine seedlings. The cell membrane represents the main cellular targets to diverse stresses. Salinity accelerates lipid peroxidation that was boosts the MP of grapevine seedlings. These results follow those recorded by Sarwar et al. [10], Abdelaal et al. [79], Dong et al. [94] in several crops. Salt stress evoked the over-production of ROS that consecutively aggravated the cell membrane damage and alternation of plasma membrane permeability [95]. The preservation of small MP% proved that the BRs-treated seedlings sustained plasma membrane integrity under salinity. The capability of the BL-treated plants to preserve plasma membrane integrity might connect to the valuable function of BL to either (i) alleviate the harmful effects of ROS, (ii) maintain membrane lipid and protein compositions, or (iii) decrease activities of lipid peroxidation and protein oxidation [57].

Proline represents the principally widespread defensive molecules within stressful conditions, i.e., salinity [7,10] as well as BL application [25,55,96]. In the present investigation, we recorded a speedy many-fold increase in the proline concentration in grapevine under salinity, with or without BL, that designated the key function of proline as a defensive substance under salinity conditions, probably to offer superior protection alongside salinity. These osmolytes, can control of the plant physio-biochemical pathways, i.e., sustaining membrane integrity, reducing the cellular water potential, facilitating continuous water uptake, preserving plant water status and cell turgidity maintaining the finest redox state, controlling salt-stress-responsive gene expression [97,98], and preserving plant water status [99]. Proline has also been recorded to participate in alleviating ROS's harmful impacts [100] and alleviating cytoplasmic pH [97]. Moreover, encouraging activation of proline assimilation in chloroplasts is a vital sink to ATP and NADPH, produced throughout the primary photosynthetic processes, thus encouraging the preservation of the electron flow among photosynthetic excitation centers, stabilization of redox equilibrium by maintaining NADPH/NADP hence preventing photoinhibition [27,101]. The hyper-assimilation of proline takes place chiefly through the motivated assimilation, the inhibition of proline oxidation, and the plant's capability to preserve the mitochondria membranes' permeability [102,103]. Several kinds of research have established that the over-expression of proline biosynthetic pathways genes Δ1-pyrroline-5-carboxylate synthetase (P5CS) exhibits improved tolerance to oxidative stress [104].

Salinity stress induces the repression of plant's metabolic pathways, including the hyper-accumulation of ROS that evoked oxidative burst [6,8,11]. Plant detoxification pathways involve the activation of antioxidant enzymes and the accumulation of antioxidant solutes [6,8,11]. The enzymatic antioxidant systems are composed of SOD, ascorbate peroxidase (APX), POD, and CAT that deactivate stress-provoked ROS production [16,79,105]. In the current study, NaCl and/or BL caused an increase in POD, PPO, and CAT activities of grapevine seedlings in both seasons. Antioxidant enzymes are part of proficient schemes for mitigating ROS and protecting plants from negative oxidative bursts [8]. Accordingly, Kaur et al. [11] also stated that salinity stress amplified antioxidant enzyme activity in chickpea genotypes. BL drastically eradicated ROS production via activation and strengthening of the antioxidant system, i.e., SOD and CAT, during salinity. Similarly, Lone et al. [26] and Arora et al. [106] found that BRs application increased antioxidant enzymes SOD, CAT, and POD.

Numerous phenolic compounds are stress-induced metabolites in plants [10,16]. It has been recorded that hyper-accumulation of phenols imparts superior radical scavenging activity so avoiding cellular oxidative rupture [91]. Soluble phenols provide an antioxidant since they have electron-donating mediators and, consequently, lessen extra ROS accumulation [107]. This production was probably provoked by eliciting the phenylpropanoid pathway and enhancing phenyl-aminolyase (PAL) gene expression [108]. In addition, a raise in PAL activity under salinity has been recorded by and is a key enzyme affecting the assimilation of plants' secondary metabolites [92].

Leaves are imperative places of essential biochemical pathways. El-Taher et al. [7], El-Banna and Abdelaal [109], and Nassar et al., [110] found that salinity decreased all leaf anatomical attributes including the thickness of the leaf blade and midvein of the mature leaf of strawberry, sweet basil, and cowpea plants, respectively. Ordinary, salt-affected plants were characterized by small cell size and declined in vascular tissue and cell wall thickness [111]. This reduction possibly resulted from the restriction of cell division and expansion plus a lessening in mesophyll parenchyma layer thickness as well as hampering procambial activity [112,113]. On the other hand, there are very few investigations related to the effect of BRs on plant anatomy. In this regard, Kulaeva et al. [114] recorded that the application of 24-EpiBL had a defensive role on cell ultra-structure in salt-affected leaves, which additionally prohibited nuclei and chloroplast deprivation, paving a way for better photosynthesis. Moreover, Ibrahim and Abo-ELwafa [115] on Thompson grape found that a high salinity level decreased the thickness of lamina and midvein of blades; the decline was more noticeable than that induced by a low salinity level, being 52.24% less than the control for thickness of lamina and 54.01% less than the control for thickness of midvein. Additionally, the thinner blade under 3000 ppm salinity could be attributed to the declines in thickness of palisade and spongy parenchyma and thickness of upper and lower epidermis by 54.62 and 56.44%, and 40.00 and 39.13%, respectively, compared with those of the control. The same authors revealed that vascular bundles of midvein displayed noticeable reduction in length by 61.29% and in width by 61.87% less than the control.

Studying the plant cell ultrastructural under salinity is possibly a practical implementation for understanding the deep strategies implicated in conferring salt tolerance. Salinity evoked the chief alterations in chloroplasts, i.e., swollen thylakoids, loose profiles of the piece of interior lamellae thylakoids, though mainly granal thylakoids were shattered. These outcomes were corroborated by Farouk and Arafa [49] and El-Banna and Abdelaal [109]. In this study, the deformations of grana stacking and swelling of thylakoids caused by salinity were possibly due to a modification in the ionic composition of the stroma. The degradation of the plastids is related to salt stress possibly provoked by ROS extra-accumulation causing oxidative anxiety [116]. The increase in the plastoglobule number evaluated in the current investigation is possibly a proper sign of ecological stress disorder [117,118]. The physical coupling among the plastoglobules and thylakoid membranes permits the free exchange of lipid molecules along with the plastoglobules and thylakoids [117]. The huge plastoglobule size and number recorded in the salt-affected plants are possibly one of the adaptive methods that avoid the oxidative injuries caused by high salinity. Conversely, boost the number of starch grains in the chloroplast (Figure 5h,i). Rahman et al. [119] reported that the raise of starch grains in chloroplast under salinity is owing to the injury of enzymes occupied in starch metabolism by alterations in ionic composition and/or the damage of the sucrose phosphate pathway biosynthesis in the cytosol leading to the triose phosphate pathway toward starch metabolism. Alternatively, BL lightened this structural injury by defending the chloroplasts from oxidative stress. Large chloroplasts with no swelling and only slight dilations of the thylakoids in BL and salt-affected plants are the existing signs of less oxidative anxiety. The relatively fewer number of plastoglobuli in chloroplasts of plants treated with salinity and BL alongside is another signal of smaller oxidative anxiety [120].

Within normal conditions, the mitochondria had well-organized cristae and an intact structure and were of similar size and appearance. Conversely, salt-affected cells had an extremely small and the largest number of mitochondria with a defeat of the integrity of the outer mitochondrial membranes. Comparable outcomes have been a statement by Zhang et al. [120]. The injury elicited by salinity in mitochondria is probably a signal of salt-associated changes in mitochondria energy status resulting in decline ATP levels [121]. Nevertheless, BL spraying enhanced the dimensions and number of mitochondria under stressed circumstances. The superior mitochondrial number and size meet the increased needs of ATP under unfavorable circumstances when photosynthesis is commonly suppressed, and these organelles respond to stress by assimilation of different precise mitochondrial stress proteins [122].

The incidence of the membrane vesicles in the grapevine mesophyll cells is believed as an adaptive mechanism for sodium ions sequestration to ease their dangerous impact on cell organelles and cytoplasm [123,124]. Moreover, the accretion of lipid droplets in the cytoplasm may be considered as a preserve of energy to be used by the cell to cover the increased requirements in metabolic energy requisite to salinity tolerance and/or effect of ROS which fast the peroxidation of membrane lipids leading to loss membrane integrity [119].

5. Conclusions

Salinity declined grapevine seedling's growth, relative water content, and mineral concentrations. Conversely, with a spray of BL, the harmful impacts of salinity were mitigated. The possible strategies consist of the following: (1) BL could boost the concentration of chlorophyll and free proline; (2) BL could control activities of key antioxidant enzymes to eradicate ROS; (3) BL enhances cell membrane stability and nutrient uptake, as well as water status; (4) BL maintains the ultrastructure of cell organelles and leaf anatomy. Hence, BL could increase grapevine seedling growth under salinity, and the most favorable concentration appears to be 2 mg L^{-1} concentration. BL spraying could present an easy application in grapevine productivity in saline soil. However, additional research is required to decide the competence of these materials.

Author Contributions: Conceptualization, M.F.E.-B., A.A.A., S.F., B.E.A.B., M.A.E.-K. and A.F.A.E.-K.; methodology, M.F.E.-B., S.F., B.E.A.B. and M.A.E.-K.; software, S.F.; validation, M.F.E.-B., A.A.A., S.F., B.E.A.B., M.A.E.-K. and A.F.A.E.-K.; formal analysis, S.F.; investigation, M.F.E.-B., A.A.A., S.F., B.E.A.B., M.A.E.-K. and A.F.A.E.-K.; resources, M.F.E.-B., A.A.A., S.F., B.E.A.B., M.A.E.-K. and A.F.A.E.-K.; data curation, M.F.E.-B., A.A.A., S.F., B.E.A.B., M.A.E.-K. and A.F.A.E.-K.; writing—original draft preparation, M.F.E.-B., A.A.A., S.F., B.E.A.B., M.A.E.-K. and A.F.A.E.-K.; writing—review and editing, S.F.; visualization, M.F.E.-B., A.A.A., S.F., B.E.A.B., M.A.E.-K. and A.F.A.E.-K.; funding acquisition, A.A.A. All authors have read and agreed to the published version of the manuscript.

Funding: This research was funded by Princess Nourah bint Abdulrahman University Researchers Supporting Project number (PNURSP2022R93), Princess Nourah bint Abdulrahman University, Riyadh, Saudi Arabia.

Institutional Review Board Statement: Not applicable.

Informed Consent Statement: Not applicable.

Data Availability Statement: Not applicable.

Acknowledgments: The authors express their gratitude to Princess Nourah bint Abdulrahman University Researchers Supporting Project number (PNURSP2022R93), Princess Nourah bint Abdulrahman University, Riyadh, Saudi Arabia, as well as Mansoura University, Mansoura, Egypt.

Conflicts of Interest: The authors declare no conflict of interest.

References

1. Senthilkumar, S.; Vijayakumar, R.M.; Soorianathasundaram, K.; Durga, D. Impact of pre-harvest sprays with gibberellic acid on yield and economics of grape (*Vitis vinifera* L.) cv. Italia. *Ann. Biol.* **2018**, *34*, 170–175.
2. Bourgaud, F.; Gravot, A.; Milesi, S.; Gontier, E. Production of plant secondary metabolites: A historical perspective. *Plant Sci.* **2001**, *161*, 839–851. [CrossRef]
3. Rahmani, M.; Bakhshi, D.; Qolov, M. Impact of pruning severity and training systems on red and white seedless table grape (*Vitis vinifera*) qualitative indices. *Aust. J. Crop Sci.* **2015**, *9*, 55–61.
4. Belal, B.E.A. Improvement of physical and chemical properties of Thompson seedless grapes (H4 Strain) by application of brassinolide and gibberellic acid. *Egypt J. Hort.* **2019**, *46*, 251–262. [CrossRef]
5. Zhou-Tsang, A.; Wu, Y.; Henderson, S.W.; Walker, A.R.; Borneman, A.R.; Walker, R.R.; Gilliham, M. Grapevine salt tolerance. *Aust. J. Grape Wine Res.* **2021**, *27*, 149–168. [CrossRef]
6. Farouk, S.; Elhindi, K.M.; Alotaibi, M.A. Silicon supplementation mitigates salinity stress on *Ocimum basilicum* L. via improving water balance, ion homeostasis, and antioxidant defense system. *Ecotoxicol. Environ. Saf.* **2020**, *206*, 111396. [CrossRef]
7. El-Taher, A.M.; Abd El-Raouf, H.S.; Osman, N.A.; Azoz, S.N.; Omar, M.A.; Elkelish, A.; Abd El-Hady, M.A.M. Effect of salt stress and foliar application of salicylic acid on morphological, biochemical, anatomical, and productivity characteristics of cowpea (*Vigna unguiculata* L.) plants. *Plants* **2022**, *11*, 115. [CrossRef]
8. Farouk, S.; AL-Huqail, A.A. Sustainable biochar and/or melatonin improve salinity tolerance in borage plants by modulating osmotic adjustment, antioxidants, and ion homeostasis. *Plants* **2022**, *11*, 765. [CrossRef]
9. Gopalakrishnan, T.; Kumar, L. Modeling and mapping of soil salinity and its impact on paddy lands in Jaffna Peninsula, Sri Lanka. *Sustainability* **2020**, *12*, 8317. [CrossRef]
10. Sarwar, M.; Anjum, S.; Ali, Q.; Alam, M.W.; Haider, M.S.; Mehboob, W. Triacontanol modulates salt stress tolerance in cucumber by altering the physiological and biochemical status of plant cells. *Sci. Rep.* **2021**, *11*, 24504. [CrossRef]
11. Kaur, H.; Hussain, S.J.; Kaur, G.; Poor, P.; Alamri, S.; Siddiqui, M.H.; Khan, I.M.R. Salicylic acid improves nitrogen fixation, growth, yield and antioxidant defense mechanisms in chickpea genotypes under salt stress. *J. Plant Growth Regul.* **2022**. [CrossRef]
12. Farhangi-Abriz, S.; Torabian, S. Antioxidant enzyme and osmotic adjustment changes in bean seedlings as affected by biochar under salt stress. *Ecotoxicol. Environ. Saf.* **2017**, *137*, 64–70. [CrossRef] [PubMed]
13. Fallah, F.; Nokhasi, F.; Ghaheri, M.; Kahrizi, D.; Beheshti Agha, A.; Ghorbani, T.; Kazemi, E.; Ansarypour, Z. Effect of salinity on gene expression, morphological and biochemical characteristics of *Stevia rebaudiana* Bertoni under in vitro conditions. *Cell Mol. Biol.* **2017**, *63*, 102–106. [CrossRef] [PubMed]
14. Saleem, M.H.; Fahad, S.; Khan, S.U.; Din, M.; Ullah, A.; Sabagh, A.E.L.; Hossain, A.; Llanes, A.; Liu, L. Copper induced oxidative stress, initiation of antioxidants and phytoremediation potential of flax (*Linum usitatissimum* L.) seedlings grown under the mixing of two different soils of China. *Environ. Sci. Pollut. Res.* **2020**, *27*, 5211–5221. [CrossRef]
15. Mumtaz, S.; Saleem, M.H.; Hameed, M.; Batool, F.; Parveen, A.; Amjad, S.F.; Mahmood, A.; Arfan, M.; Ahmed, S.; Yasmin, H. Anatomical adaptations and ionic homeostasis in aquatic halophyte *Cyperus laevigatus* L. under high salinities. *Saudi J. Biol. Sci.* **2021**, *28*, 2655–2666. [CrossRef] [PubMed]
16. Sofy, M.R.; Elhindi, K.M.; Farouk, S.; Alotaibi, M.A. Zinc and paclobutrazol mediated regulation of growth, upregulating antioxidant aptitude and plant productivity of pea plants under salinity. *Plants* **2020**, *9*, 1197. [CrossRef]
17. Hatami, S.; Pourakbar, L. Effects of manganese on physiological characters of grapevine cultivars under salinity stress. *MOJ Eco Environ. Sci.* **2020**, *5*, 62–68. [CrossRef]
18. Naveed, M.; Sajid, H.; Mustafa, A.; Niamat, B.; Ahmad, Z.; Yaseen, M.; Kamran, M.; Rafique, M.; Ahmar, S.; Chen, J.T. Alleviation of salinity-induced oxidative stress, improvement in growth, physiology and mineral nutrition of canola (*Brassica napus* L.) through calcium-fortified composted animal manure. *Sustainability* **2020**, *12*, 846. [CrossRef]
19. Saddiq, M.S.; Iqbal, S.; Hafeez, M.B.; Ibrahim, A.M.H.; Raza, A.; Fatima, E.M.; Baloch, H.; Jahanzaib Woodrow, P.; Ciarmiello, L.F. Effect of salinity stress on physiological changes in winter and spring wheat. *Agronomy* **2021**, *11*, 1193. [CrossRef]
20. Inoue, M. Salinization status and salt removal techniques. *Geotech. Eng. Manag.* **2012**, *60*, 12–15.
21. Belkheiri, O.; Mulas, M. The effects of salt stress on growth, water relations and ion accumulation in two halophyte *Atriplex* Species. *Environ. Exp. Bot.* **2013**, *86*, 17–28. [CrossRef]
22. Shams, M.; Ekinci, M.; Ors, S.; Turan, M.; Agar, G.; Kul, R.; Yildirim, E. Nitric oxide mitigates salt stress effects of pepper seedlings by altering nutrient uptake, enzyme activity and osmolyte accumulation. *Physiol. Mol. Biol. Plants* **2019**, *25*, 1149–1161. [CrossRef] [PubMed]
23. Hafeez, M.B.; Zahra, N.; Zahra, K.; Raza, A.; Khan, A.; Shaukat, K.; Khan, S. Brassinosteroids: Molecular and physiological responses in plant growth and abiotic stresses. *Plant Stress* **2021**, *2*, 100029. [CrossRef]
24. Quamruzzaman, M.; Manik, S.M.N.; Shabala, S.; Zhou, M. Improving performance of salt-grown crops by exogenous application of plant growth regulators. *Biomolecules* **2021**, *11*, 788. [CrossRef]
25. Ahmad, P.; Ahanger, M.A.; Alam, P.; Alyemeni, M.N. Modification of osmolytes and antioxidant enzymes by 24-epibrassinolide in chickpea seedlings under mercury (Hg) toxicity. *J. Plant Growth Regul.* **2018**, *31*, 309–322. [CrossRef]
26. Lone, W.A.; Majeed, N.; Yaqoob, U.; John, R. Exogenous brassinosteroid and jasmonic acid improve drought tolerance in *Brassica rapa* L. genotypes by modulating osmolytes, antioxidants and photosynthetic system. *Plant Cell Rep.* **2021**, *41*, 603–617. [CrossRef]

27. Kolomeichuk, L.V.; Efmova, M.V.; Zlobin, I.E.; Kreslavski, V.D.; Murgan, O.K.; Kovtun, I.S.; Khripach, V.A.; Kuznetsov, V.V.; Allakhverdiev, S.I. 24-Epibrassinolide alleviates the toxic effects of NaCl on photosynthetic processes in potato plants. *Photosynth. Res.* **2020**, *146*, 151–163. [CrossRef]
28. Lv, J.; Zong, X.F.; Ahmad, A.S.; Wu, X.; Wu, C.; Li, Y.P.; Wang, S.G. Alteration in morpho-physiological attributes of *Leymus chinensis* (Trin.) Tzvelev by exogenous application of brassinolide under varying levels of drought stress. *Chil. J. Agric. Res.* **2020**, *80*, 61–71. [CrossRef]
29. Mazorra Morales, L.M.; Senn, M.E.; Gergoff Grozeff, G.E.; Fanello, D.D.; Carrión, C.A.; Núñez, M.; Bishop, G.J.; Bartoli, C.G. Impact of brassinosteroids and ethylene on ascorbic acid accumulation in tomato leaves. *Plant Physiol. Biochem.* **2014**, *74*, 315–322. [CrossRef]
30. Ali, B.; Hayat, S.; Fariduddin, Q.; Ahmad, A. 24-Epibrassinolide protects against the stress generated by salinity and nickel in *Brassica juncea*. *Chemosphere* **2008**, *72*, 1387–1392. [CrossRef]
31. Krishna, P. Brassinosteroid mediated stress response. *J. Plant Growth Regul.* **2003**, *22*, 289–297. [CrossRef] [PubMed]
32. Jangid, K.; Kanwar, K.; Panwar, P.; Asiwal, R.C.; Bajya, M.; Bagdi, D.L. Effect of brassinolide in amelioration of salinity adverse effects on growth and yield of wheat. *J. Pharmacogn. Phytochem.* **2017**, *6*, 194–197.
33. Lal, B.; Bagdi, D.L.; Dadarwal, B.K. Role of brassinolide in amelioration of salinity induced adverse effects on growth, yield attributes and yield of wheat. *J. Pharmacogn. Phytochem.* **2019**, *8*, 1790–1793.
34. Su, Q.; Zheng, X.; Tian, Y.; Wang, C. Exogenous brassinolide alleviates salt stress in *Malus hupehensis* Rehd. by regulating the transcription of NHX-Type $Na^{+}(K^{+})/H^{+}$ Antiporters. *Front. Plant Sci.* **2020**, *11*, 38. [CrossRef] [PubMed]
35. Babalık, Z.; Demirci, T.; Aşcı, Ö.A.; Baydar, N.G. Brassinosteroids modify yield, quality, and antioxidant components in grapes (*Vitis vinifera* cv. Alphonse Lavallée). *J. Plant Growth Regul.* **2020**, *39*, 147–156. [CrossRef]
36. Kanwar, M.K.; Bajguz, A.; Zhou, J.; Bhardwaj, R. Analysis of brassinosteroids in plants. *J. Plant Growth Regul.* **2017**, *36*, 1002–1030. [CrossRef]
37. Vardhini, B.V.; Anuradha, S.; Rao, S.S.R. Brassinosteroids—A great potential to improve crop productivity. *Indian J. Plant Physiol.* **2006**, *11*, 1–12.
38. Cooper, T.G. *The Tools of Biochemistry*; A Wiley-Interscience Pub; John Wiley and Sons: New York, NY, USA, 1977.
39. Motsara, M.R.; Roy, R.N. *Guide to Laboratory Establishment for Plant Nutrient Analysis*; FAO Fertilizer and Plant nutrition bulletin; Food and Agriculture Organization: Rome, Italy, 2008; 219p.
40. Lichtenthaler, H.; Wellburn, A. Determinations of total carotenoids and chlorophylls a and b of leaf extracts in different solvents. *Biochem. Soc. Trans.* **1983**, *603*, 591–592. [CrossRef]
41. Dionisio-Sese, M.L.; Tobita, S. Antioxidant responses of rice seedlings to salinity stress. *Plant Sci.* **1998**, *135*, 1–9. [CrossRef]
42. Bates, L.; Waldren, R.P.; Teare, I.D. Rapid determination of free proline for water stress studies. *Plant Soil* **1973**, *39*, 205–207. [CrossRef]
43. Dong, Y.J.; He, M.R.; Wang, Z.L.; Chen, W.F.; Hou, J.; Qiu, X.K.; Zhang, J.W. Effects of new coated release fertilizer on the growth of maize. *J. Soil Sci. Plant Nutr.* **2016**, *16*, 637–649. [CrossRef]
44. Augustin, M.A.; Ghazali, H.M.; Hashim, H. Polyphenoloxidase from guava (*Psidium guajava* L.). *J. Agric. Food Chem.* **1985**, *36*, 1259–1265. [CrossRef]
45. Tian, X.Y.; He, M.R.; Wang, Z.L.; Zhang, J.W.; Song, Y.L.; He, Z.L.; Dong, Y.J. Application of nitric oxide and calcium nitrate enhances tolerance of wheat seedlings to salt stress. *Plant Growth Regul.* **2015**, *77*, 343–356. [CrossRef]
46. Tzortzakis, N.G.; Tzanakaki, K.; Economakis, C. Effect of origanum oil and vinegar on the maintenance of postharvest quality of tomato. *Food Nutr. Sci.* **2011**, *2*, 974–982. [CrossRef]
47. Rizk, M.H.; Rizk, N.A. Effect of Dormex on bud behavior, yield and rate of wood maturity of Thompson seedless grapevine. *Egypt J. Appl. Sci.* **1994**, *9*, 525–542.
48. Boursiac, Y.; Chen, S.; Luu, D.T.; Sorieul, M.; van den Dries, N.; Maurel, C. Early effects of salinity on water transport in *Arabidopsis* roots. Molecular and cellular features of aquaporin expression. *Plant Physiol.* **2005**, *139*, 790–805. [CrossRef]
49. Farouk, S.; Arafa, S.A. Mitigation of salinity stress in canola plants by sodium nitroprusside application. *Span. J. Agric. Res.* **2018**, *16*, e0802. [CrossRef]
50. Queiros, F.; Rodrigues, J.A.; Almeida, J.M.; Almeida, D.P.; Fidalgo, F. Differential responses of the antioxidant defense system and ultrastructure in a salt-adapted potato cell line. *Plant Physiol. Biochem.* **2011**, *49*, 1410–1419. [CrossRef]
51. López-Gómez, M.; Hidalgocastellanos, J.; Lluch, C.; Herreracervera, J.A. 24-Epibrassinolide ameliorates salt stress effects in the symbiosis *Medicago truncatula-Sinorhizobium meliloti* and regulates the nodulation in cross-talk with polyamines. *Plant Physiol. Biochem.* **2016**, *8*, 212–221. [CrossRef]
52. Tao, Y.Z.; Zheng, J.; Xu, Z.M.; Zhang, X.H.; Zhang, K.; Wang, G.Y. Functional analysis of ZmDWF1, a maize homolog of Arabidopsis brassinosteroids biosyntheticDWF1/DIMgene. *Plant Sci.* **2004**, *167*, 743–751. [CrossRef]
53. Nomura, T.; Nakayama, N.; Reid, J.B.; Takeuchi, Y.; Yokota, T. Blockage of brassinosteroid biosynthesis and sensitivity cause dwarfism in *Pisum sativum*. *Plant Physiol.* **1997**, *113*, 31–37. [CrossRef] [PubMed]
54. Friedrichsen, D.M.; Nemhauser, J.; Muramitsu, T.; Maloof, N.N.; Alonso, J.; Ecker, J.R. Three redundant brassinosteroids early response genes encode putative bHLH transcription factors required for normal growth. *Genetics* **2002**, *162*, 1445–1456. [CrossRef] [PubMed]

55. Hu, D.G.; Ma, Q.J.; Sun, C.H.; Sun, M.H.; You, C.X.; Hao, Y.J. Overexpression of MdSOS2L1, a CIPK protein kinase, increases the antioxidant metabolites to enhance salt tolerance in apple and tomato. *Physiol. Plant.* **2016**, *156*, 201–214. [CrossRef] [PubMed]
56. Khan, T.A.; Yusuf, M.; Ahmad, A.; Bashir, Z.; Saeed, T.; Fariduddin, Q.; Hayat, S.; Mock, H.P.; Wu, T. Proteomic and physiological assessment of stress sensitive and tolerant variety of tomato treated with brassinosteroids and hydrogen peroxide under low-temperature stress. *Food Chem.* **2019**, *289*, 500–511. [CrossRef]
57. Ahanger, M.A.; Mir, R.A.; Alyemeni, M.N.; Ahmad, P. Combined effects of brassinosteroid and kinetin mitigates salinity stress in tomato through the modulation of antioxidant and osmolyte metabolism. *Plant Physiol. Biochem.* **2020**. [CrossRef]
58. Anwar, A.; Liu, Y.; Dong, R.; Bai, L.; Yu, X.; Li, Y. The physiological and molecular mechanism of brassinosteroid in response to stress: A review. *Biol. Res.* **2018**, *51*, 46. [CrossRef]
59. Anwar, A.; Di, Q.; Yan, Y.; He, C.; Li, Y.; Yu, X. Exogenous 24-epibrassinolide alleviates the detrimental effects of suboptimal root zone temperature in cucumber seedlings. *Arch. Agron. Soil Sci.* **2019**, *65*, 1927–1940. [CrossRef]
60. Mohanabharathi, M.; Sritharan, N.; Senthil, A.; Ravikesavan, R. Physiological studies for yield enhancement in finger millet under drought condition. *J Pharm. Phytochem.* **2019**, *8*, 3308–3312.
61. Hasan, S.A.; Hayat, S.; Ahmad, A. Brassinosteroids protect photosynthetic machinery against the cadmium induced oxidative stress in two tomato cultivars. *Chemosphere* **2011**, *84*, 1446–1451. [CrossRef]
62. Gupta, P.; Srivastava, S.; Seth, C.S. 24-Epibrassinolide and sodium nitroprusside alleviate the salinity stress in *Brassica juncea* L. cv. Varuna through cross talk among proline, nitrogen metabolism and abscisic acid. *Plant Soil* **2017**, *411*, 483–498. [CrossRef]
63. Zhiponova, M.K.; Vanhoutte, I.; Boudolf, V.; Betti, C.; Dhondt, S.; Coppens, F.; Mylle, E.; Maes, S.; González-García, M.P.; Caño-Delgado, A.I.; et al. Brassinosteroid production and signaling differentially control cell division and expansion in the leaf. *N. Phytol.* **2013**, *197*, 490–502. [CrossRef] [PubMed]
64. Karlidag, H.; Yildirim, E.; Turan, M. Role of 24-epibrassinolide in mitigating the adverse effects of salt stress on stomatal conductance, membrane permeability, and leaf water content, ionic composition in salt stressed strawberry (Fragaria x ananassa). *Sci. Hortic.* **2011**, *130*, 133–140. [CrossRef]
65. Miao, S.Q.; Han, Y.F.; Xiao-Zenga, M.A.N. Nodule formation and development in soybeans (*Glycine max* L.) in response to phosphorus supply in solution culture. *Pedosphere* **2007**, *17*, 36–43. [CrossRef]
66. Rahman, A.; Nahar, K.; Hasanuzzaman, M.; Fujita, M. Calcium supplementation improves Na^+/K^+ ratio, antioxidant defense and glyoxalase systems in salt-stressed rice seedlings. *Front. Plant Sci.* **2016**, *7*, 609. [CrossRef]
67. Othman, Y.; Al-Karaki, G.; Al-Tawaha, A.R.; Al-Horani, A. Variation in germination and ion uptake in barley genotypes under salinity conditions. *World J. Agric. Sci.* **2006**, *2*, 11–15.
68. Ghonaim, M.M.; Mohamed, H.I.; Omran, A.A.A. Evaluation of wheat salt stress tolerance using physiological parameters and retrotransposon-based markers. *Genet. Resour. Crop Evol.* **2020**. [CrossRef]
69. Farouk, S.; Al-Amri, S.M. Exogenous melatonin-mediated modulation of arsenic tolerance with improved accretion of secondary metabolite production, activating antioxidant capacity and improved chloroplast ultrastructure in rosemary herb. *Ecotoxicol. Environ. Saf.* **2019**, *180*, 333–347. [CrossRef]
70. Farouk, S.; Al-Amri, S.M. Exogenous zinc forms counteract NaCl-induced damage by regulating the antioxidant system, osmotic adjustment substances and ions in canola (*Brassica napus* L. cv Pactol) plants. *J. Soil Sci. Plant Nutr.* **2019**, *19*, 887–899. [CrossRef]
71. Ren, J.; Ye, J.; Yin, L.; Li, G.; Deng, X.; Wang, S. Exogenous melatonin improves salt tolerance by mitigating osmotic, ion, and oxidative stresses in maize seedlings. *Agronomy* **2020**, *10*, 663. [CrossRef]
72. Miranda, R.S.; Mesquita, R.O.; Costa, J.H.; Alvarez-Pizarro, J.C.; Prisco, J.T.; Gomes-Filho, E. Integrative control between proton pumps and SOS1 antiporters in roots is crucial for maintaining low Na^+ accumulation and salt tolerance in ammonium-supplied *Sorghum bicolor*. *Plant Cell Physiol.* **2017**, *58*, 522–536. [CrossRef]
73. Hassanpouraghdam, M.B.; Mehrabani, L.V.; Tzortzakis, N. Foliar application of nano-zinc and iron affects physiological attributes of *Rosmarinus officinalis* and quietens NaCl salinity depression. *J. Soil Sci. Plant Nutr.* **2019**, *20*, 335–345. [CrossRef]
74. Talaat, N.B.; Shawky, B.T. 24-Epibrassinolide alleviates salt induced inhibition of productivity by increasing nutrients and compatible solutes accumulation and enhancing antioxidant system in wheat (*Triticum aestivum* L.). *Acta Physiol. Plant* **2013**, *35*, 729–740. [CrossRef]
75. Wu, C.J.; Wang, J.; Zhu, J.; Ren, J.; Yang, Y.X.; Luo, T.; Xu, L.X.; Zhou, Q.H.; Xiao, X.F.; Zhou, Y.X.; et al. Molecular characterization of Mg-chelatase CHLI subunit in pea (*Pisum sativum* L.). *Front. Plant Sci.* **2022**, *13*, 821683. [CrossRef] [PubMed]
76. Li, J.; Zhou, H.; Zhang, Y.; Li, Z.; Yang, Y.; Guo, Y. The GSK3-like Kinase BIN2 Is a molecular switch between the salt stress response and growth recovery in *Arabidopsis thaliana* developmental. *Cell* **2020**, *9*, 367–380.e6. [CrossRef] [PubMed]
77. Zhao, S.; Zhang, Q.; Liu, M.; Zhou, H.; Ma, C.; Wang, P. Regulation of plant responses to salt stress. *Int. J. Mol. Sci.* **2021**, *22*, 4609. [CrossRef]
78. Tahjib-Ul-Arif, M.; Sohag, A.A.M.; Afrin, S.; Bashar, K.K.; Afrin, T.; Mahamud, A.G.M.S.U.; Polash, M.A.S.; Hossain, M.T.; Sohel, M.A.T.; Brestic, M.; et al. Differential response of sugar beet to long-term mild to severe salinity in a soil–plot Culture. *Agriculture* **2019**, *9*, 223. [CrossRef]
79. Abdelaal, K.A.; EL-Maghraby, L.M.; Elansary, H.; Hafez, Y.M.; Ibrahim, E.I.; El-Banna, M.; El-Esaw, M.I.; Elkelish, A. Treatment of sweet pepper with stress tolerance-inducing compounds alleviates salinity stress oxidative damage by mediating the physio-biochemical activities and antioxidant systems. *Agronomy* **2020**, *10*, 26. [CrossRef]

80. Van Hoorn, J.W.; Katerji, N.; Hamdy, A.; Mastrorilli, M. Effect of salinity on yield and nitrogen uptake of four grain legumes and on biological nitrogen contribution from the Soil. *Agric. Water Manag.* **2001**, *51*, 87–98. [CrossRef]
81. Siddiqui, M.H.; Alamri, S.; Alsubaie, Q.D.; Ali, H.M.; Khan, M.N.; Al-Ghamdi, A.; Ibrahim, A.A.; Alsadon, A. Exogenous nitric oxide alleviates sulfur deficiency-induced oxidative damage in tomato seedlings. *Nitric Oxide Biol. Chem.* **2020**, *94*, 95–107. [CrossRef]
82. Szafranska, K.; Reiter, R.J.; Posmyk, M.M. Melatonin improves the photosynthetic apparatus in pea leaves stressed by paraquat via chlorophyll breakdown regulation and it's accelerated de novo synthesis. *Front. Plant Sci.* **2017**, *8*, 179. [CrossRef]
83. Liu, L.; Nakamura, Y.; Taliman, N.A.; Sabagh, A.E.; Moghaieb, R.E.; Saneoka, H. Differences in the growth and physiological responses of the leaves of *Peucedanum japonicum* and *Hordeum vulgare* exposed to salinity. *Agriculture* **2020**, *10*, 317. [CrossRef]
84. Farouk, S.; Arafa, S.A.; Nassar, R.M.A. Improving drought tolerance in corn (*Zea mays* L.) by foliar application with salicylic acid. *Int. J. Environ.* **2018**, *7*, 104–123.
85. Wang, M.; Jiang, W.; Yu, H. Effects of exogenous epibrassinolide on photosynthetic characteristics in tomato (*Lycopersicon esculentum* Mill) seedlings under weak light stress. *J. Agric. Food Chem.* **2010**, *58*, 3642–3645. [CrossRef]
86. Bajguz, A. Effect of brassinosteroids on nucleic acid and protein content in cultured cell of *Chlorella vulgaris*. *Plant Physiol. Biochem.* **2000**, *38*, 209–215. [CrossRef]
87. Li, G.; Peng, X.; Wei, L.; Kang, G. Salicylic acid increases the contents of glutathione and ascorbate and temporally regulates the related gene expression in salt-stressed wheat seedlings. *Gene* **2013**, *529*, 321–325. [CrossRef] [PubMed]
88. Deng, Z.P.; Zhang, X.; Tang, W.Q.; Oses-Prieto, J.A.; Suzuki, N.; Gendron, J.M.; Chen, H.; Guan, S.; Chalkley, R.J.; Peterman, T.K.; et al. A proteomics study of brassinosteroid response in *Arabidopsis*. *Mol. Cell Proteom.* **2007**, *6*, 2058–2071. [CrossRef]
89. Yildirim, E.; Turan, E.; Guvenc, M. Effect of foliar salicylic acid application on growth, chlorophyll and mineral content cucumber grown under salt stress. *J. Plant Nutr.* **2008**, *31*, 593–612. [CrossRef]
90. Hayat, S.; Hasan, S.A.; Hayat, Q.; Ahmad, A. Brassinosteroids protect *Lycopersicon esculentum* from cadmium toxicity applied as shotgun approach. *Protoplasma* **2010**, *239*, 3–14. [CrossRef]
91. Ahanger, M.A.; Agarwal, R.M. Potassium up-regulates antioxidant metabolism and alleviates growth inhibition under water and osmotic stress in wheat (*Triticum aestivum* L). *Protoplasma* **2017**, *254*, 1471–1486. [CrossRef]
92. Ahanger, M.A.; Qin, C.; Maodong, Q.; Dong, X.X.; Ahmad, P.; Abd-Allah, E.F.; Zhang, L. Spermine application alleviates salinity induced growth and photosynthetic inhibition in *Solanum lycopersicum* by modulating osmolyte and secondary metabolite accumulation and differentially regulating antioxidant metabolism. *Plant Physiol. Biochem.* **2019**, *144*, 1–13. [CrossRef]
93. Morillon, R.; Catterou, M.; Sangwan, R.S.; Sangwan, B.S.; Lassalles, J.P. Brassinolide may control aquaporin activities in *Arabidopsis thaliana*. *Planta* **2001**, *212*, 199–204. [CrossRef] [PubMed]
94. Dong, Y.J.; Wang, Z.L.; Zhang, J.W.; Liu, S.; He, Z.L.; He, M.R. Interaction effects of nitric oxide and salicylic acid in alleviating salt stress of *Gossypium hirsutum* L. *J. Soil Sci. Plant Nutr.* **2015**, *15*, 561–573.
95. Hu, Y.; Xia, S.; Su, Y.; Wang, H.; Luo, W.; Su, S.; Xiao, L. Brassinolide increases potato root growth in vitro in a dose-dependent way and alleviates salinity stress. *Biomed Res. Int.* **2016**, 8231873. [CrossRef]
96. Mansour, M.M.F. Plasma membrane permeability as an indicator of salt tolerance in plants. *Biol. Plant* **2013**, *57*, 1–10. [CrossRef]
97. Hare, P.; Cress, W. Metabolic implications of stress induced proline accumulation in plants. *Plant Growth Regul.* **1997**, *21*, 79–102. [CrossRef]
98. Reddy, P.S.; Jogeswar, G.; Rasineni, G.K.; Maheswari, M.; Reddy, A.R.; Varshney, R.K.; Kishor, P.B.K. Proline over-accumulation alleviates salt stress and protects photosynthetic and antioxidant enzyme activities in transgenic sorghum [*Sorghum bicolor* (L.) Moench]. *Plant Physiol. Biochem.* **2015**, *94*, 104–113. [CrossRef]
99. Farouk, S.; Omar, M.M. Sweet basil growth, physiological and ultrastructural modification and oxidative defense system under water deficit and silicon forms treatment. *J. Plant Growth Regul.* **2020**, *39*, 1307–1331. [CrossRef]
100. Miller, G.; Honig, A.; Stein, H.; Suzuki, N.; Mittler, R.; Zilberstein, A. Unraveling Δ1-pyrroline-5-carboxylate-proline cycle in plants by uncoupled expression of proline oxidation enzymes. *J. Biol. Chem.* **2009**, *284*, 26482–26492. [CrossRef]
101. Zhang, Y.P.; Zhu, X.H.; Ding, H.D.; Yang, S.J.; Chen, Y.Y. Foliar application of 24-epibrassinolide alleviates high-temperature-induced inhibition of photosynthesis in seedlings of two melon cultivars. *Photosynthetica* **2013**, *51*, 341–349. [CrossRef]
102. Huang, Z.; Zhao, L.; Chen, D.; Liang, M.; Liu, Z.; Shao, H.; Long, X. Salt stress encourages proline accumulation by regulating proline biosynthesis and degradation in Jerusalem artichoke plantlets. *PLoS ONE* **2013**, *8*, e62085. [CrossRef]
103. Meena, M.; Divyanshu, K.; Kumar, S.; Swapnil, P.; Zehra, A.; Shukla, V.; Yadav, M.; Upadhyay, R.S. Regulation of L-proline biosynthesis, signal transduction, transport, accumulation and its vital role in plants during variable environmental conditions. *Heliyon* **2019**, *5*, e02952. [CrossRef] [PubMed]
104. Kumar, V.; Shriram, V.; Kavi Kishor, P.B.; Jawali, N.; Shitole, M.G. Enhanced proline accumulation and salt stress tolerance of transgenic indica rice by over-expressing P5CSF129A gene. *Plant Biotechnol. Rep.* **2010**, *4*, 37–48. [CrossRef]
105. Caverzan, A.; Casassola, A.; Brammer, S.P. Reactive oxygen species and antioxidant enzymes involved in plant tolerance to stress. In *Abiotic and Biotic Stress in Plants-recent Advances and Future Perspectives*; Shanker, A.K., Shanker, C., Eds.; InTech: London, UK, 2016; pp. 463–480.
106. Arora, N.; Bhardwaj, R.; Sharma, P.; Arora, H.K. Effects of 28-homobrassinolide on growth, lipid peroxidation and antioxidative enzyme activities in seedlings of *Zea mays* L. under salinity stress. *Acta Physiol. Plant* **2008**, *30*, 833–839. [CrossRef]

107. Michalak, A. Phenolic compounds and their antioxidant activity in plants growing under heavy metal stress. *Pol. J. Environ. Stud.* **2006**, *15*, 523–530.
108. Rahman, M.M.; Das, R.; Hoque, M.M.; Zzaman, W. Effect of freeze drying on antioxidant activity and phenolic contents of Mango (*Mangifera indica*). *Int. Food Res. J.* **2015**, *22*, 613–617.
109. El-Banna, M.F.; Abdelaal, K.A.A. Response of strawberry plants grown in the hydroponic system to pretreatment with H_2O_2 before exposure to salinity stress. *J. Plant Prod. Mansoura Univ.* **2018**, *9*, 989–1001. [CrossRef]
110. Nassar, M.A.; Azoz, D.N.; Wessam, S.; Serag El-Din, M. Improved growth and productivity of basil plants grown under salinity stress by foliar application with ascorbic acid. *Middle East J. Agric* **2019**, *8*, 211–225.
111. Guerfel, M.; Baccouri, O.; Boujnah, D.; Chaibi, W.; Zarrouk, M. Impacts of water stress on gas exchange, water relations, chlorophyll content and leaf structure in the two main Tunisian olive (*Olea europaea* L.) cultivars. *Sci. Hortic.* **2009**, *15*, 15–20. [CrossRef]
112. Nieman, R.H. Expansion of bean leaves and its suppression by salinity. *Plant Physiol.* **1965**, *40*, 156–161. [CrossRef]
113. Rashid, P.; Karmoker, J.L.; Chakrabortty, S.; Sarker, B.C. The effect of salinity on ion accumulation and anatomical attributes in mungbean (*Phaseolus radiate* L. cv BARI-3) seedlings. *Int. J. Agric. Biol.* **2004**, *6*, 495–498.
114. Kulaeva, O.N.; Burkhanova, E.; Fedina, A.; Khokhlova, V.; Bokebayeva, G.; Vorbrodt, H.; Adam, G. Effect of brassinosteroids on protein synthesis and plant-cell ultrastructure under stress conditions. In *Brassinosteroids. Chemistry, Bioactivity and Applications*; Cutler, H.G., Yokota, T., Adam, G., Eds.; American Chemical Society: Washington, DC, USA, 1991; pp. 141–155.
115. Ibrahim, S.A.; Abo-ELwafa, T.S.A. Alleviation harmful impacts of salinity using some antioxidants substances on Thompson seedless grapevines seedlings. *J. Plant Prod. Mansoura Univ.* **2018**, *9*, 39–49. [CrossRef]
116. Sarker, U.; Oba, S. The response of salinity stress-Induced *A. tricolor* to growth, anatomy, physiology, non-enzymatic and enzymatic antioxidants. *Front. Plant Sci.* **2020**, *11*, 559876. [CrossRef]
117. Austin, J.R.; Frost, E.; Vidi, P.A.; Kessler, F.; Staehelin, L. Plastoglobules are lipoprotein subcompartments of the chloroplast that are permanently coupled to thylakoid membranes and contain biosynthetic enzymes. *Plant Cell* **2006**, *18*, 1693–1703. [CrossRef] [PubMed]
118. Arafa, A.A.; Khafagy, M.A.; El-Banna, M.F. The effect of glycinebetaine or ascorbic acid on grain germination and leaf structure of sorghum plants grown under salinity stress. *Aust. J. Crop Sci.* **2009**, *3*, 294–304.
119. Rahman, M.S.; Matsumuro, T.; Miyake, H.; Takeoka, Y. Salinity-induced ultrastructural alternations in leaf cells of rice (*Oryza sativa* L.). *Plant Prod. Sci.* **2000**, *3*, 422–429. [CrossRef]
120. Zhang, J.L.; Flowers, T.J.; Wang, S.M. Mechanisms of sodium uptake by roots of higher plants. *Plant Soil* **2010**, *326*, 45–60. [CrossRef]
121. Pareek, A.; Singla, S.L.; Grover, A. Short-term salinity and high temperature stress-associated ultrastructural alterations in young leaf cells of *Oryza sativa* L. *Ann. Bot.* **1997**, *80*, 629–639. [CrossRef]
122. Rizhsky, L.; Liang, H.J.; Shuman, J.; Shulaev, V.; Davletova, S.; Mittler, R. When defense pathways collide. The response of *Arabidopsis* to a combination of drought and heat stress. *Plant Physiol.* **2004**, *134*, 1683–1696. [CrossRef]
123. Rahman, M.S.; Miyake, H.; Takeoka, Y. Effects of exogenous glycinebetaine on growth and ultrastructure of salt-stressed rice seedlings (*Oryza sativa* L.). *Plant Prod. Sci.* **2002**, *5*, 33–44. [CrossRef]
124. Arafa, A.A.; Khafagy, M.A.; El-Banna, M.F. Role of glycinebetaine and ascorbic acid in the alleviation of salt-stress induced micro-morphological damages in sweet pepper seedlings. *J. Biol. Sci.* **2007**, *7*, 879–887. [CrossRef]

horticulturae

Article

Brassinosteroid Applications Enhance the Tolerance to Abiotic Stresses, Production and Quality of Strawberry Fruits

Ramiro N. Furio [1], Sergio M. Salazar [1,2,*], Jorge A. Mariotti-Martínez [1], Gustavo M. Martínez-Zamora [3], Yamilet Coll [4] and Juan C. Díaz-Ricci [3]

1 Instituto Nacional de Tecnología Agropecuaria, EEA Famaillá, Tucumán CP4132, Argentina; ramirofurio@gmail.com (R.N.F.); mariottimartinez.ja@inta.gob.ar (J.A.M.-M.)
2 Facultad de Agronomía y Zootecnia, Universidad Nacional de Tucumán, San Miguel de Tucumán CP4000ACS, Argentina
3 Instituto de Química Biológica, Facultad de Bioquímica, Química y Farmacia, Universidad Nacional de Tucumán, and Instituto Superior de Investigaciones Biológicas ((INSIBIO, CONICET-UNT)), San Miguel de Tucumán CPT4000ILI, Argentina; martin.martinezzamora@fbqf.unt.edu.ar (G.M.M.-Z.); juan@fbqf.unt.edu.ar (J.C.D.-R.)
4 Centro de Estudios de Productos Naturales, Facultad de Química, Universidad de La Habana, Vedado CP10400, Cuba; yaqsaida@gmail.com
* Correspondence: salazar.sergio@inta.gob.ar

Abstract: Brassinosteriods (BRs) have increasingly been used to improve the yields and quality of various crops. In this work we studied the effect of two brassinosteroids, BB16 and EP24, on the growth promotion of *Fragaria ananassa* plants under normal conditions or exposed to water or saline stress. The influence of both BRs on the plant development and fruit quality was evaluated when cultivated in semi-hydroponic conditions. A marked growth-promoting effect was observed with both compounds when plants were cultivated under normal irrigation conditions and under saline and water stresses. BB16 and EP24 yielded plants with a higher dry weight, root length and surface, a higher number and area of leaves, a higher total weight of fruits per plant, and a higher percentage of fruits of commercial quality. Additionally, a higher content of chlorophyll, number of leaves, and increased dry weight was detected in plants treated with both BRs and exposed to water and saline stresses. Finally, when evaluating the production and quality of fruits obtained under semi-hydroponic conditions, we observed that the pre-harvest treatment with both compounds induced a higher fruit production and better quality of fruits. These results suggest the potential of these compounds to achieve a more sustainable management of strawberry cultivation.

Keywords: brassinosteroids; strawberry; growth; stress; quality

Citation: Furio, R.N.; Salazar, S.M.; Mariotti-Martínez, J.A.; Martínez-Zamora, G.M.; Coll, Y.; Díaz-Ricci, J.C. Brassinosteroid Applications Enhance the Tolerance to Abiotic Stresses, Production and Quality of Strawberry Fruits. *Horticulturae* **2022**, *8*, 572. https://doi.org/10.3390/horticulturae8070572

Academic Editors: Małgorzata Majewska, Barbara Hawrylak-Nowak and Agnieszka Hanaka

Received: 1 May 2022
Accepted: 2 June 2022
Published: 24 June 2022

1. Introduction

The application of plant growth stimulators for increasing the quality of crops and yields is of great importance in agriculture due to its social and economic implications. In the early 1970s, the role of growth promoters to accelerate the germination of pollen grains was investigated and characterized. Mitchell et al. [1] reported that some extracts of the pollen of *Brassica napus* L. caused a marked elongation effect on the bean stem. The term brassinosteroids (BRs) was assigned by Mandava in 1988 [2] and since then they have been considered as a special group of endogenous steroid plant hormones essential for plant growth. In addition to stem elongation, BRs affect root and flower development, cell division, photomorphogenesis, tissue vascular matrix, proton pumps, membrane polarization, and stress modulation [3,4].

Several BRs were evaluated under field conditions showing that they can induce a significant yield increase of various crops such as: *Solanum lycopersicum* [5], *Solanum tuberosum* [6], *Opuntiaficus-indica* (L.) Mill [7], and pomegranate [8]. Wu et al. [9] showed that the level of BRs is a limiting factor for plant growth rate and the increase of the level of BRs was effective in promoting plant growth and crop yield in rice.

The growth and development of plants depend on their metabolic and physiological capacity to adapt themselves to the changes of the environmental conditions they have to face to grow; among them, drought and salinity are the most frequent adverse situations that plants must overcome. It is well known that water and salt stresses cause serious damage in plants affecting osmotic processes, the absorption of nutrients, the inhibition of the photosynthetic activity [10,11], plant growth rate, and crop productivity [12]. Strawberry plants are very sensitive to water and saline stresses, due to their large leaf surface, a shallow root system, and their production of fruits with a high water content [13,14].

The protective effect of BRs in different crops against abiotic stresses began to be widely studied in recent times [15]. It was possible to demonstrate their protective effect against saline stress in *Lactuca sativa* [16] and *Oryza sativa* L. [17], as well as the induction of tolerance against water stress in *Lycopersicon esculentum* [18]. From this background and previous studies in our work group, where we were able to demonstrate a marked protective effect against two of the main fungal pathogens of strawberry cultivation, *Colletotrichum acutatum* [19] and *Botryitis cinerea* [20], we decided to study the effect of these stresses on strawberry plants and to investigate whether the BRs can help plants to cope with these stresses. Hence, in this work we studied the effect of two BR isomers (e.g., EP24 and BB16) on the vegetative growth and fruit production of strawberry plants (*Fragaria x ananassa*) of the cultivar 'Festival'. We used a natural brassinosteroid, 24-epibrasinolide (EP24) and a formulation based on the synthetic brassinosteroid spirostanic analogue DI-31 (an active ingredient of the commercial formulation BIOBRAS 16).

After studying the effect of EP24 and BB16 on the growth promotion and fruit quality of strawberry plants grown under greenhouse conditions and exposed to water and salinity stresses, we proposed to extend the study by analyzing the effect of BRs on plants and fruits of strawberry plants grown under semi-hydroponic conditions. The hydroponic production of strawberries has increasingly been used in recent years, because it provides an interesting, valuable alternative orientated to a more sustainable crop production [21]. Hydroponic cultures have many advantages over traditional cultivation, since they use less water, allow for the production to be close to the markets, reduce the carbon footprint, minimize the use of agrochemicals, optimize the use of fertilizers by controlling the electrical conductivity of irrigation water, allow for the automation and robotization of the harvest processes, and increase crop yields [22].

Finally, in this work we also present results of the effect of EP24 and BB16 on the yield and fruit quality of strawberry plants (cv. Festival) exposed to high values of electrical conductivity during fertigation when grown under semi-hydroponic conditions

2. Materials and Methods

2.1. Plant Material

Strawberry plants (*Fragaria ananassa*) cv. 'Festival' were provided by the BGA Active Strawberry Germplasm Bank of the National University of Tucumán). Healthy seedlings were obtained from in vitro cultures in MS medium [23], rooted in pots with sterilized substrate (humus and perlite, 2:1), and maintained at 28 °C, 70% relative humidity (RH), with a light cycle of 16 h (white fluorescent, 350 µmol photons m^{-2} s^{-1}).

For the determinations of growth promotion and tolerance to water and saline stress, plants were kept under greenhouse conditions, while for determinations of fruit quality a semi-hydroponic system was used.

The semi-hydroponic system under greenhouse conditions consisted of hydroponic growing bags 0.9 m long and 25 L, in which 10 plants were placed per bag (11.11 plants/linear meter). These bags are made up of Growmix TerraFertil® inert substrate, whose components are Sphagnum moss peat, pine bark compost, and perlite. The fertilization that was carried out consisted of a nutritive solution composed of 80 cc/m^3 of phosphoric acid (85%), 130 g/m^3 of potassium nitrate, 140 cc/m^3 of nitric acid, and 30 g/m^3 of Fetrilom

Combi 2®. Drip irrigation by means of a micro-perforated tape installed inside the bags, was carried out 5 times a day, through short irrigations of 3 min.

2.2. Treatments with BRs

The aerial part of the plants was sprayed with BB16 or EP24 at a concentration of 0.1 mg L^{-1} up to the dripping point. In the trials in which the plants were exposed to abiotic stresses, the treatment was carried out 3 days before subjecting them to said conditions.

2.3. Growth Promotion

Growth promotion experiments were randomized with 10 plants and 10 replicates for the destructive and non-destructive parameters evaluated. The plants were kept in a greenhouse during the experiment and watered 3 times a week.

The treatments were carried out every 30 days by spraying the plants with BB16 or EP24 at a concentration of 0.1 mg l^{-1}, and after 8 months the number of leaves, leaf area, leaf greenness, length of the root, crown diameter, root surface, fresh and dry weights, number of stolons, and the weights of total and commercial fruits per plant, were determined. Fruits were graded into marketable (>10 g per fruit) and non-marketable (<10 g, with disease symptoms, or deformed) categories. The threshold value for marketable fruit was 10 g since fruits over this weight are sold for either fresh consumption (larger fruit sizes) or for processing (smaller fruit sizes).

Root length (cm) and crown diameters (cm) were measured using a caliper, dry weights (g) were measured by oven drying at 60 °C to constant weight, and root surface area was determined by the method of calcium nitrate [24]. In this method, the roots were gently washed with water, dried for 30 s on absorbent paper, and weighed. Then, they were immersed in saturated $Ca(NO_3)_2$ solution, drained over the same container for 30 s, and weighed, the weight difference being equivalent to the root area.

For leaf greenness determinations, relative chlorophyll content was measured using a Minolta SPAD-502 Chlorophyll Meter. These results are expressed as SPAD values. To determine the leaf area, photographs of all the leaflets were taken and the measurement was made using the ImageJ program [25]. Measurements were made in 10 replicates and experiments were performed three times.

2.4. Abiotic Stress

For salt stress, plants were irrigated with 80 mL of NaCl 100 mM every 3 days. For water stress, plants were irrigated with 32 mL of water every 3 days for 20 days, which corresponds to 40% of the amount of water used in the normally irrigated control plants.

The controls used in these trials were: (i) plants not treated with BRs and subjected to normal irrigation with 80 mL per plant (control); (ii) plants not treated with BRs and subjected to irrigation with 80 mL 100 mM NaCl (saline stress); (iii) plants not treated with BRs and subjected to irrigation with 32 mL of water (water stress).

The determinations were carried out were: the number of leaves, leaf area, leaf greenness, root length, crown diameter, root surface, fresh and dry weights. All these determinations carried out in the same way as mentioned above when evaluating growth promotion

In addition, a soil analysis of the different stress conditions evaluated was carried out, making the following determinations: pH, volumetric humidity [26], salinity by electrical conductivity, sodium by photometry of flame [27], and chlorides by volumetry [28]. The soil analyzes were evaluated on ten random samples taken 20 days after the start of the treatments.

These experiments were randomized with 10 plants and 10 replicates to evaluate destructive and non-destructive parameters, and 3 independent experiments were also performed.

2.5. Fruit Quality Measurements

The freshly harvested fruits of the semi-hydroponic crops were used to evaluate the effect of the pre-harvest treatment with BB16 or EP24 at a concentration of 0.1 mg L^{-1} or with water (control plants). At harvest time, the total soluble solids (TSS) of the strawberry juice were determined using a refractometer (Arcano REF103) and recording three readings per fruit. In addition, acidity was determined using an aliquot of 10 g of strawberry juice in 100 mL of deionized water and titrating with 0.1 N NaOH at pH 8.1 [29]. Titratable acidity is expressed in grams of citric acid per liter. The surface color was evaluated with a colorimeter (Minolta, Model CR-300, Osaka, Japan) by measuring the parameters L*, a* and b*. Negative L* indicates darkness and positive L* indicates lightness. Negative a* indicates green color, and a* high positive indicates red color. A high positive b* indicates a more yellow color and a negative b* indicates a blue color. The chroma value (C*), calculated as $C^* = (a^{*2} + b^{*2})^{1/2}$, indicates the intensity or saturation of the color [30]. The color was measured in three random positions of each fruit. Fruit firmness was evaluated using a penetrometer (Effegi, Italia) of 2 mm diameter. Finally, the weight loss of the fruits was recorded at 5 days post-harvest (dph) using a balance with a precision of 0.01 g and expressed as a percentage of the initial weight. Thirty fruits per treatment were analyzed and the experiments were performed in triplicate.

2.6. Incidence of Postharvest Diseases in Fruits

The influence of BRs on the appearance of latent natural infection by the microbiota present in fruits was evaluated. Freshly harvested fruits were stored in hermetically sealed trays, containing 3 fruits each, and kept at 25 °C and a high RH (95%). Disease progress was assessed at 5 dph. Each treatment consisted of thirty repetitions and the experiment was carried out in triplicate.

2.7. Statistical Analysis

Statistical analysis of the data was performed using the InfoStat software [31]. All data were obtained from at least three independent experiments and are expressed as mean ± standard error. Data were also analyzed by one-way analysis of variance (ANOVA) test and means separated by the Tukey's test for $p < 0.05$.

3. Results

The evaluation of the effect of BB16 and EP24 on the growth promotion of *Fragaria ananassa* cv. 'Festival' plants under greenhouse conditions showed a clear increase of fresh and dry weight, the number and area of leaves, the total weight of fruits per plant, the percentage of fruits with commercial quality, and the SPAD in plants treated with both BRs (Table 1). Likewise, when evaluating the effects of BB16 and EP24 on root parameters a clear increase of root dry weight, surface, length, and crown diameter was found (Table 2).

Figure 1 shows the aspect of the canopy and root of plants treated with BRs. In that figure we can appreciate that plants treated both BRs exhibited a greater development of the aerial part and root when compared with not treated plants, confirming the parameter values presented in Tables 1 and 2. Noteworthily, in plants treated with EP24, more stolonization was observed when compared to not treated plants and plants treated with BB16 (Figure 1).

Table 1. Growth-promoting effect on parameters of the aerial part of strawberry plants treated with BB16 and EP24.

Canopy Parameters	Control	BB16	EP24
Fresh weight (g)	21.9 ± 2.49 a	29.73 ± 2.49 ab	36.95 ± 2.49 b
Dry weight (g)	8.93 ± 0.65 a	12.4 ± 0.65 bc	14.95 ± 0.65 c
Number of leaves	11.75 ± 0.93 a	17.5 ± 0.93 bc	19.75 ± 0.93 c
Leaf area (cm^2)	11.63 ± 1.6 a	28.18 ± 1.6 b	27.77 ± 1.6 b
Number of stolons	4.98 ± 0.56 a	6.50 ± 0.56 a	9.08 ± 0.56 a
Weight of total fruits per plant (g)	49.09 ± 1.56 a	59.27 ± 1.56 b	56.67 ± 1.56 b
Commercial quality fruit (%)	18.15 ± 1.57 a	25.46 ± 1.57 b	28.15 ± 1.57 b
SPAD	32.92 ±- 0.74 a	40.34 ± 0.74 b	40.56 ± 0.74 b

Different letters represent groups which are significantly different (Tukey, α < 0.05).

Table 2. Growth promoting effect of BB16 and EP24 on root parameters of strawberry plants.

Root Parameters	Control	BB16	EP24
Fresh weight (g)	23.5 ± 4.33 a	48.23 ± 4.33 b	43.1 ± 4.33 b
Dry weight (g)	4.6 ± 0.58 a	12.83 ± 0.58 b	11.88 ± 0.58 b
Root lenght (cm)	29.04 ± 1.26 a	34.75 ± 1.26 b	32.25 ± 1.26 ab
Root surface (mg $Ca(NO_3)_2$)	3.13 ± 0.72 a	13.13 ± 0.72 b	14.1 ± 0.72 b
Crown diameter (cm)	9.38 ± 0.55 a	12.98 ± 0.55 bc	11.95 ± 0.55 bc

Different letters represent groups which are significantly different (Tukey, $\alpha < 0.05$).

Figure 1. Appearance of the aerial parts and roots of plants of *Fragaria ananassa* cv. 'Festival', either not treated (control) and treated with BB16 or EP24, grown in solid substrate and greenhouse conditions. Scale bars correspond to 5 cm.

The evaluation of the effects of salinity and water stress on the growth of strawberry plants was carried out in solid substrates that were previously prepared and analyzed. The values of the soil parameters evaluated are presented in Table 3. As shown in the Table, the substrate used to test the saline stress in plants presented a lower moisture content and a significant increase in electrical conductivity, as well as the sodium and chloride concentrations when compared to the normal (control) substrate. On the other hand, the substrate used to test the plants under water stress presented a significant reduction of moisture content when compared to the normal (control) substrate. These values let us validate that the plants would be considerably exposed to a saline and water stresses.

Table 3. Parameters of the soil used in plants subjected to salt stress (100 mM NaCl) and water stress.

Soil Analysis			
	Control	Saline Stress	Water stress
pH (1:2,5)	5.78 ± 0.18 b	5.73 ± 0.18 a	5.72 ± 0.18 ab
CE (dS/m)	2.66 ± 0.17 a	35.66 ± 2.35 b	2.73 ± 0.17 a
H_2O (%)	240.67 ± 4.9 c	101.33 ± 4.9 b	68.33 ± 4.9 a
Na (meq/l)	17.23 ± 1.05 a	339.13 ± 10.92 b	16.26 ± 1.05 a
Cl (meq/l)	17.8 ± 0.48 b	333.33 ± 15.04 c	15.77 ± 0.48 a

Different letters represent groups which are significantly different (Tukey, $\alpha < 0.05$).

After treating the plants with BB16 or EP24 and subjecting them to both types of stress, the physiological state of the plants was evaluated in comparison with control plants not treated with BRs and subjected to stress (Figure 2). A clear protective effect of both BRs was observed, since the plants did not show adverse effects, exhibiting a physiological state similar to that of the control plants not subjected to stress conditions.

Figure 2. Aspect of *Fragaria ananassa* cv. 'Festival'. Plants were either controls (**a**), or treated with BB16 (**b**) or EP24 (**c**), and subjected to saline or water stress after 20 days.

When analyzing the effect of saline stress on the chlorophyll content of plants a lower rate of SPAD decrease was observed in plants treated with BB16 or EP24, being more notorious at 20 dpt (Figure 3a). However, this effect was only observed in plants treated with BB16 in response to water stress (Figure 3b). Plants treated with BB16 and EP24 and exposed to water stress exhibited a greater number of leaves than the control plants, whereas plants exposed to saline stress showed this effect only when pretreated with BB16 (Figure 3c). When evaluating leaf area, a larger area was observed in the plants treated with EP24 or BB16 and exposed to water stress, but no significant change was observed in plants treated with the BRs and exposed to saline stress when compared to controls (Figure 3d). Additionally, when evaluating the canopy dry weight of plants exposed to abiotic stresses, plants pretreated with both BRs exhibited a higher dry weight in those exposed to water stress, whereas plants exposed to saline stress showed higher canopy dry weight only in those pretreated with BB16 (Figure 3e).

Other aspects analyzed were the chlorophyll content of the plants throughout the trial, and a marked decrease in SPAD was observed over time in the control plants, which was not observed in the plants treated with BB16 or EP24 (Figure 6a). In addition, when finishing carrying out the determinations in fruits, a marked difference could be seen in terms of the survival of the plants, since the treatment with BRs gave rise to a percentage of survival almost 50% higher with respect to the control plants (Figure 6b).

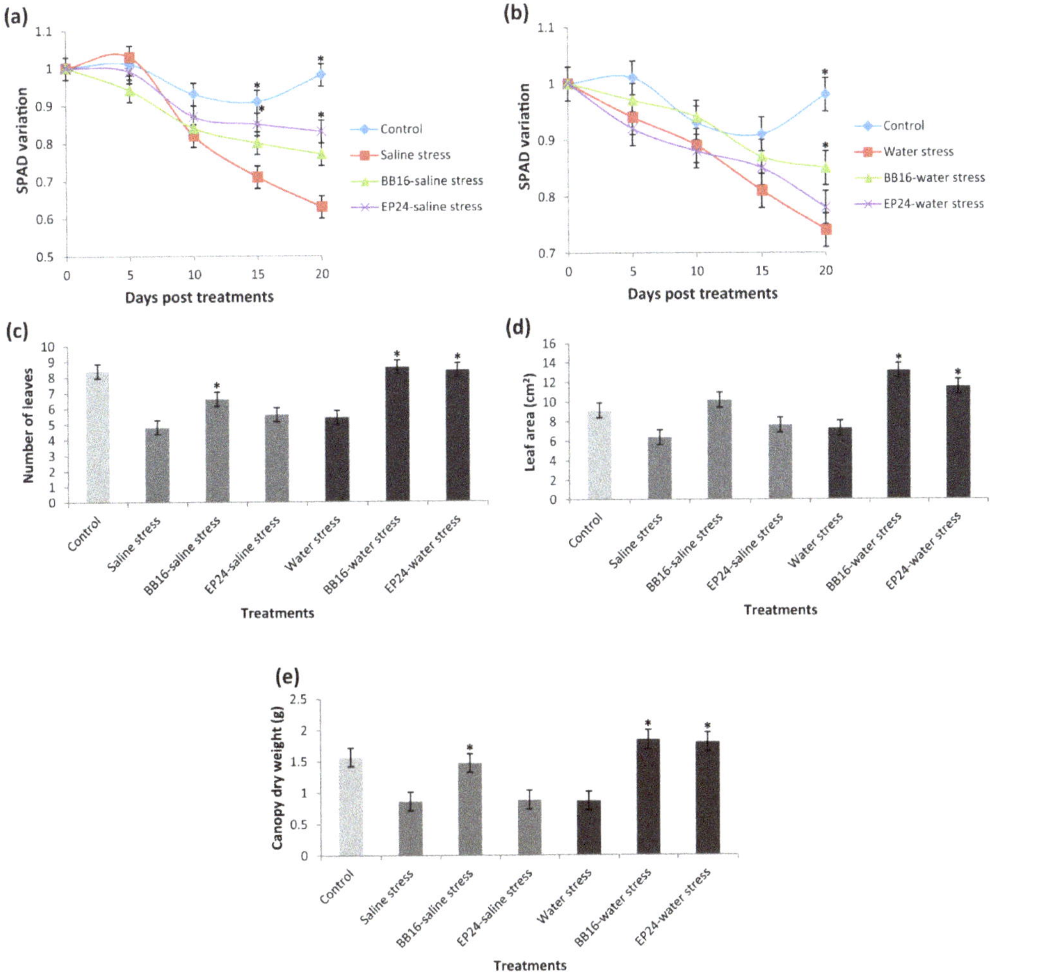

Figure 3. Effect of BB16 and EP24 treatments on the canopy growth parameters of strawberry plants (cv. 'Festival') grown on solid substrate, and exposed to water and saline stress. (**a**,**b**) Greenness index, (**c**) number of leaves, (**d**) leaf area, and (**e**) canopy dry weight. Mean values ± SE were obtained from three independent experiments ($n = 10$). Analysis of variance (ANOVA) followed by the Tukey's test was performed using InfoStat/L software ($p < 0.05$). Asterisks represent statistically significant differences.

When evaluating the effect of BRs on root morphological parameters of plants exposed to saline or water stresses, a clear increment of the root dry weight, surface, length, and crown diameter was observed in the plants pretreated with both BRs and exposed to water stress (Figure 4a–d), whereas an increment of root surface was only observed in plants treated with BB16 and exposed to saline stress (Figure 4b).

When analyzing the effect of BB16 and EP24 on fruit production of plants grown under semi-hydroponic culture, higher fruit weights (per plant and total) were observed in plants treated with both BRs (Figure 5a). However, an increase of the number of fruits with commercial value was observed only in plants treated with BB16 (Figure 5b) as compared to control non-treated plants.

When studying the effect of both compounds on the quality of fruits produced under hydroponic conditions, a higher luminosity (lightness) (Figure 7a) and red coloration (Figure 7b) was observed in plants grown under semi-hydroponic conditions that were treated with both BRs. However, a higher color intensity (chroma) was only observed in plants treated with BB16 (Figure 7c).

The treatment with both BRs also yielded fruits that presented a lower rate of weight loss after the harvest (Figure 8a), a higher firmness (Figure 8b), and an increased content of soluble solids (Figure 8c); however, only EP24 brought about fruits with a lower acidity when compared to control (not treated) plants or those treated with BB16 (Figure 8d).

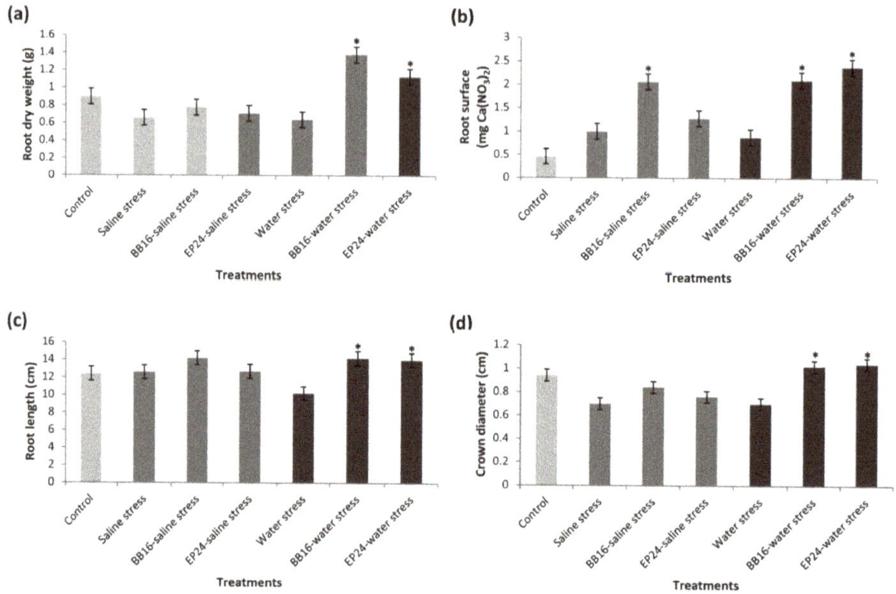

Figure 4. Effect of BB16 and EP24 treatments on the root parameters of strawberry plants (cv. 'Festival') grown in solid substrate and exposed to water and saline stress. (**a**) Root dry weight, (**b**) root surface, (**c**) root length, and (**d**) crown diameter. Mean values ± SE were obtained from three independent experiments (n = 10). Analysis of variance (ANOVA) followed by the Tukey's test was performed using InfoStat/L software ($p < 0.05$). Asterisks represent statistically significant differences.

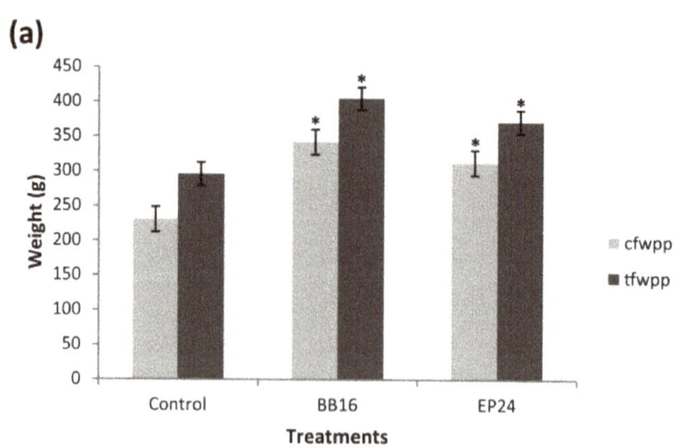

Figure 5. *Cont.*

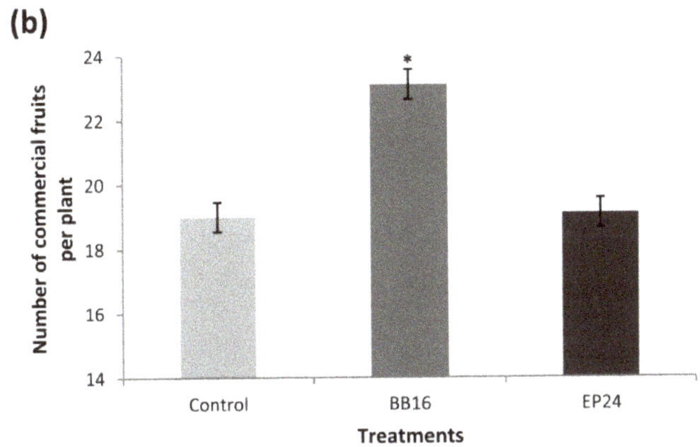

Figure 5. Effect of BRs on the fruit production of strawberry plants (cv. 'Festival') grown in semi-hydroponic conditions. (**a**) Commercial fruit weight per plant (cfwpp) and total fruit weight per plant (tfwpp), and (**b**) number of commercial fruits, obtained in response to treatment with BB16 and EP24. Asterisks correspond to statistically different values (Tukey's test, $p < 0.05$).

The incidence of diseases in fruits produced by the natural microbiota present was also evaluated. Plants treated with BB16 and EP24 exhibited a lower rate of fruit rot, displaying up to 50% fewer fruit rots when compared to untreated control plants (Figure 9).

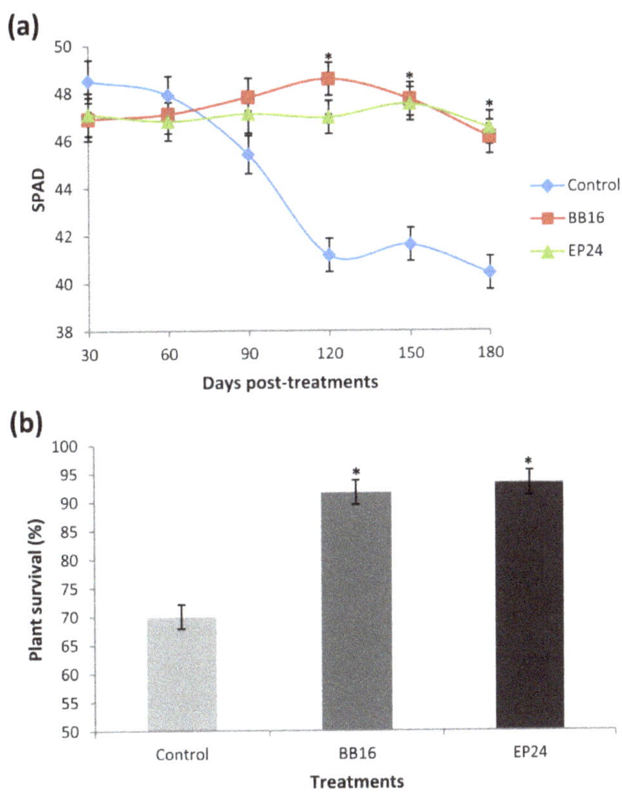

Figure 6. Effect of BRs on: (**a**) the greenness index (SPAD) of strawberry plants (cv. 'Festival'), and (**b**) on the percentage of survival of the plants grown in semi-hydroponic conditions at the end of the study. Asterisks correspond to statistically different values (Tukey's test, $p < 0.05$).

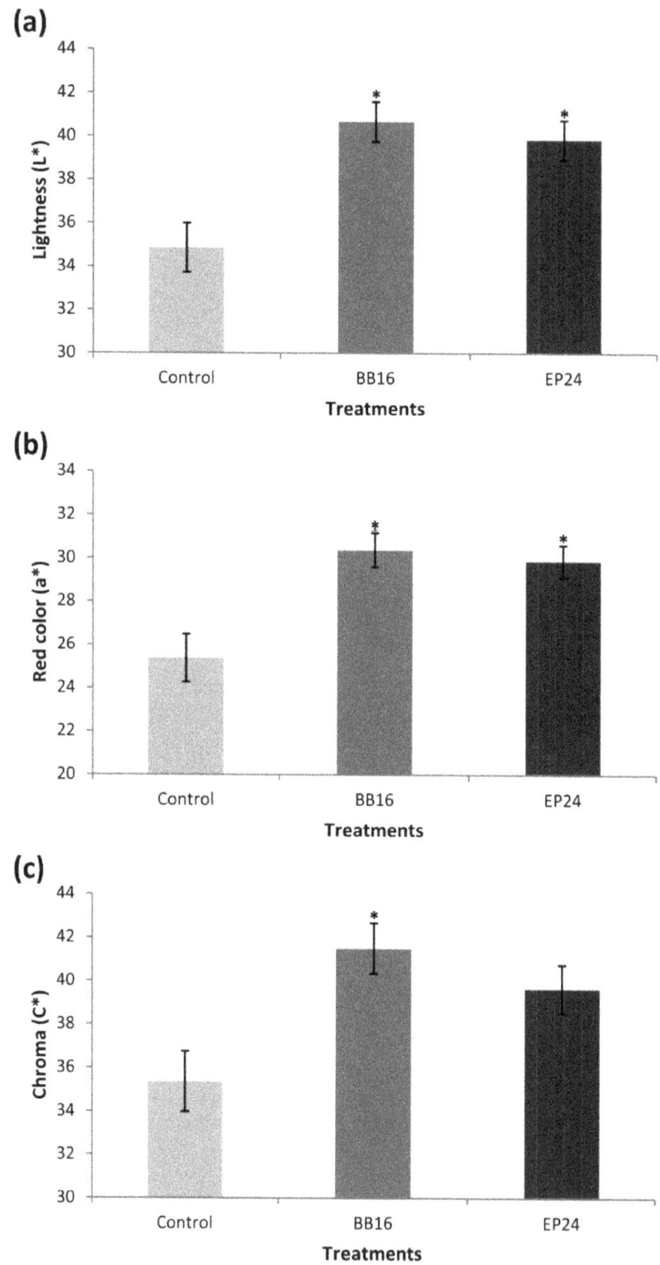

Figure 7. Evaluation of fruit color of strawberry plants (cv. 'Festival') grown in semi-hydroponic conditions after pre-harvest treatment with BRs. Lightness (**a**), red coloration (**b**), and color intensity (**c**) were evaluated. Asterisks indicate a statistically significant difference between control and BR-treated plants. Mean values ± SE were obtained from three independent experiments (n = 10). Analysis of variance (ANOVA) followed by the Tukey's test was performed using InfoStat/L software ($p < 0.05$). Asterisks represent statistically significant differences.

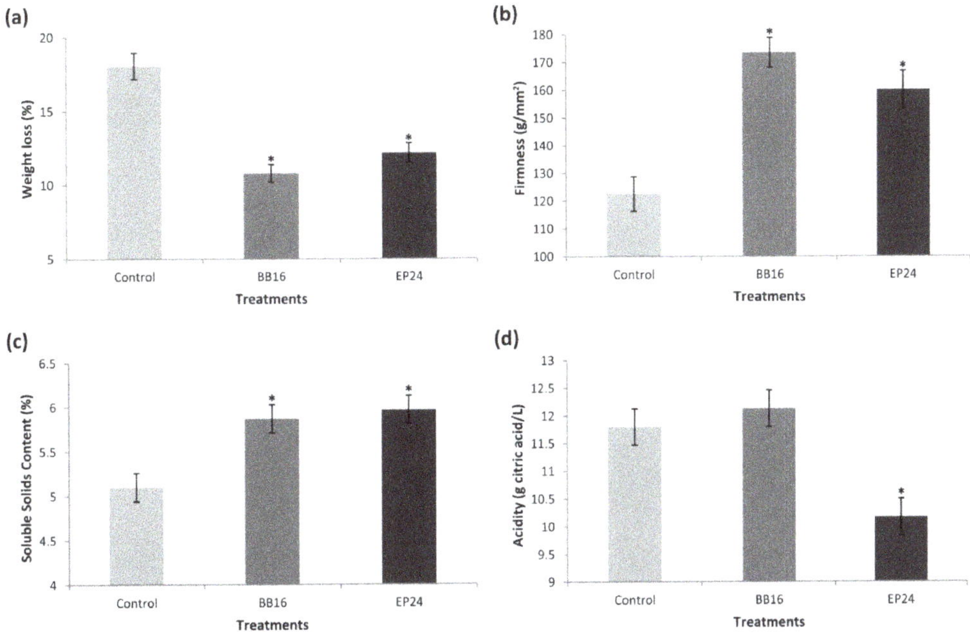

Figure 8. Effect of Bs treatment on the fruit quality parameters of strawberry plants (cv. 'Festival') grown in semi-hydroponic culture. (**a**) Weight loss expressed as a percentage (%) relative to the initial weight of strawberry fruit, (**b**) its firmness, (**c**) soluble solids' content (%), and (**d**) acidity (expressed as g citric acid/L). Fruits were kept at room temperature (25 °C) during the experiment. Mean values were obtained from ten independent samples. Vertical bars represent standard deviation (± SE). Analysis of variance (ANOVA) followed by the Tukey's test was performed using InfoStat/L software ($p < 0.05$). Asterisks indicate a statistically significant difference between control and BRs-treated plants.

Figure 9. Decay of fruits. (**a**) Representative images of the decay of fruits in response to different treatments, and (**b**) fruit rot as a percentage of the total fruits obtained of strawberry plants grown in semi-hydroponic culture. Fruit evaluation was carried out at 5 dph. Asterisks correspond to statistically different values (Tukey's test, $p < 0.05$).

4. Discussion

The potential use of brassinosteroids as plant growth stimulators to improve crop yields and quality has intensively been investigated in recent years. Various studies have shown the effect of these compounds on various crops such as: tomato [5], potato [6], yellow passion fruit [8] and rice [9].

Previously we reported that strawberry plants treated with EP24 and BB16 exhibited a marked increase in fruit yield [32]. Now, we decided to carry out a comprehensive study on the use of BRs as a feasible alternative to increase plant growth under greenhouse conditions, and to evaluate whether the BRs can provide plants tolerance against water or saline stresses.

We were also interested to study the quality of the fruits produced by using a semi-hydroponic cultivation system. In this system, the plants grow on an inert support and all the nutrients are supplied through irrigation. This type of cultivation allows the reduction of agrochemicals needs and to increase the plant density when compared to the traditional system in the field. As we mentioned earlier the semi-hydroponic system not only contributes to lower production costs, but also provides better working conditions to workers, as the substrate bags are supported in structures one meter high, hence they do not have to work at ground level. Additionally, soil disinfection is avoided by eliminating the use of methyl bromide or other contaminating products required to soil treatments.

When evaluating the effect of the BB16 and EP24 on the growth promotion of *Fragaria ananassa* cv. 'Festival' plants grown in a solid substrate, a marked growth-promoting effect was observed. Analysis of leaf and root dry weights, the number of leaves, leaf area, greenness index, root length, root surface, crown diameter, fruit production, and the percentage of commercial-quality fruits showed a clear improvement with respect to control non-treated plants. The use of BRs to enhance the growth of various plants have already been reported by other research groups. Xia et al. [33] reported that the improvement in the growth of cucumber plants (*Cucumis sativus*) after treatment with 24-epibrasinolide (EBR) was associated with a higher CO_2 assimilation and a higher quantum yield of photosystem II (PSII), and that the treatment with brassinazole (Brz), a specific inhibitor for BR biosynthesis, reduced plant growth, CO_2 assimilation, and PSII performance. There is also evidence showing that the cellular redox state controls the expression of photosynthetic genes and enzyme activities [34–36]. In this sense, it has already been shown that the exogenous application of 28-homobrassinolide increases photosynthetic activity and the antioxidant defense system [37].

Previously, it was reported that a treatment with both BRs (e.g., BB16 and EP24) induced a transient accumulation of H_2O_2 in strawberry leaves [18]. We may hypothesize that H_2O_2 functions as a second messenger that would activate a signaling pathway through MAPK to regulate the cellular redox state and photosynthetic activity [38–40].

It has also been demonstrated that BRs have a growth-promoting effect on plant roots. Yokota et al. [41] reported the existence of several BRs in tomato roots, and that an *lk* mutant, a BR-deficient pea mutant, exhibited thicker and low number of lateral roots, showing that an excess or lack of BRs have detrimental effects on the root growth and development. On the one hand, mutants that lack components of the BRs signaling pathway or their receptors exhibited short roots, indicating that the BR signaling is necessary for root growth [42,43]. Shorter roots were also observed in bes1-D mutants (gain of function), or in plants treated with high concentrations of BRs [43,44]. These results confirmed our observations, suggesting that a correct balance of BR levels is required for the normal root growth and development

Taking into account that strawberry plants are very sensitive to water and saline stresses [45–47]. we decided to evaluate whether BB16 and EP24 would provide tolerance to those stresses.

When the roots of plants treated with BB16 or EP24 and exposed to water stress were analyzed, greater dry weights, surface areas, and lengths, as well as greater crown diameters were observed (Figure 4). These results are consistent with previous reports of other authors who observed an increase in root branching, total length, and area in *Silene vulgaris* plants subjected to moderate drought stress [48,49]. We can speculate that the morpho-anatomical change observed in roots minimizes the area of the rhizosphere exposed to water depletion, and by increasing the area, this enhances water and nutrients' absorption. Interestingly, these effects were not observed in the roots of plants exposed to salt stress, in which a larger root surface was only observed in the plants treated with BB16. We may speculate that the spiroketalic ring present in structure of BB16 but absent in the EP24 structure exerts a stronger effect on the signaling of the induced morpho-anatomical structures, which is associated with the tolerance to the abiotic stresses studied, as suggested elsewhere [18].

Plants constantly regulate their physiological processes in response to various internal and external stimuli. Biological processes are integrated by multiple hormonal signals, and stress induces different hormonal signaling pathways in plants [50]. Among the external factors, salt and water stresses are those that can cause significant damage to plants, especially to strawberries [47]. It was reported that water and salt stresses can cause in strawberries a reduction in the number and area of the leaves, the dry weight of the shoot, the number of crowns, and the yield [51]. Considering the sensitivity of plants to saline conditions, the strawberry is among the most sensitive species [52,53].

High concentrations of salt in soil or irrigation water can have a devastating effect on plant metabolism, altering the level of growth regulators and uncoupling major physiological and biochemical processes [54]. Since osmotic stress is the basic cause in both stress situations, it is expected to observe a marked stomatal closure in response to these conditions to prevent water loss as reported by Furio et al. [18]. This situation causes an increase in free radicals, which leads to a degradation of lipids and proteins of the cell membrane, and also of an extremely important cell component, chlorophyll [55]. The latter would explain the loss of the green coloration [56] (observed as a marked reduction of SPAD mainly at 20 dpt) in the control non-treated plants and those exposed to saline and water stresses, while the plants previously treated with BRs exhibited greener and healthier leaves (Figure 6). As the water uptake capacity is reduced under these stressful conditions, the plants usually exhibit lower weights [57]. The lack of water availability also causes a decrease in cell turgor, a decrease in the water potential of the plant affecting the plant growth [58,59]. The positive effect exerted by BRs on plants exposed to salinity could be due to the action of osmolytes such as proline, glycine betaine, and total free amino acids, as proposed in several works [60,61].

The thin epidermis and high water content makes the strawberry fruit very perishable and susceptible to deterioration caused by physical or biological damages. It is essential therefore, to evaluate strategies to obtain fruits with lower rates of water loss and greater firmness after harvest. Accordingly, the effect of BB16 and EP24 fruit quality was evaluated on strawberry plants of the (cv. 'Festival') grown under semi-hydroponic conditions.

It is well known that after harvesting the fruit, this increases its respiratory rate, inducing a high loss of water through the thin skin of the strawberries. This water loss causes the fruit skin to wrinkle, lose shine, and cause a significant deterioration in their appearance and organoleptic quality [62]. For this reason, some parameters associated with fruit quality of strawberry plants (cv. 'Festival') grown under semi-hydroponic conditions treated with BB16 and EP 24 (or not) were evaluated.

Results obtained showed that plants treated with the BRs yielded fruits with higher luminosity (Figure 7a), redness (Figure 7b), firmness (Figure 8b), and soluble solids (Figure 8c), and a lower rate of weight loss (Figure 8a) and decay of fruits (Figure 9). The differential effect observed in fruits treated with BB16 or EP24 on some parameters of quality evaluated (see Figures 7c and 8d) can be attributed to the influence of the different chemical structure of EP24 and BB16 on the activation of the BRs' signaling pathway, as mentioned above and reported by Furio et al. [18].

Fruits with a greater luminosity and with a more intense red color are fruits that will have a greater acceptance by the consumer, as they would present a more attractive visual appearance [63]. Another important organoleptic characteristic which was markedly improved by both BRs is the soluble solids content. This characteristic, together with the decrease in acidity observed in response to EP24, implies a better and more pleasant flavor of the fruits. The improvement of quality parameters observed in strawberry fruits in response to treatment with BRs, agree with studies reported in other species such as: tomatoes [64], cherries [65], and grapes [66], among others.

When studying the influence of BRs in the reduction of fruits' decay due to the latent natural microbiota present in the fruits, a marked effect of both compounds was verified (Figure 9). We observed that after 5 days of harvest, the fruits of the untreated plants presented a clear advance of rots, characterized by the growth of a white mycelium, with more than 80% of the fruits infected. On the other hand, with the treatment with BB16 and EP24, the fruits' decay decreased markedly to 40 and 55%, respectively. Fruit rot, mainly due to gray mold (*Botrytis cinerea*), is the main source of post-harvest losses in strawberries. The results obtained in this work show that pre-harvest treatments with BRs cause a reduction in the damage caused by natural pathogens.

Finally, the results obtained let us conclude that B16 and EP24 have clear benefits on the strawberry cultivation, as they allow not only increases in plant growth under normal conditions, but also under abiotic stress conditions such as salinity or water stress. Additionally, we demonstrate that under semi-hydroponic conditions, the treatment with BB16 and EP24 improved the survival level of the plants and the fruit quality.

These results further show that the use of BRs in strawberry production not only renders higher yields, productivity, and fruit quality, but also a more sustainable and environmentally friendly management of the crop, reducing the use of toxic and contaminant agrochemicals.

Author Contributions: R.N.F. is CONICET fellowship, G.M.M.-Z. and J.C.D.-R. are member of CONICET, S.M.S. and J.A.M.-M. are member of INTA, Y.C. is member of Universidad de la Habana. R.N.F.: Conceptualization, Methodology, Software, Validation, Investigation, Writing—Original draft preparation; S.M.S.: Resources, Writing—Review & Editing; J.A.M.-M.: Methodology, Investigation; G.M.M.-Z.: Investigation; Y.C.: Resources, Methodology; J.C.D.-R.: Resources, Writing—Review & Editing. All authors have read and agreed to the published version of the manuscript.

Funding: This research was funded partially supported with grants of the Universidad Nacional de Tucumán (PIUNT 26/D621), Agencia Nacional de Promoción Científica y Tecnológica (PICT 2017-0653), CONICET (PUE-2016-0104) and INTA (2019-PD-E4-I069-001 and 2019-PE-E1-I009-001) and the APC was funded by PUE 104-2016-Díaz-Ricci.

Institutional Review Board Statement: Not applicable.

Informed Consent Statement: Not applicable.

Data Availability Statement: Data are contained within the article. The datasets generated and/or analysed during the current study are available from the corresponding author on reasonable request.

Acknowledgments: Authors are grateful to Strawberry Active Germplasm Bank (BGA) from Universidad Nacional de Tucumán (UNT) for providing strawberry plants. We thank Luis H. Montenegro and Norberto Ybarra for their help in testing the semi-hydroponic system under greenhouse conditions.

Conflicts of Interest: The manuscript has not been published and is not under consideration for publication elsewhere. All authors have approved the manuscript and agreed to submit the paper to Horticulturae. The research was conducted in the absence of any commercial relationships that could be considered as a potential conflict of interest.

References

1. Mitchell, J.W.; Mandava, N.B.; Worley, J.F.; Plimmer, J.R.; Smith, M.V. Brassins: A new family of plant hormones from rape pollen. *Nature* **1970**, *225*, 1065–1066. [CrossRef] [PubMed]
2. Mandava, N.B. Plant growth-promoting brassinosteroids. *Annu. Rev. Plant Physiol. Plant Mol. Biol.* **1988**, *39*, 23–52. [CrossRef]
3. Clouse, S.D. Brassinosteroid signal transduction: From receptor kinase activation to transcriptional networks regulating plant development. *Plant Cell* **2011**, *23*, 1219–1230. [CrossRef] [PubMed]

4. Clouse, S.D.; Sasse, J.M. Brassinosteroids: Essential regulators of plant growth and development. *Annu. Rev. Plant Biol.* **1998**, *49*, 427–451. [CrossRef] [PubMed]
5. Núñez, M.; Torres, W.; Coll, F. Effectiveness of a synthetic brassinosteroid on potato and tomato yields. *Cultiv. Trop.* **1995**, *16*, 26–27.
6. Torres, W.; Núñez, M. The application of biobras–6 and its effect on potato (*Solanum tuberosum* L.) yields. *Cultiv. Trop.* **1997**, *18*, 8–10.
7. Cortes, P.A.; Terrazas, T.; León, T.; Larqué-Saavedra, A. Brassinosteroid effects on the precocity and yield of cladodes of cactus pear (*Opuntiaficus-indica* (L.) Mill.). *Sci. Hortic.* **2003**, *97*, 65–73. [CrossRef]
8. Gomes, M.M.A.; Compostrini, E.; Rocha, N.; Pio, A.; Massi, T.; Siqueira, L. Brassinosteroid analogue effects on the yield of yellow passion fruit plants (*Passiflora edulis* f. *flavicarpa*). *Sci. Hortic.* **2006**, *110*, 235–240. [CrossRef]
9. Wu, C.Y.; Trieu, A.; Radhakrishnan, P.; Kwok, S.F.; Harris, S.; Zhang, K.; Wang, J.L.; Wan, J.; Zhai, H.; Takatsuto, S.; et al. Brassinosteroids regulate grain filling in rice. *Plant Cell* **2008**, *20*, 2130–2145. [CrossRef]
10. Stepien, P.; Klobus, G. Water relations and photosynthesis in *Cucumis sativus* L. leaves under salt stress. *Biol. Plant.* **2006**, *50*, 610–616. [CrossRef]
11. Türkan, I.; Demiral, T. Recent developments in understanding salinity tolerance. *Environ. Exp. Bot.* **2009**, *67*, 2–9. [CrossRef]
12. Rady, M.M. Effect of 24-epibrassinolide on growth, yield, antioxidant system and cadmium content of bean (*Phaseolus vulgaris* L.) plants under salinity and cadmium stress. *Sci. Hortic.* **2011**, *129*, 232–237. [CrossRef]
13. Chandler, C.K.; Ferree, D.C. Response of 'Raritan' and 'Surecrop' strawberry plants to drought stress. *Fruit Var. J.* **1990**, *44*, 183–184.
14. Klamkowski, K.; Treder, W. Morphological and physiological responses of strawberry plants to water stress. *Agric. Conspec. Sci.* **2006**, *71*, 159–165.
15. Manghwar, H.; Hussain, A.; Ali, Q.; Liu, F. Brassinosteroids (BRs) Role in Plant Development and Coping with Different Stresses. *Int. J. Mol. Sci.* **2022**, *23*, 1012. [CrossRef]
16. Serna, M.; Coll, Y.; Zapata, P.J.; Botella, M.Á.; Pretel, M.T.; Amorós, A. A brassinosteroid analogue prevented the effect of salt stress on ethylene synthesis and polyamines in lettuce plants. *Sci. Hortic.* **2015**, *185*, 105–112. [CrossRef]
17. Núñez-Vázquez, M.; Pérez-Domínguez, G.; Martínez-González, L.; Reyes-Guerrero, Y.; Coll-García, Y. Spirostanic analogues of brassinosteroids enhance the rice (*Oryza sativa* L.) cv. INCA LP-7 seedling growth under NaCl stress. *Cultiv. Trop.* **2016**, *37*, 152–159.
18. Yuan, G.F.; Jia, C.G.; Li, Z.; Sun, B.; Zhang, L.P.; Li, N.; Wang, Q.M. Effect of brassinosteroids on drought resistance and abscisic acid concentration in tomato underwater stress. *Sci. Hortic.* **2010**, *126*, 103–108. [CrossRef]
19. Furio, R.N.; Albornoz, P.L.; Coll, Y.; Martinez Zamora, G.M.; Salazar, S.M.; Martos, G.G.; Díaz Ricci, J.C. Effect of natural and synthetic Brassinosteroids on strawberry immune response against *Colletotrichum acutatum*. *Eur. J. Plant Pathol.* **2019**, *153*, 167–181. [CrossRef]
20. Furio, R.N.; Salazar, S.M.; Martínez-Zamora, G.M.; Coll, Y.; Hael-Conrad, V.; Díaz-Ricci, J.C. Brassinosteroids promote growth, fruit quality and protection against Botrytis on *Fragaria x ananassa*. *Eur. J. Plant Pathol.* **2019**, *154*, 801–810. [CrossRef]
21. de Carvalho, R.O.; Machado, M.B.; Göebel, J.T.S.; Lang, G.H.; da Luz, M.L.G.S.; Gadotti, G.I.; Silveira da Luz, C.A.; Gomes, M.C. Economical Feasibility of Strawberry Production in a Semi-Hydroponic System and Agroindustry of Jelly on a Small Property . *Agric. Eng. Int. CIGR J.* 2015, Special Issue, pp. 173–176. Available online: http://library.tuit.uz/knigiPDF/Ebsco/8-112.pdf (accessed on 14 March 2021).
22. Treftz, C.; Omaye, S.T. Comparision between hydroponic and soil systems for growing strawberries in a greenhouse. *Int. J. Agric. Ext.* **2016**, 3, 195–200.
23. Murashige, T.; Skoog, F. A revised medium for rapid growth and bio assays with tobacco tissue cultures. *Physiol. Plant.* **1962**, *15*, 473–497. [CrossRef]
24. Carley, H.E.; Watson, T.W. A new gravimetric method for estimating root-surface areas. *Soil Sci.* **1966**, *102*, 289–291. [CrossRef]
25. Schneider, C.A.; Rasband, W.S.; Eliceiri, K.W. NIH Image to ImageJ: 25 years of image analysis. *Nat. Methods* **2012**, *9*, 671–675. [CrossRef]
26. Montenegro-Gonzalez, H.; Malagón-Castro, D.; Guerrero, I. *Propiedades Físicas de Los Suelos*; CO-BAC: Bogotá, Colombia, 1990.
27. Barnes, R.B.; Richardson, D.; Berry, J.W.; Hood, R.L. Flame photometry a rapid analytical procedure. *Ind. Eng. Chem. Anal. Ed.* **1945**, *17*, 605–611. [CrossRef]
28. Kafkafi, U.; Xu, G.; Imas, P.; Magen, H.; Tarchitzky, J. *Potassium and Chloride in Crops and Soils: The Role of potassium Chloride Fertilizer in Crop Nutrition*; International Potash Institute: Basel, Switzerland, 2001.
29. Sapers, G.M. Color characteristics and stability of nonbleeding cocktail cherries dyed with carotenoid pigments. *J. Food Sci.* **1994**, *59*, 135–138. [CrossRef]
30. Agüero, J.J.; Salazar, S.M.; Kirschbaum, D.S.; Jerez, E.F. Factors affecting fruit quality in strawberries grown in a subtropical environment. *Int. J. Fruit Sci.* **2015**, *15*, 223–234. [CrossRef]
31. Di Rienzo, J.A.; Casanoves, F.; Balzarini, M.G.; Gonzalez, L.; Tablada, M.; Robledo, C.W. InfoStat Version 2013. Grupo InfoStat, FCA, Universidad Nacional de Córdoba, Argentina. Available online: http://www.infostat.com.ar (accessed on 8 November 2013).

32. Salazar, S.M.; Coll, Y.; Viejobueno, J.; Coll, F. Response of Strawberry Plants to the Application of Brassinosteroid Under Field Conditions. *Rev. Agronómica Del Noroeste Argent.* **2016**, *36*, 37–41. Available online: https://ranar.faz.unt.edu.ar/index.php/ranar/article/view/68 (accessed on 13 July 2021).
33. Xia, X.J.; Huang, L.F.; Zhou, Y.H.; Mao, W.H.; Shi, K.; Wu, J.X.; Asami, T.; Chen, Z.; Yu, J.Q. Brassinosteroids promote photosynthesis and growth by enhancing activation of Rubisco and expression of photosynthetic genes in *Cucumis sativus*. *Planta* **2009**, *230*, 1185. [CrossRef]
34. Oswald, O.; Martin, T.; Dominy, P.J.; Graham, I.A. Plastid redox state and sugars: Interactive regulators of nuclear-encoded photosynthetic gene expression. *Proc. Natl. Acad. Sci. USA* **2001**, *98*, 2047–2052. [CrossRef]
35. Pfannschmidt, T.; Allen, J.F.; Oelmüller, R. Principles of redox control in photosynthesis gene expression. *Physiol. Plant.* **2001**, *112*, 1–9. [CrossRef]
36. Zhang, N.; Kallis, R.P.; Ewy, R.G.; Portis, A.R. Light modulation of Rubisco in Arabidopsis requires a capacity for redox regulation of the larger Rubisco activase isoform. *Proc. Natl. Acad. Sci. USA* **2002**, *99*, 3330–3334. [CrossRef] [PubMed]
37. Hasan, S.A.; Hayat, S.; Ahmad, A. Brassinosteroids protect photosynthetic machinery against the cadmium induced oxidative stress in two tomato cultivars. *Chemosphere* **2011**, *84*, 1446–1451. [CrossRef]
38. Neill, S.; Desikan, R.; Hancock, J. Hydrogen peroxide signalling. *Curr. Opin. Plant Biol.* **2002**, *5*, 388–395. [CrossRef]
39. Apel, K.; Hirt, H. Reactive oxygen species: Metabolism, oxidative stress, and signal transduction. *Annu. Rev. Plant Biol.* **2004**, *55*, 373–399. [CrossRef]
40. Xia, X.J.; Wang, Y.J.; Zhou, Y.H.; Tao, Y.; Mao, W.H.; Shi, K.; Asami, T.; Chen, Z.; Yu, J.Q. Reactive oxygen species are involved in brassinosteroid-induced stress tolerance in cucumber. *Plant Physiol.* **2009**, *150*, 801–814. [CrossRef] [PubMed]
41. Yokota, T.; Sato, T.; Takeuchi, Y.; Nomura, T.; Uno, K.; Watanabe, T.; Takatsuto, S. Roots and shoots of tomato produce 6-deoxo-28-norcathasterone, 6-deoxo-28-nortyphasterol and 6-deoxo-28-norcastasterone, possible precursors of 28-norcastasterone. *Phytochemistry* **2001**, *58*, 233–238. [CrossRef]
42. Chaiwanon, J.; Wang, Z.Y. Spatiotemporal brassinosteroid signaling and antagonism with auxin pattern stem cell dynamics in Arabidopsis roots. *Curr. Biol.* **2015**, *25*, 1031–1042. [CrossRef] [PubMed]
43. González-García, M.P.; Vilarrasa-Blasi, J.; Zhiponova, M.; Divol, F.; Mora-García, S.; Russinova, E.; Caño-Delgado, A.I. Brassinosteroids control meristem size by promoting cell cycle progression in Arabidopsis roots. *Development* **2011**, *138*, 849–859. [CrossRef] [PubMed]
44. Mussig, C.; Fischer, S.; Altmann, T. Brassinosteroid-regulated gene expression. *Plant Physiol.* **2002**, *129*, 1241–1251. [CrossRef]
45. Grant, O.M.; Johnson, A.W.; Davies, M.J.; James, C.M.; Simpson, D.W. Physiological and morphological diversity of cultivated strawberry (*Fragaria* × *ananassa*) in response to water deficit. *Environ. Exp. Bot.* **2010**, *68*, 264–272. [CrossRef]
46. Arkin, G.F.; Taylor, H.M. *Modifying the Root Environment to Reduce Crop Stress*; American Society of Agricultural Engineers: St. Joseph, MI, USA, 1981; pp. 305–343.
47. Saied, A.S.; Keutgen, A.J.; Noga, G. The influence of NaCl salinity on growth, yield and fruit quality of strawberry cvs '*Elsanta*' and '*Korona*'. *Sci. Hortic.* **2005**, *103*, 289–303. [CrossRef]
48. Arreola, J.; Martínez-Sánchez, J.J.; Conesa, E.; Franco, J.A. Effect of pre-conditioning water regimes during nursery production on seedling root system characteristics of *Silene vulgaris*. *Acta Hortic.* **2008**, *782*, 287–292. [CrossRef]
49. Franco, J.A.; Arreola, J.; Vicente, M.J.; Martínez-Sánchez, J.J. Nursery irrigation regimes affect the seedling characteristics of *Silene vulgaris* as they relate to potential performance following transplanting into semi-arid conditions. *J. Hortic. Sci. Biotechnol.* **2008**, *83*, 15–22. [CrossRef]
50. Teale, W.D.; Ditengou, F.A.; Dovzhenko, A.D.; Li, X.; Molendijk, A.M.; Ruperti, B.; Paponov, I.; Palme, K. Auxin as a model for the integration of hormonal signal processing and transduction. *Mol. Plant.* **2008**, *1*, 229–237. [CrossRef] [PubMed]
51. Pirlak, L.; Esitken, A. Salinity effects on growth, proline and ion accumulation in strawberry plants. *Acta Agric. Scand.-B Soil Plant Sci.* **2004**, *54*, 189–192. [CrossRef]
52. Maas, E.V.; Hoffman, G.J. Crop salt tolerance–current assessment. *J. Irrig. Drain. Div.* **1977**, *103*, 115–134. [CrossRef]
53. Bould, C.; Hewitt, E.J.; Needham, P. The occurence and treatment of mineral disorders in the field. *Diagn. Miner. Disord. Plants* **1983**, *1*, 139–155.
54. Hasanuzzaman, M.; Nahar, K.; Fujita, M. Plant response to salt stress and role of exogenous protectants to mitigate salt-induced damages. In *Ecophysiology and Responses of Plants under Salt Stress*; Springer: New York, NY, USA, 2013. [CrossRef]
55. Makale, P.; Kontturi, M.; Pehu, E.; Somersalo, S. Photosynthetic response of drought and salt-stressed tomato and turnip rape plants to foliar applied glycinebetaine. *Physiol. Plant.* **1999**, *105*, 45–50. [CrossRef]
56. Wise, R.R.; Naylor, A.W. Chilling-enhanced photooxidation: Evidence for the role of singlet oxygen and Superoxide in the Breakdown of Pigments and endogenous antioxidant. *Plant Physiol.* **1987**, *83*, 278–282. [CrossRef] [PubMed]
57. Zapata, P.J.; Serrano, M.; Petrel, M.T.; Botella, M.A. Changes in free polyamine concentration induced by salt stress in seedling of different species. *Plant Growth Regul.* **2008**, *56*, 167–177. [CrossRef]
58. Lovisolo, C.; Perrone, I.; Carra, A.; Ferrandino, A.; Flexas, J.; Medrano, H.; Schubert, A. Drought-induced changes in development and function of grapevine (*Vitis* spp.) organs and in their hydraulic and non-hydraulic interactions at the whole-plant level: A physiological and molecular update. *Funct. Plant Biol.* **2010**, *37*, 98–116. [CrossRef]

59. Meggio, F.; Prinsi, B.; Negri, A.S.; Simone Di Lorenzo, G.; Lucchini, G.; Pitacco, A.; Espen, L. Biochemical and physiological responses of two grapevine rootstock genotypes to drought and salt treatments. *Aust. J. Grape Wine Res.* **2014**, *20*, 310–323. [CrossRef]
60. Zeng, H.; Tang, Q.; Hua, X. Arabidopsis brassinosteroid mutants *det2-1* and *bin2-1* display altered salt tolerance. *J. Plant Growth Regul.* **2010**, *29*, 44–52. [CrossRef]
61. Shahid, M.A.; Balal, R.M.; Pervez, M.A.; Garcia-Sanchez, F.; Gimeno, V.; Abbas, T.; Mattson, N.S.; Riaz, A. Treatment with 24-epibrassinolide mitigates NaCl-induced toxicity by enhancing carbohydrate metabolism, osmolyte accumulation, and antioxidant activity in *Pisum sativum*. *Turk. J. Bot.* **2014**, *38*, 511–525. [CrossRef]
62. Nunes, M.C.N.; Emond, J.P.; Rauth, M.; Dea, S.; Chau, K.V. Environmental conditions encountered during typical consumer retail display affect fruit and vegetable quality and waste. *Postharvest Biol. Technol.* **2009**, *51*, 232–241. [CrossRef]
63. Velickova, E.; Winkelhausen, E.; Kuzmanova, S.; Alves, V.D.; Moldão-Martins, M. Impact of chitosan-beeswax edible coatings on the quality of fresh strawberries (*Fragaria ananassa* cv Camarosa) under commercial storage conditions. *LWT* **2013**, *52*, 80–92. [CrossRef]
64. Liu, L.; Jia, C.; Zhang, M.; Chen, D.; Chen, S.; Guo, R.; Wang, Q. Ectopic expression of a BZR1-1D transcription factor in brassinosteroidsignalling enhances carotenoid accumulation and fruit quality attributes in tomato. *Plant Biotechnol. J.* **2014**, *12*, 105–115. [CrossRef] [PubMed]
65. Roghabadi, M.A.; Pakkish, Z.A.H.R.A. Role of brassinosteroid on yield, fruit quality and postharvest storage of 'Tak Danehe Mashhad' sweet cherry (*Prunus avium* L.). *Agric. Commun.* **2014**, *2*, 49–56.
66. Pakkish, Z.; Ghorbani, B.; Najafzadeh, R. Fruit quality and shelf life improvement of grape cv. Rish Baba using Brassinosteroid during cold storage. *J. Food Meas. Charact.* **2019**, *13*, 967–975. [CrossRef]

horticulturae

MDPI

Article

Effect of Elevated Temperature and Excess Light on Photosynthetic Efficiency, Pigments, and Proteins in the Field-Grown Sunflower during Afternoon

Antonela Markulj Kulundžić [1,*], Marija Viljevac Vuletić [2,*], Maja Matoša Kočar [1], Jasenka Antunović Dunić [3], Ivana Varga [4], Zvonimir Zdunić [5,6], Aleksandra Sudarić [1,6], Vera Cesar [3,7] and Hrvoje Lepeduš [7,8]

1 Department of Industrial Plants, Agricultural Institute Osijek, 31000 Osijek, Croatia; maja.matosa@poljinos.hr (M.M.K.); aleksandra.sudaric@poljinos.hr (A.S.)
2 Agrochemical Laboratory, Agricultural Institute Osijek, 31000 Osijek, Croatia
3 Department of Biology, J.J. Strossmayer University of Osijek, 31000 Osijek, Croatia; jantunovic@biologija.unios.hr (J.A.D.); vcesarus@yahoo.com (V.C.)
4 Faculty of Agrobiotechnical Sciences Osijek, J.J. Strossmayer University of Osijek, 31000 Osijek, Croatia; ivana.varga@fazos.hr
5 Service of General, Legal and Personnel Affairs, Agricultural Institute Osijek, 31000 Osijek, Croatia; zvonimir.zdunic@poljinos.hr
6 Centre of Excellence for Biodiversity and Molecular Plant Breeding, Faculty of Agriculture, University of Zagreb, 10000 Zagreb, Croatia
7 Faculty for Dental Medicine and Health, J.J. Strossmayer University of Osijek, 31000 Osijek, Croatia; hlepedus@yahoo.com
8 Faculty of Humanities and Social Sciences, J.J. Strossmayer University of Osijek, 31000 Osijek, Croatia
* Correspondence: antonela.markulj@poljinos.hr (A.M.K.); marija.viljevac@poljinos.hr (M.V.V.)

Citation: Markulj Kulundžić, A.; Viljevac Vuletić, M.; Matoša Kočar, M.; Antunović Dunić, J.; Varga, I.; Zdunić, Z.; Sudarić, A.; Cesar, V.; Lepeduš, H. Effect of Elevated Temperature and Excess Light on Photosynthetic Efficiency, Pigments, and Proteins in the Field-Grown Sunflower during Afternoon. *Horticulturae* **2022**, *8*, 392. https://doi.org/10.3390/horticulturae8050392

Academic Editors: Barbara Hawrylak-Nowak, Małgorzata Majewska and Agnieszka Hanaka

Received: 30 March 2022
Accepted: 27 April 2022
Published: 28 April 2022

Abstract: This study examined the photosynthetic responses of two sunflower hybrids to elevated temperatures and excess light intensity in the flowering stage by measuring the chlorophyll *a* fluorescence (ChlF) under morning and afternoon field conditions to determine the photosynthetic pigment contents and the relative accumulation of photosynthetic proteins. The morning environmental conditions were considered optimal, while the afternoon was characterised by elevated temperatures and excess light intensity. The minimum fluorescence intensity (F_0), the electron-flux-reducing end electron acceptors at the photosystem I acceptor side per reaction centre (RE_0/RC), and the D1 protein had significant, high, and positive correlations with the environmental conditions, which indicates that they were the most useful in the sunflower-stress-response research. In hybrid 7, the elevated temperatures and the excess light intensity resulted in the inactivation of the oxygen-evolving complex, which was indicated by the positive L, K, and J steps, the increase in the maximum quantum yield of PSII (TR_0/ABS), the decrease in the electron transport further than the primary acceptor Q_A ($ET_0/(TR_0\text{-}ET_0)$), the reduction in the performance index (PI_{ABS}), and the higher relative accumulation of the light-harvesting complex of the photosystem (LHCII). Hybrid 4 had smaller changes in the fluorescence curves in phases O–J and J–I, and especially in steps L, K, J, and I, and a higher PI_{ABS}, which indicates a more efficient excitation energy under the unfavourable conditions. As the tested parameters were sensitive enough to determine the significant differences between the sunflower hybrids in their photosynthetic responses to the elevated temperatures and excess light intensity in the flowering stage, they can be considered useful selection criteria. The development of more adaptable sunflower hybrids encourages sustainable sunflower production under stressful growing conditions.

Keywords: chlorophyll *a* fluorescence; temperature; solar radiation; tolerance; pigments; proteins

1. Introduction

Sunflower (*Helianthus annuus* L.) is a high-value, widespread oilseed. It is considered to be an important raw material for the production of oil and many other food and nonfood products. Sunflower production is influenced by many different environmental factors that interact, and that often cause plant stress. Because extreme temperature variations, sunlight intensity, quality variations, as well as uneven precipitation patterns that are accompanied by changes in the wind intensity and cloudiness are becoming more frequent [1], studies that investigate the relationship between plants and unfavourable weather conditions are becoming more and more important, and they enable the prediction of plant responses. The environmental factors that play major roles in plant stress are high temperatures, excess light, and drought. According to García-López et al. [2], sunflower is moderately tolerant to drought and heat stress. It is most sensitive to heat from early flowering to grain filling. Growing sunflower at elevated temperatures reduces its growth, which is reflected in the specific leaf mass, the leaf surface, and the soluble protein content [3]. Among the physiological processes, photosynthesis [4] is the most susceptible to the negative impact of heat. In addition to heat, the high light intensity can also negatively affect photosynthesis [5].

At the end of the 20th century, significant progress was made by using modern optical methods and techniques to study photosynthetic processes [6], among which one of the most commonly used methods is the determination of the chlorophyll *a* fluorescence (ChlF). The ChlF provides insight into the photosynthetic apparatus status and function, the efficiency of photosystem II (PSII), and the electron transport chain function [7]. Many authors have used ChlF parameters to study the effect of stress on plants [5,8–11].

Acclimatisation to different environments is interrelated to photosynthetic adjustment, which consequently affects the biochemical and physiological processes, the growth, and the yield [12]. Most plants show a significant ability to adjust photosynthesis to temperature and light fluctuations. Elevated temperatures negatively affect cell division and expansion, and they are one of the main stresses that stimulate protein degradation and that cause tissue senescence or death [13]. The decrease in photosynthesis occurs not only because of changes in the flow of energy through PSII, which is extremely sensitive to elevated temperatures, but also because of the reduced content of the pigments in leaves [14]. PSII damage that is caused by elevated temperatures occurs mainly at the oxygen-evolving centre (OEC), and so even slightly increased temperatures cause its deactivation [15]. Although an increase in light intensity can gradually increase the photosynthetic rate, reactive centres absorb more light when the intensity of the light is high (i.e., when there is more light than can be used in photochemistry). The remaining energy is dissipated as heat and fluorescence. If the energy is not utilised or if it is dissipated, it causes photooxidative stress and it increases the level of reactive oxygen species (ROS) [16]. High light intensity is the cause of numerous other disorders in plants besides photodamage, photoactivation, and photoinhibition. One of these disorders is the degradation of photosynthetic proteins. It has been proven that the accumulation of the light-harvesting complex of PSII (LHCII) is related to the chlorophyll *a* and *b* ratio, and that it depends on daily weather fluctuations [17]. Protein D1, which is an essential part of the photosynthetic apparatus, is also sensitive to stress [18]. Excess light causes D1 protein phosphorylation, which results in degradation, *de novo* protein synthesis, and protein incorporation into PSII [19]. The accumulation of the cytochrome *f* protein is also crucial because the *c6f* protein complex connects to PSII and to photosystem I (PSI) with cyclic and linear electron transfer [20]. Another significant protein that is directly related to temperature is Rubisco. Its abundance changes under stress [21].

As the global weather changes, extreme weather conditions that occur in the estival afternoons during the most critical sunflower developmental stages are becoming more frequent. Such conditions can often cause short-term temperature and light stress in sunflower plants, which are reflected as changes in the ChlF parameters [10]. Although plants can show a significant ability to adjust photosynthesis to temperature and light fluctuations through daily changes during plant growth, there are those that are less tolerant. Therefore, the timely determination of the stress occurrence and the stress tolerance and the elimination of stress-susceptible genotypes are of great importance for plant production. This study aimed to differentiate genotypes on the basis of their photosynthetic responses to elevated temperatures and excess light intensity in the flowering stage by measuring the ChlF, the photosynthetic pigment contents, and the relative accumulation of photosynthetic proteins in the field (i.e., in their ambient environments). The synergic effect of elevated temperatures and excess light, which are known to be correlated with sunflower leaf stress and to cause the most problems in many production areas, were determined in the afternoon hours during the sunflower flowering stage. Knowing the specific reactions of the individual hybrids to adverse environmental conditions allows breeders to better understand the characteristics of the material during selection, which also increases the breeding programme's success. Since this study is a part of the sunflower breeding programme, and since it compares the responses of different hybrids to the conditions that are known to affect the metabolism in plants, it is important to select superior material. On the basis of the abovementioned, it was hypothesised that one hybrid would be more adaptable to the elevated temperatures and excess light determined in the chosen afternoons, and that it would show minor daily changes in the tested parameters.

2. Materials and Methods

2.1. Plant Material

The experiment was conducted at the Agricultural Institute Osijek (Osijek, Croatia) on two sunflower hybrids. The hybrids differed in pedigree and in agronomic properties (plant height, head diameter, yield potential), but they had similar maturation times. The hybrids were chosen on the basis of the results of previous trials [22,23]. Hybrid 4 has been recognised and has been widely spread throughout the sunflower production in Croatia in recent years. The producers accept it because of its high seed quality and seed yields under various growing conditions, which shows its stability and wide adaptability. Hybrid 7 is an experimental material that stood out in multiyear microtrials, with good overall agronomic qualities and high oil content. A comparison of the difference in the response mechanisms of these two hybrids to unfavourable environmental conditions (elevated temperature and excess light intensity) will be beneficial for determining the direction of future breeding programmes.

The sowing was performed with manual hand planters (two seeds per hill) at a 4 cm depth in four 5 m-long rows, with a 70 cm distance between the rows, and a 23 cm distance within the rows (45°32′ N, 18°44′ E; 94 m altitude), in four replications. The final density was 6.29 plant/m^2. A randomised complete block experimental design was used. The soil analysis determined that the soil texture was silty clay loamy, and it is classified as anthropogenic Eutric Cambisol. The physical and chemical properties in the upper soil layer of the soil were: 64.7% silt; 32.5% clay; 2.8% sand; a pH in K_2O of 7.3; a pH in H_2O of 7.9; P_2O_5 > 41 mg/100 g; N: 0.16%; K_2O > 40 mg/100 g; $CaCO_3$: 0.9%; Al (mobile): 0.26 mg/100 g; and a humus content of 2.18%. During the experiment, all of the agrotechnical measures were performed by following the recommendations and requirements of sunflower cultivation.

2.2. Weather Conditions

The minimum, maximum, and mean air temperatures, the solar radiation intensity, and the precipitation for the ten days preceding the measurement and sampling are shown in Supplementary Table S1 in order to provide more insight into the environmental conditions that the photosynthetic apparatus needed to adapt to. The temperature, solar radiation, and precipitation measurements were recorded every 10 min (0–24 h), after which the average values per day were calculated.

The measurements were made on a cloudless day (4 August) during the flowering stage, in the morning (7:30–9:00 am) and in the afternoon (12:30–2:00 pm). These two measurement times were selected to quantify the photosynthetic apparatus reaction to and the changes in the physiological parameters under the elevated temperature and excess light conditions that were determined to be characteristic of the early afternoon hours, compared to the lower temperatures and low light conditions that were determined to be characteristic of the early morning hours (optimal conditions). The mean air temperature and the solar radiation during the measurement and sampling in the morning were 27.6 °C and 332.3 W/m^2, respectively. In the afternoon, they were 35.7 °C and 830.2 W/m^2, respectively. The mentioned afternoon temperature and solar radiation were previously found to be the causes of changes in the photosynthetic efficiency [5,10]. The temperatures, the solar radiation intensities, and the precipitation were recorded by a command and data-acquisition station near the experiment.

2.3. *Chlorophyll a Fluorescence (ChlF)*

The ChlF was determined during the flowering stage (according to Schneiter and Miller [24], in the R5.5 stage) by a plant efficiency analyser (Handy PEA, Hansatech, Norfolk, UK). The youngest (upper) developed sunflower leaves were used for the ChlF measurements. A leaf is considered developed if it is larger than 4 cm [24]. The measurements were carried out in the middle two rows of each hybrid in the morning (7:30–9:00 am) and afternoon hours (12:30–2:00 pm), and in their ambient environments. The ChlF was determined on 12 leaves per hybrid (three leaves × four replicates) under field conditions. Before measuring the ChlF, the sunflower leaves were adapted to the dark for 30 min and were subjected to dark conditions, during which the electron transfer in the photosynthetic electron transport chain ceases. There is no water oxidation in PSII (i.e., at the OEC), as there is no charge separation at the reaction centre in darkness. After tissue illumination, the obtained information on the intensity of the ChlF during one second is displayed on the OJIP curve. Different environmental conditions can cause the appearance of additional steps in the ChlF OJIP transients [6]. The ChlF transients were induced by using a pulse of saturating red light (peak at 650 nm, 3200 μmol m^{-2} s^{-1}). The JIP parameters that were calculated from the recorded data are shown in Supplementary Table S2.

A double normalisation of the OJIP transients was made between the O and P steps. The logarithmic time scale was used for presenting the relative variable fluorescence: $W_{OP} = (F_t - F_0)/(F_J - F_0)$. The K, L, J, and I steps were presented as the variable fluorescence: $W_{OK} = (F_t - F_0)/(F_K - F_0)$, $W_{OJ} = (F_t - F_0)/(F_J - F_0)$ $W_{OI} = (F_t - F_0)/(F_I - F_0)$, and $W_{O50} = (F_t - F_0)/(F_{50} - F_0)$ plotted with difference kinetics: $\Delta W_{OK} = W_{OK} - (W_{OK})_{ref}$, $\Delta W_{OJ} = W_{OJ} - (W_{OJ})_{ref}$, $\Delta W_{OI} = W_{OI} - (W_{OI})_{ref}$, and $\Delta W_{O50} = W_{O50} - (W_{O50})_{ref}$. The measurements under the morning conditions were used as the reference values of $(W_{OK})_{ref}$, $(W_{OJ})_{ref}$, $(W_{OI})_{ref}$, and $(W_{O50})_{ref}$.

2.4. *Laboratory Analyses*

In total, the eight youngest leaves per hybrid on which the ChlF was determined were sampled for their photosynthetic pigments and for protein analysis. Before the biochemical analyses, the composite sample was homogenised into a powder by using liquid nitrogen.

2.4.1. Photosynthetic Pigment Content Determination

About 0.05–0.1 g of plant tissue, which was previously homogenised by liquid nitrogen with the addition of magnesium hydroxide carbonate, was extracted by 1 mL of cold acetone. The extraction procedure was repeated six times until the plant tissue was completely discoloured. Supernatants were pooled and used for the spectrophotometric measurement of the absorbance at 470, 645, and 662 nm. Acetone was used for the blank. The content of the photosynthetic pigments was calculated by using the appropriate extinction coefficients, according to Lichtenthaler [25]. The chlorophyll *a*/*b* and the chlorophyll *a* + *b*/Car were calculated as well. Five replicates were performed per the condition of each genotype.

2.4.2. SDS-PAGE and Immunodetection

The relative protein accumulation was determined by sodium dodecyl sulfate-polyacrylamide gel electrophoresis (SDS-PAGE). About 0.5 g of plant tissue was extracted with 1 mL of buffer, heated to 80 °C. The buffer consisted of 0.13 M Tris/HCl (pH = 6.8), 4.6% SDS, 16% glycerol, and 0.01 M dithiothreitol. The protein concentration was determined according to Bradford [26]. The protein extract contained 10 or 30 μg of protein, depending on the protein detected. For the detection of the Rubisco large subunit (LSU), a sample containing 10 μg of protein and 1 μg of loading buffer was applied to the gel, and for the other proteins (Lhcb2, D1, and cytochrome *f*), 30 μg of protein and 2 μg of buffer were used. After the separation with 12% SDS-PAGE [27], the proteins were transferred from the gel to the nitrocellulose membrane (Bio-Rad) in semidry conditions by using Biometra Fastblot B43 [28]. For the immunodetection, the specific primary antibodies against LHCII (anti-Lhcb2, Agrisera), D1 of Photosystem II (anti-PsbA, Agrisera), cytochrome *f* (anti-Cyt *f*, Agrisera), and Rubisco LSU (anti-RbcL, Agrisera), and then secondary antibodies (Donkey anti-rabbit IgG-HRP, Santa Cruz Biotechnology, Dallas, TX, USA), were used. Primary and secondary antibodies were diluted in a buffer in ratios of 1:5000 and 1:10,000, respectively. According to the manufacturer's instructions, a commercial chemiluminescence detection substrate (Lumi-Light Western Blotting Substrate, Roche, Basel, Switzerland) was used to incubate the membranes. After that, protein bands were detected on ECL films (AGFA, Mortsel, Belgium), according to the standard procedure. ImageJ software was used for the protein band quantification. Three replicates were performed per the condition of each genotype.

2.5. Data Analyses

The ChlF parameters were calculated and visualised in Microsoft Excel, according to Strasser et al. [7]. A one-way ANOVA was used for determining the statistical differences between the hybrids under the morning and afternoon conditions of the ChlF ($n = 12$) and pigment-content ($n = 5$) measurements, which were followed by the Tukey's post hoc honest significant difference (HSD) test at $p < 0.05$. The correlations among the JIP parameters, the pigment content, the photosynthetic proteins, the sunflower hybrids, and the environmental conditions were explored by principal component analysis (PCA) at $p < 0.05$. Before the PCA analysis, the data were standardised. The PCA was performed by using a correlation matrix of the average values after autoscaling. The mean values ± standard deviations in the table are used for presenting the data.

3. Results and Discussion

3.1. Fluorescence Transient Curves

The raw fluorescence induction curves showed a high deviation between the hybrids measured in the morning and in the afternoon, where a notable change in the OJIP-curve shape occurred in hybrid 7 (Figure 1a). The double O–P normalised curves, which show the measurement values of hybrids 4 and 7 under the morning conditions, had the typical form of a normalised OJIP curve (Figure 1b), while the curves that were measured in the afternoon altered significantly. A similar shape was found in peach leaves when exposed to high (more than 35 °C) temperatures [29].

The O–J phase, which is also known as the light-dependent phase, represents the 2 ms increase in the OJIP curve. This phase provides information about the excitation energy transfer between the PSII RCs and PSI [30]. It reveals the difference between the measurements under the morning and afternoon conditions, which is seen as the rise in the fluorescence curves for both hybrids. Still, the increase was more pronounced for hybrid 7, which is a consequence of reducing the primary plastoquinone (Q_A) acceptor [31]. The J–I phase is characterised by a partial reduction in the pool of plastoquinones, which is unlike the I–P phase, which represents the reduction in the PSI's acceptor side [32]. It is evident from Figure 1b that the curves in the J step are more pronounced for hybrid 7; however, in the I step, the curves are the same for both hybrids and for both conditions. Brestic et al. [33] found a decrease in the fluorescence transient intensity in the J–I phase, followed by an I–P phase increase, which was confirmed in this investigation as well.

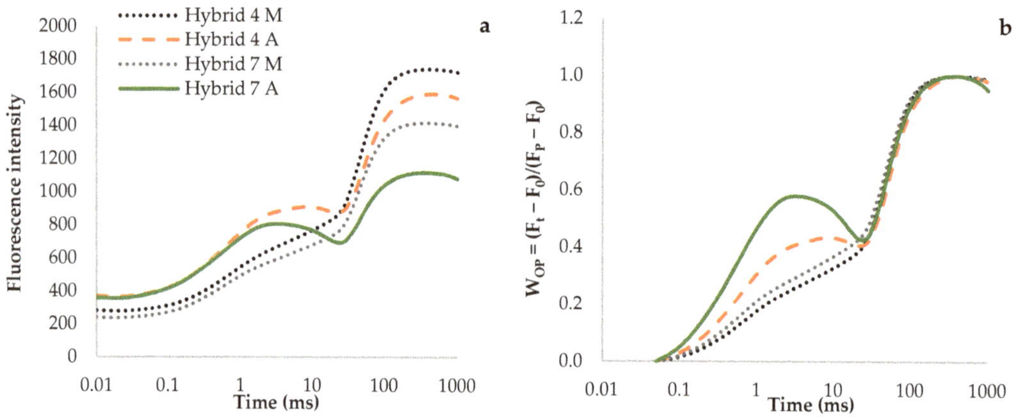

Figure 1. (**a**) Raw fluorescence induction curves, and (**b**) double O–P normalised OJIP curves of chlorophyll *a* fluorescence kinetics of dark-adapted leaves of sunflower hybrids 4 and 7 under the morning (M) and afternoon (A) conditions. $W_{OP} = (F_t - F_0)/(F_P - F_0)$ represents normalised OJIP transient data between steps O and P. Each curve represents the average kinetics of 12 replicates per condition.

A more precise image of the O–P phase's fluorescence intensity can be obtained from the individual representations of the normalised O–K, O–J, O–I, and O–50 curves. They clearly differentiate between the L, K, J, and I steps among the tested hybrids (Figure 2).

Under the synergic effect of elevated temperatures and excess light, the L step's appearance at 150 μs reflects the positive transient values in both hybrids (Figure 2a), which signify the weaker energy connectivity and stability of the PSII units [32]. Under the same conditions, the rise in the kinetic fluorescence of the K, J, and I steps has positive curve amplitudes for both hybrids; however, higher amplitudes in all the steps were determined for hybrid 7 (Figure 2b–d). The positive curve amplitudes at step K (300 μs) indicate an impaired PSII antenna function during the electron flow, which was due to an increased reduction rate of the Q_A of the primary PSII electron acceptor, which indicates impaired OEC function [34]. Furthermore, steps J (2 ms) and I (30 ms) explain the reduction in the plastoquinone pool between PSII and PSI [35]. Although many studies confirm the occurrence of the K and L steps under high temperatures [29,33] and high light intensity [36], numerous authors report their occurrence in other stress conditions as well. The described reactions of sunflower hybrids 4 and 7 in the individual steps concur with the synergic effect of the elevated temperatures and the excess light that was determined in apple cultivars [5].

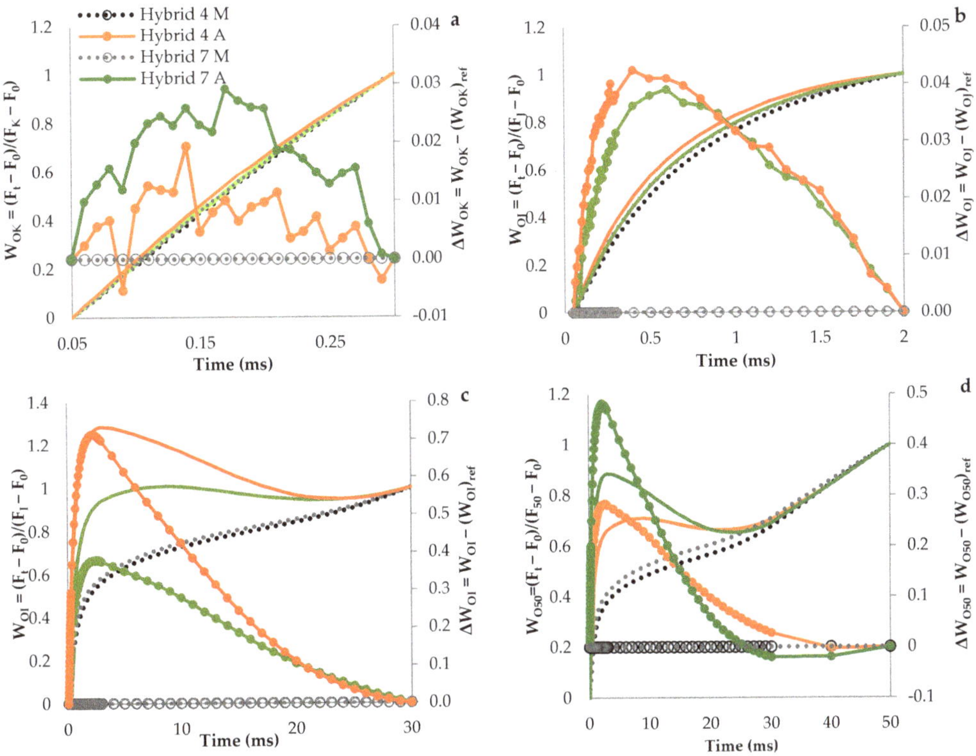

Figure 2. Chlorophyll *a* fluorescence transient curves in sunflower hybrids 4 and 7 under the morning (M) and afternoon (A) conditions. $W_{OK} = (F_t - F_0)/(F_K - F_0)$ represents normalised transient data between steps O and K (L step); (**a**) plotted as difference kinetics ($\Delta W_{OK} = W_{OK} - (W_{OK})_{ref}$) in the 0.05–0.3 ms time range. $W_{OJ} = (F_t - F_0)/(F_J - F_0)$ represents normalised transient data between steps O and J (K step); (**b**) plotted as difference kinetics ($\Delta W_{OJ} = W_{OJ} - (W_{OJ})_{ref}$) in the 0.05–2 ms time range. $W_{OI} = (F_t - F_0)/(F_I - F_0)$ represents normalised transient data between steps O and I (J step); (**c**) plotted as difference kinetics ($\Delta W_{OI} = W_{OI} - (W_{OI})_{ref}$) in the 0.05–30 ms time range. $W_{O50} = (F_t - F_0)/(F_{50} - F_0)$ represents normalised transient data between steps O and 50 (I step); (**d**) plotted as difference kinetics ($\Delta W_{O50} = W_{O50} - (W_{O50})_{ref}$) in the 0.05–50 ms time range. Each curve represents the average kinetics of 12 replicates. Measurements in morning conditions were used as reference values of $(W_{OK})_{ref}$, $(W_{OJ})_{ref}$, $(W_{OI})_{ref}$, and $(W_{O50})_{ref}$. Curve lines with marker points show W_{OK}, W_{OJ}, W_{OI}, and W_{O50} (primary axis), and curve lines without marker points show ΔW_{OK}, ΔW_{OJ}, ΔW_{OI}, and ΔW_{O50} (secondary axis).

3.2. Chlorophyll JIP-Test Parameters

The results of the JIP-test parameters and the photosynthetic pigments in the morning and afternoon conditions are shown in Table 1.

The afternoon F_0 measurements (when all the PSII RCs were open) increased in both hybrids compared to the values under the morning conditions, and with a more pronounced increase for hybrid 7. Contrary to the F_0 values, the F_m values decreased because of the synergic effect of the elevated temperatures and excess light for both hybrids, and they represented their maximal intensity when all of the PSII RCs were closed. According to Schansker et al. [37], the fluorescence increase between steps F_0 and F_m indicates Q_A reduction. These two parameters calculate the TR_0/ABS, which demonstrates the likelihood that the absorbed photon energy can be trapped by PSII RCs [38]. Under the morning conditions, the TR_0/ABS values were similar for hybrids 4 and 7; however, they decreased in the afternoon in both hybrids, which indicates PSII damage [15]. It has been proven that, under heat and light stress, the F_0 values typically increase, while the F_m and TR_0/ABS values decrease. This result was confirmed in the research by Mihaljević et al. [5]. Misra et al. [39] define the decrease in the maximum quantum efficiency of PSII, which is calculated by the TR_0/ABS as photoinhibition.

Table 1. Mean values and standard deviations of JIP parameters (n = 12) and photosynthetic pigment contents (n = 5) in sunflower hybrids 4 and 7 in morning and afternoon conditions.

	Hybrid 4		**Hybrid 7**	
Parameters	**Morning**	**Afternoon**	**Morning**	**Afternoon**
Minimum fluorescence intensity (F_0)	267.50 ± 8.51 b	348.75 ± 25.09 a	223.75 ± 7.31 c	341.83 ± 28.87 a
Maximum fluorescence intensity (F_m)	1740.92 ± 67.50 a	1591.83 ± 104.15 b	1416.75 ± 99.07 c	1115.58 ± 92.91 d
Relative variable fluorescence at 150 µs (V_L step)	0.54 ± 0.01 b	0.54 ± 0.02 b	0.54 ± 0.01 b	0.56 ± 0.02 a
Relative variable fluorescence at 300 µs (V_K step)	0.36 ± 0.02 c	0.39 ± 0.03 b	0.40 ± 0.02 b	0.44 ± 0.03 a
Relative variable fluorescence at 3 ms (V_J step)	0.24 ± 0.02 d	0.40 ± 0.03 b	0.28 ± 0.02 c	0.58 ± 0.03 a
Relative variable fluorescence at 30 ms (V_I step)	0.46 ± 0.05 ab	0.45 ± 0.03 b	0.50 ± 0.05 a	0.47 ± 0.03 ab
Maximum quantum yield of PSII (TR_0/ABS)	0.85 ± 0.01 a	0.78 ± 0.01 b	0.84 ± 0.01 a	0.69 ± 0.03 c
Density of active PSII reaction centres (RCs) per cross section (RC/CS_0)	157.84 ± 7.44 b	173.32 ± 12.43 a	119.35 ± 6.53 d	135.53 ± 10.54 c
Density of RC on chlorophyll *a* basis (RC/ABS)	0.59 ± 0.03 a	0.50 ± 0.04 b	0.53 ± 0.03 b	0.40 ± 0.04 c
Flux ratio trapping per dissipation (TR_0/DI_0)	5.51 ± 0.19 a	3.58 ± 0.30 b	5.34 ± 0.45 a	2.28 ± 0.31 c
Electron transport further than primary acceptor Q_A ($ET_0/(TR_0\text{-}ET_0)$)	3.16 ± 0.26 a	1.53 ± 0.20 c	2.61 ± 0.26 b	0.73 ± 0.10 d
Performance index (PI_{ABS})	10.26 ± 0.66 a	2.73 ± 0.48 c	7.44 ± 1.20 b	0.67 ± 0.17 d
Quantum yield for reduction in end electron acceptors at the PSI acceptor side (RE_0/ABS)	0.60 ± 0.04 c	0.72 ± 0.03 b	0.58 ± 0.04 c	0.87 ± 0.07 a
Probability that an electron from the electron transport chain is transferred to reduce end electron acceptors at the PSI acceptor side (RE_0/ET_0)	0.71 ± 0.05 c	0.92 ± 0.04 b	0.69 ± 0.06 c	1.26 ± 0.11 a
Electron-flux-reducing end electron acceptors at the PSI acceptor side per RC (RE_0/RC)	0.78 ± 0.09 b	0.87 ± 0.10 ab	0.79 ± 0.10 b	0.92 ± 0.12 a
Electron transport from PQH_2 to final PSI acceptors ($RE_0/(ET_0\text{-}RE_0)$)	2.54 ± 0.51 b	15.95 ± 11.20 a	2.33 ± 0.60 b	−5.83 ± 2.98 c
Performance index for energy conservation from exciton to the reduction in PSI end acceptors (PI_{total})	26.09 ± 5.89 b	40.21 ± 22.01 a	16.97 ± 3.80 b	−4.22 ± 3.26 c
Chlorophyll *a* (Chl *a*)	1.38 ± 0.02 b	1.49 ± 0.03 a	1.46 ± 0.02 a	1.37 ± 0.01 b
Chlorophyll *b* (Chl *b*)	0.32 ± 0.01 c	0.35 ± 0.01 b	0.37 ± 0.00 ab	0.37 ± 0.01 a
Total chlorophyll *a* + *b* (Chl *a* + *b*)	1.70 ± 0.02 c	1.85 ± 0.03 a	1.83 ± 0.02 a	1.74 ± 0.01 b
Total carotenoids (Car)	0.40 ± 0.00 bc	0.45 ± 0.01 a	0.41 ± 0.01 b	0.40 ± 0.00 c
Ratio of chlorophyll *a* and *b* (Chl *a*/*b*)	4.37 ± 0.20 a	4.21 ± 0.04 a	3.99 ± 0.06 b	3.70 ± 0.06 c
Ratio of total chlorophyll content and carotenoids (Chl *a* + *b*/Car)	4.24 ± 0.01 c	4.09 ± 0.03 d	4.47 ± 0.05 a	4.39 ± 0.03 b

According to Tukey's HSD test, means with the same letters are not significantly different at $p < 0.05$.

The values of the V_L remained unchanged in hybrid 4, and they increased in hybrid 7, while the V_K and the V_J increased in both hybrids when they were measured in the afternoon. However, hybrid 4 had lower baseline values compared to hybrid 7. The elevated V_J in both hybrids indicate that the Q_A reoxidation was limited, which resulted in reduced Q_A accumulation and decreased electron transport [7]. The described trend of the V_L, the V_J, and the V_K under heat and light stress was also observed in common fig leaves [8]. The V_I showed no significant change in both hybrids under the afternoon conditions compared to the morning conditions, which indicates that the mentioned parameter is not directly related to the changes in PSII [38]. The RC/CS_0 increased in both hybrids when they were measured in the afternoon, which indicates that the inactivation of a particular number of reaction centres did not occur. Hybrid 4 had higher RC/CS_0 values than hybrid 7, but the stress still had more impact on hybrid 7. According to other studies that have been conducted on wheat [40] and quinoa [41], stress conditions reduced the RC/CS_0 values, which directly affected the TR_0/ABS and caused its reduction. The RC/CS_0 did not affect the TR_0/ABS in this study because the TR_0/ABS values decreased despite the increased RC/CS_0. All of the above leads to the conclusion that sunflower has a partially different defence reaction of the photosynthetic apparatus to the synergic effect of elevated temperatures and excess light compared to other plant species, at least in terms of the RC/CS_0 parameter, which was concluded previously by Çiçek et al. [42] as well. Furthermore, numerous studies have shown that the most sensitive parameters of the JIP-test are the PI_{ABS} and the PI_{total} [5,43]. The PI_{ABS} and its components significantly decreased under the synergic effect of elevated temperatures and excess light. Hybrid 4 had a higher PI_{ABS}, RC/ABS, and $ET_0/(TR_0\text{-}ET_0)$ than Hybrid 7. The hybrids had similar TR_0/DI_0 and PI_{total} under the morning conditions. The PI_{total} parameter includes the PI_{ABS} with its components (RC/ABS, TR_0/DI_0, and $ET_0/(TR_0\text{-}ET_0)$) and the probability of the PSI reducing its end acceptors ($RE_0/(ET_0\text{-}RE_0)$) [32]. The $RE_0/(ET_0\text{-}RE_0)$ shows the efficiency of the processes that involve PSI and its ability to reduce its end acceptors. Contrary to the other stress parameters, the PI_{total} increased in hybrid 4 and decreased in hybrid 7, while the $RE_0/(ET_0\text{-}RE_0)$ proved to be the most sensitive component of the PI_{total}, as it had the largest changes in its values. The sensitivity of the RE_0/ET_0 was noted earlier by Pavlović et al. [9], who tested brassicas for salt stress, and by Viljevac Vuletić and Španić [43], who investigated leaf senescence in winter wheat. The results described above are similar to those observed in Zoysiagrass leaves that were exposed to cold stress [44], and in nutrient-deficient maise and tomato plants [45]. Çiçek et al. [42] obtained positive and negative PI_{total} values, and they examined the impact of drought on sensitive and tolerant sunflower hybrids. In this study, the tendency of PSI under the synergic effect of elevated temperatures and excess light is presented through the following parameters: the RE_0/ABS, the RE_0/ET_0, and the RE_0/RC. These parameters showed no significant differences between the hybrids under the morning conditions. At the same time, they increased in the afternoon in both hybrids, which reflects the electron flow from the PQH_2 to the PSI end electron acceptors. An increase in the RE_0/ET_0 values under heat stress occurs when fewer electrons are donated to reduce the PQH_2 [8,46]. Similarly, Arslan et al. [47] report that the RE_0/ABS, the RE_0/ET_0, and the PI_{total} were reduced by drought in all sunflower lines. By testing peach leaves for heat stress at three levels (25, 30, and 35 °C), Martinazzo et al. [29] proved that increasing the temperature increases the values of the RE_0/RC and the RE_0/ABS, while the RE_0/ET_0 remains the same.

3.3. Photosynthetic Pigment Content

Hybrid 4 had lower chlorophyll *a* and *b* (Chl *a* and *b*) and carotenoid (Car) contents than hybrid 7 under the morning conditions. The elevated temperatures and the excess light under the afternoon conditions increased the photosynthetic pigment content in hybrid 4, while their amount decreased (except for Chl *b*) in hybrid 7 (Table 1). Gupta et al. [48] conclude that the decrease in the chlorophyll in wheat seedlings resulted from thylakoid membrane damage that was caused by high temperatures. This study indicates that hybrids 4 and 7 have a different adaptation of the light-harvesting complex (LHC) under the synergic effect of elevated temperatures and excess light. On the other hand, a significant reduction in the Car in hybrid 7 reduced the photosynthetic efficiency because the carotenoids protect the chlorophyll from photooxidative destruction [49]. Therefore, the car-content increase in hybrid 4 as a result of the synergic effect of the elevated temperatures and excess light in the afternoon indicates the initiation of the photoprotective plant defence mechanism for the avoidance of photooxidation, which is further confirmed by the reduction in the ratio of the total chlorophyll content and carotenoids (Chl *a* + *b*/Car).

3.4. Photosynthetic Proteins

The LHCII protein is very important for the speed of the adaptation of plants to changes in the light intensity, which results in the relative accumulation of proteins and in changes in the organisation of the antenna position. In this way, plants simultaneously regulate the light absorption and the nonphotochemical dissipation of the excess excitation energy. According to Chen et al. [50], light and heat stress cause the LHCII super-complexes to disassemble, which indicates more susceptibility to stress compared to the other protein complexes in the thylakoid membrane. The antenna complex's main protein structure for the LCHII is made up of Lhcb2 proteins that bond 45–60% of the pigment molecules (Chl *a*, Chl *b*, and carotenoids). In the present experiment, the elevated temperature and the excess light resulted in an accumulation of the LHCII protein in both hybrids, which correspond to the changes in the photosynthetic pigment contents of leaves, as demonstrated by Oguchi et al. [51]. Furthermore, Tanaka and Tanaka [52] conclude that the accumulation of LHCII depends on the content of Chl *b*. In this research, the accumulation of the LHCII increased during the synergic effect of elevated temperatures and excess light, and more so for hybrid 7 than for hybrid 4 (Figure 3). Mlinarić et al. [8] and Tanaka and Tanaka [52] have determined that higher LHCII accumulation is related to higher Chl *b* content. This was also confirmed here because hybrid 7 had a higher Chl *b* content and Chl *a*/*b* and higher LHCII accumulation than hybrid 4 (Table 1 and Figure 3) under the synergic effect of elevated temperatures and excess light. In hybrid 7, the Chl *b* content remained stable despite the stress, but the LHCII increased significantly. A higher LHCII protein accumulation during heat and light stress was also confirmed in wheat [50].

One of the main reaction centres in PSII is transmembrane subunit protein D1 (coded genome psbA), which is the most susceptible to environmental stress of all of the PSII complex components [53]. Su et al. [54] claim that the D1 protein is the target place for heat and light stress action. According to Chan et al. [55], heat and/or light stress stimulate ROS synthesis in the thylakoid membranes, which damages the D1 protein. Although other studies note the decrease in the D1 protein that is due to the synergic effect of elevated temperatures and excess light [54], this was not the case here. The D1 protein was higher in the afternoon than in the morning, and especially in hybrid 4, which indicated that the ROS did not affect the D1 protein (i.e., that the cell oxidation damage was partly prevented). Hybrid 4 showed weaker dynamics between the degradation and the biosynthesis of the D1 protein compared to hybrid 7, as was evidenced by the D1 protein level in the morning, which was much lower in hybrid 4 in the post-stress period (i.e., recovery overnight). The D1 protein accumulation is more affected by the light intensity than by the temperature. According to Guo et al. [56], a higher light intensity causes increased protein phosphorylation. More recent studies indicate that the high D1 protein reactivity under unfavourable high-light-intensity conditions is not just the result of the photoinhibition processes. Its function is to protect PSI from the high flow of the electrons that are generated in PSII that could cause oxidation damage [57]. This is confirmed in the research by Vojta et al. [58]. They report the parallel existence and activity of different electron flow routes in the electron transport chain that prevent excess ROS synthesis.

Besides the D1 protein, the daily fluctuations in the temperatures and the insolation affect cytochrome *b6f* complex proteins as well. The membrane cytochrome *b6f* protein complex connects PSI to PSII through electron connections [20]. During the synergic effect of elevated temperatures and excess light, sunflower hybrid 4 accumulated more cyt *f* proteins than hybrid 7. The relative cyt *f* protein accumulation during high light intensity is one of the most sensitive components of electron transport [59]. According to Hojka [60], plants adapt to high light intensity by increasing cyt *f* protein synthesis, but the change amplitudes depend on the species. Higher cyt *f* accumulation under high-light-intensity conditions was confirmed by Yamori et al. [61] on spruce, and by Pavlovič et al. [62] on tobacco leaves.

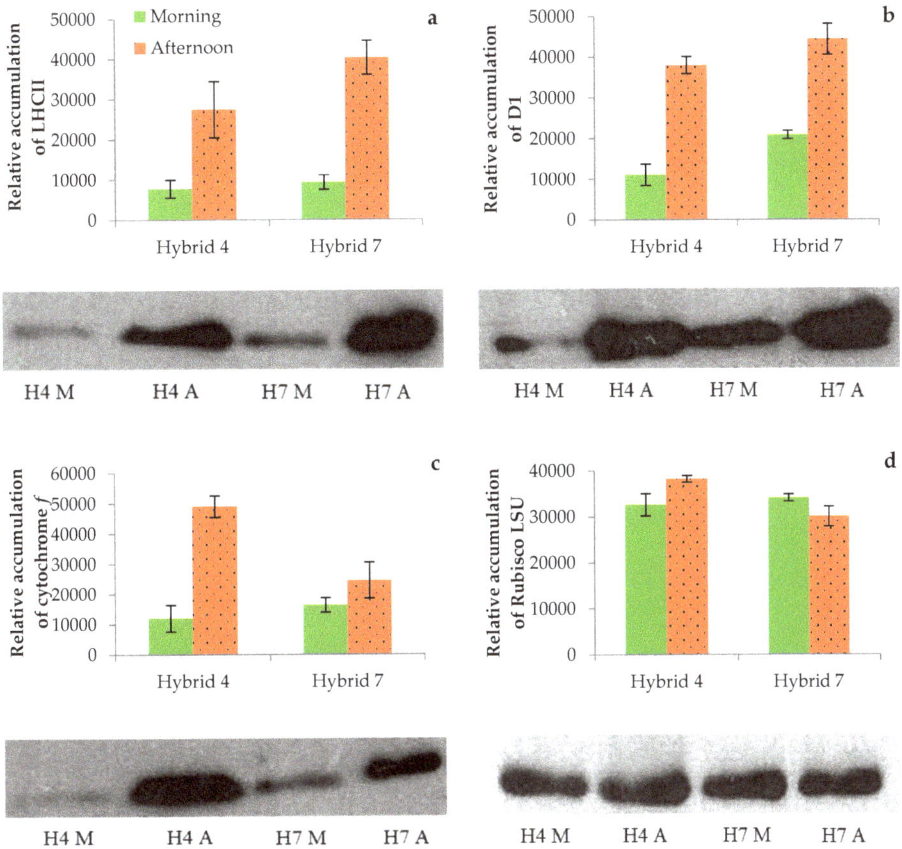

Figure 3. The relative accumulation of photosynthetic proteins: (**a**) light-harvesting complex of photosystem II—(LHCII); (**b**) D1; (**c**) cytochrome *f* (cyt *f*); and (**d**) Rubisco large subunit—(Rubisco LSU) in sunflower hybrids 4 and 7 (H4 and H7) under the morning (M) and afternoon (A) conditions. Lines in graphs represent mean values ± standard deviations of three replicates (n = 3).

The photosynthesis efficiency during heat and light stress depends on the stomatal conductivity and on the CO_2 diffusion, which affect the activation of the ribulose-1,5-biphosphate (RuBP) carboxylase/oxygenase enzyme that is known as Rubisco [63]. It is well known that the CO_2 diffusion and the Rubisco activity in RuBP carboxylation affect the photosynthetic rate. Rubisco mainly affects the efficiency of PSII and the relative electron transport through CO_2 fixation [64]. The same was confirmed in the research by Chen et al. [48], who investigated the effect of simultaneous heat and light stress conditions on wheat. In the afternoon conditions, the Rubisco LSU accumulation increased in hybrid 4, while it decreased in hybrid 7, which may indicate differences in the ROS accumulation, as ROS cause the degradation and fragmentation of the Rubisco LSU. Chen et al. [50] and Zivcak et al. [65] confirm that Rubisco LSU is deactivated under heat and light stress conditions, while Mlinarić et al. [14] report that high temperatures and excess light at noon did not affect the Rubisco LSU accumulation. Lu et al. [66] studied the impact of high temperatures on tomatoes. They confirm that high temperatures promoted the transcription of the Rubisco LSU, which was accompanied by a substantial reduction in the photosynthetic capacity, and by a slight inhibition of the Rubisco activity.

3.5. Principal Component Analysis

The presented and discussed relationships between the environmental conditions, the hybrid sensitivities, and the tested parameters were summarised and visualised with the principal component analysis (PCA) (Figure 4).

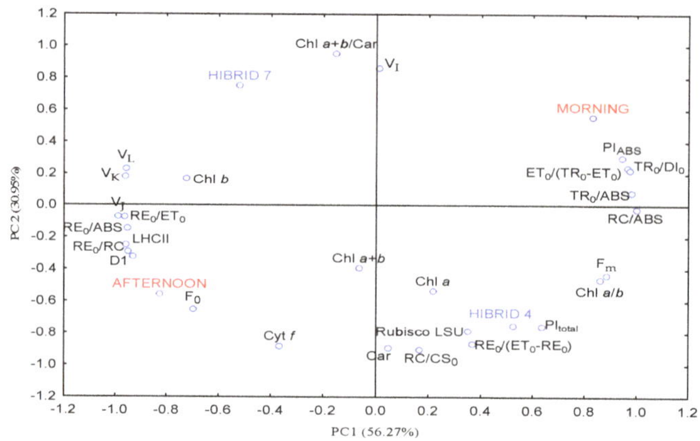

Figure 4. Biplot of principal component analysis of chlorophyll *a* fluorescence, photosynthetic pigment contents, and photosynthetic proteins in hybrids 4 and 7 (blue colour) under the morning and afternoon conditions (red colour). JIP parameters: minimum fluorescence intensity (F_0); maximum fluorescence intensity (F_m); relative variable fluorescence at 150 µs (V_L step); relative variable fluorescence at 300 µs (V_K step); relative variable fluorescence at 3 ms (V_J step); relative variable fluorescence at 30 ms (V_I step); maximum quantum yield of PSII (TR_0/ABS); density of active PSII reaction centers (RCs) per cross section (RC/CS_0); density of RC on chlorophyll *a* basis (RC/ABS); flux ratio trapping per dissipation (TR_0/DI_0); electron transport further than primary acceptor Q_A ($ET_0/(TR_0\text{-}ET_0)$); performance index (PI_{ABS}); quantum yield for reduction in the end electron acceptors at the PSI acceptor side (RE_0/ABS); probability that an electron from the electron transport chain is transferred to reduce end electron acceptors at the PSI acceptor side (RE_0/ET_0); electron-flux-reducing end electron acceptors at the PSI acceptor side per RC (RE_0/RC); electron transport from PQH_2 to final PSI acceptors ($RE_0/(ET_0\text{-}RE_0)$); and performance index for energy conservation from exciton to the reduction in the PSI end acceptors (PI_{total}). Photosynthetic pigment content: chlorophyll *a* (Chl *a*); chlorophyll *b* (Chl *b*); total chlorophyll *a* + *b* (Chl *a* + *b*); total carotenoids (Car); ratio of chlorophyll *a* and *b* (Chl *a*/*b*); and ratio of total chlorophyll content and carotenoids (Chl *a* + *b*/Car). The relative accumulation of photosynthetic proteins: light-harvesting complex of photosystem II (LHCII); D1; cytochrome *f* (cyt *f*); and Rubisco large subunit (Rubisco LSU).

The two principal components (PCs) explained 87.22% of the total variability. PC1 and PC2 were responsible for 56.27% and 30.95% of the variability. The RE_0/ABS, the TR_0/ABS, and the $ET_0/(TR_0\text{-}ET_0)$ positively contributed to PC1. A negative contribution to PC1 was determined for the V_J, the RE_0/ET_0, and the LHCII. The Chl *a* + *b*/Car, the V_I, and hybrid 7 positively contributed to PC2. According to the correlation coefficients, the synergic effect of the elevated temperatures and excess light was in a highly significant positive correlation with the F_0, the RE_0/RC, and protein D1, which means that these parameters could be useful as stress indicators for sunflowers. Furthermore, the RE_0/RC is in a significant, strong, and positive correlation with the V_J, the RE_0/ABS, the RE_0/ET_0, the LHCII, and D1, while it is in a significant, strong, and negative correlation with the TR_0/ABS, the TR_0/DI_0, the $ET_0/(TR_0\text{-}ET_0)$, and the PI_{ABS} (Supplementary Table S3), which is evident from the position of the mentioned parameters in the PCA biplot (Figure 4). D1 is in significant, strong, and positive correlation with the RE_0/RC and the LHCII, while the correlations with the TR_0/DI_0, the $ET_0/(TR_0\text{-}ET_0)$, and the PI_{ABS} were significant, strong, and negative (Figure 4 and Supplementary Table S3). These relationships indicate that increases in the

V_J, the RE_0/ABS, the RE_0/ET_0, the RE_0/RC, and the LHCII, as well as decreases in the TR_0/ABS, the TR_0/DI_0, the $ET_0/(TR_0\text{-}ET_0)$, and the PI_{ABS} can be signs that sunflower plants are under stress. Furthermore, the parameters that are in significant correlation can be used interchangeably, which simplifies and speeds up the analysis process. The association of the ChlF parameters with the environmental conditions was previously studied and confirmed by Pavlović et al. [9], Mihaljević et al. [5], and Viljevac Vuletić and Španić [43].

4. Conclusions

Although most of the tested parameters changed as expected during the elevated temperatures and excess light that were determined in the afternoon hours, only the F_0, the RE_0/RC, and D1 revealed significant, high, and positive correlations with the environmental conditions, which indicates their usefulness in sunflower-stress-response research. Some other chlorophyll fluorescence parameters (V_J, TR_0/ABS, TR_0/DI_0, $ET_0/(TR_0\text{-}ET_0)$, RE_0/ABS, RE_0/ET_0, RE_0/RC) and photosynthetic proteins (LHCII and D1) can be used as indicators of the physiologic changes that are caused by elevated temperatures and excess light as well, although they are only indirectly associated with environmental conditions.

According to the tested parameters, hybrid 4 appeared more adaptable to the elevated temperatures and excess light that were determined in the afternoon hours than hybrid 7. The better adaptability of hybrid 4 is evident from the smaller changes in the fluorescence curves in phases O–J and J–I, and especially in steps L, K, J, I, and by the higher PI_{ABS} values under the afternoon conditions. The photosynthetic apparatus of hybrid 7 can be considered to be more susceptible to the tested unfavourable weather conditions than that of hybrid 4 because of the significant impairment of its functionality, as is indicated by the positive L, K, and J steps, the increase in the TR_0/ABS, and the decrease in the $ET_0/(TR_0\text{-}ET_0)$, which caused the reduction in the PI_{ABS}. The more pronounced stress effect in hybrid 7 was confirmed by the higher relative accumulation of the LHCII potential as well. The determination of the photosynthetic efficiency, pigments, and proteins could be a useful selection criterion for the development of sunflower hybrids that are highly tolerant to elevated temperatures and excess light, which encourages sustainable sunflower production under stressful growing conditions.

Supplementary Materials: The following supporting information can be downloaded at: https://www.mdpi.com/article/10.3390/horticulturae8050392/s1, Table S1: Temperature and solar radiation ten days before measurements of chlorophyll *a* fluorescence; Table S2: Measured and calculated chlorophyll *a* fluorescence parameters according to Strasser et al. [7] and Yusuf et al. [32]; Table S3: Correlation coefficients among analysed traits and environmental conditions in sunflower hybrids.

Author Contributions: Conceptualisation, H.L., V.C. and A.S.; methodology, M.V.V., H.L. and J.A.D.; software, I.V.; validation, H.L. and M.V.V.; formal analysis, A.M.K., M.V.V. and J.A.D.; investigation, A.M.K.; resources, A.S. and Z.Z.; writing—original draft preparation, A.M.K.; writing–review and editing, M.V.V., M.M.K., J.A.D., A.S. and V.C.; visualisation, A.M.K.; supervision, V.C. and A.S. All authors have read and agreed to the published version of the manuscript.

Funding: This research was funded by the Ministry of Science and Education, Croatia, under grant numbers 073-0000000-3538 and 073-0731674-1673.

Institutional Review Board Statement: Not applicable.

Informed Consent Statement: Not applicable.

Data Availability Statement: Not applicable.

Conflicts of Interest: The authors declare no conflict of interest.

References

1. Jug, D.; Jug, I.; Brozović, B.; Vukadinović, V.; Stipešević, B.; Đurđević, B. The role of conservation agriculture in mitigation and adaptation to climate change. *Poljoprivreda* **2018**, *24*, 35–44. [CrossRef]
2. García-López, J.; Lorite, I.J.; García-Ruiz, R.; Domínguez, J. Evaluation of three simulation approaches for assessing yield of rainfed sunflower in a Mediterranean environment for climate change impact modelling. *Clim. Chang.* **2014**, *124*, 147–162. [CrossRef]
3. De la Haba, P.; De la Mata, L.; Molina, E.; Aguera, E. High temperature promotes early senescence in primary leaves of sunflower (*Helianthus annuus* L.) plants. *Can. J. Plant Sci.* **2014**, *94*, 659–669. [CrossRef]
4. Greer, D.H. Temperature-dependent responses of the photosynthetic and chlorophyll fluorescence attributes of apple (*Malus domestica*) leaves during a sustained high temperature event. *Plant Physiol. Biochem.* **2015**, *97*, 139–146. [CrossRef] [PubMed]
5. Mihaljević, I.; Lepeduš, H.; Šimić, D.; Viljevac Vuletić, M.; Tomaš, V.; Vuković, D.; Dugalić, K.; Teklić, T.; Babojelić, M.S.; Zdunić, Z. Photochemical efficiency of photosystem II in two apple cultivars affected by elevated temperature and excess light in vivo. *S. Afr. J. Bot.* **2020**, *130*, 316–326. [CrossRef]
6. Stirbet, A.; Lazár, D.; Guo, Y.; Govindjee, G. Photosynthesis: Basics, history and modelling (Review: Part of a special issue on functional-structural plant growth modelling). *Ann. Bot.* **2020**, *126*, 511–537. [CrossRef]
7. Strasser, R.J.; Tsimilli-Michael, M.; Srivastava, A. Analysis of the fluorescence transient. In *Chlorophyll Fluorescence: A Signature of Photosynthesis*; Advances in Photosynthesis and Respiration Series; Govindjee, G., Papageorgiou, G., Eds.; Springer: Dordrecht, The Netherlands, 2004; pp. 321–362.
8. Mlinarić, S.; Antunović Dunić, J.; Babojelić, M.S.; Cesar, V. Differential accumulation of photosynthetic proteins regulates diurnal photochemical adjustments of PSII in common fig (*Ficus carica* L.) leaves. *J. Plant Physiol.* **2017**, *209*, 1–10. [CrossRef]
9. Pavlović, I.; Mlinarić, S.; Tarkowská, D.; Oklestkova, J.; Novák, O.; Lepeduš, H.; Bok, V.V.; Brkanac, S.R.; Strnad, M.; Salopek-Sondi, B. Early Brassica Crops Responses to Salinity Stress: A Comparative Analysis Between Chinese Cabbage, White Cabbage, and Kale. *Front. Plant Sci.* **2019**, *10*, 450. [CrossRef]
10. Markulj Kulundžić, A.; Viljevac Vuletić, M.; Matoša Kočar, M.; Mijić, A.; Varga, I.; Sudarić, A.; Cesar, V.; Lepeduš, H. The Combination of Increased Temperatures and High Irradiation Causes Changes in Photosynthetic Efficiency. *Plants* **2021**, *10*, 2076. [CrossRef]
11. Misra, A.N.; Misra, M.; Singh, R. Chlorophyll fluorescence in plant biology. In *Biophysics*; Misra, A.N., Ed.; Intech Open: London, UK, 2012; pp. 171–192. Available online: http://www.intechopen.com (accessed on 22 April 2022).
12. Morales, A.; Kaiser, E. Photosynthetic Acclimation to Fluctuating Irradiance in Plants. *Front. Plant Sci.* **2020**, *11*, 268. [CrossRef]
13. Scheurwater, I.; Dunnebacke, M.; Eising, R.; Lambers, H. Respiratory costs and rate of protein turnover in the roots of a fast-growing (*Dactylis glomerata* L.) and a slow-growing (*Festuca ovina* L.) grass species. *J. Exp. Bot.* **2000**, *51*, 1089–1097.
14. Mlinarić, S.; Antunović Dunić, J.; Štolfa, I.; Cesar, V.; Lepeduš, H. High irradiation and increased temperature induce different strategies for competent photosynthesis in young and mature fig leaves. *S. Afr. J. Bot.* **2016**, *103*, 25–31. [CrossRef]
15. Yanhui, C.; Hongrui, W.; Beining, Z.; Shixing, G.; Zihan, W.; Yue, W.; Huihui, Z.; Guangyu, S. Elevated air temperature damage to photosynthetic apparatus alleviated by enhanced cyclic electron flow around photosystem I in tobacco leaves. *Ecotoxicol. Environ. Saf.* **2020**, *204*, 111136. [CrossRef] [PubMed]
16. Pospíšil, P. Production of Reactive Oxygen Species by Photosystem II as a Response to Light and Temperature Stress. *Front. Plant Sci.* **2016**, *7*, 1950. [CrossRef]
17. Busheva, M.; Garab, G.; Liker, E.; Tóth, Z.; Szèll, M.; Nagy, F. Diurnal fluctuations in the content and functional properties of the light harvesting chlorophyll a/b complex in thylakoid membranes. *Plant Physiol.* **1991**, *95*, 997–1003. [CrossRef]
18. Michoux, F.; Ahmad, N.; Wei, Z.Y.; Belgio, E.; Ruban, A.V.; Nixon, P.J. Testing the role of the N-terminal tail of D1 in the maintenance of photosystem II in tobacco chloroplasts. *Front. Plant Sci.* **2016**, *7*, 844. [CrossRef] [PubMed]
19. Chen, L.; Jia, H.; Tian, Q.; Du, L.; Gao, Y.; Miao, X.; Liu, Y. Protecting effect of phosphorylation on oxidative damage of D1 protein by down-regulating the production of superoxide anion in photosystem II membranes under high light. *Photosynth. Res.* **2012**, *112*, 141–148. [CrossRef]
20. Cramer, W.A.; Baniulus, D.; Yamashita, E.; Zhang, H.; Zatsman, A.I.; Hendrich, M.P. Cytochrome *b6f* complex, colon structure, spectroscopy, and function of heme cn: N-side electron and proton transfer reactions. In *Photosynthetic Protein Complexes*; Fromme, P., Ed.; Weily-Blackwell Verlag GmbH&Co.: Weinheim, Germany, 2008; pp. 155–179.
21. Helbling, E.W.; Buma, A.G.J.; Boelen, P.; Strate, H.J.; Giordanino, M.V.F.; Villafane, V.E. Increase in Rubisco activity and gene expression due to elevated temperature partially counteracts ultraviolet radiation-induced photoinhibition in the marine diatom *Thalassiosira weissflogii*. *Limnol. Oceanogr.* **2011**, *56*, 1330–1342. [CrossRef]
22. Markulj Kulundžić, A.; Kovačević, J.; Viljevac Vuletić, M.; Josipović, A.; Liović, I.; Mijić, A.; Lepeduš, H.; Matoša Kočar, M. Impact of abiotic stress on photosynthetic efficiency and leaf temperature in sunflower. *Poljoprivreda* **2016**, *22*, 17–22. [CrossRef]
23. Markulj Kulundžić, A.; Kovačević, J.; Viljevac Vuletić, M.; Jocić, S.; Cvejić, S.; Matoša Kočar, M.; Mijić, A.; Liović, I.; Sudarić, A.; Lepeduš, H.; et al. Effect of different soil water content effect on genotype expession in photosynthetic efficiency and leaf temperature in sunflower. *Genet.-Belgrade* **2016**, *48*, 971–982. [CrossRef]
24. Schneiter, A.A.; Miller, J.F. Description of sunflower growth stages. *Crop Sci.* **1981**, *21*, 901–903. [CrossRef]
25. Lichtenthaler, H.K. Chlorophylls and carotenoids: Pigments of photosynthetic biomembranes. *Methods Enzymol.* **1987**, *148*, 350–382. [CrossRef]

26. Bradford, M.M. A Rapid and sensitive method for quantitation of microgram quantities of protein utilising principle of protein-dye binding. *Anal. Biochem.* **1976**, *72*, 248–254. [CrossRef]
27. Laemmli, U.K. Cleavage of structural proteins during assembly of head of Bacteriophage-T4. *Nature* **1970**, *227*, 680–685. [CrossRef]
28. Towbin, H.; Staehelin, T.; Gordon, J. Electrophoretic transfer of proteins from polyacrylamide gels to nitrocellulose sheets: Procedure and some applications. *Proc. Natl. Acad. Sci. USA* **1979**, *76*, 4350–4354. [CrossRef]
29. Martinazzo, E.G.; Ramm, A.; Bacarin, M.A. The chlorophyll a fluorescence as an indicator of the temperature stress in the leaves of Prunus persica. *Braz. J. Plant Physil.* **2012**, *24*, 237–246. [CrossRef]
30. Tsimilli-Michael, M.; Strasser, R. The energy flux theory 35 years later: Formulations and applications. *Photosynth. Res.* **2013**, *117*, 289–320. [CrossRef] [PubMed]
31. Kalaji, H.M.; Jajoo, A.; Oukarroum, A.; Brestic, M.; Zivcak, M.; Samoborska, I.A.; Cetner, M.D.; Łukasik, I.; Goltsev, V.; Ladle, R.J. Chlorophyll fluorescence as a tool to monitor physiological status of plants under abiotic stress conditions. *Acta Physiol. Plant.* **2016**, *38*, 102. [CrossRef]
32. Yusuf, M.A.; Kumar, D.; Rajwanshi, R.; Strasser, R.J.; Tsimilli-Michael, M.; Govindjee; Sarin, N.B. Overexpression of g-tocopherol methyl transferase gene in transgenic Brassica juncea plants alleviates abiotic stress: Physiological and chlorophyll a fluorescence measurements. *Biochim. Biophys. Acta* **2010**, *1797*, 1428–1438. [CrossRef]
33. Brestic, M.; Zivcak, M.; Kalaji, H.M.; Carpentier, R.; Allakhverdiev, S.I. Photosystem II thermostability in situ: Environmentally induced acclimation and genotype-specific reactions in *Triticum aestivum* L. *Plant Physiol. Biochem.* **2012**, *57*, 93–105. [CrossRef] [PubMed]
34. Papageorgiou, G.C.; Govindjee. The Non-Photochemical Quenching of the Electronically Excited State of Chlorophyll a in Plants: Definitions, Timelines, Viewpoints, Open Questions. In *Non-Photochemical Quenching and Energy Dissipation in Plants, Algae and Cyanobacteria*; Demmig-Adams, B., Garab, G., Adams, W., III, Govindjee, Eds.; Springer: Dordrecht, The Netherlands, 2014; p. 44. [CrossRef]
35. Paunov, M.; Koleva, L.; Vassilev, A.; Vangronsveld, J.; Golstev, V. Effects Different Metals on Photosynthesis: Cadmium and Zink Affect Chlorophyll Fluorescence in Durum Wheat. *Int. J. Mol. Sci.* **2018**, *19*, 787. [CrossRef] [PubMed]
36. Desotgiu, R.; Pollastrini, M.; Cascio, C.; Gerosa, G.; Marzuoli, R.; Bussotti, F. Responses to ozone on Populus Oxford clone in an open top chamber experiment assessed before sunrise and in full sunlight. *Photosynthetica* **2013**, *51*, 267–280. [CrossRef]
37. Schansker, G.; Tóth, S.Z.; Holzwarth, A.R.; Garab, G. Chlorophyll *a* fluorescence: Beyond the limits of the Q_A model. *Photosyn. Res.* **2014**, *120*, 43–58. [CrossRef] [PubMed]
38. Kalaji, M.H.; Goltsev, V.N.; Żuk-Gołaszewska, K.; Zivcak, M.; Brestic, M. *Chlorophyll Fluorescence Understanding Crop Performance—Basics and Applications*, 1st ed.; CRC Press: Boca Raton, FL, USA, 2017.
39. Misra, A.N. Chlorophyll fluorescence: A practical approach to study ecophysiology of green plants. In *Advances in Plant Ecophysiology Techniques*; Sánchez-Moreiras, A.M., Reigosa, M.J., Eds.; Springer: Cham, Switzerland, 2018; Chapter 5; pp. 77–97.
40. Feng, B.; Liu, P.; Li, G.; Dong, S.T.; Wang, F.H.; Kong, L.A.; Zhang, J.W. Effect of Heat Stress on the Photosynthetic Characteristics in Flag Leaves at the Grain-Filling Stage of Different Heat-Resistant Winter Wheat Varieties. *J. Agro. Crop Sci.* **2014**, *200*, 143–155. [CrossRef]
41. Fghire, R.; Anaya, F.; Ali, O.I.; Benlhabib, O.; Ragab, R.; Wahbi, S. Physiological and photosynthetic response of quinoa to drought stress. *ChileanJAR* **2015**, *75*, 174–183. [CrossRef]
42. Çiçek, N.; Pekcan, V.; Arslan, Ö.; Erdal, S.E.; Balkan Nalçaiyi, A.S.; Çil, A.N.; Şahin, V.; Kaya, Y.; Ekmekçi, Y. Assessing drought tolerance in field-grown sunflower hybrids by chlorophyll fluorescence kinetics. *Braz. J. Bot.* **2019**, *42*, 249–260. [CrossRef]
43. Viljevac Vuletić, M.; Španić, V. Special issue in honour of Prof. Reto, J. Strasser–Characterisation of photosynthetic performance during natural leaf senescence in winter wheat: Multivariate analysis as a tool for phenotypic characterisation. *Photosynthetica* **2020**, *58*, 301–313. [CrossRef]
44. Gururani, M.A.; Venkatesh, J.; Ganesan, M.; Strasser, R.J.; Han, Y.; Kim, J.-I.; Lee, H.-Y.; Song, P.-S. In Vivo Assessment of Cold Tolerance through Chlorophyll-a Fluorescence in Transgenic Zoysiagrass Expressing Mutant Phytochrome A. *PLoS ONE* **2015**, *10*, e0127200. [CrossRef]
45. Kalaji, H.M.; Schansker, G.; Ladle, R.J.; Goltsev, V.; Bosa, K.; Allakhverdiev, S.I.; Brestic, M.; Bussotti, F.; Calatayud, A.; Dąbrowski, P.; et al. Frequently asked questions about in vivo chlorophyll fluorescence: Practical issues. *Photosynth. Res.* **2014**, *122*, 121–158. [CrossRef]
46. Yan, K.; Shao, H.; Shao, C.; Zhao, S.; Brestic, M. Dissection of photosynthetic electron transport process in sweet sorghum under heat stress. *PLoS ONE* **2013**, *8*, e62100. [CrossRef]
47. Arslan, Ö.; Balkan Nalçaiyi, A.S.; Çulha Erdal, S.; Pekcan, V.; Kaya, Y.; Çiçek, N.; Ekmekçi, Y. Special issue in honour of Prof. Reto J. Strasser–Analysis of drought response of sunflower inbred lines by chlorophyll a fluorescence induction kinetics. *Photosynthetica* **2020**, *58*, 348–357. [CrossRef]
48. Gupta, N.K.; Shubhi Agarwal, S.; Agarwal, V.P.; Nathawat, N.S.; Gupta, S.; Singh, G. Effect of short-term heat stress on growth, physiology and antioxidative defence system in wheat seedlings. *Acta Physiol. Plant* **2013**, *35*, 1837–1842. [CrossRef]
49. Prasad, S.M.; Dwivedi, R.; Zeeshan, M. Growth, photosynthetic electron transport, and antioxidant responses of young soybean seedlings to simultaneous exposure of nickel and UV-B stress. *Photosynthetica* **2005**, *43*, 177–185. [CrossRef]
50. Chen, Y.E.; Zhang, C.M.; Su, Y.Q.; Ma, J.; Zhang, Z.W.; Yuan, M.; Zhang, H.U.; Yuan, S. Responses of photosystem II and antioxidative systems to high light and high temperature co-stress in wheat. *Environ. Exp. Bot.* **2017**, *135*, 45–55. [CrossRef]

51. Oguchi, R.; Hikosaka, K.; Hirose, T. Does the photosynthetic light-acclimation need change in leaf anatomy? *Plant Cell Environ.* **2003**, *26*, 505–512. [CrossRef]
52. Tanaka, R.; Tanaka, A. Chlorophyll cycle regulates the construction and destruction of the light-harvesting complexes. *Biochim. Biophys. Acta* **2011**, *1807*, 968–976. [CrossRef]
53. Li, H.; Xu, H.; Zhang, P.; Gao, M.; Wang, D.; Zhao, H. High temperature effects on D1 protein turnover in three wheat varieties with different heat susceptibility. *Plant Growth Regul.* **2017**, *81*, 385–397. [CrossRef]
54. Su, X.Y.; Wu, S.; Yang, L.; Xue, R.L.; Li, H.; Wang, Y.X.; Zhao, H.J. Exogenous progesterone alleviates heat and high light stress-induced inactivation of photosystem II in wheat by enhancing antioxidant defense and D1 protein stability. *Plant Growth Regul.* **2014**, *74*, 311–318. [CrossRef]
55. Chan, T.; Shimizu, Y.; Pospisil, P.; Nijo, N.; Fujiwara, A.; Taninaka, Y.; Ishikawa, T.; Hori, H.; Nanba, D.; Imai, A.; et al. Quality Control of Photosystem II: Lipid Peroxidation Accelerates Photoinhibition under Excessive Illumination. *PLoS ONE* **2013**, *8*, e52100. [CrossRef]
56. Guo, W.-D.; Guo, Y.-P.; Liu, J.-R.; Mattson, N. Midday depression of photosynthesis is related with carboxylation efficiency decrease and D1 degradation in bayberry (*Myrica rubra*) plants. *Sci. Hortic.* **2009**, *123*, 188–196. [CrossRef]
57. Järvi, S.; Suorsa, M.; Aro, E.-M. Photosystem II repair in plant chloroplasts-Regulation, assisting proteins and shared components with photosystem II biogenesis. *Biochim. Biophys. Acta* **2015**, *1847*, 900–909. [CrossRef]
58. Vojta, L.; Carić, D.; Cesar, V.; Antunović Dunić, J.; Lepeduš, H.; Kveder, M.; Fulgosi, H. TROL-FNR interaction reveals alternative parthways of electron partitioning in photosynthesis. *Sci. Rep.* **2015**, *5*, 10085. [CrossRef]
59. Björkman, O. Response to different quantum flux densities. In *Physiological Plant Ecology I. Responses to the Physical Environment*; Lange, O.D., Nobel, P.S., Osmond, C.B., Ziegler, H., Eds.; Springer: Berlin, Germany, 1981; pp. 57–107.
60. Hojka, M.; Thiele, W.; Tóth, S.Z.; Lein, W.; Bock, R.; Schöttler, M.A. Inducible Repression of Nuclear-Encoded Subunits of the Cytochrome b6f Complex in Tobacco Reveals an Extraordinarily Long Lifetime of the Complex. *Plant Physiol.* **2014**, *165*, 1632–1646. [CrossRef]
61. Yamori, W.; Evans, J.R.; von Caemmerer, S. Effects of growth and measurement light intensities on temperature dependence of CO_2 assimilation rate in tobacco leaves. *Plant Cell Environ.* **2010**, *33*, 332–343. [CrossRef]
62. Pavlovič, A.; Stolárik, T.; Nosek, L.; Kouřil, R.; Ilík, P. Light-induced gradual activation of photosystem II in dark-grown Norway spruce seedlings. *Biochim. Biophys. Acta* **2016**, *1857*, 799–809. [CrossRef]
63. Yamori, W.; Hikosaka, K.; Way, D.A. Temperature response of photosynthesis in C3, C4, and CAM plants: Temperature acclimation and temperature adaptation. *Photosynth. Res.* **2014**, *119*, 101–117. [CrossRef]
64. Cheng, L.; Fuchigami, L.H.; Breen, P.J. The relationship between photosystem II efficiency and quantum yield for CO_2 assimilation is not affected by nitrogen content in apple leaves. *J. Exp. Bot.* **2001**, *52*, 1865–1872. [CrossRef]
65. Zivcak, M.; Brestic, M.; Kunderlikova, K.; Sytar, O.; Allakhverdiev, S.I. Repetitive light pulse-induced photoinhibition of photosystem I severely affects CO_2 assimilation and photoprotection in wheat leaves. *Photosynth. Res.* **2015**, *126*, 449–463. [CrossRef]
66. Lu, T.; Meng, Z.; Zhang, G.; Qi, M.; Sun, Z.; Liu, Y.; Li, T. Sub-high Temperature and High Light Intensity Induced Irreversible Inhibition on Photosynthesis System of Tomato Plant (*Solanum lycopersicum* L.). *Front. Plant Sci.* **2017**, *8*, 365. [CrossRef]

horticulturae

MDPI

Article

Melatonin Decreases Negative Effects of Combined Drought and High Temperature Stresses through Enhanced Antioxidant Defense System in Tomato Leaves

Mumithra Kamatchi K. Annadurai [1], Senthil Alagarsamy [1,*], Kalarani M. Karuppasami [2], Swarnapriya Ramakrishnan [3], Marimuthu Subramanian [4], Prasad R. B. Venugopal [1], Raveendran Muthurajan [5], Geethalakshmi Vellingiri [6], Vijayalakshmi Dhashnamurthi [1], Ravichandran Veerasamy [1], Boominathan Parasuraman [1], Sivakumar Rathinavelu [1] and Djanaguiraman Maduraimuthu [1,*]

1 Department of Crop Physiology, TNAU, Coimbatore 641003, India; mumithra1996@gmail.com (M.K.K.A.); prasadvenugopal@gmail.com (P.R.B.V.); vijiphysiology@gmail.com (V.D.); ravilux67@gmail.com (R.V.); boominathan.p@tnau.ac.in (B.P.); sivatnau5@gmail.com (S.R.)
2 Directorate of Crop Management, Tamil Nadu Agricultural University, Coimbatore 641003, India; kalarani.mk@tnau.ac.in
3 Floriculture Research Station, Thovalai 629302, India; swarnapriya@tnau.ac.in
4 Department of Agronomy, AC & RI, Eachangkottai, Thanjavur 631006, India; sm20@tnau.ac.in
5 Directorate of Research, TNAU, Coimbatore 641003, India; raveendrantnau@gmail.com
6 Office of the Vice-Chancellor, TNAU, Coimbatore 641003, India; geetha@tnau.ac.in
* Correspondence: senthil.a@tnau.ac.in (S.A.); jani@tnau.ac.in (D.M.)

Citation: Annadurai, M.K.K.; Alagarsamy, S.; Karuppasami, K.M.; Ramakrishnan, S.; Subramanian, M.; Venugopal, P.R.B.; Muthurajan, R.; Vellingiri, G.; Dhashnamurthi, V.; Veerasamy, R.; et al. Melatonin Decreases Negative Effects of Combined Drought and High Temperature Stresses through Enhanced Antioxidant Defense System in Tomato Leaves. *Horticulturae* **2023**, 9, 673. https://doi.org/10.3390/horticulturae9060673

Academic Editors: Małgorzata Majewska, Barbara Hawrylak-Nowak and Agnieszka Hanaka

Received: 3 May 2023
Revised: 24 May 2023
Accepted: 30 May 2023
Published: 6 June 2023

Abstract: In tomato (*Lycopersicon esculentum* L.), the effects of combined drought (D) and high temperature (HT) stress during the flowering stage had not been studied in detail. Therefore, this study was conducted with an objective of quantifying the effects of foliar spray of melatonin under individual and combined drought and HT stress. At flowering stage, D stress was imposed through withholding irrigation, while HT stress was imposed through exposing the plants to ambient temperature (AT) along with an increase of +5 °C. Under D + HT, plants were first subjected to drought followed by a + 5 °C increase in AT. The duration of individual or combined stress was ten days. At 80% available soil moisture, 100 μM melatonin was sprayed on D, HT, or D + HT treated plants. Among the stresses, D + HT stress increased the thylakoid membrane damage and decreased the photosynthetic rate and fruit yield more than D or HT stress. Foliar spray of 100 μM melatonin produced decreased thylakoid membrane damage [D: 31%, HT: 26%, and D + HT: 18%] and increased antioxidant enzyme, viz., superoxide dismutase, catalase, peroxidase, ascorbate peroxidase, and glutathione reductase, activity over stress-control plants. The photosynthetic rate [D: 24%, HT: 22%, and D + HT: 19%] and fruit yield [D: 32%, HT: 23%, and D + HT: 16%] were increased over stress-control plants. Hence, it is evident that the increased photosynthetic rate and fruit yield in D + HT and 100 μM melatonin-sprayed plants may be associated with an increased antioxidant defense system. Melatonin as a novel biostimulator has a great potential in scavenging free radicals through increased antioxidant activity, which shields the photosynthetic membrane from damage and therefore helps in stress mitigation.

Keywords: melatonin; drought; high temperature; antioxidants; free radicals; photosynthesis; lipid peroxidation; mitigation

1. Introduction

Climate variability is associated with releasing greenhouse gas emissions [1,2]. The Intergovernmental Panel on Climate Change (IPCC) indicates that the increase in air temperature from baseline should be less than 1.5 °C, and if it exceeds the threshold, it will affect crop productivity [3]. Similar to high temperature (HT), drought (D) is also an abiotic stress which is more frequent due to reduced precipitation and water vapor fluxes in the atmosphere, which affects crop productivity [4]. From 1970 to 2000, the percentage of drought-affected area was doubled [5]. The global population is projected to increase significantly by 2050, demanding increased crop production or productivity to meet food security [6]. Hence, to meet the global food demand and sustain the crop yield under a changing climate, developing a crop management solution to mitigate drought or high-temperature stress is mandatory [7].

Drought inhibits photosynthesis [8], thus decreasing the assimilate partitioning and lowering fruit yield [9]. Drought causes decreased stomatal conductance due to which diffusion of CO_2 also decreases, which in turn results in stomatal closure [10]. Additionally, high-temperature stress denatures the photosynthetic pigments involved in the light reaction and damages the thylakoid membrane responsible for producing $NADPH_2$ and ATP [11]. Under abiotic stress, increased malondialdehyde levels indicate oxidative damage in plants. Therefore, plants rely on the enzymatic antioxidants, viz., superoxide dismutase (SOD), catalase (CAT), peroxidase (POD), ascorbate peroxidase (APX), and glutathione reductase (GR), to scavenge reactive oxygen species (ROS) produced under stress, thereby protecting the membrane from damage [12].

Melatonin (N-acetyl-5-methoxytryptamine), an indoleamine compound, was discovered in the pineal gland of animals [13] and has similarities with other tryptophan derivatives [14]. Studies suggest that melatonin has various roles in the plant developmental process, namely, improved seed germination and seedling growth [15,16], photosystem activity [17], antioxidant defense system [18], osmoregulation [19], rooting depth [20], and fruit yield and quality [21,22]. In contrast, melatonin decreases the leaf senescence process [23]. It is predicted that in the current and future climate, crop yield will be affected by two or more abiotic stresses during their reproductive phase [24,25]. The effect of melatonin on drought or high-temperature stress in tomato has been studied in detail [26,27]. However, the impacts of combined drought and high-temperature stress on plants have not been quantified.

Tomato is one of the most popular and commercially grown vegetable crops and is susceptible to drought or HT stress which could cause a yield loss of 70% [28]. In tomato, the reproductive stage is more sensitive to drought or high-temperature stress because it affects the pre- and post-fertilization processes, and carbohydrate translocation from source to sink, thus, reducing fruit yield [29]. Previous research on tomato confirmed that melatonin could increase antioxidant enzymes [30]. The antioxidant molecules are used to mitigate the detrimental effects of abiotic stress through (i) decreasing thylakoid membrane damage (F_0/F_m ratio), (ii) increasing the photosynthetic activity due to less damage in photosystem II (PSII), where the initial reaction of photosynthesis take place in the thylakoid membrane, and (iii) decreasing chlorophyll degradation via protecting the chlorophyll biosynthetic enzyme [31]. In contrast, antioxidants will reduce levels of (i) malondialdehyde, (ii) free radicals, and (iii) electrolyte leakage [32,33]. The effect of melatonin on crops is presented in Supplementary Table S1 [34–47]. Based on this, we hypothesize that melatonin could increase the antioxidant defense system, resulting in increased photosynthetic rate and yield. The main aim of this study is to exploit the antioxidant potential of melatonin against drought, high-temperature, or combined drought and high-temperature induced oxidative stress; its protective role in the photosynthetic system; and its impact on membrane integrity.

2. Materials and Methods

2.1. Experimental Details

An experiment was conducted in a completely randomized block design with two factors and four replications. The first factor was the type of stress with three levels (drought, high temperature, and combined drought and high temperature), and the second factor was the foliar spray of melatonin with four levels: (i) absolute control (plants were grown in ambient temperature, maintained under 100% field capacity, and received no spray), (ii) stress control (for drought stress, plants were maintained under drought stress and received no spray; for high-temperature stress, plants were maintained under high-temperature stress and received no spray; and for combined drought and high-temperature stress, plants were drought and high-temperature stressed and received no spray), (iii) 80 µM melatonin, and (iv) 100 µM melatonin (Figure 1).

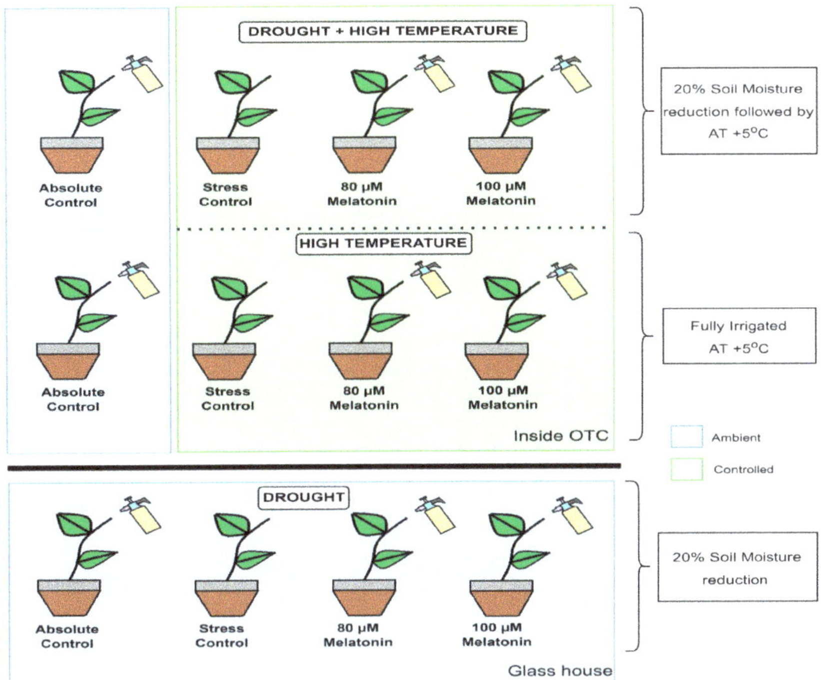

Figure 1. The image represents the stress imposition methodology and foliar treatment details used in this experiment conducted in OTC and glass house for 10 days.

The seedlings of tomato hybrid 'Shivam' were grown in portrays containing a vermicompost and coir pith. Based on uniform growth and good health, twenty-one-day-old tomato seedlings were used for transplanting. This experiment was conducted in the Glasshouse and Open Top Chamber (OTC) at the Department of Crop Physiology, Tamil Nadu Agricultural University, Coimbatore, India, from March to June 2022. The twenty-one-day-old seedlings were moved to large-sized plastic pots (46 cm in length and 60 cm in diameter) containing a mixture of red soil, sand, and vermicompost in a ratio of 3:1:1. In a pot, two plants were maintained, and the plants were watered on alternate days. All the pots were maintained under open sunlit condition. Ten days after transplanting, plants were supplied with a recommended dose of nutrients. During crop growth and development, the required crop management practices were followed as per the horticulture crop production guide [48].

2.2. Stress Imposition and Treatment Details

The plants were maintained under well-watered and ambient temperature conditions until the flower initiation stage, which coincides with the last week of April. At 50% flowering stage, plants were moved to the controlled environment facility for imposing Drought (D), high-temperature (HT), or combined drought or high-temperature stress (D + HT) for 10 days. The duration of stress imposition for ten days depends on the reduction in soil moisture content up to 60 to 70 percent under D and D + HT stress, while in case of HT stress, it depends on reduction in relative humidity up to 40 percent. Well-watered and D-stressed plants were maintained under ambient conditions, whereas the high-temperature-stressed plants were placed inside an Open Top Chamber (OTC) maintained at AT + 5 °C. In case of D + HT stress, plants were first exposed to drought (20% soil moisture reduction) and then subjected to AT + 5 °C. During the experimental period, the relative humidity ranged between 47 and 75 percent. Meteorological data of OTC are shown in Figure 2A. In D and D + HT stress experiments, soil moisture was regularly measured using a theta probe, the moisture content was calculated based on a reduction from 100% field capacity, and the soil moisture data are recorded and presented in Figure 2B. Melatonin (80 μM or 100 μM) was sprayed at 80% field capacity, and observations were recorded at the end of D, or HT, or D + HT stress. Plants exposed to HT stress were maintained at fully irrigated conditions and on the fourth day of stress, the plants under D, HT, and D + HT stress were sprayed with either 80 μM or 100 μM of melatonin.

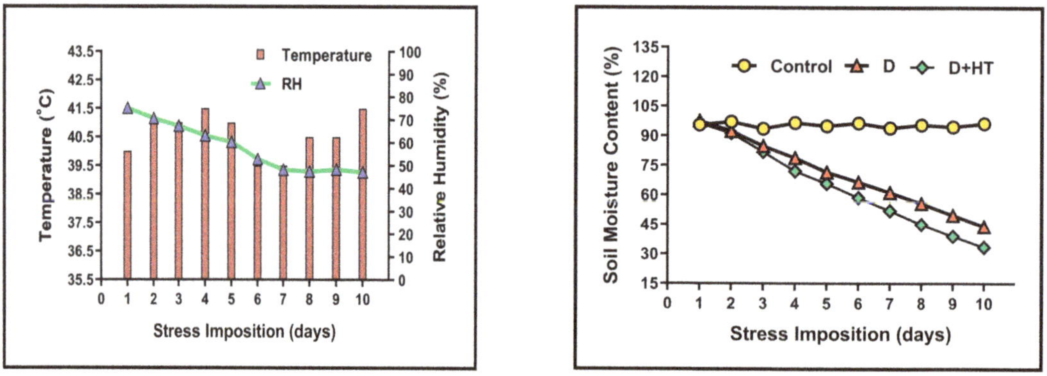

(**A**) OTC (**B**) Soil moisture

Figure 2. Temperature and soil moisture data recorded during the experiment. (**A**) Daily temperature and relative humidity under OTC during ten days of stress imposition; (**B**) soil moisture content under drought and combined stress for a period of ten days.

2.3. Preparation of Melatonin Solution

Melatonin chemical was purchased from Sigma-Aldrich Pvt. Ltd. India and stored at −20 °C. Irrespective of varieties, a previous study reported the significant results of 0.1 mM melatonin among different concentrations [49]. However, a preliminary lab study was performed on germination parameters using various concentrations of melatonin, viz., 20 μM, 40 μM, 60 μM, 80 μM, 100 μM, and 120 μM, under the PEG-induced drought and temperature-inductive response methodology. Based on the results obtained from initial screening, 80 μM and 100 μM melatonin showed significant difference among other treatments. Therefore, stock solution was prepared using the required quantity of melatonin, dissolving in 99.9% ethanol, and made to final volume using distilled water. The two final concentrations of melatonin (80 μM or 100 μM) were prepared via diluting the stock solution, and 0.25 mL of surfactant (Tween 20) was added to the melatonin solution to increase its absorption efficiency in leaves.

2.4. Sampling

The leaf samples were collected at the end of the stresses, and the collected leaf was used for physiological and biochemical analysis in one of the two plants. The yield and yield components were recorded in both plants, and the average was presented. Fresh leaves were collected at the end of the stress and immediately dipped in liquid nitrogen, grounded using liquid nitrogen, to assess biochemical parameters and enzyme activity.

2.5. Physiological Attributes

The chlorophyll index was determined in the second distal leaflet of the second and fourth leaf from the top using a chlorophyll meter (SPAD) (Minolta, Japan). The photosynthetic rate (P_n), stomatal conductance (g_s), transpiration rate (E), and intercellular CO_2 concentration (Ci) were measured in third leaf using a portable photosynthesis system (LI-6400 XT; LI-COR Inc., Lincoln, NE, USA). The leaf chlorophyll fluorescence was measured in the third leaf using a chlorophyll fluorometer [50]. Upon dark adaptation of the leaf using clips for 30 min, minimal fluorescence (F_0), maximum fluorescence (F_m), and variable fluorescence ($F_v = F_m - F_0$) were measured. The ratio of minimum fluorescence to maximum fluorescence (F_0/F_m ratio) was calculated using the data taken. The F_0/F_m ratio is referred to as thylakoid membrane damage. These observations were taken from 10:00 am to 12:30 pm simultaneously from the fully expanded leaf below the apex.

2.6. Histochemical Detection of ROS

Hydrogen peroxide (H_2O_2) and superoxide anion (O_2^-) generation rate was detected histochemically, as mentioned in Lei et al. [51] using the 3,3-diaminobenzidine (DAB) and nitro blue tetrazolium (NBT) staining method. Fresh leaves were dipped in 1 mg mL^{-1} DAB solution containing 50 mM sodium phosphate buffer (pH 3.8) and incubated for 5 h in the dark, during which brown precipitates were formed, indicating H_2O_2 accumulation. To detect superoxide anions, the leaves were immersed in 50 mM sodium phosphate buffer (pH 7.5) containing 0.2% NBT. The formation of dark blue insoluble formazan detects O_2^- accumulation. The destaining was followed with ethanol, glacial acetic acid, and glycerol in the ratio of 3:1:1, respectively, and the excess stain was removed via two to three washes using distilled water. Samples were placed in 80% glycerol, and photographs were taken.

2.7. Analysis of Hydrogen Peroxide and Superoxide Anion Content

Hydrogen peroxide content (H_2O_2) was measured as per Velikova and Loreto's method [52] through measuring the absorbance at 390 nm and expressed in µmol per gram of fresh weight. The superoxide anion (O_2^-) was estimated as per the method of Doke [53]. 0.5 g leaves was placed in the test tube containing 7 mL of 50 mM sodium azide and incubated for 5 min in the dark. From this solution, 2 mL was taken and subjected to heating at 85 °C for 15 min, then cooling on ice for 5 min. The data is expressed as an increase in absorbance at 580 nm per gram of fresh weight.

2.8. Membrane Integrity

Malondialdehyde content was estimated using the thiobarbituric acid method, according to Heath and Packer [54]. 500 mg of the leaf samples was taken and macerated with 0.1% TCA and centrifuged at 5000 rpm for 10 min, and the supernatant was collected, to which 4 mL of 20% TCA containing 0.5% TBA was added and subjected to heating at 95 °C for 30 min in a water bath followed by cooling and centrifugation. Finally, MDA content was calculated via subtracting the absorbance at 532 nm and 600 nm and expressed as µmol per gram. Leaf discs were made from the fresh leaf of drought or high-temperature stress or the combined drought-and-high-temperature-stressed plant. The leaf was immersed in distilled water and incubated for 24 h; then, the leakage was determined initially with a conductivity meter (EC_1). Then, these samples were heated at 100 °C for one hour, and the electrical conductivity of the solution was recorded (EC_2). The electrolyte leakage of the sample was expressed as a percentage [55].

2.9. Antioxidant Enzyme Activity

One gram of leaf sample was macerated with 50 mM phosphate buffer containing (pH 7.0), 0.1 mM EDTA, 0.1 mM phenyl methane sulfonyl fluoride, 1% PVP (*w*/*v*), and 0.2% (*v*/*v*) Triton X-100 using pre-chilled pestle and mortar and centrifuged at 10,000 rpm for 20 min at 4 °C. The supernatant was used to estimate the antioxidant enzyme activity as described in Camejo et al. [56].

The enzyme superoxide dismutase (SOD) was determined using the nitroblue tetrazolium (NBT) method described in Beauchamp and Fridovich [57]. The reaction mixture (3 mL) contained 0.1 mL of enzyme extract, 1.5 mL of 50 mM phosphate buffer (pH 7.8), 0.1 mL of 2 mM EDTA, 0.2 mL of 9.9 mM L-methionine, 0.1 mL of 0.02% Triton X-100, 0.1 mL of 55 μM NBT, and 0.1 mL of 1 mM riboflavin. The absorbance of control and blank was measured at 560 nm, and SOD activity was expressed as units per mg of protein. One unit of SOD is the quantity of enzymes necessary to inhibit NBT by 50% at 25 °C. According to Lowry et al. [58], the total protein was estimated using bovine serum albumin as a standard. The reaction mixture (3 mL) contained 0.1 mL enzyme extract and 2.6 mL of 50 mM potassium phosphate buffer (pH 7.0). 0.1 mL of 15 mM H_2O_2 was added, and the absorbance was recorded at 240 nm for 2 min. Catalase (CAT) activity was assessed based on the disappearance of H_2O_2 during the reaction initiation and calculated using an extinction coefficient of 43.6 mM^{-1} cm^{-1} and expressed as enzyme units per mg of proteins [59]. The peroxidase (POD) activity was measured according to the procedure of Kumar and Khan [60]. A 0.1 mL enzyme extract was added to the reaction mixture (3 mL) containing 1 mL of 100 mM phosphate buffer (pH 7.0), 0.5 mL of 10 mM pyrogallol, and 0.5 mL of 5 mM H_2O_2. Later, the solution was incubated for 5 min at 25 °C, and the reaction was terminated through adding 0.5 mL of 2.5 N H_2SO_4. The absorbance was recorded at 420 nm for 3 min at 30 s intervals, and the activity was calculated using the extinction coefficient of 12 mM^{-1} cm^{-1} and expressed in μmol of purpurogallin min^{-1} mg of $protein^{-1}$. According to Chen and Asada [61], ascorbate peroxidase (APX) activity was determined using 1 mL of the reaction mixture comprised of 0.05 mL enzyme extract, 0.85 mL of 50 mM phosphate buffer (pH 7.0), 0.05 mL of 0.1 mM ascorbate, and 0.05 mL of 0.3 mM H_2O_2, and the measure of absorbance was recorded at 290 nm for 1 min. APX activity was calculated using an extinction coefficient of 2.8 mM^{-1} cm^{-1} and expressed in units per mg of protein. Glutathione reductase (GR) was quantified as per the procedure of Smith et al. [62]. The enzyme activity was measured with 1 mL of reaction mixture containing enzyme extract, 100 mM potassium phosphate buffer (pH 7.0) containing 1 mM EDTA, 150 μM NADPH, and 500 μM oxidized glutathione. The enzyme activity was measured at an absorbance of 340 nm and expressed as enzyme units per mg of protein.

2.10. Relative Tolerance Index (RTI)

The tolerance level of plants exposed to stress and foliar spray was indirectly calculated using stomatal conductance [63]. The RTI was calculated using the formula:

$$\text{RTI (\%)} = \frac{\text{Stomatal conductance of stressed plant}}{\text{Stomatal conductance of unstressed plant}} \times 100$$

2.11. Yield

Fruit was harvested for seven pickings; the number of fruits harvested per picking was counted, and the total was represented as the total fruits per plant. The weight of tomato fruit at each harvest was recorded and expressed as fruit yield per plant.

2.12. Statistical Analysis

The experiment was laid out in a Factorial Completely Randomized Design (FCRD) with four replications. The data were statistically analyzed using SPSS for windows, version 16.0. Chicago, SPSS Inc., USA, and the graphs of observed variables were obtained using Graphpad prism software for windows, version 9.0.0. The results were presented as the mean of four replications and standard error of means (SEM). Based on analysis of variance (ANOVA), the least significant difference test ($LSD_{5\%}$) was used for means comparison. The significance was denoted using small letters, given that the means with same letters are not statistically significant at $p = 0.05$. The mean value of each trait is presented in Supplementary Tables S2–S16.

3. Results

The effect of stress, foliar spray, and the interaction of stress and foliar spray was significant ($p < 0.05$) for the chlorophyll index (Figure 3A) and thylakoid membrane damage (Figure 3B). Among the stresses, D + HT stress decreased the chlorophyll index by a greater magnitude than D or HT stresses alone. Among the foliar sprays, a higher level of chlorophyll index was observed in 100 µM melatonin-treated plants than in other treatments. Application of 100 µM melatonin to D (15%), HT (13%), and D + HT (10%) stressed plants increased the chlorophyll index more than other treatment combinations. In contrast, the thylakoid membrane damage was more remarkable in D + HT-stressed plants than D or HT-stressed plants (Figure 3B). Foliar spray of 100 µM melatonin to D + HT-stressed plants decreased the thylakoid membrane damage by 18%, which was lower than D + 100 µM melatonin (31%), and HT + 100 µM melatonin (26%) sprayed plants (Figure 3B).

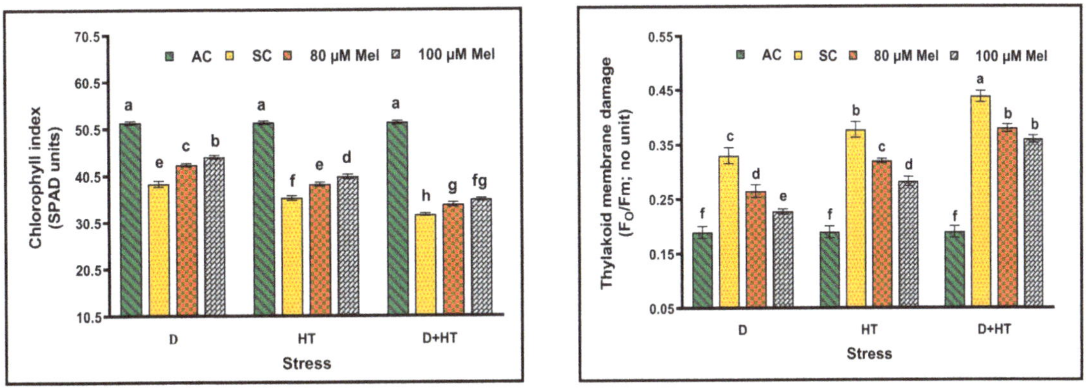

(**A**) Chlorophyll index (**B**) Thylakoid membrane damage

Figure 3. Effect of stress (drought—D, high temperature—HT, and D + HT) and foliar spray (irrigated control—AC, stress control—SC, 80 µM melatonin—80 µM Mel, and 100 µM melatonin—100 µM Mel) on (**A**) chlorophyll index (SPAD units) and (**B**) thylakoid membrane damage (F_0/F_m) in tomato on 10th day of stress. The results were presented as mean of four replications and standard error of means (SEM). Based on analysis of variance (ANOVA), the least significant difference test ($LSD_{5\%}$) was used for means comparison. The significance was denoted by small letters, given that the means with same letters are not statistically significant at $p = 0.05$.

The gas exchange parameters, viz., P_n (Figure 4A), E (Figure 4B), g_s (Figure 4C), and Ci (Figure 4D), were significantly ($p < 0.05$) influenced by stress, foliar spray, and their interactions (Figure 4A–D). Among the stresses, a higher decrease in P_n, E, and g_s was recorded in D + HT-stressed plants than D or HT stress (Figure 4A–D). Among the foliar sprays, 100 µM melatonin-treated plants showed an increased P_n, E, and g_s and decreased Ci compared to other foliar spray treatments (Figure 4A–D). A foliar spray of 100 µM melatonin on D-stressed plants yielded a higher increase in P_n (24%), E (14%), and g_s (32%) than HT + 100 µM melatonin and D + HT + 100 µM melatonin-sprayed plants (Figure 4A–D).

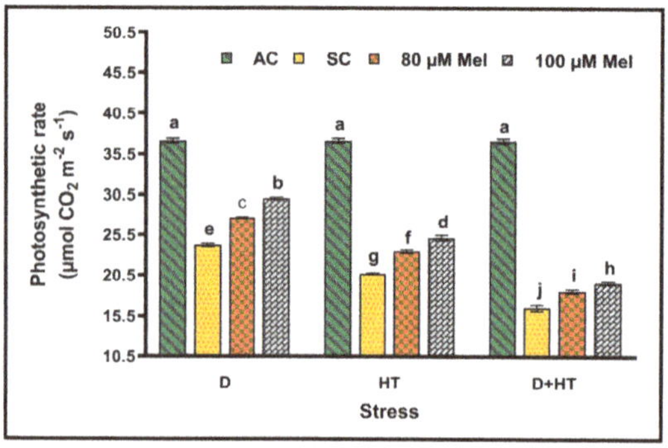

(**A**) Photosynthetic rate

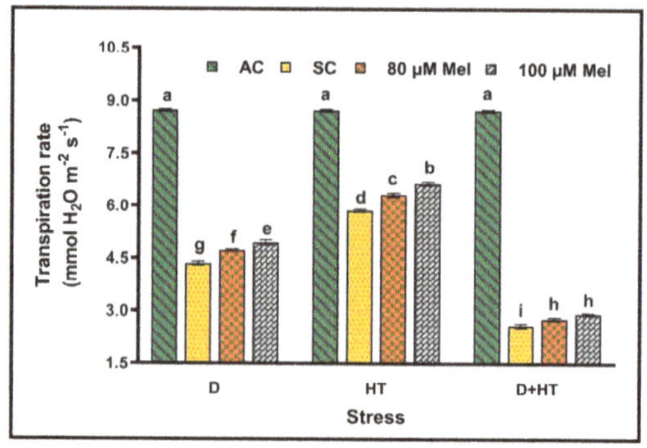

(**B**) Transpiration rate

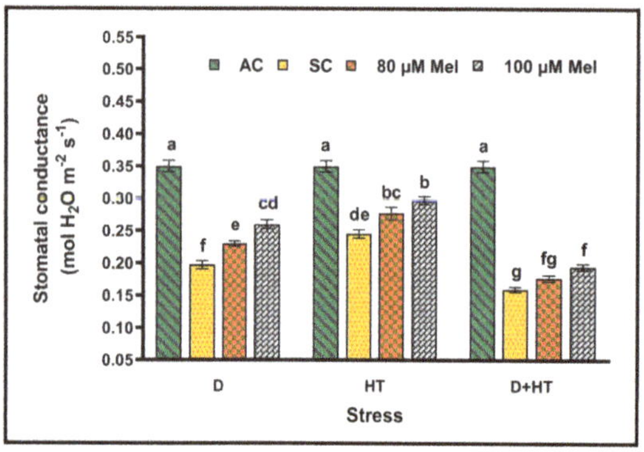

(**C**) Stomatal conductance

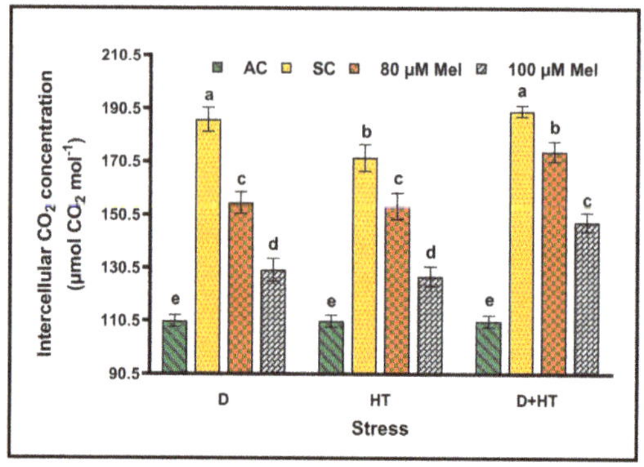

(**D**) Intercellular CO_2 concentration

Figure 4. Effect of stress (drought—D, high temperature—HT, and D + HT) and foliar spray (irrigated control—AC, stress control—SC, 80 µM melatonin—80 µM Mel, and 100 µM melatonin—100 µM Mel) on (**A**) photosynthetic rate, (**B**) transpiration rate, (**C**) stomatal conductance, and (**D**) intercellular CO_2 concentration in tomato on 10th day of stress. The results were presented as mean of four replications and standard error of means (SEM). Based on analysis of variance (ANOVA), the least significant difference test ($LSD_{5\%}$) was used for means comparison. The significance was denoted by small letters, given that the means with same letters are not statistically significant at $p = 0.05$.

The effect of stress, foliar spray, and the interaction of stress and foliar spray was significant ($p < 0.05$) for staining (Figure 5A,B) and hydrogen peroxide and superoxide anion contents (Figure 6A,B). Among the stresses, D + HT-stressed plants had a higher free radical content and staining than D or HT stress (Figures 5A,B and 6A,B). Among the foliar sprays, decreased H_2O_2 and O_2^- content and staining were observed in 100 µM melatonin-treated plants than in other treatments (Figures 5A,B and 6A,B). Drought-stressed plants sprayed with 100 µM melatonin had decreased free radical content and staining to a higher level than HT + 100 µM melatonin and D + HT + 100 µM melatonin-sprayed plants (Figures 5A,B and 6A,B).

(**A**) Hydrogen peroxide generation rate (**B**) Superoxide anion generation rate

Figure 5. Effect of stress (drought—D, high temperature—HT, and D + HT) and foliar spray (irrigated control—AC, stress control—SC, 80 μM melatonin—80 μM Mel, and 100 μM melatonin—100 μM Mel) on (**A**) histochemical detection of hydrogen peroxide generation rate via DAB staining and (**B**) histochemical detection of superoxide anion generation rate via NBT staining in tomato on 10th day of stress.

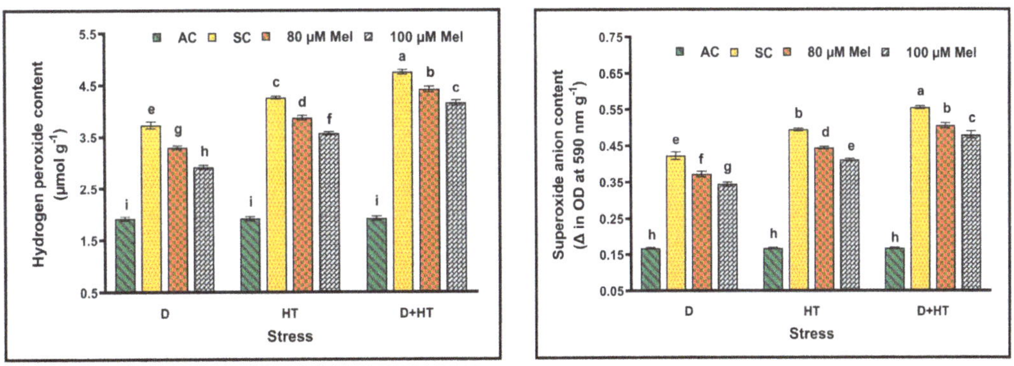

(**A**) Hydrogen peroxide content (**B**) Superoxide anion content

Figure 6. Effect of stress (drought—D, high -temperature—HT, and D + HT) and foliar spray (irrigated control—AC, stress control—SC, 80 μM mMelatonin—80 μM Mel, and 100 μM mMelatonin—100 μM Mel) on (**A**) hydrogen peroxide content, and (**B**) superoxide anion content in tomato on 10th day of stress. The results were presented as mean of four replications and standard error of means (SEM). Based on analysis of variance (ANOVA), the least significant difference test ($LSD_{5\%}$) was used for means comparison. The significance was denoted by small letters, given that the means with same letters are not statistically significant at $p = 0.05$.

The effect of stress, foliar spray, and the interaction of stress and foliar spray was significant ($p < 0.05$) for malondialdehyde (MDA) content and electrolyte leakage (EL) (Figure 7A,B). Among the stresses, D + HT-stressed plants showed increased MDA contents and electrolyte leakage to a higher level than HT or D stresses (Figure 7A,B). Among the foliar sprays, 100 μM melatonin-treated plants had decreased MDA content and electrolyte leakage level than in other treatments (Figure 7A,B). A greater decrease in MDA and electrolyte leakage was observed under D + 100 μM melatonin-sprayed plants than HT + 100 μM melatonin and D + HT + 100 μM melatonin-sprayed plants (Figure 7A,B).

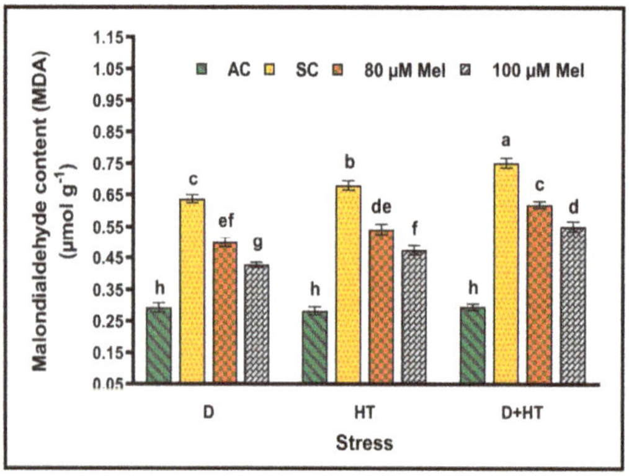

(**A**) Malondialdehyde content

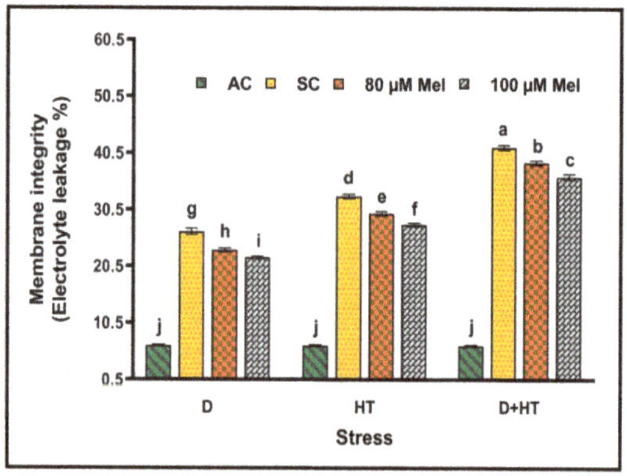

(**B**) Electrolyte leakage

Figure 7. Effect of stress (drought—D, high temperature—HT, and D + HT) and foliar spray (irrigated control—AC, stress control—SC, 80 μM melatonin—80 μM Mel, and 100 μM melatonin—100 μM Mel) on (**A**) malondialdehyde content and (**B**) electrolyte leakage in tomato on 10th day of stress. The results were presented as mean of four replications and standard error of means (SEM). Based on analysis of variance (ANOVA), the least significant difference test ($LSD_{5\%}$) was used for means comparison. The significance was denoted by small letters, given that the means with same letters are not statistically significant at $p = 0.05$.

The antioxidant enzymes, viz., SOD (Figure 8A), CAT (Figure 8B), POD (Figure 8C), APX (Figure 8D), and GR (Figure 8E), were significantly ($p < 0.05$) influenced by stress, foliar spray, and the interaction of stress and foliar spray (Figure 8A–E). Among the stresses, SOD, CAT, and POD activity was higher under D + HT stress than under D or HT stress. In contrast, the same treatment showed less activity of APX and GR (Figure 8A–E). Among the foliar sprays, increased SOD, CAT, and POD enzyme activity was recorded in 100 μM melatonin-treated plants compared to other foliar spray treatments (Figure 8A–C). D + 100 μM melatonin-sprayed plants had an increased SOD (17%), CAT (24%), and POD (27%) activity than HT + −100 μM melatonin-treated plants and D + HT +100 μM melatonin-treated plants (Figure 8A–C). A similar trend was observed for APX and GR enzyme activity (Figure 8D,E).

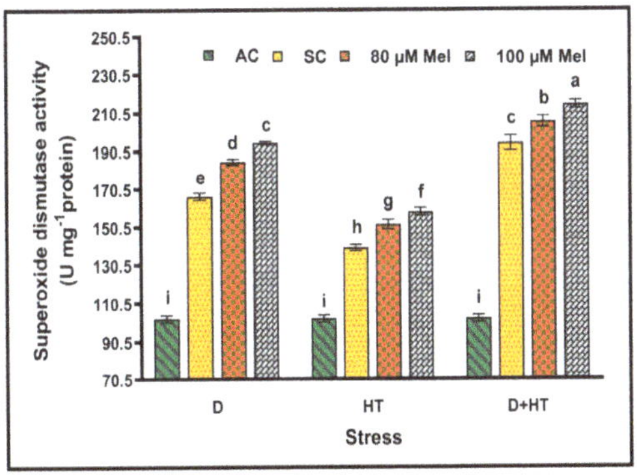

(**A**) Superoxide dismutase

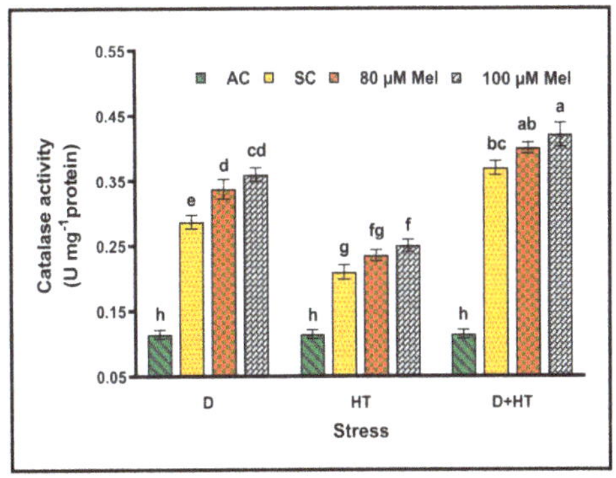

(**B**) Catalase

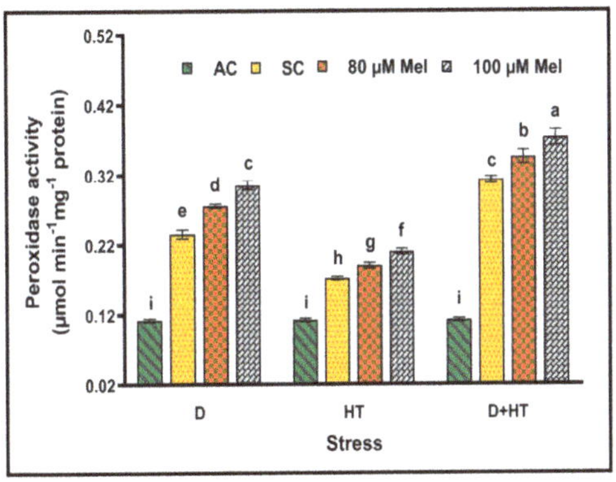

(**C**) Peroxidase

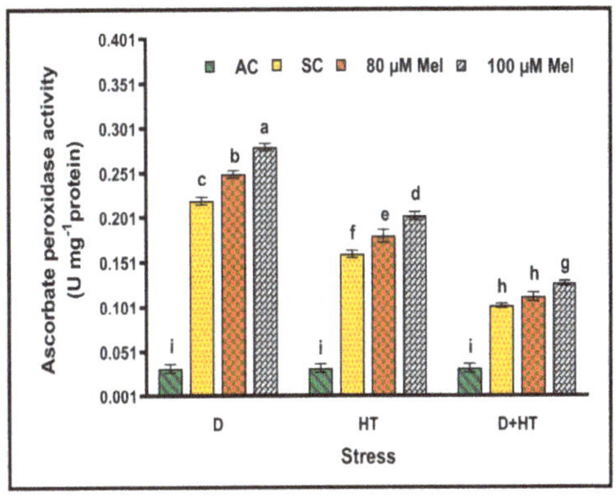

(**D**) Ascorbate peroxidase

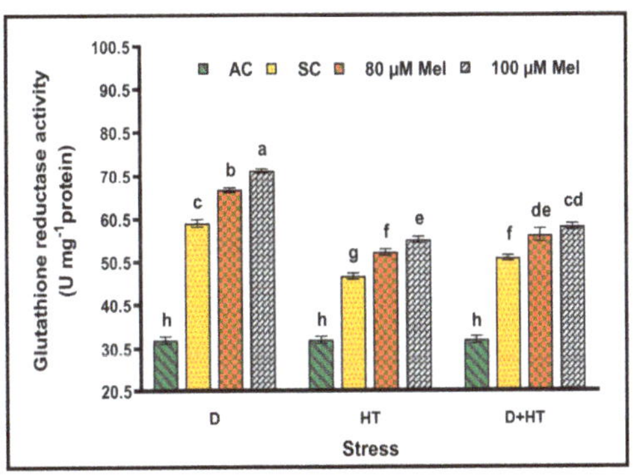

(**E**) Glutathione reductase

Figure 8. Effect of stress (drought—D, high temperature—HT, and D + HT) and foliar spray (irrigated control—AC, stress control—SC, 80 μM melatonin—80 μM Mel, and 100 μM melatonin—100 μM Mel) on (**A**) superoxide dismutase (SOD), (**B**) catalase (CAT), (**C**) peroxidase (POD), (**D**) ascorbate peroxidase (APX), and (**E**) glutathione reductase (GR) enzyme activity in tomato on 10th day of stress. The results were presented as mean of four replications and standard error of means (SEM). Based on analysis of variance (ANOVA), the least significant difference test ($LSD_{5\%}$) was used for means comparison. The significance was denoted by small letters, given that the means with same letters are not statistically significant at $p = 0.05$.

The effect of stress, foliar spray, and the interaction of stress and foliar spray was significant ($p < 0.05$) for the relative tolerance index and fruit yield (Table 1). Among the stresses, a greater decrease in relative tolerance index was observed in D + HT-stressed plants than for individual stresses (Table 1). Among the foliar sprays, a higher relative tolerance index was observed in 100 µM melatonin-treated plants than for other foliar spray treatments (Table 1). The HT + 100 µM melatonin-treated plants had an increased (86%) relative tolerance index (Table 1) compared to the D + 100 µM melatonin (74%) and D + HT and 100 µM melatonin (56%) groups.

Table 1. Effect of different stress and melatonin treatment on relative tolerance index and fruit yield in tomato.

Parameters	Treatments	D	HT	D + HT
Relative tolerance index (%)	Stress Control	55.7 ± 1.56 ᵉ	70.1 ± 2.16 ᶜᵈ	45.9 ± 2.69 ᶠ
	80 µM melatonin	65.9 ± 2.61 ᵈ	79.4 ± 2.63 ᵇ	51.5 ± 1.57 ᵉᶠ
	100 µM melatonin	73.6 ± 1.65 ᵇᶜ	85.9 ± 3.36 ᵃ	55.9 ± 2.49 ᵉ
Yield (kg plant^{-1})	Absolute Control	3.84 ± 0.08 ᵃ	3.84 ± 0.08 ᵃ	3.84 ± 0.08 ᵃ
	Stress Control	2.22 ± 0.06 ᵈ	1.65 ± 0.03 ᵍ	1.07 ± 0.04 ⁱ
	80 µM melatonin	2.55 ± 0.02 ᶜ	1.82 ± 0.05 ᶠ	1.15 ± 0.03 ⁱ
	100 µM melatonin	2.84 ± 0.05 ᵇ	2.04 ± 0.04 ᵉ	1.35 ± 0.03 ʰ

The data represent the mean of four replications and the error bars represent SEM. The means with different letters are significantly different at $p = 0.05$. The stress treatments represented as drought (D), high temperature (HT), and combined drought and high temperature (D + HT); foliar treatments represented as irrigated control (AC), stress control (SC), 80 µM melatonin (80 µM Mel), and 100 µM melatonin (100 µM Mel).

Among the stresses, compared to HT and D + HT stress, D-stressed plants had increased fruit yield (Table 1). Among the foliar sprays, 100 µM melatonin-treated plants showed increased fruit yield compared to other foliar spray treatments (Table 1). The plants treated with D + 100 µM melatonin had an increased fruit yield (32%) compared to plants treated with HT + 100 µM melatonin (23%) and D + HT + 100 µM melatonin (16%) (Table 1).

4. Discussion

Abiotic stress, viz., drought or high temperature, affects the productivity of horticultural crops to a greater extent ranging from 50% to 70% [64]. The effect of drought (D) or high temperature (HT) either individually or in combination triggers ROS production that impairs the photosynthetic membrane and thylakoid membrane due to imbalanced antioxidant activity that results in increased levels of lipid peroxidation and ion leakage [65]. As an antioxidant booster, exogenous melatonin is used in the current study to decrease the stress-induced oxidative damage [66]. Similarly, previous findings on tomatoes revealed that exogenous melatonin (100 µM) has a prominent effect on mitigating ROS-induced oxidative damage [67,68]. In addition, many investigators have reported that the individual effects of D or HT stress can be mitigated via exogenous melatonin application in maize [17], soyabean [18], tomato [20], and strawberry [21], but little information is available on the effect of melatonin under combined drought or HT stress.

The chlorophyll index measures the chlorophyll content and is directly associated with photosynthetic efficiency [69]. This study suggested that D, HT, or D + HT stress decreased the chlorophyll index, and it could be associated with thylakoid membrane damage or decreased 5-aminolevulinate dehydratase enzyme activity. Our research results were similar to the findings of Din et al. [70]. Moreover, D + HT stress-treated plants showed a more decreased chlorophyll index, which evidenced that the effects of combined stress are predominant over individual D or HT stress [71,72]. However, the findings of our study resulted that the exogenous melatonin spray under D or HT stress, individually or in combination, increased the chlorophyll index over the stress-control group, which could be associated with reduced activity of chlorophyll degradation enzymes. These results agree with Yang et al. [15].

The decreased P_n under abiotic stress could be due to damage in the site of light reaction situated in the thylakoid membrane and carbon metabolism [73]. Drought decreased P_n is mediated by a turgor-loss-induced stomatal closure mechanism, which resulted in a decrease in g_s [74]. In contrast, decreased P_n under HT stress occurs due to biochemical changes of photosynthetic enzymes [75]. Similarly, the process of photosynthesis is examined in the present study, which results in decreased stomatal conductance and photosynthetic rate and increased intercellular CO_2 concentration and transpiration rate under D, HT, or D + HT stress. Our results were corroborated by the reports of Benavides et al. [63]. However, melatonin spray under D-stressed plants increased the g_s and P_n rate more than in stress-control plants, proving that melatonin could acclimate the tomato plants to withstand the stress. Similar findings of Altaf et al. [20] reported that melatonin pretreatment in tomato restored the gas exchange parameters through reducing the negative effects of stress. The results of this study also imply that melatonin could regulate the balanced flow of electrons in PSII, which prevents chlorophyll pigment degradation and decreases thylakoid membrane damage (F_0/F_m), which could upregulate the PSII photochemistry and therefore enhance photosynthesis. Similarly, the results of Arena et al. [76] follow the same trend.

Free radical production is significantly higher under D or HT stress; in particular, increased ROS production was found to have more adverse effects under combined stress [77]. Our present study revealed that the ROS content was enhanced under D, HT, or D + HT stress, which could result in oxidative damage. Among the individual stresses, plants exposed to HT stress showed increased membrane damage, indicating that HT is more deleterious than D stress. The severity of oxidative damage caused by H_2O_2 and O_2^- was assessed via histochemical staining, and the result indicated that D + HT stress showed a tremendous increase in ROS production. The results of our study agree with the report of Hussain et al. [78] on maize. In contrast, the foliar spray of melatonin decreased ROS production more than the stress-control. Decreased ROS production would reduce the levels of MDA content and electrolyte leakage that improve membrane integrity. The results were supported by Fahad et al. [79]. Also, few results convinced that increased membrane integrity under stress could be due to increased antioxidants enzymes activity in peach [80], and pepper [81,82].

The antioxidant enzymes, viz., superoxide dismutase, catalase, peroxidase, ascorbate peroxidase, and glutathione reductase, were increased under individual and combined D or HT stress [83,84]. The results of this study indicate that under HT and D + HT stress, the activity of antioxidant enzymes, viz., SOD, CAT, and POD, was found to be increased, while APX and GR activity was found to be insufficient to scavenge free radicals. Our results are similar to Ayidin et al. [85] and Duan et al. [86] for tomato. Therefore, foliar spray of melatonin (100 µM) supplementation increased the SOD, CAT, and POD activity compared to stress-control plants. This trend is similar to the findings of Zandalinas et al. [87], suggesting that activation of antioxidant enzymes might be the reason for decreased membrane damage in citrus. In addition, APX removes H_2O_2 similar to CAT and POD, which cope to withstand combined drought and high-temperature stress [88]. In our study, APX and GR activity showed higher increases in D + 100 µM melatonin than HT + 100 µM melatonin and D + HT + 100 µM melatonin-treated plants. Although melatonin spray is effective under all stress, D + 100 µM predominantly mitigates the negative effects through increasing the antioxidant enzymes over increased ROS production, thereby maintaining redox homeostasis [89]. The results were comparable to Huang et al. [73] for maize and Raja et al. [90] for tomato, as melatonin keeps the equilibrium between ROS generation and antioxidant enzyme activity under stress.

To determine whether melatonin's foliar application could help mitigate stress, we calculated the relative tolerance index (RTI) based on the stomatal conductance in stressed and unstressed plants [63]. Plants normally depend on transpiration, a cooling mechanism, to escape drought and high-temperature stress [91]. In such conditions, responses of stomatal opening and closing under D or HT that depend on g_s were studied in detail [92,93]. Our results showed an increased RTI (70%) under HT stress compared to D (56%) and D + HT stress (46%). The trend of RTI is similar to g_s. However, the RTIs of D + 100μM melatonin, HT + 100μM melatonin, and D + HT + 100μM melatonin-sprayed plants were 74%, 86%, and 56%, respectively. The above finding proves that foliar application of melatonin can be the best crop management strategy to increase crop stress tolerance [7]. In addition, abiotic stress, viz., D or HT, adversely affects crop productivity in horticultural crops [94] and, therefore, intensive efforts were taken to improve stress tolerance to meet global food demand [95]. In recent years, melatonin-related studies also reported on the detrimental effects of D, HT, or D + HT stress on crop yield for lentil [96], moringa [97], and tomato [34]. Our study showed that foliar application of melatonin under all stresses increased the fruit yield, and this could be due to sustained photosynthesis under stressful environments through efficient activation of the antioxidant defense system.

5. Conclusions

In summary, D, HT, or D + HT stress can increase the production of ROS, which could increase membrane damage due to poor antioxidant activity. Among the stresses, D + HT stress is more detrimental than HT and D stress alone. The foliar spray of 100 μM melatonin under all stress decreased the ROS more than stress-control, proving its antioxidant potential, resulting in lower thylakoid membrane damage and increased photosynthetic rate and fruit yield in tomato. Therefore, exogenous melatonin application effectively mitigates the negative effects of D, HT, or D + HT stress through increasing the antioxidant activity which protects the photosynthetic system from oxidative damage. The current study on melatonin will help the researchers to understand how plants cope to withstand D, HT, or D + HT stress. Since a few years, melatonin is gaining interest among the researchers, although topics related to mitigation of combined stresses were recently under progress. There is a lack of ideas on how melatonin functions effectively in plant systems and how its mechanisms related to foliar uptake and translocation overcome stress Amidst difficulties, the pathways involved in melatonin biosynthesis and its associated genes, melatonin signaling and its regulation, and crosstalk with other hormones under abiotic stress need to be explored in future. Future research may also aim to focus on unexplored parts of the anisotropic or isotropic stomatal behavior and its mechanisms under stress to understand the photosynthetic process in depth, which could also be an effective strategy to improve crop productivity.

Supplementary Materials: The following supporting information can be downloaded at: https://www.mdpi.com/article/10.3390/horticulturae9060673/s1, Table S1: Role of melatonin in drought and high-temperature stress on crop yield; Tables S2–S16: Mean and ANOVA for physiological and yield traits.

Author Contributions: Conceptualization, M.K.K.A., S.A., K.M.K., S.R. (Swarnapriya Ramakrishnan), M.S., R.M., G.V. and P.R.B.V.; methodology, M.K.K.A., S.A., K.M.K., P.R.B.V., V.D., R.V., B.P. and S.R. (Sivakumar Rathinavelu); software, M.K.K.A.; validation, M.K.K.A., S.A., K.M.K., S.R. (Swarnapriya Ramakrishnan) and M.S.; statistical analysis, M.K.K.A. and S.A.; endnote, M.K.K.A.; investigation, S.A. and K.M.K.; data curation, M.K.K.A. and S.A.; writing—original draft preparation, M.K.K.A., S.A. and D.M.; review and editing, M.K.K.A., S.A., D.M., K.M.K., S.R. (Swarnapriya Ramakrishnan) and M.S.; visualization, M.K.K.A., S.A., K.M.K., S.R. (Swarnapriya Ramakrishnan) and M.S.; supervision, S.A. All authors have read and agreed to the published version of the manuscript.

Funding: This research received no external funding.

Data Availability Statement: Not applicable.

Conflicts of Interest: The authors declare no conflict of interest.

Abbreviations

D	Drought
HT	High-temperature
D + HT	Combined drought and high-temperature
AT	Ambient temperature
SOD	Superoxide dismutase
CAT	Catalase
POD	Peroxidase
APX	Ascorbic peroxidase
GR	Glutathione reductase
ROS	Reactive oxygen species
PSII	Photosystem II
OTC	Open top chamber
PEG	Polyethylene glycol
SPAD	Soil plant analysis development
P_n	Photosynthetic rate
E	Transpiration rate
g_s	Stomatal conductance
Ci	Intercellular CO_2 concentration
H_2O_2	Hydrogen peroxide
O_2^-	Superoxide anion
NBT	Nitroblue tetrazolium
DAB	3,3- diaminobenzidine
TCA	Trichloroacetic acid
TBA	Thiobarbituric acid
EC	Electrical conductivity
EL	Electrolyte leakage
PVP	Poly vinyl pyrrolidone
EDTA	Ethylene diamine tetraacetic acid
RTI	Relative tolerance index

References

1. Bouabdelli, S.; Zeroual, A.; Meddi, M.; Assani, A. Impact of temperature on agricultural drought occurrence under the effects of climate change. *Theor. Appl. Climatol.* **2022**, *148*, 191–209. [CrossRef]
2. Dos Santos, T.B.; Ribas, A.F.; De Souza, S.G.H.; Budzinski, I.G.F.; Domingues, D.S. Physiological Responses to Drought, Salinity, and Heat Stress in Plants: A Review. *Stresses* **2022**, *2*, 113–135. [CrossRef]
3. Portner, H.O.; Roberts, D.C.; Adams, H.; Adler, C.; Aldunce, P.; Ali, E.; Begum, R.A.; Betts, R.; Kerr, R.B.; Biesbroek, R. *Climate Change 2022: Impacts, Adaptation and Vulnerability*; IPCC: Geneva, Switzerland, 2022; p. 3056.
4. Basavaraj, P.; Rane, J. Avenues to realize potential of phenomics to accelerate crop breeding for heat tolerance. *Plant Physiol. Rep.* **2020**, *25*, 594–610. [CrossRef]
5. Goswami, M.; Suresh, D. Plant growth-promoting rhizobacteria-alleviators of abiotic stresses in soil: A review. *Pedosphere* **2020**, *30*, 40–61. [CrossRef]
6. Islam, M.T. Applications of nanomaterials for future food security: Challenges and prospects. *Malays. J. Halal Res.* **2019**, *2*, 6–9. [CrossRef]
7. Fahad, S.; Bajwa, A.A.; Nazir, U.; Anjum, S.A.; Farooq, A.; Zohaib, A.; Sadia, S.; Nasim, W.; Adkins, S.; Saud, S. Crop production under drought and heat stress: Plant responses and management options. *Front. Plant Sci.* **2017**, *8*, 1147. [CrossRef] [PubMed]
8. Zia, R.; Nawaz, M.S.; Siddique, M.J.; Hakim, S.; Imran, A. Plant survival under drought stress: Implications, adaptive responses, and integrated rhizosphere management strategy for stress mitigation. *Microbiol. Res.* **2021**, *242*, 126626. [CrossRef] [PubMed]
9. Muhammed, M.A.; Mohamed, A.K.S.; Qayyum, M.F.; Haider, G.; Ali, H.A. Physiological response of mango transplants to phytohormones under salinity stress. *Sci. Hortic.* **2022**, *296*, 110918. [CrossRef]
10. Barickman, T.C.; Adhikari, B.; Sehgal, A.; Walne, C.H.; Reddy, K.R.; Gao, W. Drought and elevated CO_2 impacts on photosynthesis and biochemicals of basil (*Ocimum basilicum* L.). *Stresses* **2021**, *1*, 223–237. [CrossRef]
11. Mathur, S.; Agrawal, D.; Jajoo, A. Photosynthesis: Response to high temperature stress. *J. Photochem. Photobiol. B Biol.* **2014**, *137*, 116–126. [CrossRef]

12. Das, K.; Roychoudhury, A. Reactive oxygen species (ROS) and response of antioxidants as ROS-scavengers during environmental stress in plants. *Front. Environ. Sci.* **2014**, *2*, 53. [CrossRef]
13. Cen, H.; Wang, T.; Liu, H.; Tian, D.; Zhang, Y. Melatonin application improves salt tolerance of alfalfa (*Medicago sativa* L.) by enhancing antioxidant capacity. *Plants* **2020**, *9*, 220. [CrossRef]
14. Khan, T.A.; Fariduddin, Q.; Nazir, F.; Saleem, M. Melatonin in business with abiotic stresses in plants. *Physiol. Mol. Biol. Plants* **2020**, *26*, 1931–1944. [CrossRef]
15. Yang, S.J.; Huang, B.; Zhao, Y.Q.; Hu, D.; Chen, T.; Ding, C.-B.; Chen, Y.-E.; Yuan, S.; Yuan, M. Melatonin enhanced the tolerance of *Arabidopsis thaliana* to high light through improving anti-oxidative system and photosynthesis. *Front. Plant Sci.* **2021**, *12*, 752584. [CrossRef]
16. Yu, R.; Zuo, T.; Diao, P.; Fu, J.; Fan, Y.; Wang, Y.; Zhao, Q.; Ma, X.; Lu, W.; Li, A.; et al. Melatonin enhances seed germination and seedling growth of *Medicago sativa* under salinity via a putative melatonin receptor MsPMTR1. *Front. Plant Sci.* **2021**, *12*, 702875. [CrossRef]
17. Ahmad, S.; Muhammad, I.; Wang, G.Y.; Zeeshan, M.; Yang, L.; Ali, I.; Zhou, X.B. Ameliorative effect of melatonin improves drought tolerance by regulating growth, photosynthetic traits and leaf ultrastructure of maize seedlings. *BMC Plant Biol.* **2021**, *21*, 368. [CrossRef]
18. Imran, M.; Aaqil Khan, M.; Shahzad, R.; Bilal, S.; Khan, M.; Yun, B.-W.; Khan, A.L.; Lee, I.-J. Melatonin ameliorates thermotolerance in soybean seedling through balancing redox homeostasis and modulating antioxidant defense, phytohormones and polyamines biosynthesis. *Molecules* **2021**, *26*, 5116. [CrossRef] [PubMed]
19. Li, J.; Liu, Y.; Zhang, M.; Xu, H.; Ning, K.; Wang, B.; Chen, M. Melatonin increases growth and salt tolerance of *Limonium bicolor* by improving photosynthetic and antioxidant capacity. *BMC Plant Biol.* **2022**, *22*, 16. [CrossRef] [PubMed]
20. Altaf, M.A.; Shahid, R.; Ren, M.X.; Naz, S.; Altaf, M.M.; Khan, L.U.; Tiwari, R.K.; Lal, M.K.; Shahid, M.A.; Kumar, R.; et al. Melatonin improves drought stress tolerance of tomato by modulating plant growth, root architecture, photosynthesis, and antioxidant defense system. *Antioxidants* **2022**, *11*, 309. [CrossRef]
21. Okatan, V.; Askın, M.A.; Polat, M.; Bulduk, I.; Colak, A.M.; Guclu, S.F.; Kahramanoglu, İ.; Tallarita, A.V.; Caruso, G. Effects of melatonin dose on fruit yield, quality, and antioxidants of strawberry cultivars grown in different crop systems. *Agriculture* **2023**, *13*, 71. [CrossRef]
22. Cano, A.; Giraldo-Acosta, M.; Garcia-Sanchez, S.; Hernandez-Ruiz, J.; Arnao, M.B. Effect of melatonin in broccoli post-harvest and possible melatonin ingestion level. *Plants* **2022**, *11*, 2000. [CrossRef]
23. Wang, K.; Cai, S.; Xing, Q.; Qi, Z.; Fotopoulos, V.; Yu, J.; Zhou, J. Melatonin delays dark induced leaf senescence by inducing miR171b expression in tomato. *J. Pineal Res.* **2022**, *72*, e12792. [CrossRef]
24. Zeng, W.; Mostafa, S.; Lu, Z.; Jin, B. Melatonin-mediated abiotic stress tolerance in plants. *Front. Plant Sci.* **2022**, *13*, 847175. [CrossRef]
25. Tripathi, A.; Tripathi, D.K.; Chauhan, D.; Kumar, N.; Singh, G. Paradigms of climate change impacts on some major food sources of the world: A review on current knowledge and future prospects. *Agric. Ecosyst. Environ.* **2016**, *216*, 356–373. [CrossRef]
26. Ahammed, G.J.; Xu, W.; Liu, A.; Chen, S. Endogenous melatonin deficiency aggravates high temperature-induced oxidative stress in *Solanum lycopersicum* L. *Environ. Exp. Bot.* **2019**, *161*, 303–311. [CrossRef]
27. Yang, L.; Bu, S.; Zhao, S.; Wang, N.; Xiao, J.; He, F.; Gao, X. Transcriptome and physiological analysis of increase in drought stress tolerance by melatonin in tomato. *PLoS ONE* **2022**, *17*, e0267594. [CrossRef] [PubMed]
28. Raiola, A.; Rigano, M.M.; Calafiore, R.; Frusciante, L.; Barone, A. Enhancing the health-promoting effects of tomato fruit for biofortified food. *Mediat. Inflamm.* **2014**, *2014*, 16. [CrossRef]
29. Zhou, R.; Yu, X.; Ottosen, C.O.; Rosenqvist, E.; Zhao, L.; Wang, Y.; Yu, W.; Zhao, T.; Wu, Z. Drought stress had a predominant effect over heat stress on three tomato cultivars subjected to combined stress. *BMC Plant Biol.* **2017**, *17*, 24. [CrossRef] [PubMed]
30. Tiwari, R.K.; Lal, M.K.; Naga, K.C.; Kumar, R.; Chourasia, K.N.; Subhash, S.; Kumar, D.; Sharma, S. Emerging roles of melatonin in mitigating abiotic and biotic stresses of horticultural crops. *Sci. Hortic.* **2020**, *272*, 109592. [CrossRef]
31. Martinez, V.; Nieves-Cordones, M.; Lopez-Delacalle, M.; Rodenas, R.; Mestre, T.C.; Garcia-Sanchez, F.; Rubio, F.; Nortes, P.A.; Mittler, R.; Rivero, R.M. Tolerance to stress combination in tomato plants: New insights in the protective role of melatonin. *Molecules* **2018**, *23*, 535. [CrossRef]
32. Ahammed, G.J.; Xu, W.; Liu, A.; Chen, S. COMT1 silencing aggravates heat stress-induced reduction in photosynthesis by decreasing chlorophyll content, photosystem II activity, and electron transport efficiency in tomato. *Front. Plant Sci.* **2018**, *9*, 998. [CrossRef]
33. Zhou, R.; Yu, X.; Ottosen, C.O.; Zhang, T.; Wu, Z.; Zhao, T. Unique miRNAs and their targets in tomato leaf responding to combined drought and heat stress. *BMC Plant Biol.* **2020**, *20*, 107. [CrossRef]
34. Ibrahim, M.F.M.; Elbar, O.H.A.; Farag, R.; Hikal, M.; El-Kelish, A.; El-Yazied, A.A.; Alkahtani, J.; El-Gawad, H.G.A. Melatonin Counteracts Drought Induced Oxidative Damage and Stimulates Growth, Productivity and Fruit Quality Properties of Tomato Plants. *Plants* **2020**, *9*, 1276. [CrossRef] [PubMed]
35. Sadak, M.S.; Bakry, B.A. Alleviation of drought stress by melatonin foliar treatment on two flax varieties under sandy soil. *Physiol. Mol. Biol. Plants* **2020**, *26*, 907–919. [CrossRef] [PubMed]
36. Wei, W.; Li, Q.T.; Chu, Y.N.; Reiter, R.J.; Yu, X.-M.; Zhu, D.H.; Zhang, W.K.; Ma, B.; Lin, Q.; Zhang, J.S. Melatonin enhances plant growth and abiotic stress tolerance in soybean plants. *J. Exp. Bot.* **2015**, *66*, 695–707. [CrossRef]

37. Hu, W.; Zhang, J.; Wu, Z.; Loka, D.A.; Zhao, W.; Chen, B.; Wang, Y.; Meng, Y.; Zhou, Z.; Gao, L. Effects of single and combined exogenous application of abscisic acid and melatonin on cotton carbohydrate metabolism and yield under drought stress. *Ind. Crops Prod.* **2022**, *176*, 114302. [CrossRef]
38. Hosseini, M.S.; Samsampour, D.; Zahedi, S.M.; Zamanian, K.; Rahman, M.M.; Mostofa, M.G.; Tran, L.P. Melatonin alleviates drought impact on growth and essential oil yield of lemon verbena by enhancing antioxidant responses, mineral balance, and abscisic acid content. *Physiol. Plant* **2021**, *172*, 1363–1375. [CrossRef]
39. Jafari, M.; Shahsavar, A. The Effect of Foliar Application of Melatonin on Changes in Secondary Metabolite Contents in Two Citrus Species under Drought Stress Conditions. *Front. Plant Sci.* **2021**, *12*, 692735. [CrossRef]
40. Cao, L.; Kou, F.; Zhang, M.; Jin, X.; Ren, C.; Yu, G.; Zhang, Y.; Wang, M. Effect of Exogenous Melatonin on the Quality of Soybean and Natto Products under Drought Stress. *J. Chem.* **2021**, *2021*, 8847698. [CrossRef]
41. Zou, J.; Jin, X.; Zhang, Y.; Ren, C.; Zhang, M.; Wang, M. Effects of melatonin on photosynthesis and soybean seed growth during grain filling under drought stress. *Photosynthetica* **2019**, *57*, 512–520. [CrossRef]
42. Fan, X.; Zhao, J.; Sun, X.; Zhu, Y.; Li, Q.; Zhang, L.; Zhao, D.; Huang, L.; Zhang, C.; Liu, Q. Exogenous Melatonin Improves the Quality Performance of Rice under High Temperature during Grain Filling. *Agronomy* **2022**, *12*, 949. [CrossRef]
43. Khan, M.N.; Khan, Z.; Luo, T.; Liu, J.; Rizwan, M.; Zhang, J.; Xu, Z.; Wu, H.; Hu, L. Seed priming with gibberellic acid and melatonin in rapeseed: Consequences for improving yield and seed quality under drought and non-stress conditions. *Ind. Crops Prod.* **2020**, *156*, 112850. [CrossRef]
44. Zou, J.; Yu, H.; Yu, Q.; Jin, X.; Cao, L.; Wang, M.; Wang, M.; Ren, C.; Zhang, Y. Physiological and UPLC-MS/MS widely targeted metabolites mechanisms of alleviation of drought stress-induced soybean growth inhibition by melatonin. *Ind. Crops Prod.* **2021**, *163*, 113323. [CrossRef]
45. Iman Khesali, L.; Saeed, P.; Sanam Safaei, C.; Behrooz, S. Evaluating drought stress tolerance in different *Camellia sinensis* L. cultivars and effect of melatonin on strengthening antioxidant system. *Sci. Hortic.* **2023**, *307*, 111517. [CrossRef]
46. Gholami, R.; Hoveizeh, N.F.; Zahedi, S.M.; Gholami, H.; Carillo, P. Melatonin alleviates the adverse effects of water stress in adult olive cultivars (*Olea europea* cv. Sevillana & Roughani) in field condition. *Agric. Water Manag.* **2022**, *269*, 107681. [CrossRef]
47. Hojjati, M.; Ghanbari Jahromi, M.; Abdosi, V.; Mohammadi Torkashvanda, A. Exogenous Melatonin Improved Water Status, Antioxidant Capacity, Fruit Quality, and Altered Fatty Acid Profile of Sweet Cherry (*Prunus avium* L.) under Different Irrigation Regimes. Available online: https://ssrn.com/abstract=4257898 (accessed on 25 October 2022).
48. Crop Production Guide of Horticulture Crops. Directorate of Horticulture and Plantation Crops and TNAU. Available online: https://agritech.tnau.ac.in/pdf/HORTICULTURE.pdf (accessed on 24 March 2023).
49. Liu, J.; Wang, W.; Wang, L.; Sun, Y. Exogenous melatonin improves seedling health index and drought tolerance in tomato. *Plant Growth Regul.* **2015**, *77*, 317–326. [CrossRef]
50. Dalal, V.K.; Tripathy, B.C. Water-stress induced downsizing of light-harvesting antenna complex protects developing rice seedlings from photo-oxidative damage. *Sci. Rep.* **2018**, *8*, 5955. [CrossRef]
51. Lei, C.; Ye, M.; Li, C.; Gong, M. H_2O_2 Participates in the Induction and Formation of Potato Tubers by Activating Tuberization-Related Signal Transduction Pathways. *Agronomy* **2023**, *13*, 1398. [CrossRef]
52. Velikova, V.; Loreto, F. On the relationship between isoprene emission and thermotolerance in *Phragmites australis* leaves exposed to high temperatures and during the recovery from a heat stress. *Plant Cell Environ.* **2005**, *28*, 318–327. [CrossRef]
53. Doke, N. Involvement of superoxide anion generation in the hypersensitive response of potato tuber tissues to infection with an incompatible race of *Phytophthora infestans* and to the hyphal wall components. *Physiol. Plant Pathol.* **1983**, *23*, 345–357. [CrossRef]
54. Heath, R.L.; Packer, L. Effect of light on lipid peroxidation in chloroplasts. *Biochem. Biophys. Res. Commun.* **1965**, *19*, 716–720. [CrossRef]
55. Sullivan, C.Y.; Ross, W.M. Selecting for Drought and Heat Resistance in Grain Sorghum. In *Stress Physiology in Crop Plants*; Mussell, H., Staples, R.C., Eds.; John Wiley and Sons: New York, NY, USA, 1979; Volume 21, pp. 263–281.
56. Camejo, D.; Jimenez, A.; Alarcon, J.J.; Torres, W.; Gomez, J.M.; Sevilla, F. Changes in photosynthetic parameters and antioxidant activities following heat-shock treatment in tomato plants. *Funct. Plant Biol.* **2006**, *33*, 177–187. [CrossRef]
57. Beauchamp, C.; Fridovich, I. Superoxide dismutase: Improved assays and an assay applicable to acrylamide gels. *Anal. Biochem.* **1971**, *44*, 276–287. [CrossRef]
58. Lowry, O.H.; Rosebrough, N.J.; Farr, A.L.; Randall, R.J. Protein measurement with the Folin phenol reagent. *J. Biol. Chem.* **1951**, *193*, 265–275. [CrossRef] [PubMed]
59. Aebi, H. Catalase in vitro. *Methods Enzymol.* **1984**, *105*, 121–126. [CrossRef] [PubMed]
60. Kumar, K.B.; Khan, P.A. Peroxidase and polyphenol oxidase in excised ragi (*Eleusine corocana* cv PR 202) leaves during senescence. *Indian J. Exp. Biol.* **1982**, *20*, 412–416.
61. Chen, G.-X.; Asada, K. Ascorbate peroxidase in tea leaves: Occurrence of two isozymes and the differences in their enzymatic and molecular properties. *Plant Cell Physiol.* **1989**, *30*, 987–998. [CrossRef]
62. Smith, I.K.; Vierheller, T.L.; Thorne, C.A. Assay of glutathione reductase in crude tissue homogenates using 5,5′-dithiobis (2-nitrobenzoic acid). *Anal. Biochem.* **1988**, *175*, 408–413. [CrossRef]
63. Pantoja-Benavides, A.D.; Garces-Varon, G.; Restrepo-Díaz, H. Foliar growth regulator sprays induced tolerance to combined heat stress by enhancing physiological and biochemical responses in rice. *Front. Plant Sci.* **2021**, *12*, 702892. [CrossRef]
64. Boyer, J.S. Plant productivity and environment. *Science* **1982**, *218*, 443–448. [CrossRef]

65. Hasanuzzaman, M.; Bhuyan, M.; Parvin, K.; Bhuiyan, T.F.; Anee, T.I.; Nahar, K.; Hossen, M.S.; Zulfiqar, F.; Alam, M.M.; Fujita, M. Regulation of ROS Metabolism in Plants under Environmental Stress: A Review of Recent Experimental Evidence. *Int. J. Mol. Sci.* **2020**, *21*, 8695. [CrossRef]
66. Sachdev, S.; Ansari, S.A.; Ansari, M.I.; Fujita, M.; Hasanuzzaman, M. Abiotic Stress and Reactive Oxygen Species: Generation, Signaling, and Defense Mechanisms. *Antioxidants* **2021**, *10*, 277. [CrossRef] [PubMed]
67. Mushtaq, N.; Iqbal, S.; Hayat, F.; Raziq, A.; Ayaz, A.; Zaman, W. Melatonin in Micro-Tom Tomato: Improved Drought Tolerance via the Regulation of the Photosynthetic Apparatus, Membrane Stability, Osmoprotectants, and Root System. *Life* **2022**, *12*, 1922. [CrossRef]
68. Jahan, M.S.; Shu, S.; Wang, Y.; Chen, Z.; He, M.; Tao, M.; Sun, J.; Guo, S. Melatonin alleviates heat-induced damage of tomato seedlings by balancing redox homeostasis and modulating polyamine and nitric oxide biosynthesis. *BMC Plant Biol.* **2019**, *19*, 414. [CrossRef]
69. Liu, C.; Liu, Y.; Lu, Y.; Liao, Y.; Nie, J.; Yuan, X.; Chen, F. Use of a leaf chlorophyll content index to improve the prediction of above-ground biomass and productivity. *PeerJ* **2019**, *6*, e6240. [CrossRef]
70. Din, J.; Khan, S.; Ali, I.; Gurmani, A.L. Physiological and agronomic response of canola varieties to drought stress. *J. Anim. Plant Sci.* **2011**, *21*, 78–82.
71. Dreesen, F.E.; De Boeck, H.J.; Janssens, I.A.; Nijs, I. Summer heat and drought extremes trigger unexpected changes in productivity of a temperate annual/biannual plant community. *Environ. Exp. Bot.* **2012**, *79*, 21–30. [CrossRef]
72. Yousaf, M.I.; Riaz, M.W.; Jiang, Y.; Yasir, M.; Aslam, M.Z.; Hussain, S.; Sajid Shah, S.A.; Shehzad, A.; Riasat, G.; Manzoor, M.A.; et al. Concurrent Effects of Drought and Heat Stresses on Physio-Chemical Attributes, Antioxidant Status and Kernel Quality Traits in Maize (*Zea mays* L.) Hybrids. *Front. Plant Sci.* **2022**, *13*, 898823. [CrossRef]
73. Huang, B.; Chen, Y.-E.; Zhao, Y.-Q.; Ding, C.-B.; Liao, J.-Q.; Hu, C.; Zhou, L.-J.; Zhang, Z.-W.; Yuan, S.; Yuan, M. Exogenous Melatonin Alleviates Oxidative Damages and Protects Photosystem II in Maize Seedlings under Drought Stress. *Front. Plant Sci.* **2019**, *10*, 677. [CrossRef]
74. Chaves, M.; Flexas, J.; Pinheiro, C. Photosynthesis under drought and salt stress: Regulation mechanisms from whole plant to cell. *Ann. Bot.* **2009**, *103*, 551–560. [CrossRef]
75. Havaux, M. Rapid photosynthetic adaptation to heat stress triggered in potato leaves by moderately elevated temperatures. *Plant Cell Environ.* **1993**, *16*, 461–467. [CrossRef]
76. Arena, C.; Conti, S.; Francesca, S.; Melchionna, G.; Hajek, J.; Bartak, M.; Barone, A.; Rigano, M.M. Eco-Physiological Screening of Different Tomato Genotypes in Response to High Temperatures: A Combined Field-to-Laboratory Approach. *Plants* **2020**, *9*, 508. [CrossRef]
77. Zandalinas, S.I.; Fichman, Y.; Devireddy, A.R.; Sengupta, S.; Azad, R.K.; Mittler, R. Systemic signaling during abiotic stress combination in plants. *Proc. Natl. Acad. Sci. USA* **2020**, *117*, 13810–13820. [CrossRef] [PubMed]
78. Hussain, H.A.; Men, S.; Hussain, S.; Chen, Y.; Ali, S.; Zhang, S.; Zhang, K.; Li, Y.; Xu, Q.; Liao, C. Interactive effects of drought and heat stresses on morpho-physiological attributes, yield, nutrient uptake and oxidative status in maize hybrids. *Sci. Rep.* **2019**, *9*, 3890. [CrossRef] [PubMed]
79. Fahad, S.; Hussain, S.; Saud, S.; Hassan, S.; Tanveer, M.; Ihsan, M.Z.; Shah, A.N.; Ullah, A.; Nasrullah; Khan, F.; et al. A combined application of biochar and phosphorus alleviates heat-induced adversities on physiological, agronomical and quality attributes of rice. *Plant Physiol. Biochem.* **2016**, *103*, 191–198. [CrossRef]
80. Cao, S.; Shao, J.; Shi, L.; Xu, L.; Shen, Z.; Chen, W.; Yang, Z. Melatonin increases chilling tolerance in postharvest peach fruit by alleviating oxidative damage. *Sci. Rep.* **2018**, *8*, 806. [CrossRef]
81. Korkmaz, A.; Deger, O.; Szafranska, K.; Koklu, S.; Karaca, A.; Yakupoglu, G.; Kocacinar, F. Melatonin effects in enhancing chilling stress tolerance of pepper. *Sci. Hortic.* **2021**, *289*, 110434. [CrossRef]
82. Li, Y.; Fan, Y.; Ma, Y.; Zhang, Z.; Yue, H.; Wang, L.; Li, J.; Jiao, Y. Effects of exogenous γ-aminobutyric acid (GABA) on photosynthesis and antioxidant system in pepper (*Capsicum annuum* L.) seedlings under low light stress. *J. Plant Growth Regul.* **2017**, *36*, 436–449. [CrossRef]
83. Giordano, M.; Petropoulos, S.A.; Rouphael, Y. Response and defence mechanisms of vegetable crops against drought, heat and salinity stress. *Agriculture* **2021**, *11*, 463. [CrossRef]
84. Zhang, N.; Sun, Q.; Zhang, H.; Cao, Y.; Weeda, S.; Ren, S.; Guo, Y.D. Roles of melatonin in abiotic stress resistance in plants. *J. Exp. Bot.* **2015**, *66*, 647–656. [CrossRef]
85. Aydin, S.; Buyuk, I.; Aras, E.S. Expression of SOD gene and evaluating its role in stress tolerance in NaCl and PEG stressed *Lycopersicum esculentum*. *Turk. J. Bot.* **2014**, *38*, 89–98. [CrossRef]
86. Duan, M.; Feng, H.L.; Wang, L.Y.; Li, D.; Meng, Q.W. Overexpression of thylakoidal ascorbate peroxidase shows enhanced resistance to chilling stress in tomato. *J. Plant Physiol.* **2012**, *169*, 867–877. [CrossRef]
87. Zandalinas, S.I.; Balfagón, D.; Arbona, V.; Gómez-Cadenas, A. Modulation of antioxidant defense system is associated with combined drought and heat stress tolerance in citrus. *Front. Plant Sci.* **2017**, *8*, 953. [CrossRef]
88. Koussevitzky, S.; Suzuki, N.; Huntington, S.; Armijo, L.; Sha, W.; Cortes, D.; Shulaev, V.; Mittler, R. Ascorbate peroxidase 1 plays a key role in the response of *Arabidopsis thaliana* to stress combination. *J. Biol. Chem.* **2008**, *283*, 34197–34203. [CrossRef]
89. Rabiatul Basria, S.M.N.M.; Simon, I.O. Reactive Oxygen Species, Cellular Redox Homeostasis and Cancer. In *Homeostasis—An Integrated Vision*; IntechOpen: London, UK, 2018. [CrossRef]

90. Raja, V.; Qadir, S.U.; Alyemeni, M.N.; Ahmad, P. Impact of drought and heat stress individually and in combination on physio-biochemical parameters, antioxidant responses, and gene expression in *Solanum lycopersicum*. *Biotechnology* **2020**, *10*, 208. [CrossRef]
91. Marchin, R.M.; Backes, D.; Ossola, A.; Leishman, M.R.; Tjoelker, M.G.; Ellsworth, D.S. Extreme heat increases stomatal conductance and drought-induced mortality risk in vulnerable plant species. *Glob. Chang. Biol.* **2022**, *28*, 1133–1146. [CrossRef]
92. Drake, J.E.; Tjoelker, M.G.; Vårhammar, A.; Medlyn, B.E.; Reich, P.B.; Leigh, A.; Pfautsch, S.; Blackman, C.J.; Lopez, R.; Aspinwall, M.J. Trees tolerate an extreme heatwave via sustained transpirational cooling and increased leaf thermal tolerance. *Glob. Chang. Biol.* **2018**, *24*, 2390–2402. [CrossRef] [PubMed]
93. Mott, K.A.; Peak, D. Stomatal responses to humidity and temperature in darkness. *Plant Cell Environ.* **2010**, *33*, 1084–1090. [CrossRef] [PubMed]
94. Zheng, Y.; Wang, X.; Cui, X.; Wang, K.; Wang, Y.; He, Y. Phytohormones regulate the abiotic stress: An overview of physiological, biochemical, and molecular responses in horticultural crops. *Front. Plant Sci.* **2022**, *13*, 1095363. [CrossRef] [PubMed]
95. Dhankher, O.P.; Foyer, C.H. Climate resilient crops for improving global food security and safety. *Plant Cell Environ.* **2018**, *41*, 877–884. [CrossRef]
96. Sehgal, A.; Sita, K.; Kumar, J.; Kumar, S.; Singh, S.; Siddique, K.H.M.; Nayyar, H. Effects of drought, heat and their interaction on the growth, yield and photosynthetic function of lentil (*Lens culinaris*) genotypes varying in heat and drought sensitivity. *Front. Plant Sci.* **2017**, *8*, 1776. [CrossRef] [PubMed]
97. Sadak, M.S.; Abdalla, A.M.; Abd Elhamid, E.M.; Ezzo, M.I. Role of melatonin in improving growth, yield quantity and quality of *Moringa oleifera* L. plant under drought stress. *Bull. Natl. Res. Cent.* **2020**, *44*, 18. [CrossRef]

horticulturae

MDPI

Article

Adaptogenic Preparations Enhance the Tolerance to Spring Frosts, Yield and Quality of Apple Fruits

Zoya Ozherelieva *, Pavel Prudnikov, Andrey Nikitin, Anna Androsova, Anzhelika Bolgova, Anna Stupina and Oksana Vetrova

Russian Research Institute of Fruit Crop Breeding (VNIISPK), 302530 Zhilina, Russia; prudnikov@orel.vniispk.ru (P.P.); nikitin@orel.vniispk.ru (A.N.); androsova@orel.vniispk.ru (A.A.); bolgova@orel.vniispk.ru (A.B.); stupina@orel.vniispk.ru (A.S.); vetrova@orel.vniispk.ru (O.V.)

* Correspondence: ozherelieva@orel.vniispk.ru

Abstract: The goal of the research is to study the effectiveness of the use of adaptogenic preparations of the Natural Plant Complex "White Pearl" line to increase the yield and quality of apples. The objects of the studies were the apple cultivar "Sinap Orlovsky" and adaptogenic preparations: phytomodulator "White Pearl Universal Antifreeze" and phytocorrector "White Pearl Drip Ca + Mg". The experiment used the following variants: 1—control (foliar sprays with water); 2—foliar sprays with adaptogenic preparations: 1% solution of "White Pearl Universal Antifreeze" + 1% solution of "White Pearl Drip Ca + Mg". As a result of the complex use of biological products, the yield was significantly increased by 1.7 times and the average fruit weight was increased by 20.3 g. The adaptogenic preparations increased the yield of marketable apple fruits by 14.8% and contributed to a 2.5-fold decrease in fruits affected by scald and a 2-fold decrease in bitter pitting. The adaptogenic preparations improved the taste qualities of "Sinap Orlovsky" fruits compared to the control by increasing the amount of sucrose by 25.6% and ascorbic acid by 20.5%. The conducted tests of adaptogenic preparations in apple plantations show the prospects of their use as additional techniques in traditional apple cultivation technologies.

Keywords: adaptogenic preparations; foliar sprays; tolerance of spring frosts; yield; fruit quality

Citation: Ozherelieva, Z.; Prudnikov, P.; Nikitin, A.; Androsova, A.; Bolgova, A.; Stupina, A.; Vetrova, O. Adaptogenic Preparations Enhance the Tolerance to Spring Frosts, Yield and Quality of Apple Fruits. *Horticulturae* **2023**, 9, 591. https://doi.org/10.3390/horticulturae9050591

Academic Editors: Małgorzata Majewska, Barbara Hawrylak-Nowak and Agnieszka Hanaka

Received: 28 March 2023
Revised: 2 May 2023
Accepted: 11 May 2023
Published: 17 May 2023

1. Introduction

The major challenges for agriculture scientists and experts are to improve crop quality and yield with minimal inputs, focusing on environmental sustainability. To fulfill this aim, various breeding programs have been introduced, but they are time-consuming and species-specific methods. The use of an organic substance can stimulate healthy plant metabolism and improve their growth and development functions [1].

Organic farming is generally characterized by lower crop yield as compared with conventional production systems, mainly because of the limitations imposed on fertilization (no use of chemical fertilizers) and on plant defense (no use of pesticides) [2–5].

Initially, plant biostimulants were used for organic production [6], but as its benefits have been explored, it is now being adopted in sustainable agricultural practices and integrated cropping systems [7].

Biostimulants are considered to be one of the most innovative and promising solutions for increasing the sustainability and profitability of agriculture [8]. Biostimulants are defined as "any substance or microorganism applied to plants in order to increase the efficiency of nutrition, resistance to abiotic stress and quality characteristics of the crop, regardless of the content of nutrients in it" [9]. Biostimulants are the extracts derived from organic raw substances containing bioactive compounds. Some common components of biostimulants are humic substances, mineral elements, amino acids, chitin, chitosan, vitamins, and poly- and oligosaccharides [1]. The main categories of plant biostimulants include natural substances such as humic and fulvic acids, protein hydrolysates, seaweed extracts [10–12], useful fungi (such as arbuscular mycorrhizal fungi and *Trichoderma* spp.) [13] and rhizobacteria that promote plant growth [14]. Growth regulators affect the course of physiological processes and thereby allow changes to the metabolism of the plant organism and in the soil [15–18]. Modern biological products are indispensable for increasing germination of plant seeds, as they are able to increase resistance to plant diseases [19–22], abiotic stresses [23,24] and other stressful situations [25]; accelerate flowering and fruiting; increase yield; and ensure the ecological purity of the crop [26–28]. Biostimulants could also be considered for their implementation in the post-harvest management of fruits. Biostimulants containing mineral nutrients such as zinc and silicon might contribute with calcium to the strengthening of cell wall structure [29], thereby allowing the preservation of fruit quality attributes for longer period. This is of particular interest for the organic apple production system, which is presently lacking any useful means to manage apple physiological disorders during storage. Biostimulants have been found active in promoting final crop quality and, more in detail, studies have highlighted the relevance of biostimulant applications for selected functional quality traits [30].

Despite the large and increasing number of publications dealing with biostimulants [16], science-based information on their optimal use, crop specificity, and interaction with growing conditions is many ways still incomplete. Studies on the effect of biostimulants on the growth and yield potential of plants have been conducted primarily on vegetable crops.

To further sustain the growth and profitability of the apple sector, the implementation of new agroecological means, such as biopreparations, in the management of horticultural systems is highly requested by growers. The use of these tools must nevertheless follow information derived from scientifically sound research about their effects on plant physiological and biochemical responses.

The priority in the application of adaptogenic preparations is a significant increase in the adaptive properties of plants and, as a result of their use, an increase in productivity and crop quality.

With this goal, this work aimed to investigate the effects of adaptogenic preparations of the Natural Plant Complex "White Pearl" line applications on the yield and fruit quality of apple trees belonging to the "Sinap Orlovsky" cultivar. The biopreparations were tested on apple trees, and, as far as we know, this was the first study in which the effectiveness of two adaptogenic preparations was evaluated simultaneously and during two consecutive growing seasons. In addition, their effect was also considered during the storage period of fruits by measuring the physiological disorder of "Sinap Orlovsky" apples.

2. Materials and Methods

2.1. Study Area and Research Conditions

The studies were carried out on the grounds of the laboratory of physiology of resistance of fruit plants and experimental plots of the Russian Research Institute of Fruit Crop Breeding (VNIISPK) from 2021–2022.

The institute is located in the Orel Region (53°00′ north, 36°00′ east), which is part of the Central Federal District of Russia. VNIISPK is located 368 km southwest of Moscow.

According to the data of the VNIISPK meteorological station, the winter period of 2021 in the Orel region was characterized by sharp drops in positive and negative temperatures. In spring, during the flowering period of garden crops, the minimum air temperature did not fall below 0 °C. The summer and autumn months were characterized by an uneven distribution of precipitation and temperature. Both low and high conditions of plant moisture were noted (Figure 1).

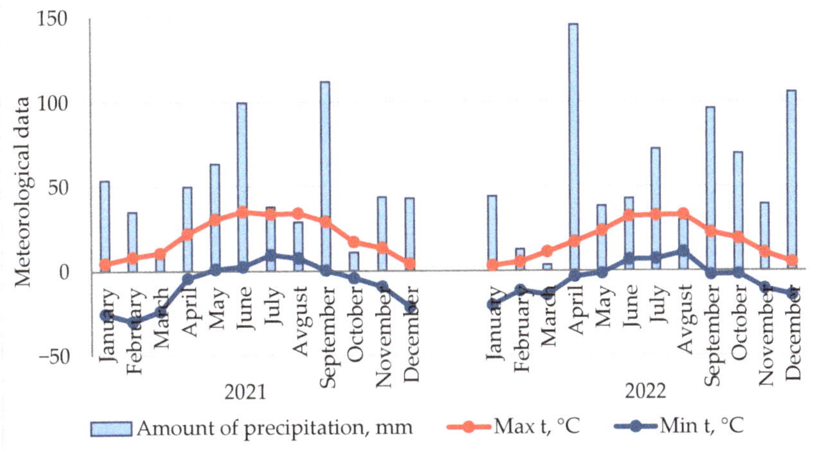

Figure 1. Meteorological conditions of the Orel region in 2021–2022 (data of the VNIISPK meteorological post).

The winter period of 2022 was characterized by moderate frosts without sudden temperature changes. In spring, a prolonged cold snap was observed in May, which led to a later flowering of fruit crops. Summer and autumn were also marked by an uneven distribution of precipitation and temperature. December was abnormally warm and rainy. The average daily air temperature of this winter month exceeded the norm by 6.4 °C, and the amount of precipitation exceeded the average annual value by 70.7 mm (Figure 1).

The experiment was performed on agro-gray soils. The agro-gray soils are formed on loess-like loams underlain by dolomitic limestone. Agrochemical indicators of the soil in the studied orchard are presented in Table 1.

Table 1. Agrochemical indicators of the soil in the experiment plot.

Depth, cm	pH_{KCl}	Humus, %	N (common), mmol/100 g	Content			
				P_2O_5	K_2O	Ca^{2+}	Mg^{2+}
				mg/kg		mmol/100 g	
0…20	5.39	4.61	3.94	177.12	75.68	14.98	4.39
20…40	5.16	3.81	4.22	129.18	58.00	15.59	4.58
40…60	6.36	2.78	2.86	128.48	56.12	14.76	4.77

2.2. Research Objects

The object of the study was apple cultivar "Sinap Orlovsky" was used from the bioresource collection of Russian Research Institute of Fruit Crop Breeding, growing on a medium-sized rootstock 54–118. "Sinap Orlovsky", a cultivar having fruit of late winter maturation, has been developed from the crossing of the "Severny Sinap" and "Pamyat Michurina". The cultivar is zoned for the Central and Central Chernozem regions of Russia. "Sinap Orlovsky" is precocious and winter-hardy. The fruits are of high quality and in terms of consumer qualities are characterized by a long shelf life. The disadvantage of the cultivar is that its fruits are predisposed to scald and bitter pitting [31].

The cultivar was planted in the experimental plot with a spacing of 6 m × 3 m in 2013. Natural tinning was used in the aisles; herbicides were used in the trunk strips. The crown shape was spindle. Agronomical practices generally accepted for apple trees were used. The soil-forming rock was loess-like loam of medium mechanical composition.

The following products from AgroPlus Group of Companies LLC (Krasnodar, Russia) were tested: phytomodulator "White Pearl Universal Antifreeze" ("WPU Antifreeze") and phytocorrector "White Pearl Drip Ca + Mg" ("WP Drip Ca + Mg"). The tested preparations are the Natural Plant Complex (NPC) "White Pearl" ("WP").

The NPC phytomodulator "White Pearl Universal Antifreeze" is a suspension of a group of minerals of natural origin containing a concentrate of extracts of spruce needles, pine and Siberian fir. The composition includes the following: mineral elements SiO_2 5.6%, N (common) 2–6%, CaO 5000 ppm, MgO 7000 ppm, K_2O B 130 ppm, Zn 150 ppm, Mo 200 ppm, Al_2O_3 1600 ppm and other micro elements; vitamins A (carotene, lutein), D (phytosterols), E, K, B1, B2, B6, PP, H; and phytoncides (volatile oils), chlorophyll, flavonoids, sugars, proteins and amino acids.

The NPC phytocorrector "White Pearl Drip Ca + Mg" is an extract of vegetative mass of oceanic bioflora on an organo-mineral basis. The composition includes the following: bioelements Ca 3490.0 ppm, Mg 2829.0 ppm, P 42.9 ppm, K 38.8 ppm, S 0.3 ppm, Fe 68.7 ppm, Mn 3.65 ppm, B 3.37 ppm, Cu 0.85 ppm, Zn 0.05 ppm, Si 0.1 ppm, Se 0.003 ppm, J 2.1 ppm and Mo 0.01 ppm; mineral elements SiO_2 5.6%, CaO 0.4%, MgO 0.4%, K_2O 0.2%, Fe_2O_3 0.4% and Al_2O_3 0.16%; vitamins A (carotene, lutein), D (phytosterols), E, K, B1, B2, B6, PP and H; and chlorophyll, sulfonic acids, humic acids, sugars, proteins and amino acids.

2.3. *Regulations for the Use of Adaptogenic Products in the Experiment*

The experiment was performed with two variations: 1—control (foliar sprays with water); and 2—foliar sprays with preparations: 1% solution of "White Pearl Universal Antifreeze" + 1% solution of "White Pearl Drip Ca + Mg". There were 3 repetitions of each experiment. In each repetition, there were 5 accounting trees.

The tested biological products are intended specifically for foliar sprays.

Foliar sprays were carried out with the RT-16LI knapsack sprayer (Patriot, Beijing, China), with a solution of the tested biological products directly on the trees of the experimental apple cultivar. This was done so that the nutrients that fall in the form of small drops on the bark and foliage of the apple tree were absorbed faster by the plants. Foliar sprays of apple trees were carried out in the morning, in calm weather. The consumption rate of the 1% solution of the tested biological products was 100 mL per 10 L of water.

In the early spring period, to prevent damage to generative organs from recurrent spring frosts, "Sinap Orlovsky" plants were treated twice with a 1% NPC "WPU Antifreeze" solution.

The first foliar sprays of the experimental cultivar trees were performed in the first ten days of April, and the second was performed at 20 days.

The next treatment was carried out in early May with a 1% solution of NPC "WPU Antifreeze" + 1% solution of NPC "WP Drip Ca + Mg".

In summer, a four-fold leaf treatment with preparations of a 1% solution of NPC "WPU Antifreeze" + 1% solution of NPC "WP Drip Ca + Mg" was carried out in order to preserve the fruits from shedding and increase yield and fruit quality. Foliar sprays were performed at specific dates: 14 days after flowering; "fruit-hazel" (Figure 2a), "fruit-walnut" (Figure 2b) and 25 days before harvesting.

2.4. *Determination of the Fractional Composition of Water*

The fractional composition of water in the bark of annual shoots and fruit buds of apple trees was determined by the Okuntsov–Marinchik method [32]. This method is based on the changing concentration of sucrose solution while immersing plant tissue therein. Prepared samples weighing 0.4 g were immersed in a 30% sucrose solution (in three replicates). Some of the water from the plant tissue passed into the solution, reducing its concentration. Based on the initial volume of the solution and its initial and final concentration, the amount of water that entered the solution from tissues was determined. The content of bound water was calculated from the difference in the total water content and the water that passed into the solution. The concentration of the sucrose solution was determined using a PAL-1 digital refractometer (Atago, Tokyo, Japan). The total water content in the bark of annual shoots and fruit buds was determined by the formula:

$$W = (m_1 - m_2)/m_1 \times 100\%, \quad (1)$$

where

W shows the total hydration from the wet mass, %;

m_1 denotes mass of crude weighed bark of annual shoots and fruit buds, g;

m_2 denotes mass of absolutely dry weighed bark of annual shoots and fruit buds, g.

To determine the absolutely dry weight, the mass of absolutely dry weighed bark of annual shoots and fruit buds was laid out in aluminum containers and dried in an oven at a temperature of 105 °C until constant weight was reached.

(a) (b)

Figure 2. Phenophases of the development of apple fruits of the "Sinap Orlovsky" cultivar: (**a**) the "fruit-hazel" d = 1.0–1.5 sm; (**b**) the "fruit-walnut" d = 2.5–3.0 sm.

2.5. Determination of Low Molecular Weight Carbohydrates and Proline Amino Acid

The amount of low molecular weight carbohydrates (sucrose, glucose) in the bark of annual shoots and fruit buds was determined in three replicates based on a resorcinol re-agent reaction at a wavelength of 520 nm. For this, 0.5 g of the material was rubbed in 10 mL of ethanol heated to 80 °C, and the tubes were heated in a UT-4301 E (Ulab, Shanghai, China) water bath (at t = 100 °C) for 10 min. The contents of the tubes were centrifuged for 10 min at 7000× *g* rpm in centrifuge B4i (Jouan, Morlaas Zone Industrielle Berlanne, Morlaàs, France). Next, 50 µL of 5 N NaOH was added to 0.5 mL of the supernatant. It was heated in a water bath UT-4301 E (Ulab, Shanghai, China) (at t = 100 °C) for 10 min. After cooling, 0.5 mL of resorcinol reagent (100 mg of resorcinol +250 mg of thiou-rea in 100 mL of ice-cold CH_3COOH) and 3.5 mL of 30% HCl were added. Test tubes were heated in the bath for 10 min. After cooling, the optical density was determined on a BioRad SmartSpec Plus spectrophotometer (BioRad, Hercules, CA, USA) at a wavelength of 520 nm. The disaccharide content was calculated using a calibration curve constructed for pure sucrose and glucose [33].

The proline content was determined in three replicates by reaction with ninhydrin reagent [34]. To do this, a 500-mg sample of bark of annual shoots and fruit buds was ground in distilled water and boiled in a UT-4301E (Ulab, Shanghai, China) water bath (at t = 100 °C) for 10 min. After this, the homogenate was centrifuged at 7000× *g* rpm in a B4i (Jouan, Morlaas Zone Industrielle Berlanne, France) centrifuge, 2 mL of the extract was taken and 2 mL of glacial acetic acid and 2 mL of ninhydrin reagent (30 mL of ice-cold CH_3COOH + 20 mL of 6 M H_3PO_4 + 1.25 g of ninhydrin) were added, followed by boiling in a UT-4301 E (Ulab, Shanghai, China) water bath (at t = 100 °C) for 1 h. The amino acid content was calculated using a calibration curve on a BioRad SmartSpec Plus spectrophotometer (BioRad, Hercules, CA, USA) constructed for pure proline at a wavelength of 520 nm. Proline content was expressed in mg per 1 kg of wet weight [35].

2.6. Modeling of Spring Frosts

Artificial freezing of apple buds and flowers was carried out in a PSL-2KPH (Espec, Osaka, Japan) climatic chamber. Spring frosts were simulated (−3°, −3.5° and −4 °C). The exposure time of the freezing was 3 h. The rate of temperature decrease was 1 °C per hour. Before evaluation, the experimental material was kept at a temperature of +22 °C until signs of damage to the buds and flowers appeared. For artificial freezing, branches of the tested cultivar were cut off in the orchard so that there were 100 flowers and 100 buds in total. After freezing, a visual assessment of damage to flowers and buds of the cultivar was carried out. Damage to the pistils and stamens was assessed by the darkening of the tissues. The number of damaged flowers and buds was calculated from the total number of flowers and buds. The degree of damage to flowers and buds was expressed as a percentage.

2.7. Yield, Average Weight, Biochemical Analysis and Keeping Quality of Apple Fruits

The weight accounting of the yield was carried out per tree in kg by weighing during the period of removable fruit maturity in accordance with the methodology [36]. The average yield from 1 accounting tree for each repetition of the experiment was calculated by dividing the total weight of the crop (harvested fruit crop + economically usable wind fallen fruits) by the number of accounting plants in the repetition.

The yield as a whole for the cultivar in the center from 1 ha was calculated by the formula:

$$Y = A/B \times 100, \quad (2)$$

where

Y—yield, c/ha;

A—average yield per 1 tree, kg;

B—nutrition area of 1 tree, m^2;

100—the coefficient of conversion of weight in kilograms to weight in hundredweight and m^2 area to hectares.

The commercial qualities, biochemical composition and keeping quality of apple fruits were studied according to the methodology [36]. The "Sinap Orlovsky" fruits were selected based on typicality of shape, color and degree of maturity.

To characterize the weight of the fruits, 100 apples were selected from each repetition of the experiment. The average weight of the fruits was determined by weighing 100 fruits and dividing the resulting weight by their number.

For biochemical analysis of fruits during harvesting, 10 fruits were selected in 3 repetitions of each variant of the experiment.

Determination of sugars (sum, monosaccharide, sucrose) was carried out according to Bertrand's method, which is based on the reducing action of sugar on alkaline solution of tartarate complex with cupric ions; the cuprous oxide formed is dissolved in a warm acid solution of ferric alum. The ferric alum is reduced to $FeSO_4$ which is titrated against standardized $KMnO_4$; Cu equivalence is correlated with the table to obtain the amount of reducing sugar. This is based on the alkaline solution of tartarate complex of cupric ion [37].

Ascorbic acid was determined by titration of oxalic acid extracts with Tilman's paint (2,6-dichlorophenolindophenol). All determinations, starting from taking the sample and ending with titration, were carried out within 1 h. The titer of the Tilman's paint was determined by the method of S.M. Prokoshev.

In total, 40 apples in 3 repetitions of each experiment variant were selected for fruit storage. The fruits were stored in a CV114-S (Polair, Volzhsk, Russia) refrigerator at a temperature of +2 °C. After storage, the degree of damage by scald, bitter pitting, wilting and rotting of fruits was determined.

2.8. Statistical Analysis

The obtained data were evaluated using mathematical statistics using single-factor analysis of variance ANOVA (Version 22, SPSS Statistics). The critical significance level between control and treatment was assumed to be 5%. The results were presented in the form of M ± m.

3. Results

3.1. The Effect of Adaptogenic Preparations on the Fractional Composition of Water in Annual Shoots and Fruit Buds of Apple Trees in Spring

The degree of hydration of plants is one of the essential indicators of the water regime. The concentration of cell juice and the water potential of individual plant organs are associated with the water content [38]. It is known that the mobility and activity of water directly depends on its state in the plant cell. From this point of view, it is customary in plant physiology to distinguish between free and bound water. Free water moves easily, enters into various biochemical reactions, evaporates during transpiration and freezes at low temperatures. Bound water, which plays a structure-forming role, is not a solvent and it has a reduced freezing point, which will significantly affect the resistance to low-temperature stress of the protoplast of the cell and the plant as a whole [39].

Thus, in early April, the foliar sprays of "Sinap Orlovsky" trees with a 1% solution of NPC "WPU Antifreeze" increased the hydration of the bark annual shoots by 1.5% compared to the control. The content of free water in annual apple shoots was significantly reduced, by 4.5% in the variant with the treatment. The level of colloidal water at the same time significantly increased by 7.0% in treated trees against control ones (Figure 3). This section may be divided by subheadings. It should provide a concise and precise description of the experimental results and their interpretation, as well as the experimental conclusions that can be drawn.

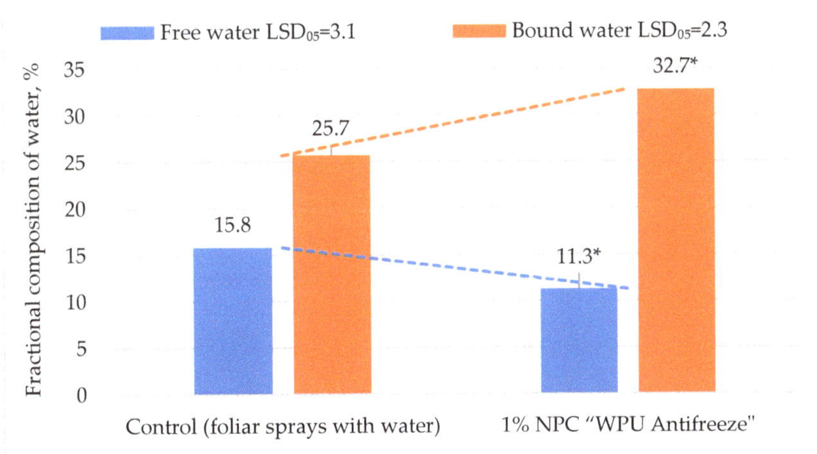

Figure 3. Fractional composition of water of the bark annual shoots of "Sinap Orlovsky" after foliar sprays (average for 2021–2022). * Significant differences with the control at the 5% significance level.

After 20 days, a second set of foliar sprays was carried out. After the second foliar sprays of "Sinap Orlovsky" with a 1% solution of NPC "WPU Antifreeze", an insignificant decrease in the free water content in annual shoots was noted compared to the control. The amount of bound water in the bark annual shoots was at the same level in both variants. In the fruit buds of treated trees, compared with the control ones, the level of bound water significantly increased by 8.4% against the background of a significant decrease of 4.2% of free water (Figure 4).

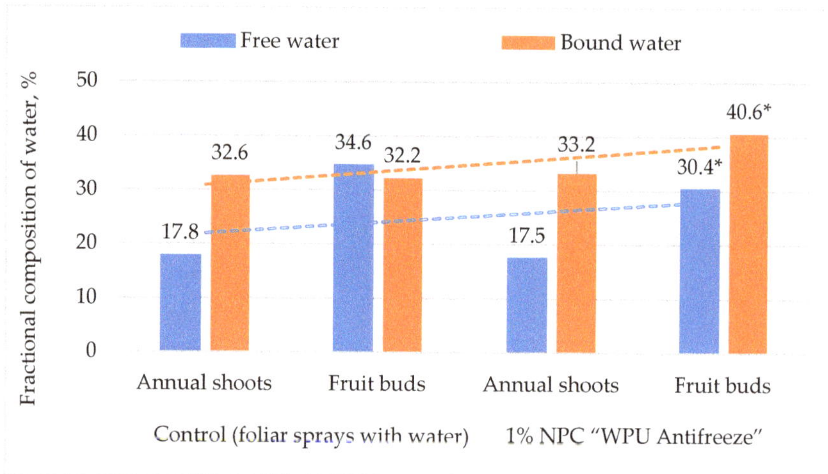

Figure 4. Fractional composition of water of the bark annual shoots and fruit buds of "Sinap Orlovsky" after foliar sprays (average for 2021–2022). * Significant differences with the control at the 5% significance level.

The subsequent foliar sprays of the experimental plants in early May with a tank mixture (1% solution of NPC "WPU Antifreeze" + 1% solution of NPC "WP Drip Ca + Mg") also significantly affected the proportion of bound water in fruit buds. In the treated of "Sinap Orlovsky" trees, fruit buds contained 5.7% more colloidally bound water than the controls. The content of free water in the fruit buds was significantly reduced, by 4.5% in the variant with treatments (Figure 5). The resistance of plants to adverse environmental conditions is determined by the state of intracellular water. It is bound water that affects the resistance to low-temperature stress of plants, since it has a reduced freezing point.

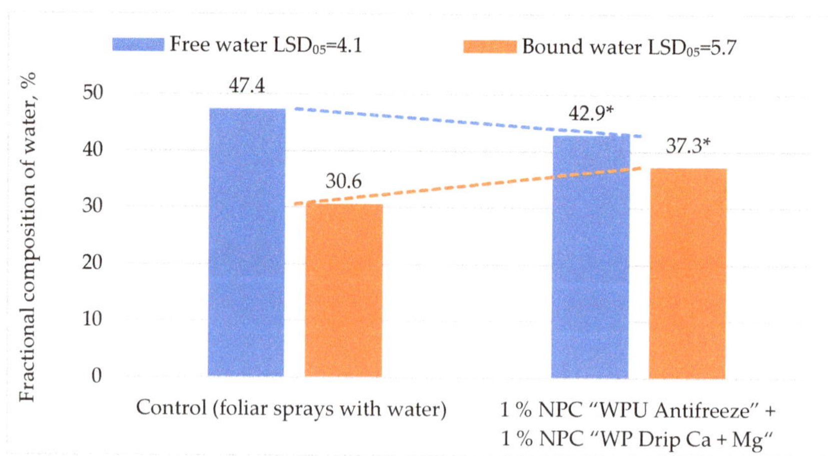

Figure 5. Fractional composition of water of fruit buds of "Sinap Orlovsky" after foliar sprays (average for 2021–2022). * Significant differences with the control at the 5% significance level.

Thus, the 3-repetition foliar sprays with adaptogenic preparations carried out in spring increased the resistance to dehydration of the bark annual shoots and fruit buds of "Sinap Orlovsky" trees against the background of an increase in bound water, which would help reduce the freezing temperature of water inside the plant cells in spring frosts.

3.2. The Effect of Adaptogenic Preparations on the Level of Low-Molecular Osmoprotectors in the Bark of Annual Shoots and Fruit Buds of Apple Trees in Spring

The spring foliar sprays of apple trees with a 1% solution of NPC "WPU Antifreeze" increased the content of free proline both in the bark of annual shoots and in the fruit buds of the plants (Table 2). The maximum effect of an increase in amino acid compared to the control in the bark of annual shoots (67.0% more than in the control) and in fruit buds (12.7% more than in the control) was noted in early April. At the same time, the maximum content of free proline was noted in the fruit buds of the plants in the variant with treatments. The proline amino acid, in addition to its building function in protein biosynthesis, performs an antioxidant and osmoregulatory role. Therefore, as an antioxidant, this amino acid is able to "extinguish" an excessive amount of the active oxygen form during the development of oxidative stress against the background of adverse environmental factors. As an osmoprotector, proline increases the concentration of cell juice, which prevents the formation of intracellular ice and prevents damage to cell membranes. By increasing the content of free proline in the cells and tissues of apple plants, NPC "WPU Antifreeze" prevents the development of not only oxidative stress, but also the formation of intracellular ice, which increases the resistance of cells and tissues to both negative and low positive temperatures.

Table 2. The content of free proline in the bark of annual shoots and fruit buds of "Sinap Orlovsky" after foliar sprays, mg/kg (average for 2021–2022).

Experiment Option	Timing of Non-Root Treatments	
	I Ten Days of April	**after 20 Days**
Bark of annual shoots		
Control (foliar sprays with water)	29.41 ± 7.44	31.00 ± 0.58
1% NPC "WPU Antifreeze"	49.02 ± 2.26 *	37.91 ± 1.06 *
LSD_{05}	6.87	1.89
Fruit buds		
Control (foliar sprays with water)	25.93 ± 0.73	81.42 ± 0.82
1% NPC "WPU Antifreeze"	29.23 ± 1.10 *	84.32 ± 0.51
LSD_{05}	3.03	2.97

* Significant differences with the control at the 5% significance level.

As in the case with proline and bound water, foliar sprays with NPC "WPU Antifreeze" contributed to an increased level of sugars, both in the bark of annual apple shoots and in the fruit buds (Table 3). At the same time, at the beginning of April, the amount of sugars in the experimental version did not significantly differ from the control, which may be due to the onset of vegetation and intensification of physiological and biochemical processes. This assumption is supported by a further decrease in the level of sugars at the end of April. It is known that sugars are a substrate for respiration, as a result of which both energy and plastic equivalents are formed for the growth and development of plants. However, it should be noted that in the variant with treatment, the intensity of sugar re-duction was significantly lower compared to the control. Therefore, during the interval of the first ten days of April till the end of the third ten days of April, the amount of sugars in the bark of annual shoots in control plants decreased by a factor of 1.76, in experimental plants by a factor of 1.74; in the fruit buds by a factor of 4.1 versus 3.3 in experimental plants, respectively (Table 3). The lower intensity of the reduction in the amount of sugars under the influence of treatment with a phytomodulator is probably associated with some inhibition of the expenditure of sugars on life support processes, and above all with the inhibition of respiration processes when plants exit the state of forced dormancy.

Table 3. The content of free sugars in the bark and fruit buds of annual shoots of "Sinap Orlovsky" after foliar sprays, mg/g (average for 2021–2022).

Experiment Option	Timing of Non-Root Treatments	
	I Ten Days of April	after 20 Days
	Bark of annual shoots	
Control (foliar sprays with water)	2.95 ± 0.09	1.68 ± 0.10
1% NPC "WPU Antifreeze"	2.96 ± 0.19	1.70 ± 0.12
	Fruit buds	
Control (foliar sprays with water)	2.52 ± 0.24	0.61 ± 0.18
1% NPC "WPU Antifreeze"	2.71 ± 0.12	0.83 ± 0.12

An increase in the amino acid proline in the bark of annual shoots by 16.5% and in fruit buds by 22.7% was noted in the experiment with the treatment with biological preparations in comparison with the control. Along with this, there was an in-crease in the amount of sugars both in the bark of annual shoots by 1.2 times, and in fruit buds by 1.6 times in "Sinap Orlovsky" under the action of adaptogenic preparations (Table 4). The experimental plants treated with a tank mixture were generally characterized by an increased background of the amount of sugars compared to the control trees, which is of importance for the protective effect of low-molecular carbohydrates in conditions of sudden spring frosts.

Table 4. The content of osmoprotectors in the bark of annual shoots and fruit buds of "Sinap Orlovsky" after foliar sprays (average for 2021–2022).

Experiment Option	Free Proline Content, mg/kg	Free Sugars Content, mg/g
	Bark of annual shoots	
Control (foliar sprays with water)	16.85 ± 0.61	1.25 ± 0.20
1% NPC "WPU Antifreeze" + 1% NPC "WP Drip Ca + Mg"	19.62 ± 0.57 *	1.48 ± 0.22
LSD_{05}	2.65	$F_f < F_t$
	Fruit buds	
Control (foliar sprays with water)	51.81 ± 3.81	0.27 ± 0.04
1% NPC "WPU Antifreeze" + 1% NPC "WP Drip Ca + Mg"	63.57 ± 2.35 *	0.42 ± 0.05
LSD_{05}	4.67	$F_f < F_t$

* Significant differences with the control at the 5% significance level.

Along with this, there was an increase in the amount of sugars both in the bark of annual shoots by 1.2 times, and in fruit buds by 1.6 times in "Sinap Orlovsky" under the action of adaptogenic preparations (Table 4). The experimental plants treated with a tank mixture were generally characterized by an increased background of the amount of sugars compared to the control trees, which is of importance for the protective effect of low molecular carbohydrates in conditions of sudden spring frosts.

Thus, adaptogenic preparations also reduced the risk of damage to the fruit buds of apple trees by negative temperature by increasing low-molecular osmoprotectors at the beginning of the growing season.

3.3. The Effect of Adaptogenic Preparations on the Resistance of Apple Buds and Flowers to Spring Frosts

The evaluation of the results of artificial freezing showed a positive effect of foliar sprays with adaptogenic preparations on the resistance of apple flowers and buds to spring frosts. After exposure to negative temperatures of −3°, −3.5° and −4 °C, a decrease in the proportion of dead buds in the variant with foliar sprays was noted by 8.6%, 7.8% and 3.4%, respectively, although statistically the differences between the variants were not confirmed (Figure 6a). Foliar sprays with a tank mixture with preparations of the "White Pearl" line significantly affected the resistance of "Sinap Orlovsky" flowers to spring frosts of −3.5 °C and −4 °C. The proportion of damaged flowers at temperatures of −3.5 °C and −4 °C significantly decreased by 6.3% and 8.2%, respectively, compared with the control (Figure 6b).

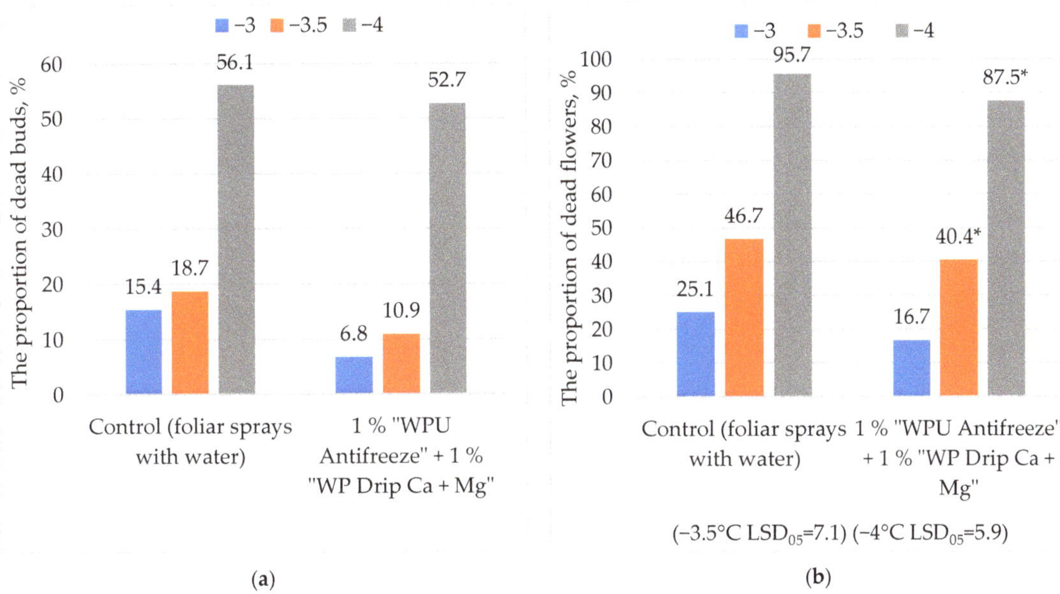

Figure 6. The effect of the preparations of NPC "White Pearl" on the resistance of buds (**a**) and flowers (**b**) of "Sinap Orlovsky" to spring frosts (average for 2021–2022). * Significant differences with the control at the 5% significance level.

Thus, the preparations of the NPC "White Pearl" line prevented the destruction of cell membranes and dehydration of plant cells by increasing bound water and the most powerful osmolytically active substance-free proline and sugars during low-temperature stress, thereby reducing the risk of damage to buds and flowers of garden crops by spring frosts.

3.4. The Effect of Adaptogenic Preparations on the Water Regime of Apple Leaves in Summer

All physiological processes in the plant normally proceed only with sufficient water supply to the plant. Water is a necessary component and an important factor in the structure of the cytoplasm of living cells. It participates in cell metabolism, in hydrolytic and synthetic processes, promotes the interaction of molecules [38].

In the summer of 2021, foliar sprays with preparations of 1% NPC "White Pearl Antifreeze" + 1% NPC "White Pearl Drip Ca + Mg" contributed to an increase in free water in the "Sinap Orlovsky" leaf apparatus by 2.1% compared to the control (Figure 7). At the same time, the proportion of bound water in the variant with foliar sprays was 2.3% lower. The overall hydration of the apple tree leaf apparatus was at the same level in both versions of the experiment. Probably, foliar sprays contributed to the strengthening of metabolic processes during the period of active growth and development of apple fruits against the background of an increase in the proportion of free water, which contributed to the intensive outflow of organic substances accumulated in the leaves during photosynthesis to the fruits.

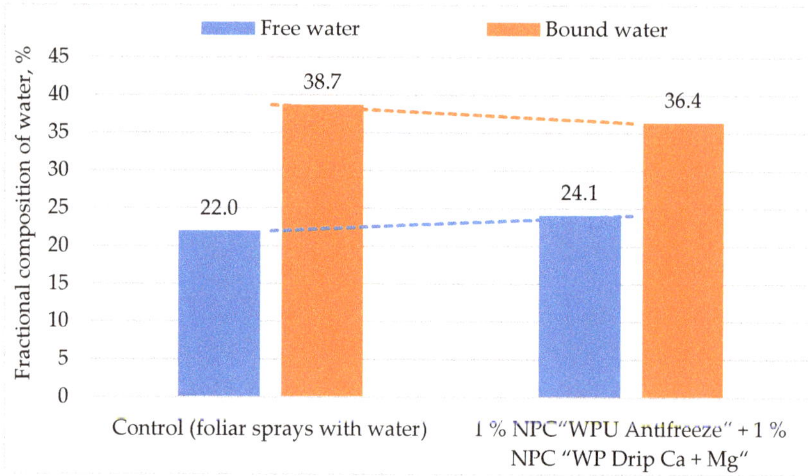

Figure 7. Fractional composition of water in the leaves of "Sinap Orlovsky" after foliar sprays in summer of 2021, %.

In the summer period of 2022, as well as of 2021, after the foliar sprays with a tank mixture of 1% NPC "WPU Antifreeze" + 1% NPC "WP Drip Ca + Mg", a 2.1% increase in free water was noted in the leaf apparatus of "Sinap Orlovsky" trees compared to the control (Figure 8). At the same time, the proportion of bound water in the variant with foliar sprays was 1.1% lower than in the control. Free water, being a solvent and the main transporter of organic substances, will presumably contribute to their intensive outflow from leaves to fruits.

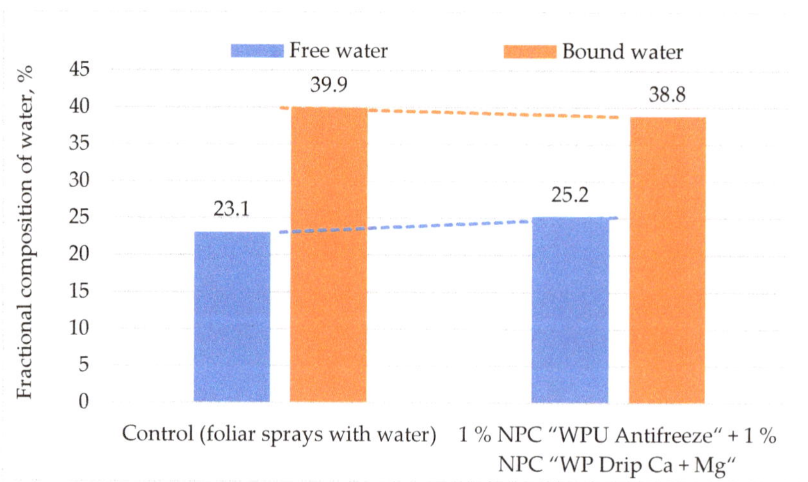

Figure 8. Fractional composition of water in the leaves of "Sinap Orlovsky" after foliar sprays in summer of 2022, %.

Thus, foliar sprays with adaptogenic preparations had a positive effect on the water regime of apple leaves, contributing to the intensive transition of bound water into a more mobile form, which was necessary for the normal functioning of plants during the formation and ripening of fruits, which subsequently affected the increase in the average fruit weight and yield of apple.

3.5. The Effect of Adaptogenic Preparations on the Proline Accumulation in Apple Leaves and Fruits in Summer

In the first ten days of June, the treatments reduced the content of free proline in leaf tissue by 43.7% and did not significantly affect its level in fruits (Table 5). However, in the third ten days of June, the analysis showed an increase in the amino acid content in both leaves (by 62.8%) and fruits (by 12.4%) compared to the control. This increase in proline under the influence of treatments may be associated with the urgent need for a building material, i.e., protein, for more intensive fruit growth. This assumption is supported by the fact that in the second ten days of July, a decrease in the level of free proline was noted in the fruits of the experimental variant, while its growth was still continuing in the leaf apparatus. In the first ten days of August, when the need for building material had virtually completely disappeared, an even greater decrease in the amount of proline was noted both in the leaf apparatus and in fruits, and in the variant with treatment it was to a greater extent. Thus, in the treated plants, the proline content in the fruits was 23.01% less than in the control indicators. Apparently, this indicates that the applied drugs initially contributed to a more intensive growth of apple fruits.

Table 5. The content of free proline in the leaves and fruits of "Sinap Orlovsky" after foliar sprays in summer of 2021, mg/kg.

Experiment Option	Date of Non-Root Treatment			
	14 Days after Flowering	"Fruit-Hazel"	"Fruit-Walnut"	25 Days before Harvesting
	Leaf			
Control (foliar sprays with water)	5.89 ± 0.18	6.51 ± 0.20	11.77 ± 0.40	0.76 ± 0.03
1% NPC "WPU Antifreeze" + 1% NPC "WP Drip Ca + Mg"	4.10 ± 0.12	10.6 ± 0.46 *	11.10 ± 0.33	0.91 ± 0.03 *
	Fruit			
Control (foliar sprays with water)	3.29 ± 0.11	5.79 ± 0.19	5.37 ± 0.20	1.28 ± 0.04
1% NPC "WPU Antifreeze" + 1% NPC "WP Drip Ca + Mg"	3.55 ± 0.13 *	6.51 ± 0.23 *	2.96 ± 0.09	1.04 ± 0.03

* Significant differences with the control at the 5% significance level.

At the same time, foliar sprays did not practically affect the content of free proline in the leaf apparatus and only during the "fruit-hazel" phenophase did it reduce its level by 32.7% compared to the control (Table 6). The amount of free proline in fruits was affected by foliar sprays only by the middle of the growing season. Thus, during the "fruit-walnut" phenophase, the study showed that the level of proline in the experimental version was 23.0% higher than in the control. However, in general, in both variants, the amount of proline decreased by the middle of the growing season, which can be explained by the intensive growth of fruits and the expenditure of amino acids on the biosynthesis of protein substances necessary for growth processes.

Thus, foliar sprays of plants with adaptogenic preparations contributed to the regulation of donor–acceptor leaf–fruit relations against the background of the accumulation of the amino acid proline in summer, which subsequently affected the increase in the average fruit weight and yield of apples.

Table 6. The content of free proline in the leaves and fruits of "Sinap Orlovsky" after foliar sprays in summer of 2022, mg/kg.

Experiment Option	Date of Non-Root Treatment			
	14 Days after Flowering	"Fruit-Hazel"	"Fruit-Walnut"	25 Days before Harvesting
	Leaf			
Control (foliar sprays with water)	3.12 ± 0.16	7.30 ± 0.37	11.60 ± 0.58	11.80 ± 0.47
1% NPC "WPU Antifreeze" + 1% NPC "WP Drip Ca + Mg"	3.00 ± 0.16	5.50 ± 0.23 *	10.90 ± 0.55	10.50 ± 0.44
	Fruit			
Control (foliar sprays with water)	5.70 ± 0.24	10.70 ± 0.37	10.50 ± 0.32	7.00 ± 0.35
1% NPC "WPU Antifreeze" + 1% NPC "WP Drip Ca + Mg"	5.10 ± 0.24	11.20 ± 0.45 *	9.90 ± 0.40	8.60 ± 0.34 *

* Significant differences with the control at the 5% significance level.

3.6. The Effect of Adaptogenic Preparations on the Accumulation of Glucose in the Leaves and Fruits of Apple Trees in Summer

An analysis of the glucose content during the growing season showed that in June, under the influence of treatments, lower carbohydrate values were noted, both in the leaf apparatus and in the fruits (Table 7). However, starting from July, in the treated version, there was an increase in glucose biosynthesis compared to June: in the leaf apparatus, it was 4.2 times more intense in the control and 5.9 times more intense in the experiment; in the fruit, it was 2.5 times more intense in the control and 4.6 times more in the experiment. In this regard, in the second ten days of July, against the background of a decrease in the level of proline in fruits, the glucose content under the influence of treatments was 9.8% higher in leaves and 11.6% higher in fruits than in the control. In August, when the fruits were ripening, a decrease in glucose biosynthesis was noted in the leaf apparatus, with a continued increase of this carbohydrate in apples. It should be noted that in the leaves of the experimental plants, the glucose content was 20.0% higher than in the control, and in the fruits it was higher by 11.3%. The higher carbohydrate content under the action of treatments in experimental plants compared to the control is explained by the improvement of both photosynthetic activity and donor-acceptor relations between the leaf apparatus and the ripening fruit.

Table 7. Glucose content in leaves and fruits of "Sinap Orlovsky" after foliar sprays in summer of 2021, mg/g.

Experiment Option	Date of Non-Root Treatment			
	14 Days after Flowering	"Fruit-Hazel"	"Fruit-Walnut"	25 Days before Harvesting
	Leaf			
Control (foliar sprays with water)	0.77 ± 0.02	0.46 ± 0.02	1.93 ± 0.07	0.76 ± 0.03
1% NPC "WPU Antifreeze" + 1% NPC "WP Drip Ca + Mg"	0.58 ± 0.02	0.36 ± 0.01	2.12 ± 0.08 *	0.91 ± 0.02 *
	Fruit			
Control (foliar sprays with water)	0.13 ± 0.006	0.44 ± 0.02	1.10 ± 0.04	1.15 ± 0.05
1% NPC "WPU Antifreeze" + 1% NPC "WP Drip Ca + Mg"	0.17 ± 0.008 *	0.27 ± 0.012	1.25 ± 0.06 *	1.28 ± 0.04 *

* Significant differences with the control at the 5% significance level.

The analysis of the low molecular weight carbohydrate content during the growing season showed an ambiguous effect of the foliar sprays on glucose levels, both in the fruits and leaves of plants. Thus, in all the dates of the conducted studies, the amount of glucose in the leaf apparatus in the experimental version did not significantly differ from the control (Table 8). Conversely, in fruits, the prevalence or decrease in the amount of the studied carbohydrate varied in the experimental variants compared to the control depending on the time of the growing season. Therefore, 14 days after flowering, under the influence of treatments, the amount of glucose exceeded the control values by 87.5%, in the "fruit-hazel" phenophase it did not significantly differ from the control, and in subsequent periods it was lower than in the control variant by 18.0–76.2%. Apparently, a gradual decrease in the amount of glucose in fruits by the middle of the growing season in the experimental variant is associated with an intense load on plants by the future yield and, as a consequence, a biological dilution of glucose concentration.

Table 8. Glucose content in leaves and fruits of "Sinap Orlovsky" after the foliar sprays in summer of 2022, mg/g.

Experiment Option	Date of Non-Root Treatment			
	14 Days after Flowering	**"Fruit-Hazel"**	**"Fruit-Walnut"**	**25 Days before Harvesting**
	Leaf			
Control (foliar sprays with water)	0.48 ± 0.020	0.29 ± 0.01	0.57 ± 0.03	0.72 ± 0.04
1% NPC "WPU Antifreeze" + 1% NPC "WP Drip Ca + Mg"	0.44 ± 0.020	0.26 ± 0.01	0.57 ± 0.03	0.68 ± 0.03
	Fruit			
Control (foliar sprays with water)	0.16 ± 0.006	0.41 ± 0.02	0.92 ± 0.04	1.11 ± 0.04
1% NPC "WPU Antifreeze" + 1% NPC "WP Drip Ca + Mg"	0.30 ± 0.014 *	0.39 ± 0.02	0.78 ± 0.03	0.63 ± 0.03

* Significant differences with the control at the 5% significance level.

Thus, foliar sprays of apple plants with adaptogenic preparations during the summer period definitely did not affect the carbohydrate metabolism of the donor–acceptor leaf–fruit relationship, and therefore it is necessary to continue the studies to draw reliable conclusions.

3.7. The Effect of Adaptogenic Preparations on the Fruit Qualities and Apple Yield

Due to a biological feature of apple trees (the frequency of fruiting) in the reporting year of 2021, the low yield of the experimental cultivar "Sinap Orlovsky" was noted, but the effectiveness of the applied biological preparations was traced at the same level as in 2022.

With the onset of harvest maturity of "Sinap Orlovsky" fruits in 2021, a significant increase in the average yield from the tree by 1.8 times was recorded against the background of foliar sprays of plants with the preparations of the NPC "White Pearl" line (Figure 9). The average weight of the fruit of the studied cultivar increased by 10.0 g compared to the control, which affected its productivity.

In 2022, a significant 1.7-fold increase in the average yield from a tree was also noted after foliar sprays of plants with biological preparations (Figure 9). Based on the treatment with the tested drugs, the average weight of the fruit of the studied cultivar increased by 30.6 g compared to the control.

When taking into account the weight of the yield of "Sinap Orlovsky" apples (Figures 10 and 11), a significant increase in the yield by 1.7 times was noted in the variant with treatment with preparations of the NPC "White Pearl" line. The increase in the yield amounted to 86.6 c/ha (Figure 10). The adaptogenic preparations contributed to the improvement of fruit quality. The average weight of the fruit of the studied cultivar significantly increased by 20.3 g in the variant with treatments compared to the control.

The results of the biochemical analysis of apple fruits showed that foliar sprays with preparations of the NPC "White Pearl" line improved the taste qualities of the fruits of "Sinap Orlovsky" compared to the control by increasing the amount of sucrose by 25.6% and ascorbic acid by 20.2% (Table 9).

Thus, foliar sprays with preparations of the NPC "White Pearl" line significantly increased the yield and quality of fruits, as well as favorably affecting the consumer and commodity qualities of "Sinap Orlovsky" fruits.

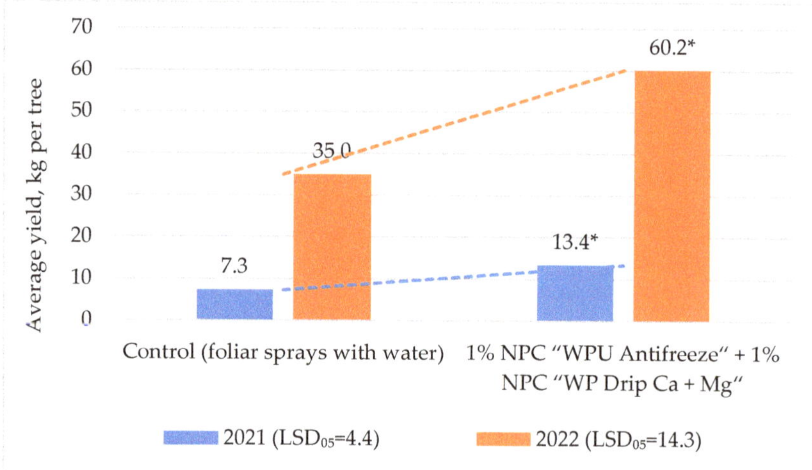

Figure 9. The effect of foliar sprays with preparations of the NPC "White Pearl" line on the average yield from a tree of "Sinap Orlovsky", kg per tree. * Significant differences with the control at the 5% significance level.

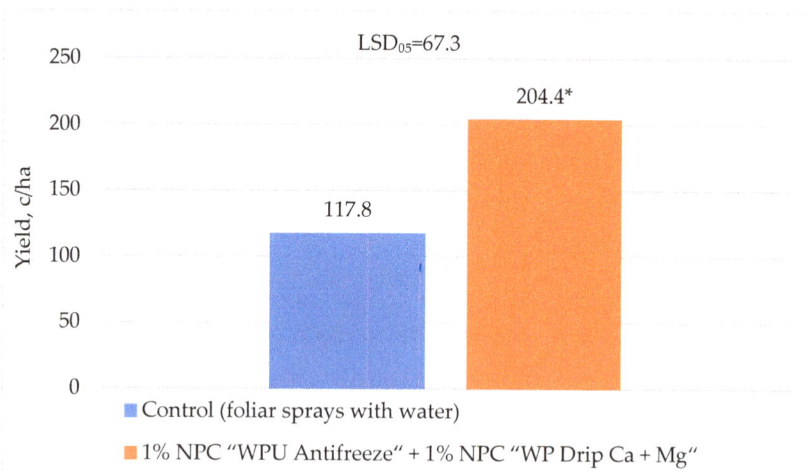

Figure 10. The effect of foliar sprays with preparations of the NPC "White Pearl" line on the yield of "Sinap Orlovsky", c/ha (average for 2021–2022). * Significant differences with the control at the 5% significance level.

Figure 11. The fruits of the "Sinap Orlovsky" in removable maturity.

Table 9. The content of sugars and ascorbic acid of the fruits of "Sinap Orlovsky" after foliar sprays (average for 2021–2022).

Experiment Option	Monosaccharum, % of Raw Weight	Sucrose, % of Raw Weight	Ascorbic Acid, mg/100 g of Raw Mass
Control (foliar sprays with water)	6.86	2.42	11.58
1% NPC "WPU Antifreeze" + 1% NPC "WP Drip Ca + Mg"	6.62	3.04 *	13.95 *

* Significant differences with the control at the 5% significance level.

3.8. The Results of Fruits of Apple Cultivar "Sinap Orlovsky" Storage

After harvesting, the apples were laid in for storage on 13 September 2021 and 9 September 2022 at a temperature of +2 °C. The fruits were removed from storage in early April. The storage duration of the studied variants was 211 days. According to the yield of commercial fruits in the experiment, significant differences were revealed between the variants at the 5% significance level. "Sinap Orlovsky" apples treated with adaptogenic preparations of the NPC "White Pearl" line had a ratio of 92.8% marketable fruits to a 7.2% waste (Figures 12b and 13b), and untreated control fruits had a marketable yield and waste of 78.0% and 22.0% (Figures 12a and 13a), respectively.

The fruits of the studied cultivar predisposed to scald were least affected by this functional disorder after being treated with adaptogenic preparations. The degree of damage was 4.7% (Figure 12b). The fruits of the control variant had a percentage of damage by scald of 11.7 (Figure 12a), although statistically the differences between the variants were not confirmed, as well as for another physiological disorder, i.e., bitter pitting (this indicates an imbalance of mineral composition). In the control variant, 5% of fruits (Figure 12b) were identified as having bitter pitting, and in the variant with treatments with preparations of the NPC "White Pearl" line only 2.5% (Figure 12a) were identified.

In addition to the above disorders, the control variant had overripe and browned fruits in the amounts of 2.7 and 1.2% (Figure 12a), respectively. The treated fruits with adaptogenic preparations did not have similar damage.

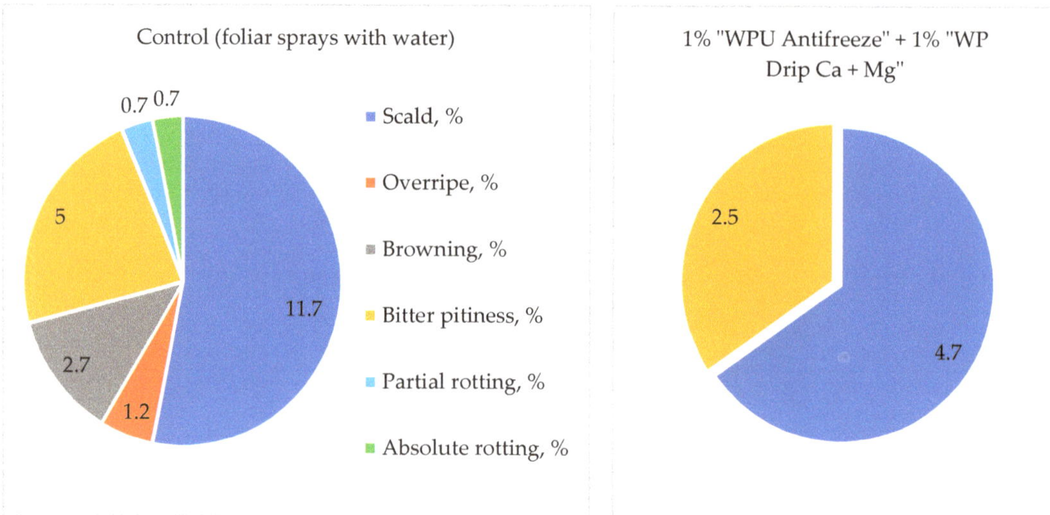

(a) (b)

Figure 12. The percentage of fruit waste during storage (+2 °C) of "Sinap Orlovsky" after on foliar sprays (2021–2022). (**a**) control of the folias sprays with water. (**b**) WPU Antifreeze.

(a) (b)

Figure 13. "Sinap Orlovsky" fruits after 211 days of storage at a temperature of +2 °C: (**a**) control (foliar sprays with water); (**b**) foliar sprays with 1% NPC "WPU Antifreeze" + 1% NPC "WP Drip Ca + Mg".

Regarding microbiological diseases on the fruits of the control variant, minor damage (0.7% with partial and 0.7% with absolute rotting) to fruits was recorded, mainly by fruit rot moniliosis (*Monilia fruktigena* Pers.). In the treated variant, the fruits had absolutely no microbiological damage (Figure 12a). At the same time, an average withering of the fruits in the control variant was recorded, and in the variant with foliar sprays with adaptogenic preparations there was a slight withering of the apples.

Thus, foliar sprays with preparations of the NPC "White Pearl" line favorably affected the consumer and commodity qualities of "Sinap Orlovsky" fruits during storage.

4. Discussion

As a result of the conducted studies, the "Synap Orlovsky" apple trees treated with adaptogenic preparations of the NPC "White Pearl" line were generally characterized by an increased content of low molecular osmolytics and bound water compared to the control trees, which is of great importance in conditions of recurrent spring frosts. It is known that sugars, along with proline, increase the concentration of cell juice by increasing bound water, which has a protective effect under conditions of negative temperatures. In our experiment, the adaptogenic preparations of 1% solution of NPC "White Pearl Universal Antifreeze" + 1% solution of NPC "White Pearl Drip Ca + Mg" presumably prevented the destruction of cell membranes and dehydration of plant cells by increasing the most powerful osmolytically active substances—free proline and sugars—during low-temperature stress, thereby reducing the degree of freezing in the generative organs of apple trees. Thus, foliar sprays with a phytomodulator NPC "White Pearl Universal Antifreeze" and a phytocorrector NPC "White Pearl Drip Ca + Mg" significantly reduced the freezing of flowers by 6.3% and 10.4% at temperature conditions of −3.5 °C and −4 °C, respectively, in comparison with the control. Earlier, V.P. Popova and the authors [40] reported the effectiveness of organomineral foliar sprays containing macro- and microelements, a complex of amino acids, with an increase in the adaptive properties of apple trees. Other studies showed that foliar sprays with the "Regalis" growth regulator enhanced resistance to adverse environmental factors in the Krasnodar Territory by increasing the intensity of metabolic processes [41].

Summer foliar sprays with adaptogenic preparations of 1% solution of NPC "White Pearl Universal Antifreeze" + 1% solution of NPC "White Pearl Drip Ca + Mg" contributed to more intensive growth and ripening of apple fruits against the background of regulation of protein-carbohydrate metabolism, water regime and donor–acceptor leaf–fruit relations, which subsequently positively affected the increase in yield by 1.7 times and the average weight of the fruit by 20.3 g. Other researchers also report an increase in the yield of apple trees with the use of growth regulators. Thus, the silicon-containing growth regulator "Mival-Agro" increased the yield of Gala apples by 3.4 t/ha against the control [42]. The use of the "Regalis" growth regulator also contributed to the active course of photosynthetic activity of apple plants and increased yield by 3.2–13.9 t/ha compared to the control variant [41]. The use of "Albit" steadily increased the yield of apple cultivars, both in comparison with the control and with the production scheme of protection [43]. Other authors, when using biostimulants based on alfalfa protein hydrolysate, seaweed extracts and B vitamins, noted an improvement in the quality and appearance of apple fruits. This study has also shown that biostimulants containing zinc are effective in reducing physiological disorders in apples during storage [30]. In our experiment, foliar sprays with adaptogenic preparations increased the yield of commercial apple fruits by 14.8% compared to the control. Treatment with bioregulators containing progexadion-Ca and paclobutrazole increased the weight of fruits and the yield of pear cultivars. At the same time, a strong relationship was shown between the illumination of the pear crown and the yield [44]. Biostimulants based on alfalfa protein hydrolysate, seaweed extracts and B vitamins also improved the taste and color of apple fruits [30].

Physiological diseases of fruits that occur during storage can seriously affect the quality of apples and, consequently, lead to significant economic losses [45]. At the same time, foliar sprays with adaptogenic preparations contributed to a decrease in fruits affected by scald by 2.5 times and bitter pitting by 2 times against the control. In addition, in the variant of the experiment with foliar sprays, there were 2.3 times fewer overripe and browned fruits. The combined use of calcium chloride with seaweed extract and with a Zn-containing product (Siliforce®) was effective in reducing the spotting of Jonathan fruit during storage [30,46]. The combined application of Ca and Zn led to a higher concentration of these elements at the level of the fruit skin [46], which may have strengthened the cell membranes [47], while reducing the development of spotting during storage. Our results on apple storage are consistent with the results of other studies. Growth regulators "Buton" and "Mival-Agro" significantly increased the yield of standard products during the storage of apples. In addition, the authors recorded an increase in fruit weight and some acceleration of the ripening process of apples [48].

Our tests of adaptogenic preparations of the NPC "White Pearl" line in apple plantations show the prospects of their use as additional techniques in traditional technologies of cultivation of this crop to regulate plant growth processes, increase resistance to spring frosts, improve fruit quality and yield, as well as to preserve consumer and commodity qualities of apples during storage.

5. Conclusions

In the soil and climatic conditions of the Orel region, the effect of adaptogenic preparations of the NPC "White Pearl Universal Antifreeze" and "White Pearl Drip Ca + Mg" on resistance to spring frosts, yield and quality of apple fruits was studied. The research results showed that the phytomodulator "White Pearl Universal Antifreeze" and the phytocorrector "White Pearl Drip Ca + Mg" increased the resistance of apple blossoms and buds to spring frosts. Moreover, adaptogenic preparations significantly increased fruit quality and yield against the background of optimization of the water regime and protein metabolism of apple plants in summer. The usage of drugs enhanced the nutritional value of fruits by increasing the content of sucrose and ascorbic acid in apples. The study also showed the effectiveness of the adaptogenic preparations used in reducing physiological disorders in apples. Thus, the application of adaptogenic preparations of the NPC "White Pearl" line is characterized by a complex effect that allows enhancing the yield and quality of products and increasing the resistance of plants to low-temperature stress in the spring period.

Author Contributions: Formal analysis, Z.O. and P.P.; investigation, Z.O., P.P. and A.N.; data curation, Z.O.; writing—original draft preparation, Z.O.; writing—review and editing, Z.O.; visualization, A.A., A.B., A.S. and O.V.; project administration, Z.O.; funding acquisition, Z.O. All authors have read and agreed to the published version of the manuscript.

Funding: This work was supported by the Ministry of Science and Higher Education of the Russian Federation under agreement No. 075-15-2021-1050 at 28 September 2021.

Institutional Review Board Statement: Not applicable.

Informed Consent Statement: Not applicable.

Data Availability Statement: In this study, the apple cultivar "Synap Orlovsky" was used from the bioresource collection of Russian Research Institute of Fruit Crop Breeding (VNIISPK). Information about this cultivar is available on the website of the Russian Research Institute of Fruit Crop Breeding (VNIISPK).

Acknowledgments: The authors thank AgroPlus Group of Companies LLC for the gratuitous provision of adaptogenic preparations the phytomodulator "White Pearl Universal Antifreeze" and the phytocorrector "White Pearl Drip Ca + Mg" for testing.

Conflicts of Interest: The authors declare no conflict of interest.

References

1. Bulgari, R.; Cocetta, G.; Trivellini, A.; Vernieri, P.; Ferrante, A. Biostimulants and crop responses: A review. *Bioljgical Agric. Hortic.* **2015**, *31*, 1–17. [CrossRef]
2. Amarante, C.V.T.; do Steffens, C.A.; Mafra, Á.L.; Albuquerque, J.A. Yield and fruit quality of apple from conventional and organic production systems. *Pesqui. Agropecuária Bras.* **2008**, *43*, 333–340. [CrossRef]
3. de Ponti, T.; Rijk, B.; van Ittersum, M.K. The crop yield gap between organic and conventional agriculture. *Agric. Syst.* **2012**, *108*, 1–9. [CrossRef]
4. Seufert, V.; Ramankutty, N.; Foley, J.A. Comparing the yields of organic and conventional agriculture. *Nature* **2012**, *485*, 229–232. [CrossRef] [PubMed]
5. Orsini, F.; Maggio, A.; Rouphael, Y.; De Pascale, S. "Physiological quality" of organically grown vegetables. *Sci. Hortic.* **2016**, *208*, 131–139. [CrossRef]
6. De Pascale, S.; Rouphael, Y.; Colla, G. Plant biostimulants: Innovative tool for enhancing plant nutrition in organic farming. *Eur. J. Hortic. Sci.* **2017**, *82*, 277–285. [CrossRef]

7. Rouphael, Y.; Colla, G. Toward a sustainable agriculture through plant biostimulants: From experimental data to practical applications. *Agronomy* **2020**, *10*, 1461. [CrossRef]
8. Povero, G.; Mejia, J.F.; Di Tommaso, D.; Piaggesi, A.; Warrior, P. A systematic approach to discover and characterize natural plant biostimulants. *Front. Plant Sci.* **2016**, *7*, 435. [CrossRef]
9. du Jardin, P. Plant biostimulants: Definition, concept, main categories and regulation. *Sci. Hortic.* **2015**, *196*, 3–14. [CrossRef]
10. Battacharyya, D.; Babgohari, M.Z.; Rathor, P.; Prithiviraj, B. Seaweed extracts as biostimulants in horticulture. *Sci. Hortic.* **2015**, *196*, 39–48. [CrossRef]
11. Canellas, L.P.; Olivares, F.L.; Aguiar, N.O.; Jones, D.L.; Nebbioso, A.; Mazzei, P.; Piccolo, A. Humic and fulvic acids as biostimulants in horticulture. *Sci. Hortic.* **2015**, *196*, 15–27. [CrossRef]
12. Colla, G.; Hoagland, L.; Ruzzi, M.; Cardarelli, M.; Bonini, P.; Canaguier, R.; Rouphael, Y. Biostimulant action of protein hydrolysates: Unraveling their effects on plant physiology and microbiome. *Front. Plant Sci.* **2017**, *8*, 2202. [CrossRef] [PubMed]
13. Rouphael, Y.; Franken, P.; Schneider, C.; Schwarz, D.; Giovannetti, M.; Agnolucci, M.; De Pascale, S.; Bonini, P.; Colla, G. Arbuscular mycorrhizal fungi act as biostimulants in horticultural crops. *Sci. Hortic.* **2015**, *196*, 91–108. [CrossRef]
14. Ruzzi, M.; Aroca, R. Plant growth-promoting rhizobacteria act as biostimulants in horticulture. *Sci. Hortic.* **2015**, *196*, 124–134. [CrossRef]
15. Ertani, A.; Cavani, L.; Pizzeghello, D.; Brandellero, E.; Altissimo, A.; Ciavatta, C.; Nardi, S. Biostimulant activity of two protein hydrolyzates in the growth and nitrogen metabolism of maize seedlings. *J. Plant Nutr. Soil Sci.* **2009**, *172*, 237–244. [CrossRef]
16. Colla, G.; Rouphael, Y. Biostimulants in horticulture. *Sci. Hortic.* **2015**, *196*, 1–134. [CrossRef]
17. Lucini, L.; Rouphael, Y.; Cardarelli, M.; Canaguier, R.; Kumar, P.; Colla, G. The effect of a plant-derived biostimulant on metabolic profiling and crop performance of lettuce grown under saline conditions. *Sci. Hortic.* **2015**, *182*, 124–133. [CrossRef]
18. Rouphael, Y.; Cardarelli, M.; Bonini, P.; Colla, G. Synergistic action of a microbial-based biostimulant and a plant derived-protein hydrolysate enhances lettuce tolerance to alkalinity and salinity. *Front. Plant Sci.* **2017**, *8*, 131. [CrossRef]
19. Hermosa, R.; Viterbo, A.; Chet, I.; Monte, E. Plant-beneficial effects of Trichoderma and of its genes. *Microbiology* **2012**, *158*, 17–25. [CrossRef]
20. Studholme, D.J.; Harris, B.; Le Cocq, K.; Winsbury, R.; Perera, V.; Ryder, L.; Ward, J.L.; Beale, M.H.; Thornton, C.R.; Grant, M. Investigating the beneficial traits of Trichoderma hamatum GD12 for sustainable agriculture—Insights from genomics. *Front. Plant Sci.* **2013**, *4*, 258. [CrossRef]
21. Hussein, M.A.M.; Hassan, M.H.A.; Abo-Elyousr, K.A.M. Biological control of Botrytis allii by Trichoderma viride on onion Allium cepa. *World Appl. Sci. J.* **2014**, *32*, 522–526. [CrossRef]
22. Mendoza-Mendoza, A.; Zaid, R.; Lawry, R.; Hermosa, R.; Monte, E.; Horwitz, B.A.; Mukherjee, P.R. Molecular dialogues between Trichoderma and roots: Role of the fungal secretome. *Fungal Biol. Biotechnol.* **2018**, *32*, 62–85. [CrossRef]
23. Fontenelle, A.D.B.; Guzzo, S.D.; Lucon, C.M.M.; Harakava, R. Growth promotion and induction of resistance in tomato plant against Xanthomonas euvesicatoria and Alternaria solani by *Trichoderma* spp. *Crop Prot.* **2011**, *30*, 1492–1500. [CrossRef]
24. Lorito, M.; Woo, S.L. Trichoderma: A multi-purpose tool for integrated pest management. In *Principles of Plant-Microbe Interactions*; Lugtenberg, B., Ed.; Springer International Publishing: Cham, Switzerland, 2015; pp. 345–353. [CrossRef]
25. Vinocur, B.; Altman, A. Recent advances in engineering plant tolerance to abiotic stress: Achievements and limitations. *Curr. Opin. Biotechnol.* **2005**, *416*, 123–132. [CrossRef]
26. Sharma, H.S.S.; Fleming, C.; Selby, C.; Rao, J.R.; Martin, T. Plant biostimulants: A review on the processing of macroalgae and use of extracts for crop management to reduce abiotic and biotic stresses. *J. Appl. Phycol.* **2014**, *26*, 465–490. [CrossRef]
27. Fedulov, Y.P.; Kotlyarov, V.V.; Dotsenko, K.A. *Plant Resistance to Adverse Environmental Factors*; Kuban State Agro University: Krasnodar, Russia, 2015.
28. Van Oosten, M.J.; Pepe, O.; De Pascale, S.; Silletti, S.; Maggio, A. The role of biostimulants and bioeffectors as alleviators of abiotic stress in crop plants. *Chem. Biol. Technol. Agric.* **2017**, *4*, 5. [CrossRef]
29. Ferguson, I.; Volz, R.; Woolf, A. Preharvest factors affecting physiological disorders of fruit. *Postharvest Biol. Technol.* **1999**, *15*, 255–262. [CrossRef]
30. Soppelsa, S.; Kelderer, M.; Casera, C.; Bassi, M.; Robatscher, P.; Andreotti, C. Use of Biostimulants for Organic Apple Production: Effects on Tree Growth, Yield, and Fruit Quality at Harvest and During Storage. *Front. Plant Sci.* **2018**, *9*, 1342. [CrossRef]
31. Sedov, E.N.; Sedysheva, G.A.; Krasova, N.G.; Serova, Z.M.; Gorbacheva, N.G.; Galasheva, A.M.; Yanchuk, T.V.; Pikunova, A.V.; Van de Veg, E. Origin, economical and cytoembryological characteristics of triploid apple cultivar 'Sinap Orlovsky'. *Russ. Agric. Sci.* **2017**, *1*, 14–18.
32. Ozherelieva, Z.E.; Prudnikov, P.S.; Zubkova, M.I.; Krivushinaa, D.A.; Knyazev, S.D. *Determination of Frost Resistance of Strawberries in Controlled Conditions*; VNIISPK: Orel, Russia, 2019.
33. Turkina, M.V.; Sokolova, S.V. Study of membrane transport of sucrose in plant tissue. *Plant Physiol.* **1972**, *1*, 912–919.
34. Bates, L.S.; Waldren, R.P.; Teare, I.D. Rapid determination of free proline for water-stress studies. *Plant Soil* **1973**, *39*, 205–207. [CrossRef]
35. Prudnikov, P.S.; Ozherelieva, Z.E. *Physiological and Biochemical Methods for Diagnosing the Resistance of Fruit Crops to Drought and Hyperthermia*; VNIISPK: Orel, Russia, 2019.
36. Sedov, E.N.; Ogoltsova, T.P. (Eds.) *Program and Methodology of Fruit, Berry and Nut Variety Study*; VNIISPK: Orel, Russia, 1999.

37. Chandraju, S.; Kumar, C.S.C.; Venkatesh, R. Estimation of reducing sugar by acid hydrolysis of black grape (*Vitis vinifera* L.) peels by standard methods. *J. Chem. Pharm. Res.* **2014**, *6*, 862–866.
38. Yurtaeva, N.M. *Physiology of the Plant Cell. Water Regime of Plants*; NNGASU: Nizhniy Novgorod, Russia, 2014.
39. Ozherelieva, Z.E.; Prudnikov, P.S.; Kurashev, O.V.; Krivushina, D.A. Tolerance of gooseberry varieties to maximally low temperature in the middle of winter. *Bulg. J. Agric. Sci.* **2018**, *24*, 1053–1058.
40. Popova, V.P.; Sergeeva, N.N.; Fomenko, T.G.; Yaroshenko, O.V.; Nenko, N.I. Principles of increasing the resistance of garden cenoses to stress factors and changes in the level of soil fertility. *Sci. Work. North Cauc. Fed. Sci. Cent. Hortic. Vitic. Winemak.* **2019**, *23*, 89–99. [CrossRef]
41. Fomenko, T.G.; Popova, V.P.; Nenko, N.I.; Shadrina, Z.A. Development of regulations for the use of Regalis growth regulator in intensive apple plantations. *Agrochem. Bull.* **2018**, *3*, 51–55. [CrossRef]
42. Kalmykova, O.V. The effectiveness of using biological products in an apple orchard in the conditions of the Lower Volga region. *Bull. Altai State Agrar. Univ.* **2014**, *4*, 20–23.
43. Pantiya, G.G.; Yanushevskaya, E.B.; Mikhailova, E.V.; Karpun, N.N. Evaluation of the effectiveness of immunoinducers in increasing the nonspecific resistance of apple trees to scab. *Plant Prot. Quar.* **2019**, *7*, 33–35.
44. Kaur, S.; Gill, M.S.; Gill, P.P.S.; Singh, N.P. Effect of plant bio-regulators on photosynthesis, growth and yield efficiency of peartrained on Y-trellis system. *J. Agrometeorol.* **2020**, *22*, 140–144. [CrossRef]
45. Gago, C.M.L.; Guerreiro, A.C.; Miguel, G.; Panagopoulos, T.; da Silva, M.M.; Antunes, M.D.C. Effect of calcium chloride and 1-MCP (SmartfreshTM) postharvest treatment on 'Golden Delicious' apple cold storage physiological disorders. *Sci. Hortic.* **2016**, *211*, 440–448. [CrossRef]
46. Soppelsa, S.; Kelderer, M.; Testolin, R.; Zanotelli, D.; Andreotti, C. Effect of Biostimulants on Apple Quality at Harvest and After Storage. *Agronomy* **2020**, *10*, 1214. [CrossRef]
47. Guerriero, G.; Hausman, J.-F.; Legay, S. Silicon and the plant extracellular matrix. *Front. Plant Sci.* **2016**, *7*, 463. [CrossRef] [PubMed]
48. Drevin, V.E.; Kalmykova, E.V.; Kalmykova, O.V. The influence of growth regulators on the keeping quality of apples in the conditions of the Lower Volga region. *Storage Process. Agric. Raw Mater.* **2016**, *1*, 12–15.

horticulturae

MDPI

Article

Copper Stress Enhances the Lignification of Axial Organs in *Zinnia elegans*

Anastasia Tugbaeva [1,*], Alexander Ermoshin [1,2], Hada Wuriyanghan [3], Maria Maleva [1,*], Galina Borisova [1] and Irina Kiseleva [1]

[1] Department of Experimental Biology and Biotechnology, Institute of Natural Sciences and Mathematics, Ural Federal University, 620000 Ekaterinburg, Russia; alexander.ermoshin@urfu.ru (A.E.); galina.borisova@urfu.ru (G.B.); irina.kiseleva@urfu.ru (I.K.)

[2] Inner Mongolia Academy of Agriculture and Animal Husbandry Science, Hohhot 232813, China

[3] Key Laboratory of Forage and Endemic Crop Biology, Ministry of Education, School of Life Sciences, Inner Mongolia University, Hohhot 010070, China; nmhadawu77@imu.edu.cn

* Correspondence: anastasia.tugbaeva@urfu.ru (A.T.); maria.maleva@urfu.ru (M.M.)

Abstract: *Zinnia elegans* Jacq. is an ornamental plant, widely used in landscaping. Heavy-metal pollution in urban and rural areas is still increasing, which determines the actuality of studying plants' reactions to pollutants. Zinnia was not sufficiently studied in this regard, so the aim of our research was to identify morphophysiological changes in this species under excess copper concentration in the soil. For this, we treated a growth substrate with 200 μM $CuSO_4$ solution for 20 days. At the end of the treatment, several morphological, biochemical, and molecular genetic traits were evaluated: the root and the shoot size; the concentration of H_2O_2 and malondialdehyde (MDA), as indicators of stress; the amount of the phenolic compounds and lignin; and the level of the expression of genes, which encoded their biosynthesis. The Cu amount in the substrate and zinnia organs was quantified using atomic-absorption spectroscopy; hydrogen peroxide, MDA, and phenolic compounds were determined spectrophotometrically, while the amount of lignin was determined according to Klason. Real-time PCR was used for estimation of the gene-transcription level. Lignin in tissues was visualized by fluorescent microscopy. In experimental plants, Cu accumulation was higher in the root than in the stem. This caused an increase in stress markers and a decrease in the root and stem lengths. For the first time for zinnia, it was shown that for several genes—4-coumarate-CoA ligase (*4CL*), cinnamoyl alcohol dehydrogenase (*CAD*), and class III peroxidase (*PRX*)—the level of expression increased under copper treatment. The rise of the transcripts' amount of these genes was accompanied by a thickening and lignification of the cell walls in the metaxylem vessels. Thus, the adaptation of zinnia to the excess Cu in the growth medium was associated with the metabolic changes in the phenylpropanoid pathway. As a result, the lignification increased in the root, which led to the accumulation of Cu in this organ and limited its translocation through the xylem to the stem, which provided plant growth.

Keywords: zinnia; redox-active metal; cell-wall lignification; phenolics; phenylpropanoid metabolic-pathway genes; landscaping of urban areas

Citation: Tugbaeva, A.; Ermoshin, A.; Wuriyanghan, H.; Maleva, M.; Borisova, G.; Kiseleva, I. Copper Stress Enhances the Lignification of Axial Organs in *Zinnia elegans*. *Horticulturae* **2022**, *8*, 558. https://doi.org/10.3390/horticulturae8060558

Academic Editors: Agnieszka Hanaka, Małgorzata Majewska and Barbara Hawrylak-Nowak

Received: 30 May 2022
Accepted: 18 June 2022
Published: 20 June 2022

1. Introduction

Pollution of garden and agricultural lands with heavy metals (HMs), in particular copper, is a common problem. The widely used fungicides and phosphate fertilizers can be sources of excess copper in garden farms [1,2]. The high amount of this element (up to 2000 mg kg^{-1}) in the soil of certain regions can also be a consequence of the mining and processing of copper ore, as well as natural soil-forming processes [1–3].

Copper is one of the trace elements essential for plant life and plant growth; 4–15 mg kg^{-1} of copper in dry matter is considered sufficient for the synthesis of chlorophylls, cytochromes, nitrogen, carbon metabolism, respiration, and photosynthesis [1,3–5].

In its ion form, copper is a part of the electron-transport-chain proteins in chloroplasts and mitochondria as well as copper-containing enzymes (Cu/Zn-superoxide dismutase, laccase, etc.) [4,5]. Copper is a redox active metal; its excess in plant tissues leads to the development of oxidative stress, which manifests itself in the growth of the reactive oxygen species and products of lipid peroxidation [4–6]. Copper toxicity is also associated with the ability to bind the SH-groups of proteins, disrupting their conformation and loss of functions [6]. Phenotypically, copper stress limits the growth of the root and aboveground plant organs, causing browning of the root and chlorosis [3–6].

Plants have several protective mechanisms against the impact of HMs and copper in particular, for example, chelation, sequestration in cell walls and vacuoles, deposition in root tissues, limitation of translocation to the shoot, and activation of the non-enzymatic and enzymatic systems of antioxidant defense [3–6]. These reactions are well-studied; however, the data on plant tolerance to copper excess in specific taxa and horticultural crops are limited. Plants that accumulate heavy-metal ions in aboveground organs are promising for phytoremediation [7]. Those that keep heavy-metal ions in their roots could be used for reclamation or re-cultivation of the disturbed territories.

For plants, growing in conditions contaminated by HMs, the root cell wall is the first barrier preventing the penetration of ions into cells. It plays a key role in the absorption, immobilization, and translocation of heavy-metal ions [8,9]. The cell wall consists of polysaccharides (cellulose, pectin, hemicellulose), phenolic compounds (lignin, suberin), and proteins, the ratio of which may vary depending on the tissue type, ontogenetic state, intensity, and duration of the stress factor [10,11]. Lignin defines the rigidity and hydrophobicity of cell walls, and it is also involved in the binding of copper ions by the carboxyl and hydroxyl groups [9,10]. Increased lignification is suggested as a nonspecific reaction of plants to the excess of HMs in the medium [8,9]. Lignin precursors, monolignols (coniferyl, synapyl, *p*-coumaryl alcohols), are synthesized in the cytosol via the phenylpropanoid pathway, then transported to the apoplast, where they are oxidized by class III peroxidases (*PRX*, EC 1.11.1.7) and laccases (*LAC*, EC 1.10.3.2) with the formation of radicals. The polymerization of lignin occurs according to the free-radical mechanism [12,13].

The first three reactions of the phenylpropanoid pathway are the sequential conversion of phenylalanine to *p*-Coumaroyl-CoA [14,15]. Phenylalanine ammonium lyase (*PAL*, EC 4.3.1.24) deaminates phenylalanine to cinnamic acid. Then, cinnamate-4-hydroxylase (*C4H*, EC 1.14.13.11) catalyzes the hydroxylation of cinnamic acid to *p*-coumaric acid. Moreover, 4-coumarate-CoA ligase (*4CL*, EC 6.2.1.12) catalyzes the formation of *p*-Coumaroyl-CoA, which is a precursor of oxycinnamic alcohols, flavonoids, lignins, and isoflavonoids [14,15]. Further, cinnamoyl-CoA reductase (*CCR*, EC 1.2.1.44) synthesizes hydroxycinnamaldehydes from hydroxycinnamoyl-CoA, and then they are converted to cinnamyl alcohols by cinnamoyl-alcohol dehydrogenase (*CAD*, EC 1.1.1.195), which are the precursors of *p*-hydroxyphenyl (H)-, guaiacyl (G)-, and syringyl (S)-lignin monomers [14,15].

Zinnia elegans Jacq. is a model object for studying lignification as well as *Arabidopsis* sp. and *Populus* sp. [14–16]. Zinnia is an annual, fast-growing plant with a long flowering period (about 90 days). It is an ornamental, widely cultivated flower culture, used in landscape design [17].

We suggest that under excess copper in the soil, zinnia like other plants will enhance the lignification of the axial organs and, thereby, prevent the translocation of this element from the root to the shoot, and allow the plant to form aboveground organs. The excess deposition of lignin under copper stress could be associated with a modified expression of the genes that are involved in phenylpropanoid metabolism and lignin synthesis. The aim of our study was to identify the morphophysiological and biochemical changes in zinnia plants, grown under conditions of excess copper in the substrate. For this purpose, the anatomical and morphological characteristics of zinnia axial organs, the deposition of lignin in cell walls, the level of stress markers, and some traits of the phenolics metabolism in the control and experimental plants were evaluated.

2. Materials and Methods

2.1. Plant-Growth Conditions

Zinnia (*Zinnia elegans* Jacq.) is an annual plant from the Asteraceae family, which is often used for decoration of parks and gardens, but rarely studied in biochemical and molecular genetic experiments.

Zinnia plants (cv. Rotkappchen) were cultivated on a pre-autoclaved substrate—a mixture of soil (neutralized peat, pH 6.5, containing total nitrogen 1500 mg kg^{-1} per dry weight (DW), phosphorus 2500 mg kg^{-1} per DW, and potassium 3000 mg kg^{-1} per DW) and coco substrate (3:1, *v*/*v*) in 0.2 L vegetative vessels. An aqueous 200 μM $CuSO_4$ solution (30 mL) was added to the experimental plants every 5 days. Control plants were poured by water. Plants were grown for 20 days under a 16 h (day):8 h (night) photoperiod; 23 ± 2 °C temperature; and $65 \pm 5\%$ humidity.

The concentration of copper and the duration of the treatment were selected to avoid acute toxicity and to assess the long-term response of plants to the stressor, according to previous study—200 μM $CuSO_4$ worsened seed germination and inhibited seedling growth [18].

2.2. Quantification of Copper

The substrate, dried to a constant weight, was ground thoroughly and sieved to remove large fragments (more than 2 mm); the total amount of copper was determined by digesting 0.25 g soil with HNO_3:$HClO_4$:HF (5:1:1, *v*/*v*/*v*) on a hot plate, followed by filtration through a Whatman filter No. 42. The extraction of mobile forms of copper ions was carried out by treating the soil sample with 4 mM Na_2EDTA in a ratio of 1:25 (*w*/*v*) (shaken at 150 rpm for 24 h, pH 4.5); then the extract was acidified with 1% HNO_3 [19]. The amount of copper ions in the substrate was expressed in mg copper kg^{-1}. The analysis was performed in 5 independent replicates.

To determine the copper amount in the zinnia organs (μg g^{-1} DW), 50 mg of dried to constant-weight biomass (separately root and stem) was ashed in HNO_3. All measurements were done using atomic-emission spectroscopy (ICP-AES, iCAP 6500 Duo, Thermo Fisher, Waltham, MA, USA). The analysis was performed in 5 independent replicates, and each replicate was formed from 3 plants.

The bioconcentration factor (BCF) was calculated as the ratio of the Cu concentration in the organ (μg g^{-1} DW) to the amount of available Cu in the substrate (μg g^{-1}), performed in relative units. The translocation factor (TF) was determined as the ratio of Cu concentration in the stem (μg g^{-1} DW) to its concentration in the root (μg g^{-1} DW).

2.3. Biochemical Characteristics

The H_2O_2 concentration was assessed in a crude extract of root and stem tissues (0.1 M Tris-HCl buffer, pH 7.8) using a method based on the oxidation of xylenol orange chelates with iron (III) by peroxide, according to Bellincampi et al. [20], and expressed in μmol of hydrogen peroxide g^{-1} fresh weight (FW). The intensity of lipid peroxidation was estimated spectrophotometrically as the production of malondialdehyde (MDA) in the reaction with thiobarbituric acid and expressed in μmol MDA g^{-1} FW [21].

Phenolic compounds were extracted by 70% ethanol, and their concentration was determined using the Folin–Ciocalteu reagent and performed in μg g^{-1} FW in terms of gallic acid [22]. The content of Klason lignin (KL) and acid-soluble lignin (ASL) were determined in the dry ground roots or stems by the sulfuric acid method [23] and expressed in percentages (%). The optical density of the samples was measured on a Tecan Infinite M200 Pro spectrophotometer (Tecan Austria GmbH, Grödig, Austria). The analysis was performed in 3 biological and 15 analytical replicates.

2.4. Quantitative Real-Time PCR (qRT-PCR) Analysis

Total RNA was isolated using Trizol (TransGen Biotech, Beijing, China) [24]. The concentration and the quality of the isolated RNA was assessed spectrophotometrically using a NanoDrop ND-1000 instrument (ThermoScientific, Waltham, MA, USA). In total, 100 ng of total RNA was used for each sample with Oligo(dT)23VN and Random Hexamer primers to obtain the first strand of c-DNA, in accordance with the instructions of the manufacturer (HiScriptII 1st standard cDNA synthesis kit, Vasyme, Nanjing, China). Gene expression was assessed by qRT-PCR in a qTOWER 2.0 96-well optical amplifier (Analytikjena, Jena, Germany) using TransStrat® Tip Green qPCR SuperMix (TransGenBiotech, Beijing, China, Cat#AQ141).

The forward and reverse primers for the reaction were selected using the Blast Primer designee online program (www.ncbi.nlm.nih.gov/tools/primer-blast, accessed on 1 May 2022). Gene-specific primers are performed in Table 1. Amplification was carried out under standard conditions (1 cycle: 30 s at 94 °C; 40 cycles: 5 s at 94 °C, 15 s at 60 °C, and 10 s at 72 °C; 5 s at 60 °C). The relative expression level was calculated using the $2^{-\Delta\Delta Ct}$ method [25]. The data were normalized to the gene encoding the *18S rRNA*. The analysis was performed in 3 biological and 3 analytical replicates.

Table 1. Forward and reverse primers for qRT-PCR.

Gene, GenBank Access No.	Forward Primer Sequence (5′ → 3′)	Reverse Primer Sequence (5′ → 3′)
PAL FM879196	GTCACCAGGCGAAGAGTTTG	CGGAACACCATCCCATCCTT
C4H FM880082	GAACTTTGAGCTGTTGCCGC	TGAAAAACCCACAAACAACAATCC
4CL AU294519	ACGTCACCTTCCGTTACACC	CGTCAGCGATTATCGACGGT
CCR FM881365	CCTCGGCTTCTGGTCGATAC	TGTATGGCTTTGCTCGTGGT
CAD FM881026	CCGTAAACCATCCTCTTGCG	CAAGCTTCCTCCCCACAATC
PRX AB023959	TCGCAGCTTCAATGGTCAAAC	TCTCTTCTTCTTTCATACTTCCCTT
LAC AU286008	AATAAGGACGGGTTGGGCTG	AGGGTAAGGGATACCACGCT
18S rRNA AB089282	ATGTGGTAGCCGTTTCTCAGG	TGCCCGTTGCTGCGAT

2.5. Biometric and Anatomical Analysis

At the end of experiment, 30 plants from each variant were used for the determination of biometric characteristics. The length of the shoot and root were measured for each plant. Then, the samples were oven-dried at 85 °C for 48 h, and the dry weight (DW) was determined.

Plant axial organs (the root in the mature zone, the hypocotyl, and the first internode above the cotyledons) were fixed in a mixture of ethyl alcohol and acetic acid (3:1, v/v) [26] for the investigation of their anatomy. The fixation time was 48 h at 4 °C. Then, the plant material was washed and stored in 96% ethanol. Cross sections of the axial organs were made by hand using a razor and placed in glycerin. The transverse sections were visualized on a wide-field microscope Leica DM5500 ("Leica Microsystem", Wetzlar, Germany). Autofluorescence of lignin was detected with a standard GFP filter. The diameter of the root and stem, the thickness of the cortex and stele, and the cross-sectional area and cell-wall thickness of the metaxylem vessels were measured in cross-sectional photographs using SIMAGIS® Meso-Plant™ software for Windows XP. The number of measurements was 50 for each characteristic.

2.6. Statistical Analysis

The experiment was repeated three times. The data are presented as the arithmetic mean and the standard error. Statistical data processing was carried out in the STATISTICA 13 program for Windows 10 using Student's t-test ($p < 0.05$), Mann–Whitney U-test ($p < 0.05$), and Spearman's r-test to calculate correlations ($p < 0.01$).

3. Results

3.1. Copper Amount

The amount of the available forms of copper in the substrate was estimated as 0.56% of its total amount, in the case of treatment with 200 µM CuSO4 (Table 2). The total copper amount increased by 21.6 times compared to the untreated substrate.

Table 2. Copper amount in substrate and zinnia organs; BCF and TF on the 20th day of growth.

Treatment	Copper Amount, µg g^{-1} Dry Substrate		Copper Amount, µg g^{-1} DW		BCF		TF
	Available	Total	Root	Stem	Root	Stem	
Control (water)	n.d.	7.75 ± 0.65	9.80 ± 0.52	11.36 ± 0.70	n.d.	n.d.	1.16 ± 0.07
200 µM $CuSO_4$	1.12 ± 0.17 * [1]	167.63 ± 1.87 *	26.32 ± 1.32 *	12.04 ± 0.69	23.5 ± 1.2 *	10.75 ± 0.65 *	0.46 ± 0.02 *

[1] Result is presented as mean ± standard error (n = 5); n.d.—the copper amount below detection limit. Asterisks represent significant differences ($p < 0.05$, U-test).

The strong accumulation of copper in the root and its weak redistribution into the shoot were detected under the treatment with 200 µM $CuSO_4$ (Table 2). In the roots of the experimental plants, the copper amount increased by 168% in comparison with the control, but in the stems it did not change. The BCF for copper was higher in the root than in the stem (Table 2).

In untreated plants, the TF for copper was greater than one, i.e., the plants absorbed it from the substrate and translocated it from the root to the shoot as an essential element. In the case of the copper excess in the substrate, the TF was less than one, which proves the barrier role of the root system in the long-distance transport and allows for classifying zinnia as a copper-excluder plant (Table 2).

3.2. Concentration of Hydrogen Peroxide and MDA Products

The levels of lipid peroxidation (LPO) and hydrogen-peroxide concentration were determined as the stress markers. In the case of substrate treatment by the copper solution, the development of oxidative stress reached 120% compared to the control in the root and 150% in the stem (Table 3). There was a 5.4-fold increase in the H_2O_2 concentration in the roots of treated plants and a 2.1-fold decrease in the stems, compared to untreated plants (Table 3).

Table 3. The concentration of hydrogen peroxide and MDA in zinnia organs on the 20th day of growth.

Treatment	H_2O_2, µmol g^{-1} FW		MDA, µmol g^{-1} FW	
	Root	Stem	Root	Stem
Control (water)	29.5 ± 3.6 [1]	120.5 ± 9.1	0.56 ± 0.03	0.31 ± 0.02
200 µM $CuSO_4$	159.61 ± 5.88 *	61.50 ± 3.8 *	0.67 ± 0.02 *	0.48 ± 0.02 *

[1] Result is presented as mean ± standard error (n = 15). Asterisks represent significant differences ($p < 0.05$, U-test) from control.

3.3. Concentration of Phenolics and Lignin

A strong decrease (37.3%) in the phenolics amount was observed in the roots and in the stems (19.6%), in the case of the soil treatment with 200 µM $CuSO_4$, compared to the control (Table 4).

Table 4. The concentration of phenolic compounds and lignin (KL, ASL, and total lignin) in zinnia organs on the 20th day of growth.

Treatment	Phenolics, mg g^{-1} FW		KL, %		ASL, %		Total Lignin, %	
	Root	Stem	Root	Stem	Root	Stem	Root	Stem
Control (water)	0.43 ± 0.01 [1]	0.61 ± 0.03	9.29 ± 0.08	8.02 ± 0.44	4.90 ± 0.23	6.43 ± 0.11	14.19 ± 0.31	14.19 ± 0.31
200 μM $CuSO_4$	0.27 ± 0.01 *	0.50 ± 0.01 *	12.27 ± 0.20 *	10.58 ± 0.56 *	4.35 ± 0.42	4.47 ± 0.42 *	16.62 ± 0.76 *	15.05 ± 0.98

[1] Result is presented as mean ± standard error (n = 15). Asterisks represent significant differences ($p < 0.05$, U-test) from control.

A statistically significant increase in the KL quantity (Table 4) was found in the zinnia organs, which contributed to the increase in the total lignin amount. Both in the root and the stem, the KL content under the copper treatment increased, respectively, by 32.1% and 31.8% (Table 4).

The amount of ASL in the root of treated plants was the same as in the control group of plants (Table 4). In the stem it significantly decreased by 30.5% compared to the control (Table 4).

3.4. Expression of the Genes of Phenylpropanoid Metabolic Pathway and Lignin Biosynthesis

Under $CuSO_4$ treatment, the relative number of transcripts of *PAL* and *CCR* genes decreased in the root, and the expression of the *C4H*, *4CL*, and *CAD* genes increased (Figure 1a).

The relative quantity of transcripts of the *C4H* and *CCR* gene in the stem did not differ significantly both in the control and experimental plants, but transcripts of the *PAL*, *4CL*, and *CAD* increased in number compared to the untreated plants (Figure 1b).

The expression level of the genes involved in lignin biosynthesis has changed under stress conditions. The relative number of transcripts of *PRX* gene increased by 1.8 times in the root, and by 2.5 times in the stem, compared to the control. The expression level of the *LAC* gene tended to decrease in both organs.

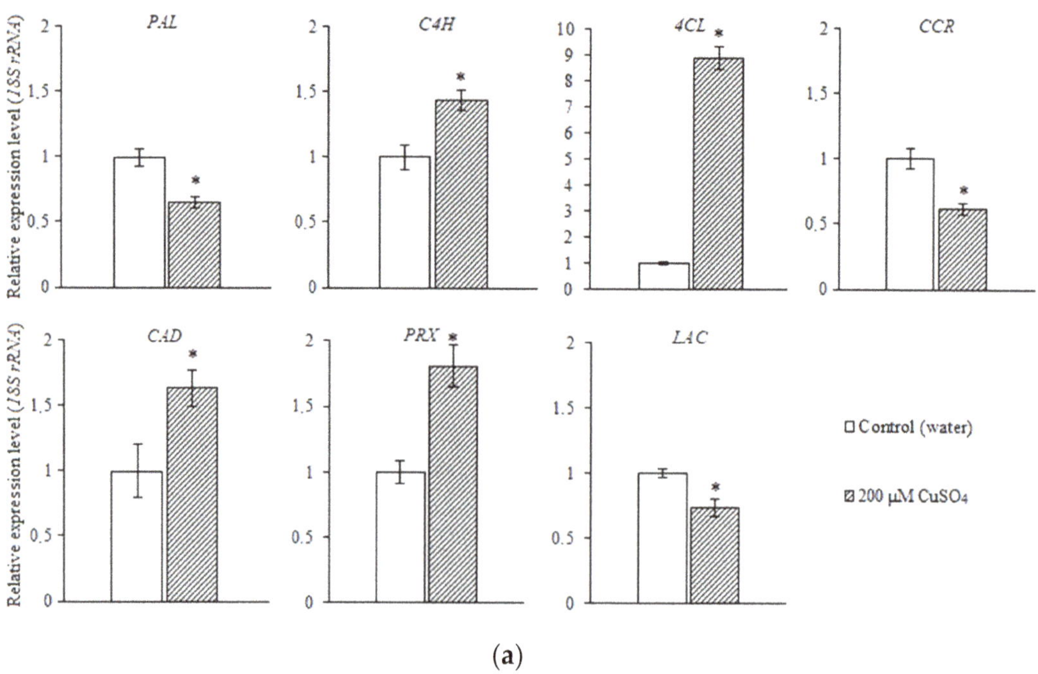

(a)

Figure 1. *Cont.*

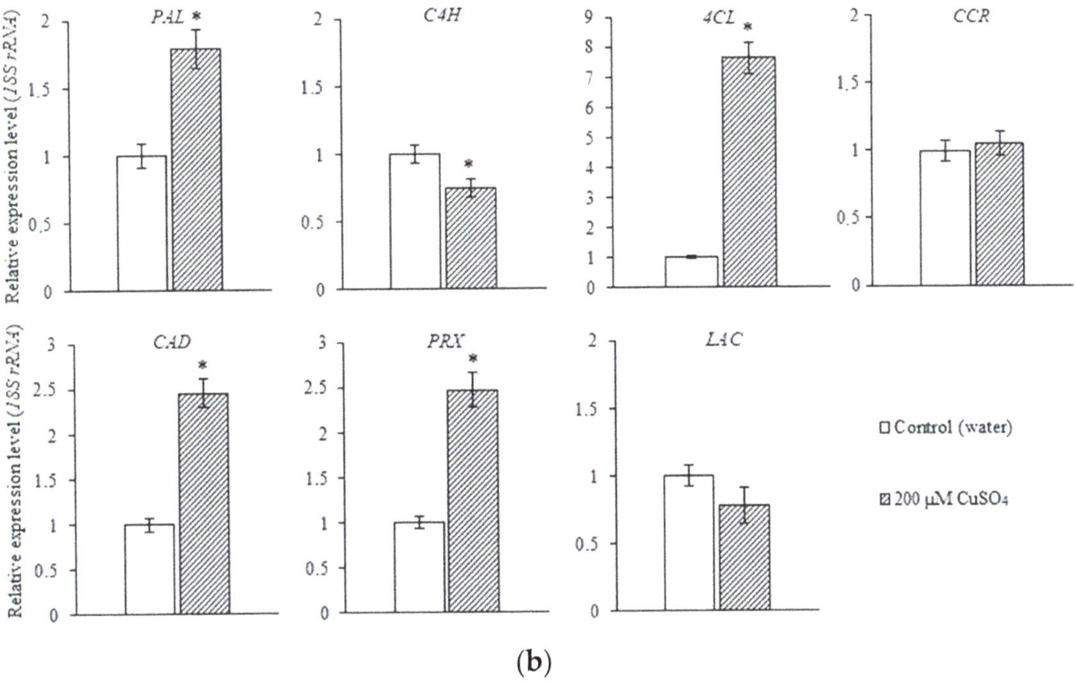

Figure 1. Relative expression levels of phenylpropanoid biosynthesis pathway genes in: (**a**) the root and (**b**) the stem in zinnia on the 20th day of growth; *18S rRNA* was used as the reference gene. Result is presented as mean ± standard error ($n = 9$). Asterisks represent significant differences ($p < 0.05$, *U*-test) from control.

3.5. Anatomical and Morphological Characteristics of Zinnia Plants

The shoot growth of zinnia plants delayed under the excess copper in the medium (Figure 2a). The length of the main root decreased by 17.2% (Figure 2c), and the root transection area increased by 31.0% (Table 5) compared to the control. The stem of treated plants was shorter by 12.7% compared to the control (Figure 2c), and the hypocotyl and stem diameter decreased by 8.6% and 22.9%, respectively (Table 5). The proportion of the cortex in the transection area increased in the root, while in the hypocotyl and stem it changed insignificantly compared to the control (Figure 2b). In response to copper stress, the thickness of the metaxylem vessels cell walls increased by 7.2% in the mature zone of the root compared to the control (Table 5). In the hypocotyl and internodes, this characteristic did not change. The cross-sectional area of the metaxylem vessels decreased by 12.2% in the root, 18.6% in the hypocotyl, and 28.9% in the stem compared to the control (Table 5, Figure 3).

Table 5. Anatomy characteristics of zinnia organs on the 20th day of growth.

Treatment	Cross-Sectional Diameter of Organ, mm			Metaxylem Cell-Wall Thickness, µm			Cross-Sectional Area of Metaxylem Vessels, µm²		
	Root	Hypocotyl	Stem (1st Internode)	Root	Hypocotyl	Stem (1st Internode)	Root	Hypocotyl	Stem (1st Internode)
Control (water)	1.74 ± 0.06 [1]	2.44 ± 0.05	2.57 ± 0.08	2.90 ± 0.06	2.62 ± 0.05	2.94 ± 0.08	51.34 ± 1.43	42.19 ± 2.17	42.97 ± 1.59
200 µM $CuSO_4$	2.24 ± 0.06 *	2.23 ± 0.05 *	1.98 ± 0.13 *	3.11 ± 0.06 *	2.60 ± 0.06	2.90 ± 0.07	45.06 ± 1.88 *	34.35 ± 1.41 *	30.55 ± 0.59 *

[1] Result is presented as mean ± standard error ($n = 50$). Asterisks represent significant differences ($p < 0.05$, *t*-test) from control.

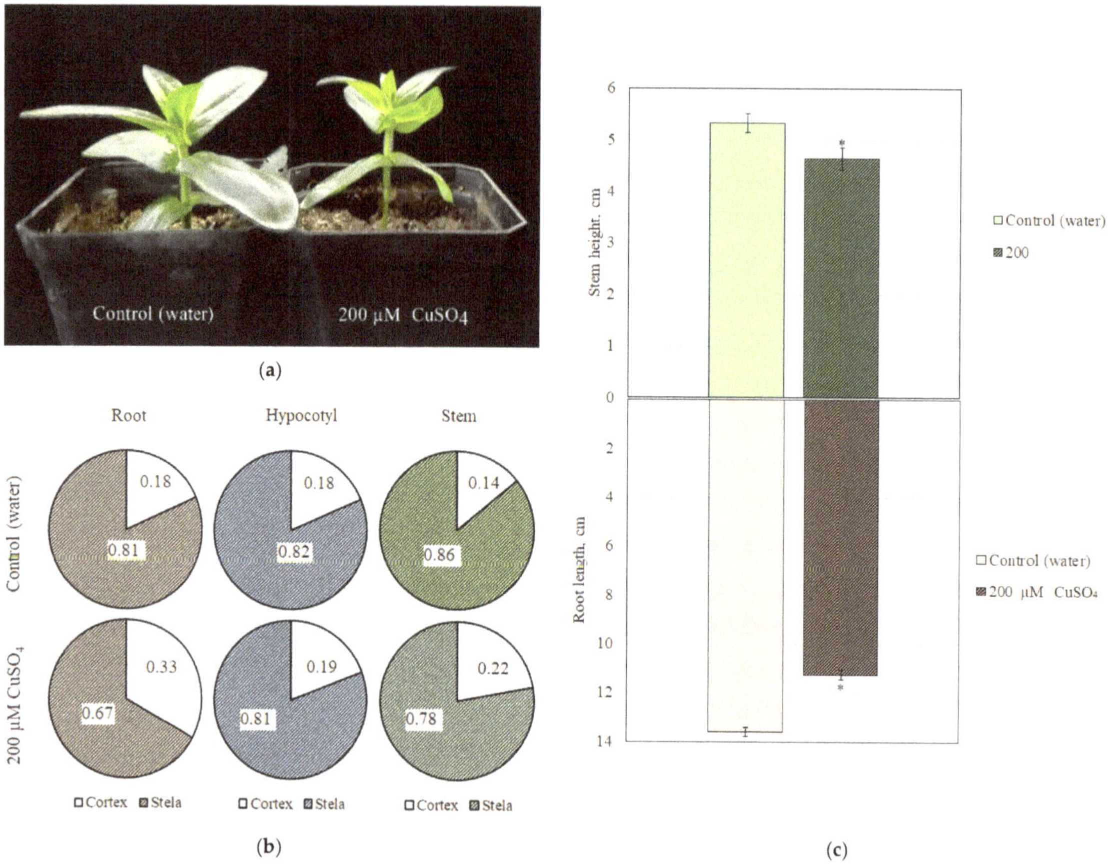

Figure 2. Morphological characteristics of zinnia plants: (**a**) zinnia plants on the 20th day of growth; (**b**) the ratio of cortex and stele in transverse sections of zinnia organs; (**c**) the length of main root and stem height (n = 30). Asterisks represent significant differences ($p < 0.05$, t-test) from control.

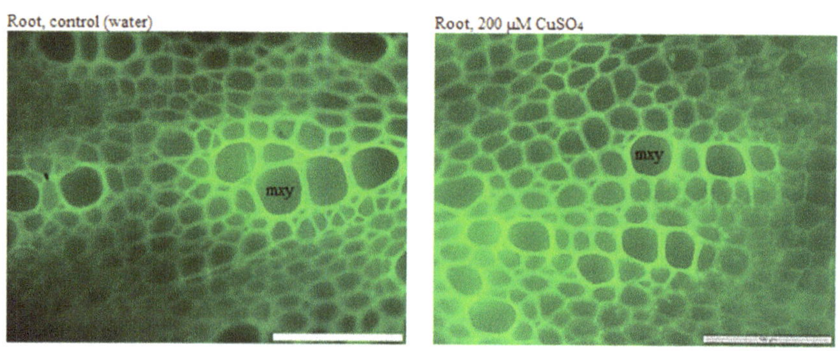

Figure 3. *Cont.*

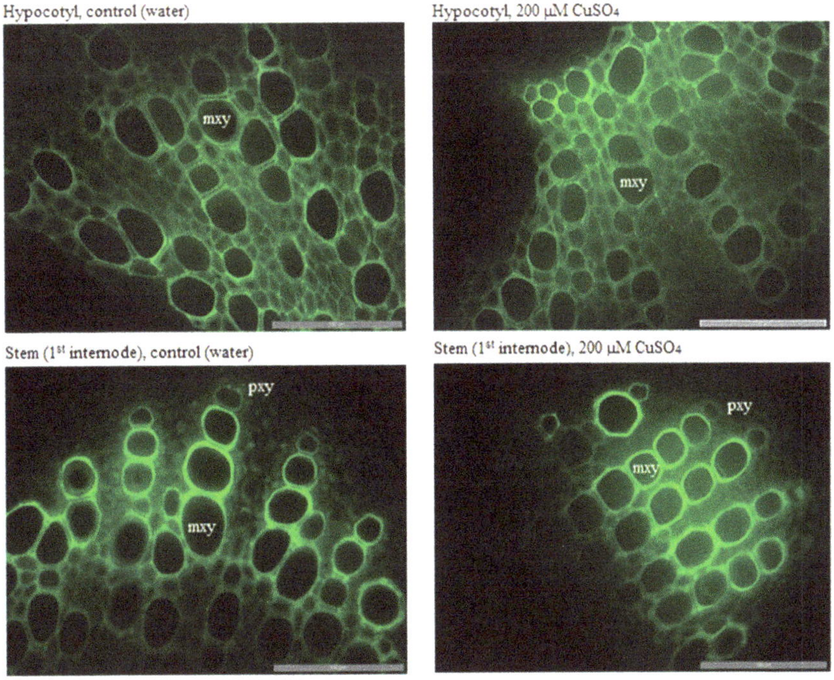

Figure 3. Transverse sections of roots and stems in zinnia plants (*Z. elegans*) visualized by wide-field fluorescence (green autofluorescence of lignin). Bars = 100 μm; mxy, metaxylem; pxy, protoxylem.

4. Discussion

Copper as both an essential element and a heavy metal revealed a complicated action on plants. In our study, the treatment of the growth substrate with 200 μM $CuSO_4$ led to an increase in the available copper forms and its total amount. Since copper is a redox active metal, it induced oxidative stress through Fenton and Haber–Weiss reactions [2,3], which manifested in an increase in the levels of LPO and hydrogen peroxide (Table S1), more significant in the roots of zinnia than in the shoots, which was also shown in other species, for example, *Helianthus annuus* L. [5], *Salvinia auriculata* Aubl. [27], *Oryza sativa* L. [28], etc. The development of stress is also evidenced by the suppression of zinnia growth, which is a typical response to copper stress [5,27,29,30]. At the same time, leaf chlorosis did not appear in our experiment, as was shown in [2,5,6], which was probably due to a slight increase in the copper concentration in the shoots. The roots contacted directly with copper ions in the substrate; therefore, the toxic effects of HMs were more pronounced in them than in the above-ground organs. The root accumulated a greater amount of copper compared to the stem, thereby performing a barrier function. The same results were obtained on *Arabidopsis thaliana* L. and *Oreganum vulgare* L., treated with copper [29,30]. The translocation factor for copper was less than one, which makes it possible to attribute zinnia to copper excluders. The reason for the limitation of copper translocation from the root to the shoot could be increased root lignification. The deposition of lignin into the cell walls could be considered as a protective mechanism, which was enhanced by the production of hydrogen peroxide under excess copper in plant tissues [8,9]. Increased lignification of cell walls is a nonspecific plant response to HMs. Cu stimulated lignin biosynthesis in the roots of many plants: *A. thaliana* [29], *Raphanus sativus* L. [31], *Glycine max* L. [32], *Panax ginseng* C.A. Meyer [4], etc. Some authors also noted the higher level of lignification in the roots of plant varieties sensitive to copper. It is known that lignin amount and composition can change under different kinds of stress. In our experiment, both in the root and in the stem the amount of KL increased, while ASL did not change in the root, it decreased in the stem in experimental plants compared to the control. KL is based on H-, S-, and G-units of lignin, while S-units predominate in ASL [33], so our results demonstrate the qualitative changes in lignin under copper stress.

Phenolic compounds are known as the precursors of lignin biosynthesis and as antioxidants, which may be involved in the quenching of reactive oxygen species (ROS) [27,34]. In our study, the increase in the copper amount in the zinnia root was accompanied by a decrease in the amount of phenolics (Table S1), that could be explained by their use for lignin synthesis and ROS deactivation [27,35], which led to a decrease in their detectable amount. The drop of phenolics was shown in response to both short-term treatment with 0.01–10 mM copper ions [27] in *S. auriculata* and two-week treatment in *Lycopersicon esculentum* L., in the case of with 10 ppm, 20 ppm, and 50 ppm of Cu [34]. The changes in lignin were the result of a complex and time-coordinated regulation of the phenylpropanoid-pathway enzymes. We have found changes in *4CL* and *CAD* transcription in the root as well as in *4CL*, *CAD*, and *PAL* in the stem, in treated plants compared to the control. Therefore, the *4CL* gene is responsible for the synthesis of G-units [36] and KL (Table S1). Our data on the increase in KL content in zinnia root in response to copper stress are likely due to this fact. An increase in the number of transcripts of the *4CL*, *CAD*, and *PAL* genes was also shown in response to Al stress in *Oryza sativa* L. [37] and in *Gossypium hirsutum* L. under Cd stress [38].

At the final stages of lignin formation, its composition and amount depend on the activity of the III class peroxidases and laccases. It was shown that under both normal and stress conditions, lignin biosynthesis was stimulated by an increase in the activity of class III peroxidases, which use hydrogen peroxide as an electron acceptor and phenolics as a substrate [11–13]. In *A. thaliana*, the level of *AtPRX62* gene transcripts in roots increased under the treatment with Cd^{2+} [39]. The induction of *PRX7* and *PRX8* gene expression was observed in the roots of *Hordeum vulgare* L., in response to 1 mM Cu^{2+}. The excess transcription of these genes led to the inhibition of growth [40]. The *POD* gene, encoding anionic peroxidase, which is involved in lignification, increased in *Paeonia ostii* T. Hong and J.X. Zhang plants, cultivated with high Cu^{2+} concentrations [41]. In our study, the enhanced expression of the zinnia *PRX* gene under copper stress was also shown. It was shown that promoters of many genes contain *cis*-regulatory elements, which are associated with transcription factors, the work of which is modified by the H_2O_2 generated during the stress caused by HMs [42,43].

When plants are exposed to metal stress, the biosynthesis of phenylpropanoids and lignin is activated, resulting in thickening of the cell walls [9,44]. In zinnia root, we also found an increase in the cell walls' thickness in the metaxylem vessels under copper stress; in the stem, it did not change. The thickening of cell walls and their lignification could provide the deposition of copper ions in the apoplast, limiting its translocation to the shoot [29,40].

Cell walls and root diameter both increased, and the root length decreased in the experimental zinnia plants treated with copper. The same effects were shown in *A. thaliana* [29] and *O. vulgare* [30]. The root cortex also thickened under stress. In maize treated with Cd, the same changes were shown [45]. The increase in cortical cell sizes led to the deposition of heavy-metal ions in vacuoles, limiting their entry into the stele [44,45]. Stress from HMs affected water uptake from the soil and, in turn, reduced the root water content [46]. Increased lignification can also limit the apoplast transport of water and minerals and cause their loading into the symplast [10,11,46]. Probably, the violation of mineral nutrition [47], the water regime [46], and the development of oxidative stress led to disorders in the development of vascular tissues in the zinnia stem, therefore, the number of xylem vessels in the vascular bundle decreased compared to the untreated plants. As a result, the radial growth of the hypocotyl and zinnia stem was limited.

5. Conclusions

Like most cultivated plants, zinnia could survive under excess of heavy metals, copper in particular, in the soil. One of the mechanisms that provide its tolerance to the moderate $CuSO_4$ concentration is the increased lignification of axial organs. For the first time for zinnia, it was shown that this process is associated with the enhanced expression of several phenylpropanoid metabolic-pathway genes (*4CL*, *CAD*) and the *PRX* gene that participates in lignin biosynthesis. Another mechanism of zinnia adaptation to excess copper is the increase in the transcript amount of *PAL*, *4CL*, and *CAD* genes that provide biosynthesis of phenolics, involved in the quenching of the reactive oxygen species in stress conditions. These changes allow zinnia to tolerate moderate copper stress.

Zinnia is an ornamental plant, characterized by rapid growth and abundant flowering. This species could be recommended for the reclamation of contamination by copper urban and industrial areas. The use of flowering plants will also increase the aesthetic value of the disturbed territories.

Supplementary Materials: The following supporting information is available online at https://www.mdpi.com/article/10.3390/horticulturae8060558/s1. Table S1: Spearman's rank correlation coefficients between copper amount, biochemical traits, and relative gene expression.

Author Contributions: Conceptualization, A.T. and I.K.; methodology, A.T., A.E., H.W. and I.K.; software, A.T. and A.E.; validation, A.T. and A.E.; formal analysis, A.T. and H.W.; investigation, A.T.; resources, A.E., H.W. and I.K.; data curation, A.E. and I.K.; writing—original draft preparation, A.T.; writing—review and editing, A.T., A.E., M.M., G.B. and I.K.; visualization, A.T.; supervision, I.K.; project administration, A.E.; funding acquisition, A.T. and A.E. All authors have read and agreed to the published version of the manuscript.

Funding: This research was supported by the Russian Science Foundation, project no. 22-24-00817, https://rscf.ru/project/22-24-00817/, accessed date: 17 July 2022.

Institutional Review Board Statement: Not applicable.

Informed Consent Statement: Not applicable.

Data Availability Statement: Not applicable.

Conflicts of Interest: The authors declare no conflict of interest.

References

1. Ballabio, C.; Panagos, P.; Lugato, E.; Huang, J.-H.; Orgiazzi, A.; Jones, A.; Fernández-Ugalde, O.; Borrelli, P.; Montanarella, L. Copper distribution in European topsoils: An assessment based on LUCAS soil survey. *Sci. Total Environ.* **2018**, *636*, 282–298. [CrossRef] [PubMed]
2. Shabbir, Z.; Sardar, A.; Shabbir, A.; Abbas, G.; Shamshad, S.; Khalid, S.; Ghulam Murtaza, N.; Dumat, C.; Shahid, M. Copper uptake, essentiality, toxicity, detoxification and risk assessment in soil-plant environment. *Chemosphere* **2020**, *259*, 127436. [CrossRef]
3. Adriano, D.C. *Trace Elements in Terrestrial Environments Biogeochemistry, Bioavailability, and Risks of Metals*; Springer: New York, NY, USA, 2001; 867p. [CrossRef]
4. Ali, M.B.; Singh, N.; Shohael, A.M.; Hahn, E.J.; Paek, K.Y. Phenolics metabolism and lignin synthesis in root suspension cultures of *Panax ginseng* in response to copper stress. *Plant Sci.* **2006**, *171*, 147–154. [CrossRef]
5. Jouili, H.; Ferjani, E. Changes in antioxidant and lignifying enzyme activities in sunflower roots (*Helianthus annuus* L.) stressed with copper excess. *Comp. Rend. Biolog.* **2003**, *326*, 639–644. [CrossRef]
6. Elleuch, A.; Chaâbene, Z.; Grubb, D.C.; Drira, N.; Mejdoub, H.; Khemakhem, B. Morphological and biochemical behavior of fenugreek (*Trigonella foenum-graecum*) under copper stress. *Ecotoxicol. Environ. Saf.* **2013**, *98*, 46–53. [CrossRef]
7. Aioub, A.A.A.; Zuo, Y.; Aioub, A.A.A.; Hu, Z. Biochemical and phytoremediation of *Plantago major* L. to protect tomato plants from the contamination of cypermethrin pesticide. *Environ. Sci. Pollut. Res.* **2021**, *28*, 43992–44001. [CrossRef] [PubMed]
8. Loix, C.; Huybrechts, M.; Vangronsveld, J.; Gielen, M.; Keunen, E.; Cuypers, A. Reciprocal interactions between cadmium-induced cell wall responses and oxidative stress in plants. *Front. Plant Sci.* **2017**, *8*, 1867. [CrossRef] [PubMed]
9. Bouazizi, H.; Jouili, H.; Geitmann, A.; Ferjani, E. Cell wall accumulation of Cu ions and modulation of lignifying enzymes in primary leaves of bean seedlings exposed to excess copper. *Biol. Trace Elem. Res.* **2010**, *139*, 97–107. [CrossRef]
10. Tenhaken, R. Cell wall remodeling under abiotic stress. *Front. Plant Sci.* **2015**, *5*, 771. [CrossRef]

11. Vaahtera, L.; Schulz, J.; Hamann, T. Cell wall integrity maintenance during plant development and interaction with the environment. *Nat. Plants* **2019**, *5*, 924–932. [CrossRef]
12. Berthet, S.; Thevenin, J.; Baratiny, D.; Demont-Caulet, N.; Debeaujon, I.; Bidzinski, P.; Leple, J.C.; Huis, R.; Hawkins, S.; Gomez, L.D.; et al. Role of plant laccases in lignin polymerization. *Adv. Bot. Res.* **2012**, *61*, 145–172. [CrossRef]
13. Jovanovic, S.V.; Kukavica, B.; Vidovic, M.; Morina, F.; Menckhoff, L. Class III peroxidases: Functions, localization and redox regulation of isoenzymes. In *Antioxidants and Antioxidant Enzymes in Higher Plants*; Gupta, D., Palma, J., Corpas, F., Eds.; Springer: Cham, Switzerland, 2018; pp. 269–300. [CrossRef]
14. Fraser, C.M.; Chapple, C. The phenylpropanoid pathway in *Arabidopsis*. *Arab. Book.* **2011**, *9*, e0152. [CrossRef]
15. Barros, J.; Serk, H.; Granlund, I.; Pesquet, E. The cell biology of lignification in higher plants. *Ann. Bot.* **2015**, *115*, 1053–1074. [CrossRef]
16. Pesquet, E.; Zhang, B.; Andras Gorzsas, A.; Puhakainen, T.; Serk, H.; Escamez, S.; Barbier, O.; Gerber, L.; Courtois-Moreau, C.; Alatalo, E.; et al. Non-cell-autonomous postmortem lignification of tracheary elements in *Zinnia elegans*. *Plant Cell* **2013**, *25*, 1314–1328. [CrossRef]
17. Ehsan, N.; Nawaz, R.; Ahmad, S.; Arshad, M.; Umair, M.; Sarmad, M. Remediation of heavy metal-contaminated soil by ornamental plant Zinnia (*Zinnia elegance* L.). *Asian J. Chem.* **2016**, *28*, 1338–1342. [CrossRef]
18. Plotnikov, D.S.; Tugbaeva, A.S.; Ermoshin, A.A.; Kiseleva, I.S. Response reactions of *Zinnia elegans* seedlings to the impact of different copper ions concentrations. *AIP Conf. Proc.* **2021**, *2388*, 030034. [CrossRef]
19. Roccotiello, E.; Marescotti, P.; Di Piazza, S.; Cecchi, G.; Mariotti, M.G.; Zotti, M. Biodiversity in Metal-Contaminated Sites—Problem and Perspective. In *A Case Study, Biodiversity in Ecosystems—Linking Structure and Function*; Blanco, J., Ed.; IntechOpen: London, UK, 2015; 1865p. [CrossRef]
20. Bellincampi, D.; Dipierro, N.; Salvi, G.; Cervone, F.; De Lorenzo, G. Extracellular H_2O_2 induced by oligogalacturonics is not involved in the inhibition of the auxin-regulated *rolB* gene expression in tobacco leaf explants. *Plant Physiol.* **2000**, *122*, 1379–1385. [CrossRef]
21. Uchiyama, M.; Mihara, M. Determination of malonaldehyde precursor in tissues by thiobarbituric acid test. *Anal. Bioch.* **1978**, *86*, 287–297. [CrossRef]
22. Ermoshin, A.A.; Kiseleva, I.S.; Nsengiyumva, I.V.; Nsengiumva, D.S.; Duan, S.; Ma, C.; Kurchenko, V.P. Antioxidant activity and chemical composition of extracts from fruiting bodies of xylotrophic fungi growing on birch. *J. Sib. Fed. Univ. Biol.* **2021**, *14*, 339–353. [CrossRef]
23. Bajpai, P. Pulp Bleaching. In *Biermann's Handbook of Pulp and Paper*; Bajpai, P., Ed.; Elsevier: New York, NY, USA, 2018; pp. 465–491. [CrossRef]
24. Kuluev, B.; Mikhaylova, E.; Ermoshin, A.; Veselova, S.; Tugbaeva, A.; Gumerova, G.; Gainullina, K.; Zaikina, E. The ARGOS-LIKE genes of *Arabidopsis* and tobacco as targets for improving plant productivity and stress tolerance. *J. Plant Physiol.* **2019**, *242*, 153033. [CrossRef]
25. Livak, K.J.; Schmittgen, T.D. Analysis of relative gene expression data using real-time quantitative PCR and the $2^{-\Delta\Delta Ct}$ method. *Methods* **2001**, *25*, 402–408. [CrossRef] [PubMed]
26. Berlin, G.P.; Miksche, J.P. *Botanical Microtechnique and Cytochemistry*; Iowa State University Press: Ames, IA, USA, 1976; 321p.
27. Bizzo, A.L.; Intorne, A.C.; Gomes, P.H.; Susuki, M.S.; Esteves, B.D. Short-term physiological responses to copper stress in *Salvinia auriculata* Aubl. *Acta Limnol. Bras.* **2014**, *26*, 268–277. [CrossRef]
28. Thounaojam, T.C.; Panda, P.; Choudhury, S.; Patra, H.K.; Panda, S.K. Zinc ameliorates copper-induced oxidative stress in developing rice (*Oryza sativa* L.) seedlings. *Protoplasma* **2014**, *251*, 61–69. [CrossRef]
29. Lequeux, H.; Hermans, C.; Lutts, S.; Verbruggen, N. Response to copper excess in *Arabidopsis thaliana*: Impact on the root system architecture, hormone distribution, lignin accumulation and mineral profile. *Plant Physiol. Biochem.* **2010**, *48*, 673–682. [CrossRef]
30. Panou-Filotheou, H.; Bosabalidis, A.M. Root structural aspects associated with copper toxicity in oregano (*Origanum vulgare* subsp. *hirtum*). *Plant Sci.* **2004**, *166*, 1497–1504. [CrossRef]
31. Chen, E.L.; Chen, Y.A.; Chen, L.M.; Liu, Z.H. Effect of copper on peroxidase activity and lignin content in *Raphanus sativus*. *Plant Physiol. Biochem.* **2002**, *40*, 439–444. [CrossRef]
32. Lin, C.-C.; Chen, L.-M.; Liu, Z.-H. Rapid effect of copper on lignin biosynthesis in soybean roots. *Plant Sci.* **2005**, *168*, 855–861. [CrossRef]
33. Yasuda, S.; Fukushima, K.; Kakehi, A. Formation and chemical structures of acid-soluble lignin I: Sulfuric acid treatment time and acid-soluble lignin content of hardwood. *J. Wood Sci.* **2001**, *47*, 69–72. [CrossRef]
34. Michalak, M. Phenolic compounds and their antioxidant activity in plants growing under heavy metal stress. *Pol. J. Environ. Stud.* **2006**, *15*, 523–530.
35. Kısa, D.; Öztürk, L.; Doker, S.; Gökçe, İ. Expression analysis of metallothioneins and mineral contents in tomato (*Lycopersicon esculentum*) under heavy metal stress. *J. Sci. Food Agric.* **2017**, *97*, 1916–1923. [CrossRef]
36. Lee, D.; Meyer, K.; Chapple, C.; Douglas, C.J. Antisense suppression of 4-Coumarate: Coenzyme A Ligase activity in *Arabidopsis* leads to altered lignin subunit composition. *Plant Cell* **1997**, *9*, 1985–1998. [CrossRef]
37. Mao, C.; Yi, K.; Yang, L.; Zheng, B.; Wu, Y.; Liu, F.; Wu, P. Identification of aluminium-regulated genes by cDNA-AFLP in rice (*Oryza sativa* L.): Aluminium-regulated genes for the metabolism of cell wall components. *J. Exp. Bot.* **2004**, *55*, 137–143. [CrossRef]

38. Chen, H.; Li, Y.; Ma, X.; Guo, L.; He, Y.; Ren, Z.; Kuang, Z.; Zhang, X.; Zhang, Z. Analysis of potential strategies for cadmium stress tolerance revealed by transcriptome analysis of upland cotton. *Sci. Rep.* **2019**, *9*, 86. [CrossRef]
39. Weber, M.; Trampczynska, A.; Clemens, S. Comparative transcriptome analysis of toxic metal responses in *Arabidopsis thaliana* and the Cd^{2+}-hypertolerant facultative metallophyte *Arabidopsis halleri*. *Plant Cell Environ.* **2006**, *29*, 950–963. [CrossRef]
40. Tamas, L.; Durcekova, K.; Huttova, J.; Mistrik, I. Expression of defence-related peroxidases *Prx7* and *Prx8* during abiotic stresses in barley roots. *Acta Physiol. Plant.* **2009**, *31*, 139. [CrossRef]
41. Wang, Y.; Dong, C.; Xue, Z.; Jin, Q.; Xu, Y. *De novo* transcriptome sequencing and discovery of genes related to copper tolerance in *Paeonia ostii*. *Gene* **2016**, *576*, 126–135. [CrossRef]
42. Li, G.-Z.; Zheng, Y.-X.; Chen, S.-J.; Liu, J.; Wang, P.-F.; Wang, Y.-H.; Guo, T.-C.; Kang, G.-Z. TaWRKY74 participates copper tolerance through regulation of TaGST1 expression and GSH content in wheat. *Ecotoxicol. Environ. Saf.* **2021**, *221*, 112469. [CrossRef]
43. Gómez-Ros, L.V.; Gabaldón, C.; Núñez-Flores, M.J.L.; Gutiérrez, J.; Herrero, J.; Zapata, J.M.; Sottomayor, M.; Juan Cuello, J.; Barceló, A.R. The promoter region of the *Zinnia elegans* basic peroxidase isoenzyme gene contains *cis*-elements responsive to nitric oxide and hydrogen peroxide. *Planta* **2012**, *236*, 327–342. [CrossRef]
44. Maksimovic, I.; Kastori, R.; Krstic, L.; Lukovic, J. Steady presence of cadmium and nickel affects root anatomy, accumulation and distribution of essential ions in maize seedlings. *Biol. Plant.* **2007**, *51*, 589–592. [CrossRef]
45. Yadav, V.; Arif, N.; Kováč, J.; Singh, V.P.; Tripathi, D.K.; Chauhan, D.K.; Vaculík, M. Structural modifications of plant organs and tissues by metals and metalloids in the environment: A review. *Plant Physiol. Biochem.* **2021**, *159*, 100–112. [CrossRef]
46. Tewari, R.K.; Yadav, N.; Gupta, R.; Kumar, R. Oxidative stress under macronutrient deficiency. *J. Soil Sci. Plant Nutr.* **2021**, *21*, 832–859. [CrossRef]
47. Rucińska-Sobkowiak, R. Water relations in plants subjected to heavy metal stresses. *Acta Physiol. Plant.* **2016**, *38*, 257. [CrossRef]

horticulturae

MDPI

Article

Adaptive Redox Reactions Promote Naturalization of Rare Orchid *Epipactis atrorubens* on Serpentine Dumps Post Asbestos Mining

Maria Maleva [1,*,†], Galina Borisova [1,†], Elena Filimonova [2], Natalia Lukina [2], Nadezhda Chukina [1], Alexander Ermoshin [1,3], Anastasia Tugbaeva [1] and Olga Voropaeva [1]

1 Department of Experimental Biology and Biotechnology, Ural Federal University, 620002 Ekaterinburg, Russia; g.g.borisova@urfu.ru (G.B.); nadezhda.chukina@urfu.ru (N.C.); alexander.ermoshin@urfu.ru (A.E.); anastasia.tugbaeva@urfu.ru (A.T.); olga.voropaeva@urfu.ru (O.V.)
2 Laboratory of Anthropogenic Dynamics of Ecosystems, Ural Federal University, 620002 Ekaterinburg, Russia; elena.filimonova@urfu.ru (E.F.); natalia.lukina@urfu.ru (N.L.)
3 Inner Mongolia Academy of Agriculture and Animal Husbandry Science, Hohhot 232813, China
* Correspondence: maria.maleva@urfu.ru
† These authors contributed equally to this work.

Citation: Maleva, M.; Borisova, G.; Filimonova, E.; Lukina, N.; Chukina, N.; Ermoshin, A.; Tugbaeva, A.; Voropaeva, O. Adaptive Redox Reactions Promote Naturalization of Rare Orchid *Epipactis atrorubens* on Serpentine Dumps Post Asbestos Mining. *Horticulturae* **2022**, *8*, 603. https://doi.org/10.3390/horticulturae8070603

Academic Editors: Agnieszka Hanaka, Małgorzata Majewska and Barbara Hawrylak-Nowak

Received: 30 May 2022
Accepted: 1 July 2022
Published: 4 July 2022

Publisher's Note: MDPI stays neutral with regard to jurisdictional claims in published maps and institutional affiliations.

Abstract: *Epipactis atrorubens* (Hoffm.) Besser. is a regionally rare orchid species with highly ornamental properties due to its very beautiful bright flowers, therefore it is of considerable interest as a horticultural plant for use in botanical gardens and greenhouses. The objective of the research was to assess metal accumulation and some pro- and antioxidant reactions in *E. atrorubens*, colonizing serpentine dumps post asbestos mining. Additionally, some physicochemical properties of substrates, microbiotic characteristics and water status were investigated in orchids growing on two serpentine dumps and in a natural forest habitat of the Middle Urals, Russia. The dump substrates were characterized by the strong stoniness and the high content of Mg, Ni, Cr and Co (by 1.8 times on average) compared to the natural habitat. In these sites, *E. atrorubens* was characterized by increased mycorrhization. In the rhizome and roots of *E. atrorubens* the concentrations of most metals studied were considerably higher (more than 4 times on average) than in the leaves. It was found that orchids colonizing serpentine dumps produced more lipid peroxidation products (by 1.4 times on average) in the leaves which was accompanied by the more active synthesis of such non-enzymatic antioxidants as ascorbate, free proline, soluble phenolic compounds (including flavonoids) and non-protein thiols. The study suggests that non-enzymatic antioxidants increased the adaptive potential of *E. atrorubens* and contributed to its naturalization on serpentine dumps post asbestos mining.

Keywords: Orchidaceae; ornamental plant introduction; serpentine outcrops; stressful conditions; adaptive responses; plant water status; redox balance; non-enzymatic antioxidants

1. Introduction

At present, the problem of preserving biological diversity is becoming increasingly important. Changes in natural habitats have lead to the extinction of many species, including a number of orchid plants [1]. At the same time, in recent decades, some species of rare orchids were found in anthropogenically disturbed habitats, including industrial dumps [2–9]. One of the representatives of the Orchidaceae family found in disturbed habitats in Russia is *Epipactis atrorubens* (Hoffm.) Besser. This species is listed on the European Red List of Vascular Plants under the category of 'Least Concern' [10] and is included in the Convention on International Trade in Endangered Species Protection Status (CITIS) [11]. It is also listed in the Red Book of many Russian regions, including the Sverdlovsk region under category III, 'Rare plant' [12].

Epipactis atrorubens is a short-rhizome herbaceous perennial, calcephilus, xeromesophyte that grows in dry and well-lit habitats [5]. This species is widely distributed in boreal, temperate and submeridional zones. It is usually found in deciduous, coniferous and mixed forests [5,13,14]. This orchid has highly ornamental properties due to its beautiful bright flowers (Figure 1a,b). Therefore, it is interesting as a horticultural plant for growing in botanical gardens and greenhouses.

(a) (b)

Figure 1. *Epipactis atrorubens* on serpentine dump: (**a**) Flowering plant; (**b**) Orchid flower.

The conservation of rare plant species is often preceded by the study of their adaptive abilities under natural conditions. In this regard, it is necessary to have a complete understanding of the orchid's adaptive reactions that increase their resistance to stressful conditions. It is well known that *E. atrorubens* has a great colonization potential on different industrial dumps [2,4,15]. It was found also on the dumps of serpentine rocks formed during the development of asbestos deposits in the Middle Urals [6,7].

Serpentine substrates are commonly unfavorable for plant growth due to their negative physicochemical properties [16–19]. It is known that under the impact of the unfavorable effect of environmental factors, the number of reactive oxygen species (ROS) in plant cells can increase, which leads to the activation of prooxidant processes and the development of oxidative stress [20,21]. The accumulation of malondialdehyde (MDA) and other products of lipid peroxidation can be on the one hand an indicator reaction reflecting the degree of stress exposure [22] and, on the other hand, a signal for the gene expression of some antioxidant enzymes and non-enzymatic antioxidants [23,24]. In higher plants among non-enzymatic antioxidants, ascorbate and glutathione are the most abundant soluble forms which play a vital role as electron donors and scavenge ROS directly through the glutathione–ascorbate cycle [21]. Other antioxidants play an equally important role. For example, many phenolic compounds, including flavonoids, have a great potential to scavenge free radicals and reduce cell damage from lipid peroxidation [21,25,26]. Free proline has a multifunctional effect, it plays an important role in both osmoregulation and antioxidant protection, providing cellular homeostasis and facilitating plant adaptation to stressful conditions [27]. In this regard, studies of plant antioxidant status and methods of stress mitigation are becoming increasingly important.

To date, some anatomical and morphological features, chemical composition, mycorrhizal associations and bacterial microflora of several representatives of the genus *Epipactis*, growing in natural and transformed ecosystems have been wellstudied [2,4,6,7,28–31]. However, the adaptive redox reactions of orchids growing abundantly on serpentine outcrops, have not been practically explored.

The aim of the research was to study metal accumulation and to assess some pro- and antioxidant reactions in *E. atrorubens*, colonizing serpentine dumps post asbestos mining. Additionally, the physicochemical properties of the rhizosphere substrate, rhizospheric microbiota and water status of this orchid were studied. The comparative analysis of *E. atrorubens*, growing in transformed and natural habitats will make it possible to identify the adaptive responses for stress mitigation contributing to rare orchid species naturalization and conservation.

2. Materials and Methods

2.1. Study Area

The study area is located within the Tagilo-Nevyansk hyperbasite massif on the eastern slope of the Middle Urals (Sverdlovsk region, Russia), belonging to the taiga zone, southern taiga subzone. Three naturally colonized plant populations of *E. atrorubens* were selected for the present study (Figure 2).

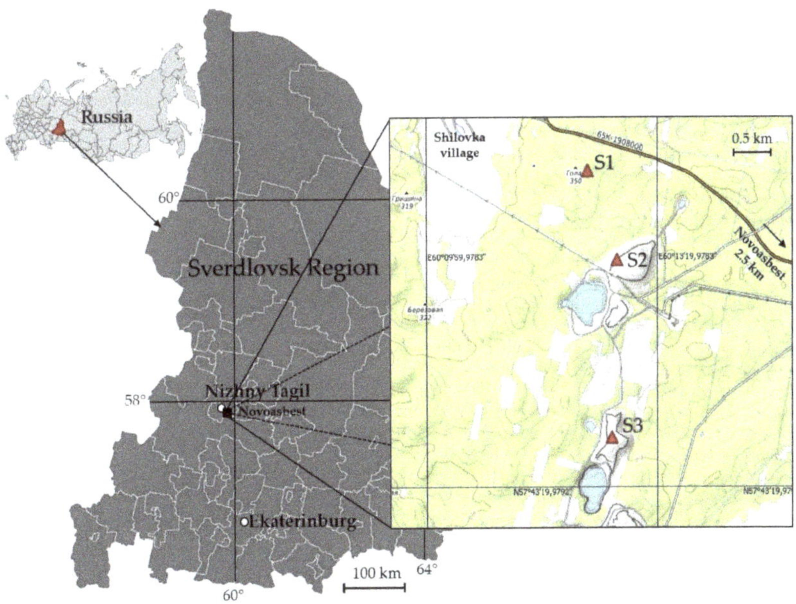

Figure 2. The studied sites of *E. atrorubens* populations (S1–S3) in the Sverdlovsk region, Russia.

The study area climate is characterized by a continental subarctic climate with an average annual air temperature of +1.0 °C; the average July temperature is +17.2 °C and the average January temperature is −16.0 °C (according to the weather station in Nizhny Tagil) [32]. The average annual rainfall is 628 mm. The snow cover lasts from the second half of November to early April, and its average thickness is 0.7 m. The depth of soil freezing is 0.8–1.7 m.

As a reference control (S1), the population of *E. atrorubens* in a pine forest on the Golaya Mount slope, located between Shilovka and Novoasbest villages, about 25 km from Nizhny Tagil, Sverdlovsk region was chosen (Figure 2, Table 1).

Table 1. Studied sites for sampling of rhizospheric substrate and *E. atrorubens* plants.

Sites	Study Area	Coordinate	Plant Community
S1	Forest area on serpentine rocks	57°45′42.93″ N 60°12′52.82″ E	Natural pine forest community on the Golaya Mount slope
S2	Shilovsky dump of serpentine rocks of the Anatol'sko-Shilovsky asbestos deposit	57°44′54.78″ N 60°12′37.55″ E	Emerging pine forest community with sparse undergrowth of trees and shrubs on the lower tier
S3	Anatol'sky dump of serpentine rocks of the Anatol'sko-Shilovsky asbestos deposit	57°43′32.92″ N 60°12′39.44″ E	Pine forest community on the second tier ledge and gentle slope

Two other populations studied were found at serpentine dumps (S2 and S3) of the Anatol'sko Shilovsky deposit, located about 2.0–4.0 km from S1 (Table 1). The area is confined to lenticular deposits of talc-chlorite-carbonate rocks and has been exploited for the extraction of fibrous asbestos ($Na_3(Mg,Fe^{2+})4Fe^{3+}Si_8O_{22}(OH)_2$) as an open cast mine from 1952 to 1992 [7]. Site S2 was represented by an emerging forest community located on the berm of the first tier on the southern side of the Shilovsky dump, while S3 was represented by a forest community formed on the eastern side of the Anatol'sky dump (Table 1). The total estimation of the forest communities formed on the study sites was carried out according to the generally accepted geobotanical methods [7].

The forest communities formed in the natural habitat (S1) and on the Anatol'sko-Shilovsky asbestos deposit (S2 and S3) showed uneven formation of vegetation. In reference site (S1) the average age of the stand of *Pinus sylvestris* L. was 70 years and the tree crown density was 0.6. The range of afternoon light intensity measured at the sampling point varied between 6 and 15 klx. Besides the dominant *P. sylvestris*, other species such as *Larix sibirica* Ledeb., *Betula pendula* Roth and *Picea obovata* Ledeb. were rarely found in the stand; sparsely isolated individuals of *Juniperus communis* L., *Rosa acicularis* Lindl., *Sorbus aucuparia* L. and *Chamaecytisus ruthenicus* (Fisch. ex Wol.) Klásk. grew in the undergrowth. *Vaccinium myrtillus* L., *V. vitis-idaea* L. and *Linnaea borealis* L. were represented in the herb-shrub layer. Herbaceous species were dominated by *Calamagrostis arundinacea* (L.) Roth, *Brachypodium pinnatum* (L.) Beauv., *Geranium sylvaticum* L., *Rubus saxatilis* L., *Fragaria vesca* L., *Potentilla erecta* (L.) Raeusch., etc. The total projective cover of the herb-shrub layer varied from 40 to 80%. *Epipactis atrorubens* occurred in groups of 3 to 14 individuals. The total number of orchid plants was about 43, the average density was about 8 individuals per 100 m^2. Flowering plants dominated the age spectrum (90%). The local population's self-maintenance was ensured through seed and vegetative propagation.

Site S2 was represented by an emerging forest community located on the berm of the first tier on the southern side of the Shilovsky dump. There, on a flat surface, rolled by road transport, there was a sparse undergrowth of *B. pendula* and *P. sylvestris* as well as some species of the *Populus* and *Salix* genera. There was no crown closure in the forest area and the afternoon light intensity in the sampling point was very high (varying between 80 and 105 klx). The herb-shrub layer was dominated by *Dendranthema zawadskii* (Herbich) Tzvel., *E. atrorubens*, *Thymus talijevii* Klok. & Des.-Shost., *Solidago virgaurea* L. and *Calamagrostis epigeios* (L.) Roth. The total projective cover of this layer varied from 0 to 15%. The distribution of *E. atrorubens* was uneven and the orchids grew along the roads, between stones and under the trees. The number of *E. atrorubens* in the local population was 189; the average density was about 32 individuals per 100 m^2. The age spectrum was dominated by pregenerative plants (58%).

Site S3 was represented by a forest community formed on the eastern side of the Anatol'sky dump. The tree layer was dominated by *P. sylvestris*. The age of the trees was between 10 and 35 years; the tree crown density was 0.5–0.6. The range of light intensity measured at the sampling point varied between 48 and 56 klx. *Calamagrostis arundinacea* dominated in the herb-shrub layer. The total projective cover of this layer varied from 0 to 20%. *Epipactis atrorubens* plants were found both as individuals and in groups of up to three individuals. The number of *E. atrorubens* in the local population was 163; the average density was about 33 individuals per 100 m^2. Individuals of the pregenerative age state prevailed (65%).

2.2. Collecting and Preparation of Plants and Substrates

Both plants and rhizospheric substrate were collected from studied habitats of *E. atrorubens* (S1–S3) over a two-year period (mid-July 2019 and 2020). All individuals were collected at the same phenological state (a fully developed inflorescence) under similar weather conditions (temperature during the daytime was about 24 ± 3 °C and the relative humidity was about 60 ± 5%).

Four generative orchid plants with 3–5 individual inflorescences (40–50 cm in length) were randomly selected from each site. The plants were carefully dug up together with the underground organs (rhizome + roots) and part of the soil (up to 15 cm in depth), placed in sterile plastic bags and transferred to the laboratory. The plant samples were cleaned of soil particles and washed first with running tap water, then with distilled and deionized water. One part of the fresh plant material (leaves and rhizome with roots) was fixed at 105 °C for 2 h and dried at 75 °C for 24 h for further metal analysis. The other part of the plant material (weighted fresh leaf cuttings) was partly used for the immediate determination of the *E. atrorubens* water status and partly frozen in liquid nitrogen and stored at −80 °C for further biochemical analysis. For the estimation of dry weight (DW), weighted fresh leaves were dried in a hot air oven at 75 °C for 24 h and the ratio of FW/DW was calculated.

The substrate samples were collected close to the orchid root zone from each studied site. The samples were air-dried for five days, oven-dried at 75 °C for 24 h and then used for granulometric and physicochemical analyses. Independently, part of the rhizospheric substrate from each site was used to determine some microbiological characteristics.

2.3. Physicochemical Characterization of Substrates

Part of the substrate was used for the determination of the percentage of different particle sizes which was performed by a standard sieve analysis (stones: >10 mm; gravel large: 5–10 mm; gravel small: 2–5 mm; sand large: 1–2 mm; sand average: 0.25–1 mm; dust and clay: <0.25 mm) as was described previously [7]. The second part of the soil was destoned, homogenized and passed through a sieve (<2 mm) [33] and a composite sample for each site was used to determine pH, electrical conductivity (EC) and available macronutrients (nitrogen, phosphorus, potassium), as well as total and available metal content.

The pH and EC of the substrate–water suspensions (1:2.5; *w*/*v*) were measured using a portable multivariable analyzer HI98129 Combo (Hanna Instruments GmbH, Graz, Austria). The alkaline-hydrolyzed nitrogen content and the available forms of phosphorus were measured as described by Filimonova et al. [7]. Subsequently, dried plant material (leaves and rhizome + roots) and soil samples were weighed and digested with concentrated nitric acid (analytical grade) using MARS 5 Digestion Microwave System (CEM, Matthews, NC, USA) for the determination of total metal concentration.

The substrate moisture content was determined using the thermostatic weight method [34] and expressed as a percentage of DW of the substrate.

The available form of metals was analyzed after mixing the substrate samples with 0.4 mM Na_2EDTA [35]. All the samples were prepared using double deionized Millipore water (Milli-Q system, Millipore, Molsheim, France). The Mg, Ca, K, Fe, Zn, Cu, Mn, Ni, Cr, Pb and Co concentrations in all samples were determined using a flame atomic absorption spectrometer AA240FS (Varian Australia Pty Ltd., Mulgrave, Victoria, Australia). Standard

Reference Materials [JSC Ural Plant of Chemical Reagents, Russia; GSS 7681-99 for Mg(II), GSS 7682-99 for Ca(II), GSS 8092-94 for K(I), GSS 7766-2000 for Fe(III), GSS 7256-96 for Zn(II), GSS 7998-93 for Cu(II), GSS 7266-96 for Mn(II), GSS 7265-96 for Ni(II), GSS 8035-94 for Cr(VI), GSS 7012-93 for Pb(II), and GSS 8089-94 for Co(II)] were used for the preparation and calibration of each analytical batch. Calibration coefficients were maintained at a high level of not less than 0.99.

2.4. Assessment of the Rhizospheric Microbiota of E. atrorubens

The quantity of mesophilic aerobic and facultative anaerobic microorganisms (QMAFAnM) in the rhizospheric soil of *E. atrorubens* from each studied site was determined by plating on Luria–Bertani (LB) agar medium. About 10 g of substrate was mixed with 90 mL of phosphate buffer (pH 6.5) and shaken in an orbital shaker at 180 rpm for 20 min at 28 °C. A series of dilutions of each sample was made and 100 µL was added to a Petri dish with LB agar nutrient medium supplemented with cycloheximide (75 mg L^{-1}) to suppress the growth of fungi. Inoculations on a nutrient medium were carried out with 2–5 dilutions, and 2 parallel inoculations were made from each dilution. For the growth of cultured bacteria, the plates were incubated for 3 days at 28 °C in a bacterial incubator (TSO-1/80 SPU, Smolensk, Russia). Colonies of bacteria were counted on the 3rd and 5th days of incubation, ignoring the dishes on which the number of colonies was less than 10 or more than 300. QMAFAnM was expressed in colony-forming units (CFU) per g of DW of soil.

The enzymatic activity of *E. atrorubens* rhizospheric soil was assessed by the activity of cellulose-degrading microorganisms (bacteria and fungi) [36].

To assess mycorrhizal colonization in the root system of *E. atrorubens*, root tips 1.0–1.5 cm were cross-sectioned to 20 µm with a freezing microtome MEP-01 (Technom, Ekaterinburg, Russia). Root sections (50 samples from each studied site) were analyzed using the light microscope Meiji MT 4300L (Meiji Techno, Saitama, Japan) at 100-x magnification [7]. The presence of pelotons or intracellular hyphal coils was determined within the root cortical cells. The percentage colonization was assessed as the proportion of sections containing pelotons compared to the total number of sections per plant.

2.5. Assessment of Plant Water Status

The transpiration rate was measured on the middle tier leaves (6–7 from the top) of *E. atrorubens* using an LI-6400XT portable infrared gas analyzer (LI-COR, Lincoln, NE, USA) [8].

Relative water content (RWC) and water saturation deficit (WSD) were measured using the floating disc method and calculated according to Hellmuth [37]. The fresh leaf cuttings (6 discs 0.9 cm^2 in diameter) were immediately weighed to obtain fresh weight (FW) and then saturated by submerging the sample in distilled water for 2 h. Afterwards, the surface water was blotted carefully and the discs were weighed to obtain the saturated weight (SW) and later dried for 24 h at 75 °C to determine the DW.

RWC was calculated using Equation (1):

$$\mathrm{RWC}\ (\%) = (\mathrm{FW} - \mathrm{DW})/(\mathrm{SW} - \mathrm{DW}) \times 100 \quad (1)$$

WSD was calculated using Equation (2):

$$\mathrm{WSD}\ (\%) = (\mathrm{SW} - \mathrm{FW})/(\mathrm{SW} - \mathrm{DW}) \times 100 \quad (2)$$

2.6. Assessment of Lipid Peroxidation and Non-Enzymatic Antioxidants

The lipid peroxidation was assessed by the content of oxidation products (malondialdehyde, MDA) according to Heath and Packer [38] in fresh leaves (0.3 g) homogenized with 4 mL of the reaction medium containing 0.25% thiobarbituric acid (TBA) in 10% trichloroacetic acid (TCA) (v/v). Then the extract was boiled for 30 min, cooled on ice and centrifuged at 12,000× g for 15 min. The supernatant absorbance was measured at 532 and 600 nm. The TBA-reactive product concentration was calculated using the extinction coefficient (155 mM^{-1} cm^{-1}) and expressed in nmol MDA per g of DW.

The free proline and ascorbate content was determined as described previously with slight modifications [39]. The amount of proline was measured after leaf extraction (0.4 g) in 10 mL of boiling water (100 °C) for 10 min; then the reaction medium containing the prepared filtered extract and a mixture of ninhydrin reagent with glacial acetic acid (1:1:1; *v*/*v*) were placed in a boiling water bath for 30 min for staining and then cooled in ice rapidly. The proline content was quantified spectrophotometrically by PD-303 UV (Apel, Saitama, Japan) at 520 nm and calculated in mg per g of DW. The proline standard curve (Sigma-Aldrich Chemie GmbH, Taufkirchen, Germany) was made following the same protocol.

The ascorbate content was measured after the homogenization of leaves (0.3 g) in 3 mL of 2% (*w*/*v*) metaphosphoric acid. The homogenate was transferred into 15 mL Falcon tubes and the volume was made up with a mixture containing 2% metaphosphoric acid:0.21 M trisodium phosphate (3:2, *v*/*v*), pH 7.3–7.4. The extract was centrifuged for 3 min at 3000× *g* and the absorbance was measured at 265 nm against a blank containing the metaphosphoric acid–trisodium phosphate mixture. If necessary, the extract was further diluted. The ascorbate concentration was calculated in mg g^{-1} of DW using an extinction coefficient (1.655×10^4 M^{-1} cm^{-1}) and the molecular weight of ascorbate (176.1 g M^{-1}).

For the determination of the total content of phenolic compounds and flavonoids, the fresh leaves (0.3 g) were crushed and extracted with 10 mL of 80% ethanol for 24 h (in the dark). Then the resulting extract was filtered through the filter paper and used for the analysis. The total phenolic content was determined with the Folin–Ciocalteu reagent [40]. Briefly, 0.1 mL of the extract sample was reacted with 0.5 mL of 0.2 M Folin–Ciocalteu reagent for 5 min and then 0.4 mL 7.5% sodium carbonate solution (*w*/*v*) was added to the reaction mixture. The absorbance readings were measured with a multimode plate reader Infinite 200 PRO (Tecan, Grödig, Austria) at 760 nm after incubation at room temperature for 1 h. Gallic acid (Sigma-Aldrich Chemie GmbH, Taufkirchen, Germany) was used as a reference standard, and the results were expressed as mg gallic acid per g of DW.

The amount of flavonoids was determined using a modified method [41] after the reaction of the extract sample with an equivalent amount of 10% aluminum chloride ethanolic solution and then incubated at room temperature for 15 min, and the absorbance of the reaction mixture was measured at 415 nm. The amount of 10% aluminum chloride was substituted by the same amount of 80% ethanol in the blank. Similarly, the standard solution of rutin (Sigma-Aldrich Chemie GmbH, Taufkirchen, Germany) was reacted with aluminum chloride to generate a calibration curve and the results were expressed as mg rutin per g of DW.

The extraction and determination of soluble protein and non-protein thiols were carried out as described by Borisova et al. [42]. The total content of soluble thiols was determined after reaction with Elman's reagent (5.5′-dithiobis (2-nitrobenzoic) acid) at 412 nm. The content of protein thiols was calculated by subtracting the amount of non-protein thiols previously obtained by precipitation of proteins with 50% trichloroacetic acid from the total soluble fraction. Reduced glutathione (Sigma-Aldrich Chemie GmbH, Taufkirchen, Germany) was used as a standard. The content of soluble protein was determined at 595 nm according to Bradford [43]. Bovine serum albumin was used as a standard.

2.7. Statistics

The study of the physicochemical characteristics of substrates, including analyses of macronutrients and heavy metals, was carried out on composite samples in quadruplicate. The granulometric analysis was performed in 4 independent replicates for each studied site. The physiological and biochemical analyses were performed in 6 replicates; data obtained over a two-year period were averaged. The tables and figures present the mean values (Means) and standard errors (SE).

After checking the normality using the Shapiro–Wilk's test and the homogeneity of variance using Levene's test, the differences between the studied orchid populations were determined with the non-parametric Kruskal–Wallis ANOVA by ranks and Mann–Whitney

U-test ($p < 0.05$). Different alphabetical letters in the tables and figures indicate a significant difference between the studied parameters.

3. Results

3.1. Physicochemical Characteristics of Substrates

The pH of the substrate on serpentine dumps (S2 and S3) was slightly alkaline, while the reference site (S1) was circumneutral (Table 2). The maximum value of ES was noted at Anatol'sky dump (S3); it was higher by 33% and 17% than in natural forest area (S1) and Shilovsky dump (S2), respectively. At the same time, the substrate moisture content was 2.3 times lower in S2 in comparison with S1 and S3 (Table 2).

Table 2. The pH values, electrical conductivity and nutrient content in substrates from studied sites.

Parameters	Sites		
	S1	S2	S3
pH	7.10 ± 0.01 b [1]	7.89 ± 0.02 a	7.65 ± 0.02 a
Electrical conductivity, µS cm^{-1}	166.80 ± 4.37 c	189.00 ± 0.58 b	222.00 ± 4.73 a
Substrate moisture content,% DW	19.50 ± 1.05 a	8.35 ± 0.47 b	19.20 ± 1.25 a
Available nutrients, mg kg^{-1} DW			
Nitrogen (N)	191.80 ± 27.04 a	32.20 ± 6.83 c	84.00 ± 17.81 b
Phosphorus (P_2O_5)	25.20 ± 8.80 b	31.70 ± 6.30 b	58.40 ± 11.70 a
Potassium (K)	136.18 ± 3.47 b	168.34 ± 2.46 a	127.24 ± 2.83 b

[1] Data is presented as Means ± SE (n = 4). Different alphabetical letters indicate a significant difference between the studied sites at $p < 0.05$.

The available nitrogen level in the soil from the natural forest community (S1) was considerably higher (by 3.3 times on average) compared to other sites (Table 2). At the same time, the differences in the content of phosphorus were not so significant. In the serpentine substrate of dumps, it was higher, especially in S3 (by 2.3 times). The maximum potassium concentration was observed in S2, while there were no differences between S1 and S3.

All the studied substrates were formed on serpentine rocks. However, in terms of particle size distribution, the following differences were revealed between them (Table 3).

Table 3. Granulometric composition of substrates from studied sites.

Site	Particle Size Distribution, %					
	>10 mm	5–10 mm	2–5 mm	1–2 mm	0.25–1 mm	<0.25 mm
S1	9.5 ± 0.3 c [1]	5.8 ± 0.2 c	3.8 ± 0.1 b	0.8 ± 0.2 c	14.6 ± 0.9 c	65.5 ± 2.4 a
S2	29.4 ± 1.1 a	16.6 ± 1.3 a	13.1 ± 0.9 a	9.3 ± 0.5 a	17.2 ± 1.2 b	14.4 ± 0.9 b
S3	26.3 ± 0.8 b	13.9 ± 1.2 b	12.6 ± 0.7 a	7.6 ± 0.3 b	22.7 ± 0.8 a	16.9 ± 1.3 b

[1] Data is presented as Means ± SE (n = 4). Different alphabetical letters indicate a significant difference between the studied sites at $p < 0.05$.

Substrates on Shilovsky (S2) and Anatol'sky (S3) dumps were very stony; fractions of large size (crushed stone and gravel) prevailed. Fractions <0.25 mm in size on S2 and S3 dumps accounted for only 15.7% on average, while in the natural forest community (S1), the proportion of small fractions was 4.2 times higher (Table 3).

3.2. Metal Content in Substrates and Plants

Data on the total content of metals in the studied substrates show that sites S1–S3 are distinguished by an increased content of many metals which is due to their serpentine nature. The total content of Mg in the substrate of serpentine dumps (S2 and S3) was on average 1.5 times higher than in the soil of the natural forest community (S1, Figure 3).

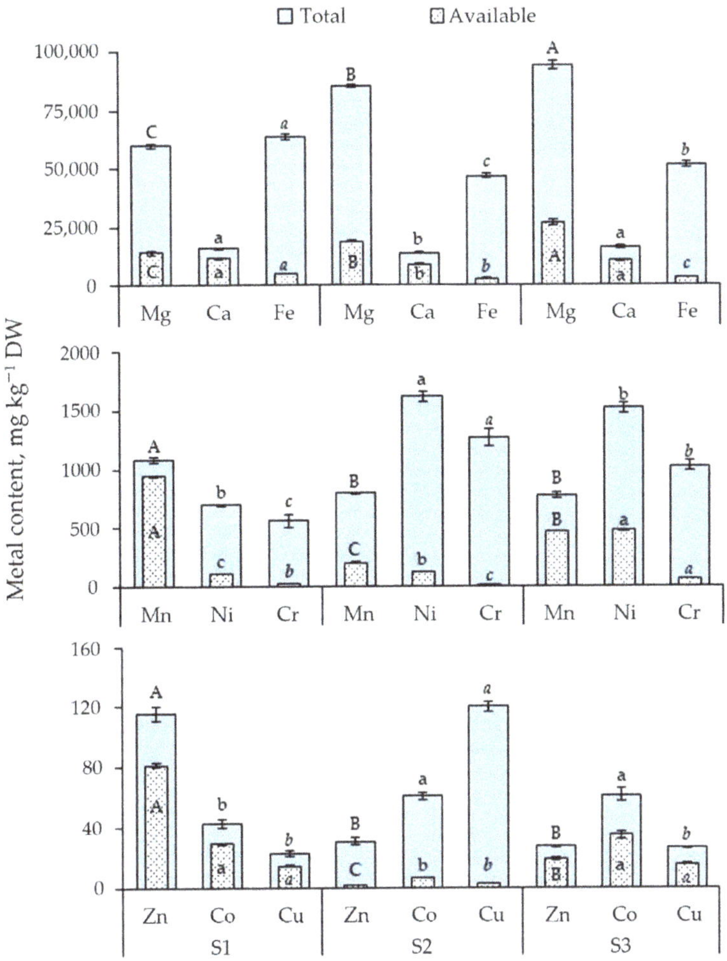

Figure 3. Total and available metal content in soil substrates from studied sites. Data is presented as Means ± SE (n = 4). Different alphabetical letters indicate a significant difference between the studied sites at $p < 0.05$.

The total content of Ni, Cr and Co in the dump substrate was also higher (by 2 times on average) than at the reference site (Figure 3). The reverse trend was noted for Fe, Mn and Zn; their maximum concentration was found in S1. In terms of the total Cu concentration, S2 was distinguished; the concentration being 5 times higher compared to other studied sites (Figure 3). The ratio of the total Mg:Ca on the dumps averaged 5.9, while in the natural community it was 3.7. The differences between the sites of the available Mg:Ca were not so considerable and on average it was 1.9.

The proportion of available metals in their total concentrations varied from 1.3% to 87% (Figure 3). The most significant part of the available form in all sites was found for Ca (75% on average). At two sites (S1 and S3), the proportion of available Mn, Zn, Co and Cu was more than 50% of the total concentration, while Fe and Cr in all sites were predominant in an inaccessible form. The Shilovsky dump (S2) was also distinguished by a low content of available Zn, Co and Cu (by 4.5% on average) (Figure 3).

The content of Fe, Mn, Ni, Zn and Cu in the rhizome + roots of *E. atrorubens* was higher on average by 2 times, while Cr and Co concentrations were higher by 4 and 7 times, respectively, than in the leaves, excepting the macronutrients Ca and Mg: their average rhizome + roots to leaves ratio was close to 1 (Figure 4).

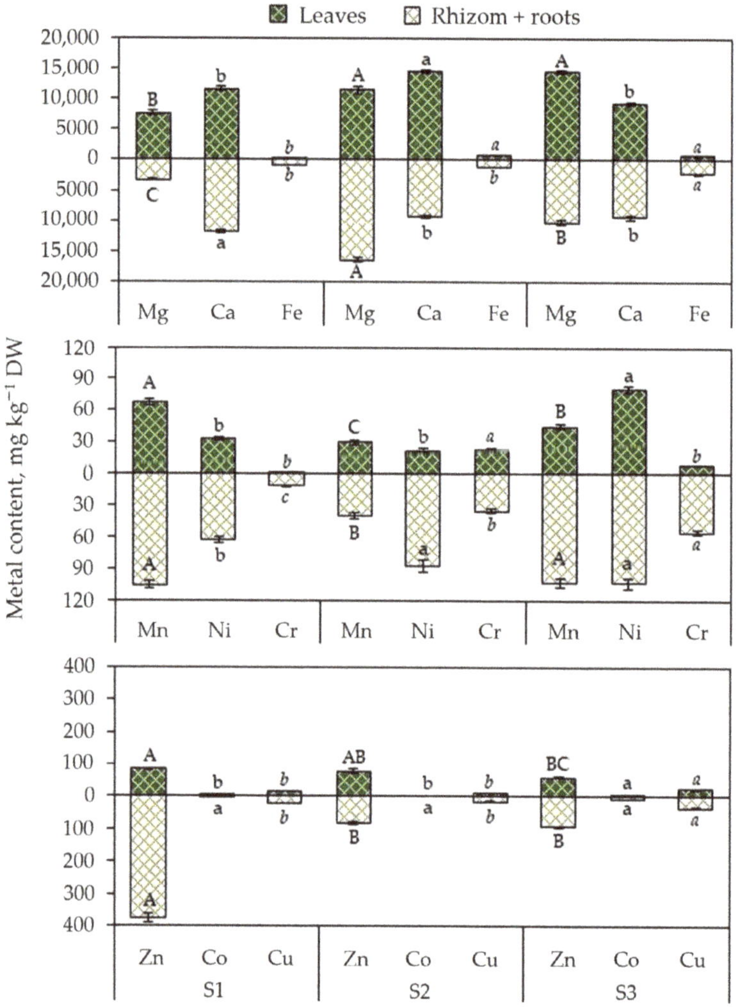

Figure 4. Metal content in the aboveground and underground organs of *E. atrorubens* growing on serpentine substrates. Data is presented as Means ± SE (n = 4). Different alphabetical letters indicate a significant difference between the studied sites at $p < 0.05$. kg^{-1} DW.

The content of Fe, Mg, Ni, Cr, Co and Cu in the leaves of *E. atrorubens* growing on dumps (S2 and S3) was on average 2 times higher than that in the natural community (S1). The opposite trend was noted for Mn and Zn (Figure 4).

3.3. The Rhizospheric Microbiota of E. atrorubens

The analysis of serpentine substrates collected in the root zone of *E. atrorubens* showed that the lowest density of bacterial cells (QMAFAnM) in the orchid rhizosphere was found on the Shilovsky dump (S2), while in the natural forest community (S1) and on Anatol'sky dump (S3), the number of rhizospheric bacteria was an order of magnitude higher (Table 4). The lowest enzymatic activity, estimated by the cellulose-decomposing microflora, was also noted in S2, which was 21 and 12 times higher than in S1 and S3, respectively (Table 4).

The degree of mycotrophy in *E. atrorubens* roots was the highest on the serpentine dumps (S2 and S3) in comparison with the natural forest community (Table 4).

Table 4. The characteristics of rhizospheric microbiota of *E. atrorubens* from studied sites.

Characteristics	Sites		
	S1	**S2**	**S3**
QMAFAnM, CFU g^{-1} DW	6.1×10^5	6.8×10^4	9.2×10^5
Enzymatic activity, %	6.5	0.3	3.6
Degree of mycotrophy, %	78	90	97

3.4. Plant Water Status

The intensity of transpiration in *E. atrorubens* leaves varied insignificantly compared to the reference site (S1). However, the highest values were noted in S2 plants (Shilovsky dump) which increased by 11 and 18% compared to S1 and S3 (Figure 5a). The RWC and WSD indexes entered the range of values of most plants and differed significantly only in S2 orchids (Figure 5b). In Shilovsky dump plants (S2) the water deficit increased by an average of 1.4 times compared to other orchid populations (S1 and S3).

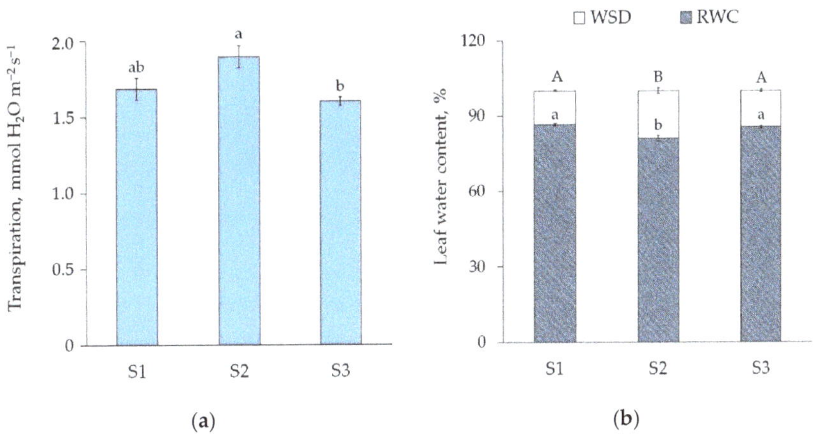

Figure 5. The water status parameters of *E. atrorubens* growing on serpentine substrates: (**a**) Transpiration rate; (**b**) Relative water content (RWC) and water saturation deficit (WSD). Data is presented as Means ± SE (n = 12). Different alphabetical letters indicate a significant difference between the studied populations at $p < 0.05$.

3.5. Redox Reactions of Plants

The study has shown that foliar MDA content in *E. atrorubens* growing on technogenic substrates was 1.5 and 1.2 times higher on the Shilovsky (S2) and Anatol'sky (S3) dumps, respectively, compared to S1 (Figure 6a).

The free proline content was 1.5 times higher in the leaves of S2 plants in comparison with S1 (Figure 6b). A slight increase (by 1.2 times) in the proline content was found in S3 plants (S3) compared to S1. The ascorbate amount in S2 plants was increased by 1.7 times in comparison with the reference site, however, there were no significant differences between S3 and S1 plants (Figure 6c).

The level of total soluble phenols in S2 orchids was higher by 1.5 and 1.6 times than in S3 and S1 plants, respectively (Figure 6d). In the leaves of plants growing on both dumps, the content of flavonoids was higher than in plants growing in natural conditions. The soluble protein thiols content was the same in all studied sites (Figure 6e). The content of non-protein thiols was higher by 1.8 times in plants growing on technogenic substrates compared to natural conditions.

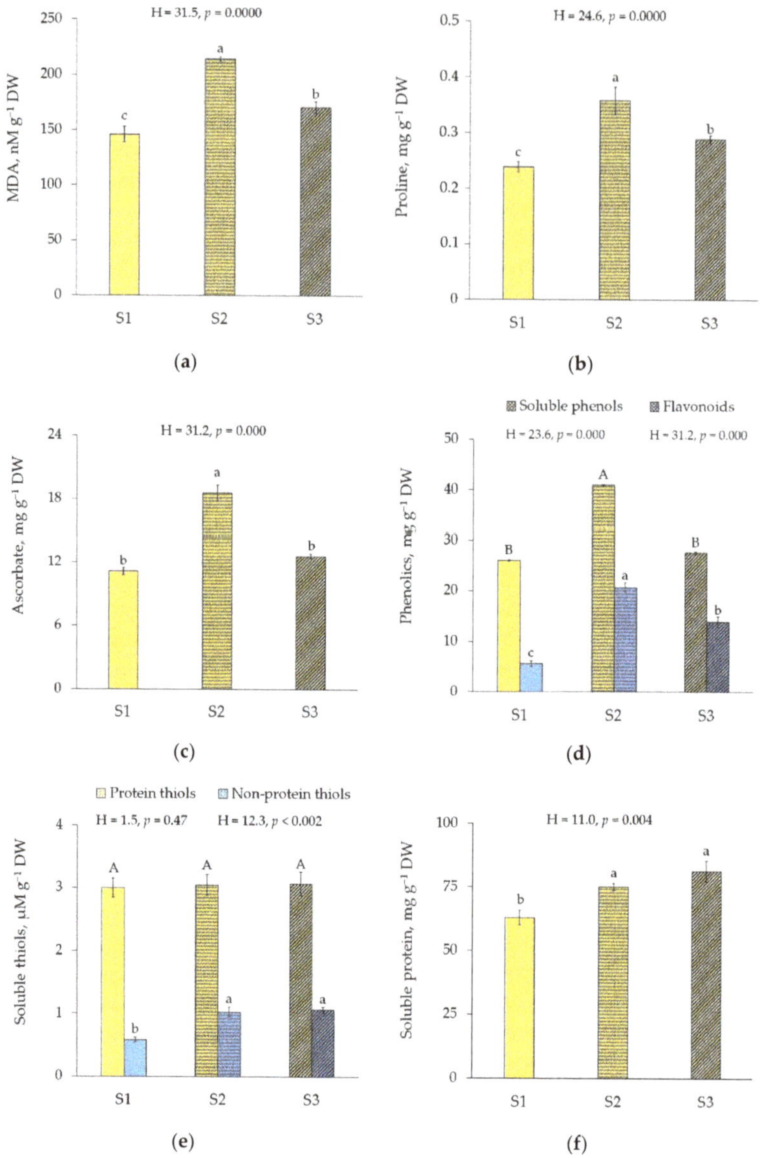

Figure 6. Pro- and antioxidant compounds in the leaves of *E. atrorubens* growing on serpentine substrates: (**a**) Malondialdehyde (MDA); (**b**) Free proline; (**c**) Ascorbate; (**d**) Soluble phenolics; (**e**) Soluble thiols; and (**f**) Soluble protein content. Data is presented as Means ± SE (n = 12). Different alphabetical letters indicate a significant difference between the studied populations at $p < 0.05$.

The content of soluble protein in *E. atrorubens* growing on technogenic substrates was increased by 1.3 times in comparison with the reference site (Figure 6f), however, there were no significant differences between S2 and S3 plants.

4. Discussion

The study of adaptive responses to abiotic stress in rare and endangered plants is an important prerequisite for their successful introduction [26]. The adaptive redox reactions of orchid *E. atrorubens* were investigated on serpentine dumps post asbestos mining in comparison with a natural forest community. The serpentine substrates are unfavorable for plant growth due to their poor physicochemical properties: high stoniness; minimum amount of silty and clay particles; high content of iron, magnesium, nickel, chromium and cobalt, which are toxic to most plants and bacterial communities; and a low content of some macro- and micronutrients [16–18].

Although numerous studies have been carried out on plants in serpentine areas [6,7,9,19,44,45], there is little information concerning the adaptations of orchids abundantly growing on serpentine outcrops.

It was found that the substrates on the studied dumps (S2 and S3) were slightly alkaline and were characterized by low concentrations of available nitrogen and phosphorus, which is unfavorable for plant growth [46]. Moreover, the substrates on the dumps were characterized by high stoniness, which was associated with the mining of asbestos by the open method, after which overburdened rocks were mixed with loose rocks and enrichment wastes on the dumps [7].

The previous comparative study of *E. atrorubens* in two different forest plant communities of the Middle Urals (on serpentine and granite rocks) has shown that serpentine substrate differed by an extremely high concentration of total Mg which was 79-fold higher than in soil on granites. The high concentration of total Ni, Cr, Co and Fe was also found in serpentine substrate, where Ni, Cr, Co and Fe were 94, 59, 17 and 4 times higher than in granite ones, respectively [7]. The comparison of metal concentrations in the serpentine substrate on dumps with the averaged data on the serpentine rocks in the Urals is of particular interest. It was found that the total content of Mg, Fe, Ni and Cr in the serpentine dumps of the Anatol'sko-Shilovsky asbestos deposit was on average three times higher compared to the averaged data on serpentine substrates in the region reported by Teptina and Paukov [44]. However, a comparison of the obtained data with the information presented by Kierczak et al. [18] shows that the total and available nickel content was within the range of values for the serpentine soils of Lower Silesia (southwestern Poland). At the same time, the content of Cr and Co was slightly lower.

The previously conducted study showed that the possible reason for the increased *E. atrorubens* tolerance to high concentrations of metals in the serpentine soil was its ability to accumulate a higher amount of metal in its root system and prevent its transfer to the aboveground organs [7]. A similar trend was also found in this study. The content of all studied heavy metals (HMs) in *E. atrorubens* leaves was lower than in the roots. Nevertheless, the content of some metals (Ni and Cr) in the leaves of plants on serpentine dumps was higher than their critical concentrations reported by Kabata-Pendias and Mukherjee [47].

The high tolerance of *E. atrorubens*, as well as of some other species of the *Epipactis* genus, to elevated concentrations of HMs was evidenced by the reported results of other authors [4,9,29]. Many researchers attributed this to a well-developed mycorrhiza in the roots of orchids [15,29]. Orchid mycorrhiza, represented by pelotons, was found in the root cells of *E. atrorubens* from all studied sites. The percentage of mycorrhiza occurrence was higher in the dumps in comparison with the natural environment. This fact allows us to assume that mycorrhiza in this orchid plant make it more tolerant to the high concentrations of HMs in serpentine soils. It is also important to point out that the soil bacteria, along with micromycetes, are the most important components of the system of symbiotic relationships between orchid plants and microorganisms [48]. It is known that many endophytic and rhizospheric bacteria have the ability to stimulate plant growth through various mechanisms [31,48,49]. However, data on the ability of bacterial communities in the rhizosphere to promote the growth of orchids and increase their viability under abiotic stress conditions are fragmentary [30].

It should be noted that, despite the complex action of unfavorable edaphic factors on orchids colonizing serpentine dumps (S2 and S3), their local populations turned out to be more numerous than in the natural forest community (S1). Furthermore, it was found that *E. atrorubens* grows well on serpentine sites without showing any detrimental effects.

Obviously, this is partly due to the fact that in the natural ecotope the orchids experienced more pronounced phytocoenotic stress. The significance of serpentines as suitable habitats for the light-demanding orchids having low competitive ability was reported previously by Djordjević et al. [15].

On the whole, *E. atrorubens* demonstrated high tolerance to adverse abiotic factors. Therefore, it is important to identify the physiological and biochemical features that contribute to the colonization of this orchid on technogenic substrates. The study showed that *E. atrorubens* growing on serpentine dumps (S2 and S3) differed by the increased MDA content in the leaves compared to the ones in the natural forest community. An imbalance between the production of ROS and their neutralization often leads to oxidative stress in plants. At the same time, the orchid in transformed habitats demonstrated high tolerance to adverse abiotic factors. An important role in ensuring plant resistance to abiotic stress is played by the antioxidant system components, including non-enzymatic antioxidants [21,25]. It was found that lipid oxidative damage in orchid plants was accompanied by the synthesis of free proline, ascorbate, soluble phenolic compounds and soluble non-protein thiols in leaves. We suggested that *E. atrorubens* has adapted to the technogenic substrates by increasing the content of non-enzymatic antioxidants.

Proline is a proteinogenic heterocyclic amino acid which plays an important role in plant cells [27]. Osmotic adjustment is an important mechanism which alleviates some of the detrimental effects of water stress due to the accumulation of osmolytes [50]. The orchid plants from the Shilovsky dump (S2) had the highest content of free proline which was correlated with water deficiency ($r = 0.66$, $p < 0.05$). The increased water loss is possibly related to the characteristics of this technogenic habitat (strong substrate stoniness and intense insolation due to the lack of crown closure).

Water deficit stress tolerance is the result of the coordination of physiological and biochemical alterations at the organ, cellular and molecular levels [50]. Obviously, the increased accumulation of proline as the most important osmolyte contributed to the survival of *E. atrorubens* in unfavorable environmental conditions. It is known that proline is involved not only in osmoregulation, but also in the stabilization of proteins, membranes and subcellular structures. Proline is also able to chelate HMs, maintain cellular redox potential and participate in ROS neutralization [27,50].

The ascorbate represents a key molecule in plant metabolism. It plays a central role in several physiological processes such as photosynthesis, photo-protection, cell division, plant growth and stress responses [21,25]. Moreover, it performs a number of antioxidant functions and participates in the termination of chain reactions of oxidation organic compounds, preventing lipid peroxidation. In addition, ascorbate can directly react with ROS and participate in their neutralization, as well as restore other important antioxidants (phenolic compounds, tocopherol, etc.). [25,51]. The S2 *E. atrorubens* plants were characterized by an increased accumulation of not only free proline but also ascorbate content compared to other habitats, which indicates their active role in the adaptation to the abiotic stress.

Many phenolic compounds are known to have antioxidant properties. Phenolic substances readily interact with ROS. Initially, they are oxidized to phenoxyl radicals, the further oxidation of which leads to the formation of quinones. They can also chelate HMs and stabilize membranes, which limits the diffusion of free radicals and reduces the rate of lipid peroxidation [51,52]. As a rule, the synthesis of phenolic compounds is enhanced under stress. Previously, it was shown that the level of antioxidant activity in *Phalaenopsis* orchid hybrids was proportional to the concentration of phenolic compounds [53]. In *E. atrorubens* plants colonizing technogenic substrates (S2 and S3), the content of phenolic compounds in leaves directly correlated with the content of MDA ($r = 0.98$; $p < 0.05$). Particularly, many phenolic compounds, including flavonoids, were synthesized in S2 plants. It is important to point out that the content of flavonoids in S2 orchids was higher (by 4 times) than that in the hybrids of orchid *Phalaenopsis* reported by Minh et al. [53].

It should be noted that redox reactions, such as the increase in MDA content and the more active synthesis of free proline, ascorbate and phenolics, were manifested most clearly in S2 plants (Shilovsky dump). Furthermore, the substrate on this dump was characterized by the lowest density of bacterial cells and lowest enzymatic activity compared to other sites. Perhaps this is due to the fact that S2 was represented by an emerging forest community without crown closure, while in S1 and S3 tree crown density was about 0.6.

The cell compounds containing SH-groups (thiols), which can be divided into protein and non-protein types, play an important role in the antioxidant protection of plants. Thiols can bind HMs and act as antioxidants, participating in the neutralization of ROS formed during oxidative stress [24,54]. An increased level of non-protein thiols was found in the plants growing on technogenic substrates (S2 and S3) compared to the natural conditions, which confirms their participation in protective reactions of plants to the action of HMs and other stressors. Additionally, many soluble proteins are involved in the antioxidant defense system of plants. They are capable of both directly chelating HMs and acting as enzymes, catalyzing the reactions of ROS neutralization [55].

Thus, the present study shows that an increased level of non-enzymatic antioxidant accumulation determines not only the ability of *E. atrorubens* to withstand negative environmental factors, but also to adapt to them successfully.

5. Conclusions

The overgrown dumps of the Anatol'sko-Shilovsky deposit today are areas with an infertile serpentine substrate containing large amounts of crushed stone and gravel, as well as higher concentrations of such heavy metals as Ni, Cr and Co, compared to natural habitat (2.0 times on average). Most of the studied metals predominantly accumulated in *E. atrorubens* rhizomes and roots compared to leaves (4 times on average). It was found that orchids colonizing serpentine dumps had more lipid peroxidation products (1.4 times on average), which demonstrated chronic oxidative stress. A comparative study of this orchid in the natural and transformed habitats allowed us to identify some compensatory adaptive reactions that contribute to its naturalization and distribution in technogenic substrates. The orchid plants demonstrated fairly high resistance to stressful conditions, probably due to the increased mycorrhization and more active synthesis of such non-enzymatic antioxidants as ascorbate, free proline, soluble phenolics including flavonoids and non-protein thiols. Further research will be aimed at studying the bacterial communities of the *E. atrorubens* rhizosphere and the estimation of the rhizobacteria characteristics that contribute to the growth and vital activity of the orchid under adverse environmental conditions.

Author Contributions: Conceptualization, M.M. and G.B.; methodology, M.M., G.B., E.F., N.L. and O.V.; software, M.M. and N.C.; validation, M.M. and G.B.; formal analysis, M.M. and N.C.; investigation, M.M., G.B., E.F., N.L., N.C., A.E., A.T. and O.V.; resources, M.M., G.B., E.F. and O.V.; data curation, M.M., G.B. and E.F.; writing—original draft preparation, M.M. and G.B.; writing—review and editing, M.M., G.B., E.F., N.L., N.C., A.T. and A.E.; visualization, M.M. and G.B.; supervision, M.M. and G.B.; project administration, M.M.; funding acquisition, M.M., G.B., E.F., N.L., A.E., A.T. and O.V. All authors have read and agreed to the published version of the manuscript.

Funding: The reported study was partly funded by RFBR and the Government of Sverdlovsk region, project number 20-44-660011 and the Ministry of Science and Higher Education of the Russian Federation as part of state task of the Ural Federal University, FEUZ-2020-0057.

Institutional Review Board Statement: Not applicable.

Informed Consent Statement: Not applicable.

Data Availability Statement: Not applicable.

Acknowledgments: The authors are very grateful to Belyaeva I.V. and Kiseleva I.S. (UrFU, Ekaterinburg, Russia) for linguistic support and editing of the English language.

Conflicts of Interest: The authors declare no conflict of interest.

References

1. Swarts, D.N.; Dixon, W.D. Terrestrial orchid conservation in the age of extinction. *Ann. Bot.* **2009**, *104*, 543–556. [CrossRef] [PubMed]
2. Jurkiewicz, A.; Turnau, K.; Mesjasz–Przybylowicz, J.; Przybylowicz, W.; Godzik, B. Heavy metal localisation in mycorrhizas of *Epipactis atrorubens* (Hoffm.) Besser (Orchidaceae) from zink mine tailings. *Protoplasma* **2001**, *218*, 117–124. [CrossRef] [PubMed]
3. Adamowski, W. Expansion of native orchids in anthropogenous habitats. *Pol. Bot. Stud.* **2006**, *22*, 35–44.
4. Shefferson, R.; Kull, T.; Tali, K. Mycorrhizal interactions of orchids colonizing Estonian mine tailings hills. *Am. J. Bot.* **2008**, *95*, 156–164. [CrossRef]
5. Vakhrameeva, M.G.; Tatarenko, I.V.; Varlygina, T.I.; Torosyan, G.K.; Zagulskii, M.N. *Orchids of Russia and Adjacent Countries (Within the Borders of the Former USSR)*; A.R.G. Gantner Verlag: Ruggell, Liechtenstein, 2008.
6. Filimonova, E.; Lukina, N.; Glazyrina, M.; Borisova, G.; Tripti; Kumar, A.; Maleva, M. A comparative study of *Epipactis atrorubens* in two different forest communities of the Middle Urals, Russia. *J. For. Res.* **2020**, *31*, 2111–2120. [CrossRef]
7. Filimonova, E.I.; Lukina, N.V.; Glazyrina, M.A.; Borisova, G.G.; Maleva, M.G.; Chukina, N.V. Endangered orchid plant *Epipactis atrorubens* on serpentine and granite outcrops of Middle Urals, Russia: A comparative morphophysiological study. *AIP Conf. Proc.* **2019**, *2063*, 040016. [CrossRef]
8. Maleva, M.; Borisova, G.; Chukina, N.; Sinenko, O.; Filimonova, E.; Lukina, N.; Glazyrina, M. Adaptive morphophysiological features of *Neottia ovata* (Orchidaceae) contributing to its natural colonization on fly ash deposits. *Horticulturae* **2021**, *7*, 109. [CrossRef]
9. Djordjević, V.; Tsiftsis, S. The role of ecological factors in distribution and abundance of terrestrial orchids. In *Orchids Phytochemistry, Biology and Horticulture*; Reference Series in Phytochemistry; Mérillon, J.M., Kodja, H., Eds.; Springer: Cham, Switzerland, 2022; pp. 3–72. [CrossRef]
10. Bilz, M.P.; Kell, S.P., Maxted, N.; Lansdown, R.V. *European Red List of Vascular Plants*; Publications Office of the European Union: Luxembourg, 2011; p. 130.
11. Tashev, A.; Tashev, N. Plant species of Bulgarian flora included in the CITES Convention. *Exp. Biol.* **2021**, *1*, 4–13. [CrossRef]
12. Zolotareva, N.V.; Knyazev, M.S.; Podgaevskaya, E.N.; Pustovalova, L.A. *Red Book of the Sverdlovsk Region: Animals, Plants, Mushrooms; Part II. Plants; Section 1. Angiosperms*; Korytin, N.S., Ed.; Mir LLC: Yekaterinburg, Russia, 2018; pp. 159–344. ISBN 978-5-60-427511-5. (In Russian)
13. Buttler, P.K. *Field Guide to Orchids of Britain and Europe*; The Crowood Press: Swindon, UK, 1991; p. 288.
14. Delforge, P. *Orchids of Europe, North Africa and the Middle East*; A. & C. Black: London, UK, 2006; p. 640.
15. Djordjević, V.; Tsiftsis, S.; LakuŠić, D.; Stevanović, V. Niche analysis of orchids of serpentine and non-serpentine areas: Implications for conservation. *Plant Biosyst.* **2016**, *150*, 710–719. [CrossRef]
16. Alexander, E.B.; Coleman, R.G.; Keeler-Wolf, T.; Harrison, S. *Serpentine Geoecology of Western North America*; Oxford University Press: New York, NY, USA, 2007; p. 512.
17. Rajakaruna, N.; Boyd, R.S. Serpentine soils. In *Oxford Bibliographies in Ecology*; Gibson, D., Ed.; Oxford University Press: New York, NY, USA, 2014. [CrossRef]
18. Kierczak, J.; Pędziwiatr, A.; Waroszewski, J.; Modelska, M. Mobility of Ni, Cr and Co in serpentine soils derived on various ultrabasic bedrocks under temperate climate. *Geoderma* **2016**, *268*, 78–91. [CrossRef]
19. Galey, M.L.; van der Ent, A.; Iqbal, M.C.M.; Rajakaruna, N. Ultramafic geoecology of South and Southeast Asia. *Bot. Stud.* **2017**, *58*, 18. [CrossRef] [PubMed]
20. Gill, S.S.; Tuteja, N. Reactive oxygen species and antioxidant machinery in abiotic stress tolerance in crop plants. *Plant Physiol. Biochem.* **2010**, *48*, 909–930. [CrossRef] [PubMed]
21. Hasanuzzaman, M.; Bhuyan, M.H.M.B.; Zulfiqar, F.; Raza, A.; Mohsin, S.M.; Mahmud, J.A.; Fujita, M.; Fotopoulos, V. Reactive oxygen species and antioxidant defense in plants under abiotic stress: Revisiting the crucial role of a universal defense regulator. *Antioxidants* **2020**, *9*, 681. [CrossRef]
22. Gajić, G.; Djurdjević, L.; Kostić, O.; Jarić, S.; Mitrović, M.; Pavlović, P. Ecological potential of plants for phytoremediation and ecorestoration of fly ash deposits and mine wastes. *Front. Environ. Sci.* **2018**, *6*, 124. [CrossRef]
23. Weber, H.; Chetelat, A.; Reymond, P.; Farmer, E.E. Selective and powerful stress gene expression in *Arabidopsis* in response to malondialdehyde. *Plant J.* **2004**, *37*, 877–888. [CrossRef] [PubMed]
24. Pradedova, E.V.; Nimaeva, O.D.; Salyaev, R.K. Redox processes in biological systems. *Russ. J. Plant Physiol.* **2017**, *64*, 822–832. [CrossRef]
25. Cavaiuolo, M.; Cocetta, G.; Ferrante, A. The antioxidants changes in ornamental flowers during development and senescence. *Antioxidants* **2013**, *2*, 132–155. [CrossRef]
26. Masouleh, S.S.S.; Sassine, Y.N. Molecular and biochemical responses of horticultural plants and crops to heat stress. Review Article. *Ornam. Hortic.* **2020**, *26*, 148–158. [CrossRef]
27. Zhang, L.; Becker, D.F. Connecting proline metabolism and signaling pathways in plant senescence. *Front. Plant Sci.* **2015**, *6*, 552. [CrossRef]
28. Těšitelová, T.; Těšitel, J.; Jersáková, J.; Říhová, G.; Selosse, M.S. Symbiotic germination capability of four *Epipactis* species (Orchidaceae) is broader than expected from adult ecology. *Am. J. Bot.* **2012**, *99*, 1020–1032. [CrossRef]

29. De Agostini, A.; Caltagirone, C.; Caredda, A.; Cicatelli, A.; Cogoni, A.; Farci, D.; Guarino, F.; Garau, A.; Labra, M.; Lussu, M.; et al. Heavy metal tolerance of orchid populations growing on abandoned mine tailings: A case study in Sardinia Island (Italy). *Ecotoxicol. Environ. Saf.* **2020**, *189*, 110018. [CrossRef] [PubMed]
30. Böhmer, M.; Ozdín, D.; Račko, M.; Lichvár, M.; Budiš, J.; Szemes, T. Identification of bacterial and fungal communities in the roots of orchids and surrounding soil in heavy metal contaminated area of mining heaps. *Appl. Sci.* **2020**, *10*, 7367. [CrossRef]
31. Jakubska-Busse, A.; Kedziora, A.; Cieniuch, G.; Korzeniowska-Kowal, A.; Bugla-Płoskon´ska, G. Proteomics-based identification of orchid-associated bacteria colonizing the *Epipactis albensis*, *E. helleborine* and *E. purpurata* (Orchidaceae, Neottieae). *Saudi J. Biol. Sci.* **2021**, *28*, 4029–4038. [CrossRef]
32. Shakirov, A.V. *Physical-Geographical Zoning of the Urals*; UB RAC: Ekaterinburg, Russia, 2011; p. 617, (In Russian with English summary).
33. Maiti, S.K. *Ecorestoration of the Coalmine Degraded Lands*; Springer: New York, NY, USA, 2013; p. 361.
34. Arinushkina, E.V. *Chemical Soil Analysis Guide*; Moscow State University: Moscow, Russia, 1970. (In Russian)
35. Tandy, S.; Bossart, K.; Mueller, R.; Ritschel, J.; Hauser, L.; Schulin, R.; Nowack, B. Extraction of heavy metals from soils using biodegradable chelating agents. *Environ. Sci. Technol.* **2004**, *38*, 937–944. [CrossRef] [PubMed]
36. Mishustin, E.N. *Microorganisms and Agricultural Productivity*; Nauka: Moscow, Russia, 1972; p. 344. (In Russian)
37. Hellmuth, E.O. Measurement of leaf water deficit with particular reference to the whole leaf method. *J. Ecol.* **1970**, *58*, 409–417. [CrossRef]
38. Heath, R.L.; Packer, L. Photoperoxidation in isolated chloroplasts: I. Kinetics and stoichiometry of fatty acid peroxidation. *Arch. Biochem. Biophys.* **1968**, *125*, 189–198. [CrossRef]
39. Maleva, M.; Borisova, G.; Chukina, N.; Kumar, A.; Prasad, M.N.V. High dose of urea enhances the nickel and copper toxicity in Brazilian elodea (*Egeria densa* Planch. Casp.). *Braz. J. Bot.* **2016**, *39*, 965–972. [CrossRef]
40. Singleton, V.L.; Orthofer, R.; Lamuela-Raventos, R.M. Analysis of total phenols and other oxidation substrates and antioxidants by means of Folin–Ciocalteu reagent. *Method. Enzymol.* **1999**, *299*, 152–178. [CrossRef]
41. Chang, C.-C.; Yang, M.-H.; Wen, H.-M.; Chern, J.-C. Estimation of total flavonoid content in propolis by two complementary colorimetric methods. *J. Food Drug Anal.* **2002**, *10*, 178–182.
42. Borisova, G.; Chukina, N.; Maleva, M.; Kumar, A.; Prasad, M.N.V. Thiols as biomarkers of heavy metal tolerance in the aquatic macrophytes of Middle Urals, Russia. *Int. J. Phytorem.* **2016**, *18*, 1037–1045. [CrossRef]
43. Bradford, M.M. A rapid and sensitive method for the quantitation of microgram quantities of protein utilizing the principle of protein-dye binding. *Anal. Biochem.* **1976**, *72*, 248–254. [CrossRef]
44. Teptina, A.Y.; Paukov, A.G. Nickel accumulation by species of *Alyssum* and *Noccaea* (Brassicaceae) from ultramafic soils in the Urals, Russia. *Aust. J. Bot.* **2015**, *63*, 78–84. [CrossRef]
45. Tripti; Kumar, A.; Maleva, M.; Borisova, G.; Chukina, N.; Morozova, M.; Kiseleva, I. Nickel and copper accumulation strategies in *Odontarrhena obovata* growing on copper smelter-influenced and non-influenced serpentine soils: A comparative field study. *Environ. Geochem. Health* **2021**, *43*, 1401–1413. [CrossRef] [PubMed]
46. Marschner, H. *Mineral Nutrition of Higher Plants*, 2nd ed.; Academic Press: San Diego, CA, USA, 1995.
47. Kabata-Pendias, A.; Mukherjee, A.B. *Trace Elements from Soil to Human*; Springer: Heidelberg, Germany, 2007; p. 550.
48. Herrera, H.; Fuentes, A.; Soto, J.; Valadares, R.; Arriagada, C. *Orchids Phytochemistry, Biology and Horticulture*; Reference Series in Phytochemistry; Mérillon, J.M., Kodja, H., Eds.; Springer: Cham, Switzerland, 2022; pp. 175–200. [CrossRef]
49. Hanaka, A.; Majewska, M.; Jaroszuk-Scisel, J. Study of the influence of abiotic and biotic stress factors on horticultural plants. *Horticulturae* **2022**, *8*, 6. [CrossRef]
50. Vijayalakshmi, T.; Varalaxmi, Y.; Jainender, S.; Yadav, S.K.; Vanaja, M.; Jyothilakshmi, N.; Maheswari, M. Physiological and biochemical basis of water-deficit stress tolerance in pearl millet hybrid and parents. *Am. J. Plant Sci.* **2012**, *3*, 1730–1740. [CrossRef]
51. Blokhina, O.; Virolainen, E.; Fagerstedt, K.V. Antioxidants, oxidative damage and oxygen deprivation stress: A review. *Ann. Bot.* **2003**, *91*, 179–194. [CrossRef]
52. Michalak, A. Phenolic compounds and their antioxidant activity in plants growing under heavy metal stress. *Pol. J. Environ. Stud.* **2006**, *15*, 523–530.
53. Minh, T.N.; Khang, D.T.; Tuyen, P.T.; Minh, L.T.; Anh, L.H.; Quan, N.V.; Ha, P.T.T.; Quan, N.T.; Toan, N.P.; Elzaawely, A.A.; et al. Phenolic compounds and antioxidant activity of *Phalaenopsis* orchid hybrids. *Antioxidants* **2016**, *5*, 31. [CrossRef]
54. Cobbett, C.S. Phytochelatins and their roles in heavy metal detoxification. *Plant Physiol.* **2000**, *123*, 825–832. [CrossRef]
55. Kulaeva, O.A.; Tsyganov, V.E. Molecular-genetic basis of cadmium tolerance and accumulation in higher plants. *Russ. J. Genet. Appl. Res.* **2011**, *1*, 349–360. [CrossRef]

horticulturae

MDPI

Review

Plant Defensive Responses Triggered by *Trichoderma* spp. as Tools to Face Stressful Conditions

Jaime Pacheco-Trejo, Eliazar Aquino-Torres, Ma Isabel Reyes-Santamaría, Margarita Islas-Pelcastre, Sergio Rubén Pérez-Ríos, Alfredo Madariaga-Navarrete and Mariana Saucedo-García *

Instituto de Ciencias Agropecuarias, Universidad Autónoma del Estado de Hidalgo, Tulancingo-Santiago Tulantepec, Tulancingo 43600, Hidalgo, Mexico
* Correspondence: saucedo@uaeh.edu.mx

Abstract: The current agriculture is facing various challenges to produce enough food to satisfy the need of the human population consumption without having a negative impact on the environment, human health and ecosystems. The exploitation of bioinoculants has been a crucial alternative for green agriculture. Bioinoculants have two great benefits: to promote plant growth by making essential nutrients available to crops and, to increase the tolerance to biotic and abiotic stresses by inducing a long-lasting defense. Certain members of genus *Trichoderma* have been recognized as biocontrol agents, biofertilizers and stress alleviators for the plants. The use of *Trichoderma* spp. has also been extended to protect and stimulate growth of horticultural crops. Elucidating the plant signaling events triggered by *Trichoderma* is of high importance in order to understand the molecular basis involving plant protection against stresses. In this review, the signaling elements of the plants from *Trichoderma* perception through late defensive responses is discussed. Enhanced understanding how *Trichoderma* spp. activate defense will lead to improvement in the use of species of this genus to increase crop production with the consequent benefits for human health and care for the environment.

Keywords: priming of defense; G proteins; calcium signaling; mitogen-activated protein kinase; phytohormones; SA signaling; JA signaling; reactive oxygen species; antioxidant proteins; defense genes

Citation: Pacheco-Trejo, J.; Aquino-Torres, E.; Reyes-Santamaría, M.I.; Islas-Pelcastre, M.; Pérez-Ríos, S.R.; Madariaga-Navarrete, A.; Saucedo-García, M. Plant Defensive Responses Triggered by *Trichoderma* spp. as Tools to Face Stressful Conditions. *Horticulturae* **2022**, *8*, 1181. https://doi.org/10.3390/horticulturae8121181

Academic Editors: Agnieszka Hanaka, Małgorzata Majewska and Barbara Hawrylak-Nowak

Received: 23 November 2022
Accepted: 6 December 2022
Published: 10 December 2022

1. Introduction

Stress in plants can be defined as any external condition that limits the photosynthetic rate and reduces the energy conversion ability of a plant to biomass, affecting its growth, development or productivity [1,2]. Plant stress can be classified as abiotic or biotic. The abiotic stress refers to any environmental factor that negatively affects the plant growth and development. Abiotic stress (e.g., extreme temperatures, drought, salinity, radiation and toxic metals) causes serious losses of major crop plants around the world [3]. On the other hand, the presence of plant pathogenic living organisms, especially viruses, fungi, bacteria, nematodes, and herbivores are the causes of plants biotic stress [3].

Plants attempt to adapt and resist the stresses by adjusting their metabolism, signal transduction, gene expression, etc.; however, the plant survival under these stress conditions will depend on the intensity, frequency and exposure time [2].

Population growth as well as current climate change and the crop losses caused by emergence of plant pathogenic microorganisms are challenges that require immediate action to ensure food security and safety in coming years. It has been estimated that agricultural food production needs to increase by about 70% by 2050 to feed an expected world population of 9.1 billion of people [4].

Human food production must focus particularly on sustainable agriculture by the means of ecological practices that maximize food productivity and minimize negative consequences on the environment [5]. In recent years, bio-priming agents are receiving large attention as a promising approach to mitigate the environmental and disease threats in agriculture [6–10].

Bio-priming has been recognized as a low-cost and eco-friendly technology that promotes growth and induces stress tolerance to achieve desired crop yield [11]. Bio-priming consists in the use of beneficial microorganisms [e.g., plant-growth-promoting bacteria (PGPB), fungi, etc.] or materials of biological origin (e.g., humus, chitosan, etc.). These materials can be used in the seeds or the whole plants to promote growth or to improve stress responses. Among these microorganism are included fungi, especially arbuscular mycorrhizal and *Trichoderma* spp. [12].

Trichoderma is mostly an asexual genus of filamentous fungi (the teleomorphic forms are *Hypocrea*) that usually are among the most common saprophytic microorganisms living in the rhizosphere [13]. *Trichoderma* genus contains 375 species that have been described by molecular phylogenetic analysis based on DNA sequencing data [14]. The drastic increase in the number of *Trichoderma* species has several explanations that are related to the technologies and applications used for identification [14].

Although *Trichoderma* was isolated for the first time in 1794 from soil and decomposing organic matter [15], it was not until the early 20th century that some *Trichoderma* species were found to have importance for biofuel industries and plant protection against pathogens by the use of mycoparasitism and/or antibiosis mechanisms [14,16,17]. In the years to follow, many strains of *Trichoderma* have been described as biocontrol agents [18]. Among *Trichoderma* species commercially available for agricultural use are *T. harzianum*, *T. virens*, *T. viride*, *T. asperellum* and *T. atroviride* [19].

The mechanism by which *Trichodema* spp. function as biocontrol agents is complex, and the mentioned biocontrol effect varies with the specie of *Trichoderma* and host plant involved in the interaction [18]. Clearly, environmental conditions (e.g., temperature, pH, salinity and nutrient availability) also influence the biocontrol mechanism [19].

Trichoderma spp. are considered as opportunistic and avirulent plant symbionts [20]. During interaction with host plants, *Trichoderma* spp. secrete several classes of chemical molecules (e.g., proteins, peptides, oligosaccharides and antibiotics) [10,21]. Some of these compounds may act as hormones that stimulate plant growth and development, or can also act as elicitors, activating defense responses in the host plant [22].

The activation of defense induced by *Trichoderma* spp. not only reduces plant diseases. It has also been proved that *Trichoderma* spp. application to the plant increases the tolerance to abiotic stress, such as drought [23–25], low temperatures [24,26], salinity [27,28], and can be used to reduce the presence of toxic metals [29,30]. This wide range of beneficial traits to their hosts is due to bio-priming, and is attributed to the induction of long plant basal resistance that improves the defensive capacity of the plants for subsequent stresses [31]. The application of bio-priming agents prepares the plant for a faster and more effectively response against future stresses [32].

Due to the ability of *Trichoderma* spp. to rapidly produce spores and antibiotic compounds, these fungi have been used for the massive production of commercial formulations that can be stored by months maintaining the beneficial effect for the crop [33]. The most widely used *Trichoderma* spp. products are formulated in a wettable powder or granules [19]. Ninety percent of various *Trichoderma* strains are applied to crops, within many horticulture species (e.g., Poaceae, Solanaceae and Cucurbitaceae) specially for the control of plant diseases due to the antagonistic characteristic against phytopathogens (see [34] for review).

The long-lasting dialogue established between plants and *Trichoderma* is one of the major gaps in the understanding of how this relationship works. In this review, we will focus on the plant signal elements underlying the priming function of *Trichoderma* spp. that may trigger plant adaptation to stress conditions.

2. Defense Responses at Early Stages of Plant–*Trichoderma* Interaction

Little is known about the plant host mechanisms that connect the perception of *Trichoderma* root colonization to the downstream signaling pathways leading to activation of defense and developmental responses [35]. It is assumed that plant defense triggered by *Trichoderma* spp. is initiated by the perception of microbial-associated molecular patterns (MAMPs) by pattern recognition receptors (PPRs), which are localized on the surface of plant cells [36]. This first phase defense induction is called MAMP-triggered immunity (MTI) [37]. MTI activated by *Trichoderma* spp. includes defense responses such as oxidative burst, callose deposition, Ca^{2+} and reactive oxygen species (ROS) signaling as well as the induction of phytoalexins and other secondary metabolites because, at that point, the plant does not recognize that it is a friendly attack [35,38,39].

2.1. Heterotrimeric G Proteins in Trichoderma Recognition

G proteins are membrane-associated, heterotrimeric, and composed of subunits α, β and γ. When GDP is bound, the subunit α associates with the βγ dimer to form an inactive heterotrimer that binds to a G-protein-coupled receptor (GPCR) [40]. When a GPCR detects an extracellular signal, α subunit decreases the GDP affinity and the leaving GDP is replaced with GTP. Once GTP is bound, the α subunit is activated and dissociated both from the GPCR and from βγ dimer [40]. Following activation, both the GTP-bound α subunit and the free βγ complex can bind to downstream effector molecules and mediate a variety of responses in the target cell, including adaptations to environmental and biotic stresses [41,42]. There is one report about the involvement of plant G-proteins after inoculation with *Trichoderma*. Pea roots inoculated only with *T. asperellum* showed a transcript accumulation of the Gα1 subunit of the heterotrimeric G protein [43]. This suggests G-proteins play an important role in the *Trichoderma* recognition by the plant and suggests that the Gα1 subunit (in its active form), could activate downstream signaling elements. Among the roles of Gα1 signaling, activation of plant plasma membrane Ca^{2+} channels and ROS accumulation have also been widely reported [44–46] (Figure 1).

2.2. Calcium Mediated Signalling in Trichoderma Bio-Priming

Calcium is a second messenger by which plants modulate signaling pathways to respond to a particular stress. The increase in intracellular calcium concentrations ($[Ca^{2+}]_i$) is one of the earliest signaling events when plants are challenged with biotic and abiotic stimulus [47,48]. Changes in $[Ca^{2+}]_i$ are commonly found during interaction between plants and beneficial microorganisms. This is the case for metabolites secreted by *T. atroviride* which increase $[Ca^{2+}]_i$ and defense responses in the first minutes after the treatment in soybean cells [49]. Also, the elicitor HYTOL1 (a hydrophobin abundantly secreted by *T. longibrachiatum* strain MK1 [50]) may be involved in adhesion of fungal hyphae to the root surface [51], inducing a transient increase of cytosolic Ca^{2+} in *Lotus japonicus* cells [52]. These results indicate that the induction of intracellular Ca^{2+} changes represents an early step during *Trichoderma*–plant interaction that primes defense mechanisms (Figure 1).

2.3. Early ROS Accumulation

One of the earliest responses during the plant defense strategy is a fast and transient production of intracellular ROS [53]. Plasma membrane NADPH oxidases, known as respiratory burst oxidase homologues (RBOHs), are one of the many sources of ROS that have been implicated in several essential processes in plants [54]. Growing lines of evidence from plants suggest the involvement of NADPH oxidase-generated oxidative burst in extracellular signaling to regulate a wide range of physiological functions in plants [55,56].

A networking between cytosolic concentrations of Ca^{2+} and RBOH-mediated ROS production has been shown in several studies [57–59]. Plant–*Trichoderma* systems have also demonstrated that these fungi or their metabolites can trigger transient increases in ROS and calcium levels in the first minutes of interaction, activating enhanced immune defense [49,60]. Additionally, tight connections of NADPH oxidases and mitogen-activated protein kinases (MAPKs) are recognized to regulate various biological processes, wherein NADPH oxidase-originated oxidative burst can act upstream to activate the MAPKs cascade [61]. It has been demonstrated that association of *T. viride* Tv-1511 and peppermint plants produces the activation of a MAPK cascade via NADPH oxidase [61]. All these findings suggest that NADPH oxidase-dependent ROS production plays vital roles in the root colonization (Figure 1).

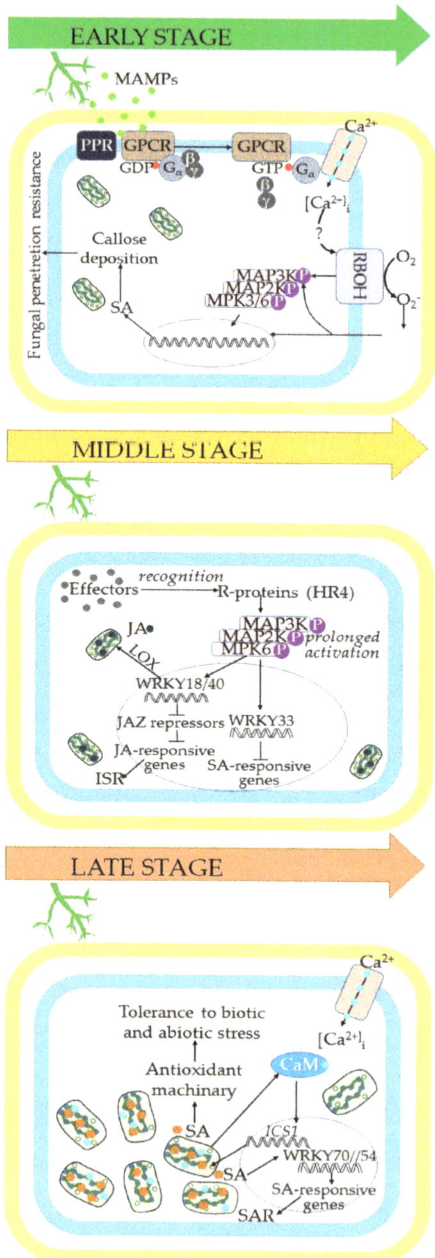

Figure 1. Mechanism of plant responses according to the time of interaction with *Trichoderma*. The model is divided into three stages. The earlier stage comprises the first hours of interaction, wherein the plant is avoiding fungal root colonization due to SA phytohormone and consequently the callose deposition. This first stage is initiated by the recognition of MAMPs secreted by *Trichoderma*, which

can trigger early defense responses mediated by Ca^{2+} and reactive oxygen species and by a rapid but transient activation of MAPK cascades through G heterotrimeric proteins. In the second stage, *Trichoderma* effectors are recognized by R-proteins to promote JA signaling by sustained MAPK activation, and to suppress SA signaling. Consequently, it is established a beneficial interaction. In the later stage, a second peak in the amount of SA is observed, which may induce antioxidative enzyme activities to reduce the oxidative damage to biomolecules and cells.

2.4. Salicylic Acid Restricts Trichoderma Invasion of Vascular System

The interactions between *Trichoderma* spp. and plant roots involve recognition, attachment, penetration, colonization and nutrient transfer [62]. It is well known that *Trichoderma* spp. grow on the outer layer of the roots of the plants [63,64].

During root colonization of *Trichoderma* spp., salicylic acid (SA) seems to be involved in preventing this fungus from entering the vascular system of the roots as well as in avoiding detrimental effects on plant growth and development of the host plants [65]. SA plays a key role in plant cell wall reinforcement (via callose synthesis) responsible for the limitation of *Trichoderma* colonization to the outer layers of roots [65]. Endogenous increase in SA levels has been reported in tomato plants inoculated with *T. virens* and *T. harzianum* T22 at 24 and 48 hpi, respectively [66,67]. The temporary induction of SA confirms a possible role in avoiding excessive *Trichoderma* penetration within the roots [65] and underlines the importance of SA in the first steps of the *Trichoderma*–plant interaction (Figure 1). It has also been shown that *T. atroviride* and *T. cremeum* induce changes in the composition of wheat seedlings roots [68]. These species promote lignin deposition and rearrangements of pectins after 14 days of incubation with *Trichoderma* spp., suggesting that modifications of wheat seedlings roots can be used as a tool against to pathogens [68].

3. Induction of Systemic Plant Defense by *Trichoderma* spp. Plays Key Role in the Crosstalk between Biotic and Abiotic Stress Responses

After MTI, *Trichoderma* spp. seem to activate a second layer of defense. In this stage, effectors secreted by fungi species prevent plant recognition and activate the plant systemic resistance to biotic and abiotic stress [36]. The second line of plant defense induction is called effector-triggered immunity (ETI), which is activated by plant resistance protein (R) and it is frequently associated with hypersensitive response (HR) [37]. Despite ETI and PTI involving a similar set of downstream defense responses, including calcium-mediated signaling, activation of MAPK cascades, production of ROS, transcriptional reprogramming, and biosynthesis of antimicrobial compounds [69–72], the responses during ETI have a longer duration and higher magnitude [73].

Induced resistance is the term used for the induced state of resistance in plants triggered by a biological or chemical inducer. This protects nonexposed plant parts from stresses [74]. Systemic acquired resistance (SAR) and induced systemic resistance (ISR) are two types of induced resistance wherein plant defenses are preconditioned by prior infection or treatment that results in resistance against subsequent challenge by a pathogen or parasite [75]. Plants, in response to virulent, avirulent and nonpathogenic microbes, elicit SAR. For the activation of SAR, the molecule SA and the accumulation of PR proteins are required. In contrast, ISR is triggered by the infection of pathogens, response to insects, herbivores, or upon root colonization by beneficial microbes in the rhizosphere (such as *Pseudomonas* spp., *Bacillus* spp. and *Trichoderma* spp.). Typically ISR is regulated by jasmonic acid and ethylene (JA/ET) [75,76], in some particular cases, ISR can requiere SA accumulation [77].

The first evidence of TISR was published in 1997 by Bigirimana et al. [78], who demonstrated that soil treated with *T. harzianum* made the leaves of bean plants resistant to diseases that are caused by the fungal pathogens *Botrytis cinerea* and *Colletotrichum lindemuthianum*, even though *T. harzianum* was present only on the roots and not on the foliage. Similar results have been reported for a wide range of host plants with different strains and species of *Trichoderma* and various classes of plant pathogen including fungi, bacteria, viruses and nematodes [39,78–80].

3.1. *Trichoderma* spp. *Induce a Prolonged Activation of Plant MAPK Cascades*

Mitogen-activated protein kinase (MAPK) cascades are well conserved signaling proteins in all eukaryotes [81]. Each cascade is minimally constituted of three proteins that are sequentially activated: a MAP kinase kinase kinase (MAPKKK or MAP3K), a MAP kinase kinase (MAPKK or MAP2K) and a MAP kinase (MAPK or MPK) [81,82].

MAPKs are intracellular proteins that can be activated by various stimuli [81]. MAPKs cascades transduce extracellular signals to cellular responses, including the biosynthesis of phytohormones, ROS generation, changes in gene expression, among others [83]. Activation of MAPK cascades is one of the earliest signaling events after plant sensing of PAMPs/MAMPs [84–87]; however sustained activation of MAPK confers a robust innate immunity [73,88]. *Arabidopsis thaliana* MPK3 and MPK6, as well as their orthologs in other species, such as tobacco SA-induced protein kinase (SIPK) and wounding-induced protein kinase (WIPK), are involved in plant responses to biotic and abiotic stresses [84,86,87,89,90]. Some studies have found the activation of MPKs associated with plant defense during plant–*Trichoderma* interactions [35,91–95]. For instance, xylanase, an elicitor from the cell walls of *T. virens* (TvX), induces the slow and prolonged activation of SIPK in tobacco [91]. Similarly, inoculation with *T. atroviride* (a specie known to promote root growth by producing auxine-like compounds [94]) in *Arabidopsis* roots induces the MPK6 activation [95]. Since the modulation of MPK6 is also responsive to auxin-like compounds, it has been suggested that *T. atroviride* alters root-system architecture modulating MPK6 and auxin action [95]. In addition, the activation of an analog of *Arabidopsis* MPK6 in peppermint by *T. viridae* is related with the modulation of essential oil metabolism at the transcriptional level and for enzymatic activation [61]. Interestingly enough, menthol, which is the main terpenoid of peppermint oil, exhibits potential abilities as plant defense potentiator in agriculture and horticulture [92].

Besides activating MPKs through posttranslational modifications, bio-priming with *Trichoderma* spp. induce expression of plant *MPK* genes. For example, the inoculation of cucumber (*Cucumis sativus*) roots with *T. asperellum* leads to a long-term expression of a *Trichoderma*-induced MPK (*TIPK*) gene, which is an ortholog of *WIPK* and *MPK3* [93], while the elicitor HYTOL1 also up-regulates the early and transient expression of *MPK3* in *L. japonicus* [52]. After inoculation in *Arabidopsis*, *T. hamatum* induces the expression of *MPK3* after 48 h [96] and *T. asperelloides* of *MPK11* at 24 h [35]. It is noteworthy that this last MPK is also responsive to PAMPs/MAMPs [97,98].

It is known that sustained activation of MPK3/6 elicits a massive reprogramming of the defense metabolome, with an accumulation of camalexin and indole glucosinolate derivatives [99]. The activation of both MPKs is also accompanied by many defense-related phytohormones such as SA, JA, and ET [99], suggesting that extended MPK activation could be involved in the modulation of the robustness of the immune signaling during plant–*Trichoderma* interactions (Figure 1).

3.2. *Hormone Signalling Pathways Involved in Systemic Resistance induced by* Trichoderma spp.

Plant hormones play a crucial role in the immune signaling networks in response to pathogens and beneficial microbes [100]. Among the most relevant hormones related to the modulation of defense responses are SA, JA, ET and abscisic acid (ABA), however auxin, gibberellic acid (GA), cytokinin (CK), brassinosteroids and peptide hormones could also be implicated in plant defense signaling pathways [101].

Several studies have shown that *Trichoderma* species induce the production of phytohormones in the host plants such as JA, SA and ET. Since *Trichoderma* spp. can also produce small amounts of phytohormones such as auxins, GA, SA and ABA [102–105], it makes difficult to discern the origin of hormones detected in some plant–*Trichoderma* spp.

interactions. The role of specific hormones during plant–*Trichoderma* interaction seems to be dependent on the experimental condition and organisms involved [66,96,106].

3.2.1. Salicylic Acid

SA plays a key role in plant defense against biotrophic pathogens [107]. The accumulated evidence shows that partial suppression of SA-dependent responses in plants is necessary for the occurrence of the symbiotic association between beneficial microbes and plants [108–112], including *Trichoderma* spp. [35,113,114]. In this regard, evidence shows a down-regulation of *PR-1,* a useful marker for the SAR response, in the first hours of various plant–*Trichoderma* spp. interactions [52,115].

SAR is a long lasting defense modulated by SA. Recently, it has been found that systemic resistance in maize plants primed with *T. atroviride* at seedling stage is detected until two months later with an increase of SA levels, suggesting SA is a key component of a regulatory network controlling the immunity of silks during systemic resistance [116]. Similarly, *Arabidopsis* seedlings exposed to *T. asperellum* Ism T5 volatile for 9 days, stimulate SA accumulation [117]. In recent years, experimental studies have found that application of exogenous SA induces biotic and abiotic stress [118–122], possibly by modulating antioxidative enzyme activities, thereby potentially reducing the damaging levels of ROS [120,123,124]. It is thus possible that alleviation of biotic or abiotic stress observed in plant–*Trichoderma* systems would involve SA in later stages of the interactions (Figure 1).

3.2.2. Jasmonic Acid

JA is synthesized from the α-linolenic acid of chloroplast membranes by the octadecanoid pathway. JA is a phytohormone involved in diverse physiological processes including plant growth and development [125], and also actively participates in the mediation of plant responses and defenses against herbivore attack, pathogen infection and abiotic stresses, including ozone, ultraviolet radiation, high temperatures, and freezing [125–128].

Multiple reports have confirmed that *Trichoderma* spp. can increase the levels of JA in host plants. For instance, during the interaction between tomato plants with *T. virens*, an increase in endogenous JA levels at 24 hpi has been observed [67]. Similarly to the content of SA, JA significantly increase in *T. longibrachiatum* H9-inoculated cucumber plants at 96 hpi [114], and in A*rabidopsis* co-cultivated with both *T. virens* and *T. atroviride* 8 days after interaction [129], implying that SA and JA play important roles in regulating the plant response and enhancing plant defense in plants (Figure 1).

3.3. *Induction of Plant Defense Gene Expression in Response to Trichoderma* spp.

Reprogramming of a cell in response to the perception of an external stimulus involves complex changes in gene expression. The expression of genes appears to be regulated by intracellular signal transduction pathways. For instance, the interaction of plants with a variety of microorganisms results in changes in the level of SA, JA and ET, which are positive regulators of transcription factors (WRKYs), defense genes (PRs), and receptors (*R* genes) [130].

To link particular pathways with actual defense responses, some molecular tools, such as qPCR, allow the use of the expression of several marker genes as indicators of the activation of specific pathways [9]. Expression studies on defense/stress-related genes suggested that *Trichoderma*-induced systemic resistance (TISR) might involve both SA- and JA-related pathways.

Comparing plants treated with *Trichoderma* spp. with mock-treated controls, hundreds of genes that are differentially expressed during ISR-prime have been identified [35,66,96] (Tables 1–3). The products of the genes are related to defense responses, signal transduction, systemic acquired resistance, antioxidant systems, programmed cell death, etc. [96]. It is difficult to establish a specific time point when early defense responses end, but it has been proposed that 48 hpi would indicate the moment of transition when the plant reprograms its transcriptional machinery mainly towards redox and defense processes, fully accepting that *Trichoderma* is not an enemy [38,66].

3.3.1. WRKY Transcription Factors

WRKY is a family of transcription factors found exclusively in plants [131]. They bind W-box and/or other *cis*-elements located in the promoter of their target genes [131]. Most *WRKY* genes are responsive to pathogens, elicitors, and defense-related phytohormones such as SA or JA, implying a major role for the *WRKY* gene family in plant immunity [132], but also, the WRKY transcription factors play an important role in the alleviation of abiotic stresses [131,133].

The WRKY proteins regulate the gene expression directly or indirectly by modulating the downstream target genes, by activating or repressing the other genes (encoding transcription factors) or by self-regulating their own expression [131].

Molecular studies have revealed that *Arabidopsis* plants under interaction with *T. atroviride* induces the expression of *WRKY8*, *WRKY33*, *WRKY38*, *WRKY42* and *WRKY60*, all of which are considered as positive regulators in JA pathway, while *WRKY70* and *WRKY54*, regulated by the SA pathway, could be activated at later stages of the interaction, when the fungus is fully established in the plant roots [134]. Similarly, the treatment of *L. japonicus* with hydrophobin HYTOL from *T. longibrachiatum*, or the inoculation of the common bean (*Phaseolus vulgaris* L.) with *T. velutinum*, lead to the expression of *WRKY33*, but not *PR-1* [52,115], suggesting that expression of *WRKY33* induced by *Trichoderma* spp. negatively regulates the SA pathway to evade the plant immunity and to establish a prolonged mutualistic association (Table 1).

On the other hand, the expression of *WRKY18*, *WRKY40* and *WRKY60* transcription factors genes in *Arabidopsis* inoculated with *T. asperelloides* is observed as early as 9 h. The three WRKY show redundant function in negatively regulating PTI in *Arabidopsis* [135]. In response to *T. asperelloides*, these transcription factors negatively regulate the induction of transcript levels of SA marker genes *FMO1*, *PAD3* and *CYP71A13*, but positively regulate the expression of *LOX2* and *AOS* related to the JA pathway through inhibition of expression of the jasmonate ZIM domain (JAZ) repressors (Figure 1). Because FMO1 negatively regulates root colonization, *WRKY18* and *WRKY40* could negatively regulate *FMO1* to allow a moderate level of colonization [35].

3.3.2. PR Proteins

Pathogenesis-related proteins (PRs) are a structurally diverse group of plant proteins that are induced by various types of pathogens. They are widely distributed in host plants in trace amounts, but are produced in much higher concentration following pathogen attack or stress conditions [136]. PR proteins impede pathogen invasion but also helps in growth and metabolism of the host plants. The PR proteins are grouped according to their properties and functions, and include β-1,3-glucanases, endochitinases, proteinases, proteinase inhibitors, peroxidases, RNases, inhibitors of pathogen hydrolases, and others [137]. Chitinases and β-1,3-glucanases are the major hydrolytic enzymes abundant in plants after fungal pathogen infection [138]. An earlier report showed that cucumber roots induced the activity of peroxidase, β-1,3-glucanase and chitinase, which are apparently of plant origin, 72 h post-inoculation with *T. harzianum* [13], suggesting that *Trichoderma* association could reduce disease through activation of both enzymes by hydrolyzing the main constituents of the structural barrier of pathogenic cell wall fungi.

Likewise, induction of *PR* gene expression is also essential for the development of induced resistance and can require the molecules SA or JA/ET. In *Arabidopsis PR-1* that inhibits fungal growth, *PR2* also called β-1,3-glucanase and *PR-5* are considered to be markers for SAR, while *PR-3* (chitinase), *PR-4* (chitinase) and *PR-12* (plant defensin) are used as markers for JA pathway. Transcriptomic analyses have shown the expression of *PR* genes in response to *Trichoderma* spp. (Table 2). The rhizosphere colonization by *Trichoderma* spp. can support the transcription of some defense-related genes for a relatively long period [139,140]. This effect is particularly strong for those inducible by SA (Table 2), suggesting that the long-term response to *Trichoderma* in plants may involve SA signaling.

Table 1. Expression of *WRKY* genes up-regulated by *Trichoderma* species.

Signaling Pathways Related	Gene	Host Plant (Full Name in the Legend)	*Trichoderma* Specie or Elicitor	Time after Inoculation	Reference
JA/abiotic stress	*WRKY33*	*A. thaliana*	*T. atroviride*	96–144 h	[134]
			T. asperelloides T203	9–24 h	[35]
		L. japonicus	Hydrophobin HYTOL from *T. longibrachiatum*	2 h	[52]
		P. vulgaris	*T. velutinum* T028	45 days	[115]
		S. lycopersicum	*T. erinaceum*	24–48 h	[141]
JA/ET	*WRKY8*	*A. thaliana*	*T. atroviride*	24–48 h	[134]
			T. asperelloides T203	24–48 h	[35]
	WRKY38	*A. thaliana*	*T. atroviride*	96 h	[134]
	WRKY42			96–144 h	
	WRKY60			72–144 h	
	WRKY41	*A. thaliana*	*T. asperelloides* T203	9–24 h	[35]
	WRKY53			24 h	
	WRKY55			24 h	
	WRKY18			9–24 h	
	WRKY60			9–24 h	
	WRKY40			9–48 h	
	WRKY1	*V. vinifera*	*T. harzianum* T39	4 days	[142]
	WRKY-C10 (*WRKY transcription factor 6*)	*V. vinifera*	*T. harzianum* T39	4 days	[142]
Negatively regulated by JA/ET. Represses plant basal defense mechanisms	*WRKY48*	*A. thaliana*	*T. asperelloides* T203	9–24 h	[35]
SA	*WRKY30*	*A. thaliana*	*T. asperelloides* T203	9 h	[35]
	WRKY54			9 h	
	WRKY15			9–24 h	
	WRKY46			9–24 h	
	WRKY70			48 h	
	WRKY54	*A. thaliana*	*T. atroviride*	144 h	[134]
	WRKY70 *			144 h	
Involved in plant defense	*WRKY37*	*S. lycopersicum*	*T. erinaceum*	24–48 h	[141]

WRKY70 is an Arabidopsis gene that is upregulated by two different strains of Trichoderma: *T. asperelloides* and *T. atroviride*. *Arabidopsis thaliana*, *Lotus japonicus*, *Phaseolus vulgaris*, *Solanum lycopersicum*, *Vitis vinifera*.

Table 2. Induction of *Pathogenesis Related (PR)* genes expression by *Trichoderma* species.

Marker for	Gene	Protein Function	Host Plant (Full Name in the Legend)	*Trichoderma* Specie	Time after Inoculation	References
JA/ET	*PR-3*	Chitinase Class 1. Hydrolytic enzymes that disrupt mycelial cell wall Antifungal properties	*A. thaliana*	*T. asperelloides* T203	24 h	[35]
			O. sativa	*T. harzianum; T.erinaceum; T. atriviride; T. hebeiensis; T. parareesei; T. longibrachiatum; T. resei*	NR *	[8]
			S. lycopersicum	*T. erinaceum*	24–48 h	[141]
	Acidic endochitinase 3 (Chit3)	Chitinases	*V. vinifera*	*T. harzianum* T39	4 days	[142]
	PR-4	Basic Chitinases	*A. thaliana*	*T. asperelloides* T203	24–48 h	[35]
	PR-P2	It is a pathogenesis related 4 (*PR4*) gene	*S. lycopersicum*	*T. atroviride; T. harzianum*	2 months	[139]
	PDF1	Plant defensin. Membrane permeabilizing functions.	*A. thaliana*	*T. asperelloides* T203	24 h	[35]
	PDF1.2				24 h	[35]
	PDF1.2c				24 h	[35]
	PDF1.2		*S. lycopersicum* cv. Oogata-fukuju	*T. virens*	4–24 h	[67]
	Defensin		*O. sativa*	*T. harzianum; T.erinaceum; T. atriviride; T. hebeiensis; T. parareesei; T. longibrachiatum; T. resei*	NR *	[8]
SA	*PR-1*	Antimicrobial function and defense signal amplification.	*A. thaliana*	*T. virens; T. atroviride;*	6–8 days	[129]
				T. hamatum T382	48–72 h	[96]
			S. lycopersicum	*T. atroviride; T. harzianum*	2 months	[139]
			S. lycopersicum cv. Oogata-fukuju	*T. virens*	4–24 h	[67]
	PR-2	Beta-1,3-endoglucanase. Hydrolytic enzymes that disrupt mycelial cell wall	*A. thaliana*	*T. hamatum* T382	48–72 h	[96]
			S. lycopersicum	*T. erinaceum*	24–48 h	[141]
	β-1,4-glucanase	Hydrolytic enzyme that disrupts mycelial cell wall	*C. sativus*	*T. asperellum*	48 h	[143]
	PR-5	Osmotins. Membrane permeabilizing proteins.	*A. thaliana*	*T. hamatum* T382	48–72 h	[96]
				T. asperelloides T203	24 h	[35]
			S. lycopersicum	*T. hamatum*	5 weeks	[140]
	OSM2	*Trichoderma*-induced osmotin 2	*V. vinifera*	*T. harzianum* T39	4 days	[142]

* NR = Not reported. *Arabidopsis thaliana, Oryza sativa, Solanum lycopersicum, Vitis vinifera, Cucumis sativus.*

3.3.3. Other Defense Gene Markers

The expression of *PR* genes can be transitory, but strongly potentiates the expression of defense-related proteins when plants are affected with biotic stress. Proteins encoded by resistance genes (*R*) are found among them. The R proteins recognize effectors from beneficial and pathogenic microorganisms to activate a stronger defense. The *HR4* gene that codifies an R protein is induced 96 h after of the *Arabidopsis–T. atroviride* interaction, suggesting the fungus is activating the recognition system and promoting a beneficial interaction establishment in the plant [130], however, little is known about *R* genes in beneficial interactions.

Additionally, there is evidence of a link between the accumulation of the phytohormones and changes in the expression of marker genes, which have been identified by analysing their expression patterns after exogenous application of single or combined phytohormone solutions [144]. The evidence has demonstrated that *Trichoderma* spp. can simultaneously or separately induce ISR and SAR associated with the biosytnthesis of SA, JA and ET according with the induction of the expression of specific resistance marker genes, which are summarized in Table 3.

Table 3. Expression of gene markers positively regulated by *Trichoderma* species.

Marker for	Gene	Protein Function	Host Plant (Full Name in the Legend)	*Trichoderma* Specie	Time after Inoculation	Reference
JA/ET	*Lox1*	Lipoxygenase enzyme involved in JA synthesis	*C. sativus*	*T. asperellum*	24 h	[61]
			A. thaliana	*T. harzanium*	72 h	[65]
			A. thaliana	*T. asperelloides* T203	24 h	[35]
			S. lycopersicum	*T. parareesei*	6 days	[145]
	Lox2		*A. thaliana*	*T. virens*, *T. atroviride*	8 days	[129]
	Lox3		*A. thaliana*	*T. asperelloides* T203	24 h	[35]
	Lox4		*A. thaliana*	*T. asperelloides* T203	24 h	[35]
	LoxA		*S. lycopersicum.*	*T. atroviride*, *T. harzianum*	2 months	[139]
	HPL	Hydroperoxide lyase	*C. sativus*	*T. asperellum*	24–48 h	[146]
	hGS	Homoglutathione synthetase related with oxidative stress	*P. vulgaris*	*T. velutinum* T028	45 days	[115]
ET	*CTR1*	Ethylene signal-associated serine/threonine protein kinase	*C. sativus*	*T. asperellum*	24 h	[143]
	ETR1					
	EIN2	Key component in ethylene signaling	*A. thaliana*	*T. asperelloides* T203	48 h	[35]
	EIN4					
	ERF-A2	Ethylene-responsive transcription factor	*S. lycopersicum*	*T. parareesei*, *T. asperellum*, *T. harzianum*	4 weeks	[147]
	CH5b	Endochitinase precursor related to ethylene signaling	*P. vulgaris*	*T. velutinum* T028	45 days	[115]
SA	*PAL1*	Phenylalanine and histidene ammonia-lyase. Enzyme involved in the production of antimicrobial compounds	*C. sativus*	*T. asperellum*	24 h	[143,146]
			A. thaliana	*T. asperelloides* T203	9–24 h	[35]
			O. sativa	*T. harzianum*, *T.erinaceum*, *T. atriviride*, *T. hebeiensis*, *T. parareesei*, *T. longibrachiatum*, *T. resei*	NR *	[8]
	PAL2		*A. thaliana*	*T. asperelloides* T203	24 h	[35]
	ICS1	Isochorismate synthase is involved in SA biosynthesis	*A. thaliana*	*T. harzanium*	72 h	[65]
	Cals	Callose synthase, involved in callose biosynthesis	*A. thaliana*	*T. harzanium*	72 h	[65]

* NR = Not reported. *Cucumis sativus, Arabidopsis thaliana, Solanum lycopersicum, Phaseolus vulgaris, Oryza sativa.*

3.4. Induction of Antioxidant Enzyme Activity Is Modulated by *Trichoderma* spp.

As noted before, one of the common responses under stress conditions is the generation of ROS. Overproduction of ROS could result in damage to macromolecules such as lipids, proteins and DNA, via oxidation, and in severe cases, leads to cell death. So it is crucial to overcome these effects either by enhancing the intrinsic antioxidant defense or by repairing the damage [148].

Stress-induced ROS accumulation can be counteracted by plant antioxidative defense that consist of enzymatic or nonenzymatic systems. Superoxide dismutase (SOD), ascorbate peroxidase (APX), catalase (CAT), peroxidases (POD) and glutathione peroxidase (GPX) are the main enzymatic scavengers of superoxide (O_2^-) and hydrogen peroxide (H_2O_2) [149], while glutathione (GSH) and ascorbic acid (ASA) are the major non-enzymatic antioxidants that, among other vital functions, maintain cellular redox homeostasis [150]. Keeping ASA and GSH in reduced form is critical for redox homeostasis and cellular vitality [151]. The activity of the enzymes that regenerate these molecules is correlated with resistance to abiotic stresses. These enzymes include glutathione reductase (GR) (which regenerates oxidized GSH), monodehydroascorbate reductase (MDHAR) and dehydroascorbate reductase (DHAR), which regenerate ASA from monodehydroascorbate (MDHA) and dehydroascorbate (DHA) [149].

Recent literature has revealed that *Trichoderma* spp. reduce the negative effects of plants stressed with biotic and abiotic stimuli through the modulation of the ROS by inducing antioxidant enzymes [24,152,153]. For instance, in the presence of *T. harzianum* T22, the ratios of reduced to oxidized forms of the molecules for ascorbate and glutathione, and the activity of SOD, APX, MDHAR, DHAR and GR in tomato seedlings are higher than non-inoculated plants. This indicates that *T. harzianum* T22 enhances systems of ROS scavenging and redox maintenance [151]. Also, *T. erinaceum* bioprimed tomato plants increase the activities of SOD and CAT compared to a control, and T. hamatum enhances the activity of enzymes CAT, POD, APX, GR and SOD in Ocharenus baccatus [154]. Similarly, the inoculation of maize and rice seeds as well as wheat seedlings with *T. harzianum* or its metabolites extracts increases SOD and CAT antioxidant enzymes activity [155,156]. This demonstrates that the pre-treatment of biocontrol Trichoderma results in increased activities of the antioxidant enzymatic pool [141].

Trichoderma strains also increase the activity of antioxidative defense through enhanced expression of genes encoding the component enzymes [148]. Transcriptional reprogramming of the oxidative stress response may also influence *Trichoderma* spp. bio-priming to overcome oxidative damage in stressed plants. Some examples of overexpression of antioxidant-related genes induced by *Trichoderma* spp. are summarized in Table 4.

Table 4. Induction of expression of plant antioxidant genes by Trichoderma species.

Gene	Host Plant (Full Name in the Legend)	*Trichoderma* Specie or Elicitor	Time after Inoculation	Reference
CAT	*C. sativus*	*T. asperelloides* T203	24 h	[35]
CAT	*O. sativa*	*T. harzianum; T.erinaceum; T. atriviride; T. hebeiensis; T. parareesei; T. longibrachiatum; T. resei*	* NR	[8]
CAT	*T. aestivum* cv.'Yongliang 4	*T. longibrachiatum T6*	* NR	[157]
GPX	*S. lycopersicum*	*T. erinaceum*	24–48 h	[141]
POD	*T. aestivum* cv.'Yongliang 4	*T. longibrachiatum T6*	* NR	[157]
POD	*O. sativa*	*T. harzianum; T.erinaceum; T. atriviride; T. hebeiensis; T. parareesei; T. longibrachiatum; T. resei*	* NR	[8]
SOD	*O. sativa*	*T. harzianum; T.erinaceum; T. atriviride; T. hebeiensis; T. parareesei; T. longibrachiatum; T. resei*	* NR	[8]
SOD	*S. lycopersicum*	*T. erinaceum*	24–48 h	[141]
SOD	*T. aestivum* cv.'Yongliang 4	*T. longibrachiatum T6*	* NR	[157]
SOD (Mn)	*C. sativus*	*T. asperelloides* T203	24 h	[35]
SOD (Cu)	*C. sativus*	*T. asperelloides* T203	24 h	[35]

* NR = Not reported. *Cucumis sativus, Oryza sativa, Triticum aestivum, Solanum lycopersicum.*

SA has been widely recognized as a promoter of antioxidant defense, including CAT, SOD, and APX, as well as non-enzymatic antioxidants, to alleviate oxidative stress in plants [124,158–160], so the late increase of endogenous SA observed in bioprimed plants with *Trichoderma* spp., might be responsible for the antioxidant enzymatic mechanism pathway improving the performance of plants under stress conditions (Figure 1). Thus, growing evidence suggests that application of strains of *Trichoderma* spp. may be an ecological strategy to help plants to recover from biotic and abiotic stress-induced oxidative damage to continue the metabolic and physiological activities in a better way.

3.5. Effects of Trichoderma on Chloroplasts

Chloroplasts are key organelles of the higher plants in which photosynthesis takes place. The chloroplasts are also the major production site of defense molecules including hormones (such as SA, JA, ABA) and secondary messengers like Ca^{2+} and ROS [161].

The effect of Trichoderma interaction on chloroplast has been poorly explored. Recently, it was observed that T. asperellum and *T. harzianum* consortium at 10^8 CFU/mL concentration increased the number and size of chloroplasts in spongy parenchyma of Passiflora caerulea after 60 days [162]. These chloroplasts also showed a reduction of starch grains, which could be related to starch degradation and the translocation of monosaccharides from chloroplasts to the rest of the cell and/or to the phloem [162] (Figure 1).

Additionally, it has been proved that some Trichoderma strains enhance photosynthetic capacity compared to uninoculated controls (see [163] for review) by increasing the photosynthetic pigment content or the expression of genes regulating the biosynthesis of chlorophyll, proteins of the light-harvesting complex, or components of the Calvin cycle [164]. Chloroplasts are considered as sensors and regulators of plant responses to biotic and abiotic stresses [165]. When plants are exposed to stress, they usually lose their photosynthetic capability by an overproduction of ROS formed during excitation of chlorophyll in photosynthesis, causing an oxidative stress in chloroplasts [166]. However, it has been shown that plants inoculated with certain strains of Trichoderma and then challenged by a stress overcome the reduction of photosynthetic capability [26,39,148,167]. This might be due to the protection against ROS levels described previously, but also to the increase in the content of carotenoids detected in the interaction of some plants with *Trichoderma* spp. [164,168–171], since carotenoid pigments act as antioxidants that quench singlet oxygen and trap peroxyl radicals [172].

Since chloroplasts produce ROS during cellular stress and ROS act as promoters of programmed cell death (PCD), *Trichoderma* spp. may be preventing cell death in plants subsequently exposed to stress. Moscatiello et al. [52] demonstrated that despite the fact that HYTOL induced the expression of defense genes, it did not affect cell viability and ultrastructure of L. japonicus cells after treatment. However, other studies have reported some markers of PCD (e.g., caspase 3-like caspase protease activity and by chromatin condensation) in soybean and tobacco cells treated with metabolite mixtures from T. atroviride or xylanase, respectively [49,173].

4. Conclusions and Future Perspectives

The negative consequences of climate change on living organisms and the environment are already forcing us to search for alternative ways of reducing these catastrophic events. Eco-friendly practices for food production have been highlighted to achieve sustainability. In horticultural crops, plant biostimulants have been proposed as agronomic tools to mitigate environmental/abiotic stress effects. However, since our knowledge about the mechanism involved during plant–biostimulant interaction is currently limited, more research is needed to understand exactly what is taking place during interactions. The elucidation of the mechanisms of action will allow us to develop new methods that involve beneficial microorganisms with better performance for the solution of agricultural problems.

Trichoderma spp. induce multiple beneficial effects on plants by reducing the severity of diseases, but also by alleviating abiotic stress-induced damage in plants. These promising results are opening the door for sustainable agriculture to exploit the potential of Trichoderma in a safe way for crop plants, agroecosystems, and humans.

Further research into the molecular bases of dialogue in plant–Trichoderma interactions should predict the impact of certain species of this genus on crops or cultivars performance to ensure their effective use.

Author Contributions: Conceptualization, M.S.-G.; investigation, J.P.-T., E.A.-T. and M.I.R.-S.; writing—original draft preparation, M.S.-G. and J.P.-T.; writing—review and editing, M.S.-G., M.I.-P., S.R.P.-R. and A.M.-N.; supervision, M.S.-G. and A.M.-N. All authors have read and agreed to the published version of the manuscript.

Funding: This research received no external funding.

Institutional Review Board Statement: Not applicable.

Informed Consent Statement: Not applicable.

Data Availability Statement: Not applicable.

Conflicts of Interest: The authors declare no conflict of interest.

Abbreviations

Abscisic acid (ABA); allene oxide synthase (AOS); ascorbate peroxidase (APX); catalase (CAT); cytochrome P450 family 71 polypeptide (CYP71A13); dehydroascorbate reductase (DHAR); ethylene (ET); effector-triggered immunity (ETI); flavin monoxygenase 1 (FMO1); gibberellic acid (GA); guanosine diphosphate (GDP); G-protein-coupled receptor (GPCR); glutathione peroxidase (GPX); glutathione reductase (GR); glutathione (GSH); heterotrimeric G-protein α (Gα1); hypersensitive response (HR); hydrophobin secreted by T. longibrachiatum strain MK1 (HYTOL1); induced systemic resistance (ISR); jasmonic acid (JA); jasmonate ZIM domain (JAZ); lipoxygenase 2(LOX2); microbe-associated molecular patterns (MAMPs); mitogen-activated protein kinases (MAPKs); monodehydroascorbate reductase (MDHAR); MAMP-triggered immunity (MTI); nicotinamide adenine dinucleotide phosphate (NADPH); phytoalexin deficient3 (PAD3); pathogen-associated molecular patterns (PAMPs); programmed cell death (PCD); plant-growth-promoting bacteria (PGPB); peroxidases (POD); pattern recognition receptors (PPRs); pathogenesis-related protein (PR); pattern-triggered immunity (PTI); resistance proteins (R); respiratory burst oxidase homologues (RBOHs); reactive oxygen species (ROS); salicylic acid (SA); systemic acquired resistance (SAR); tobacco SA-induced protein kinase (SIPK); superoxide dismutase (SOD); Trichoderma-induced MPK (TIPK); Trichoderma-induced systemic resistance (TISR); wounding-induced protein kinase (WIPK); transcription factors with the domain WRKYs (WRKY).

References

1. Verma, S.; Nizam, S.; Verma, P. Biotic and abiotic stress signaling in plants. In *Stress Signaling in Plants: Genomics and Proteomics Perspectives*; Sarwat, M., Ahmad, A., Abdin, M.Z., Eds.; Springer Science: New York, NY, USA, 2013; pp. 25–49.
2. Umar, O.B.; Ranti, L.A.; Abdulbaki, A.S.; Bola, A.L.; Abdulhamid, A.K.; Biola, M.R.; Victor, K.O. Stresses in plants: Biotic and abiotic. In *Current Trends in Wheat Research (Internet)*; Mahmood-ur-Rahman, A., Ed.; InterchOpen: London, UK, 2021. [CrossRef]
3. Gull, A.; Lone, A.A.; Wani, N.U.I. Biotic and abiotic stresses in plants. In *Abiotic and Biotic Stress in Plants (Internet)*; de Oliveira, A.B., Ed.; IntechOpen: London, UK, 2019. [CrossRef]
4. Raney, T.; Steinfeld, H.; Skoet, J. *The State of Food and Agriculture 2009: Livestock in the Balance*; FAO: Rome, Italy, 2009. Available online: https://www.fao.org/publications/card/en/c/3aa4f41c-4316-5ddd-a656-22a00ef5d414/ (accessed on 22 September 2022).
5. FAO. *Sustainable Agriculture*; FAO: Rome, Italy, 2016. Available online: https://www.fao.org/family-farming/detail/en/c/423952/ (accessed on 22 September 2022).
6. Fadiji, A.E.; Santoyo, G.; Yadav, A.N.; Babalola, O.O. Efforts towards overcoming drought stress in crops: Revisiting the mechanisms employed by plant growth-promoting bacteria. *Front. Microbiol.* **2022**, *13*, 962427. [CrossRef]
7. Mashabela, M.D.; Tugizimana, F.; Steenkamp, P.A.; Piater, L.A.; Dubery, I.A.; Mhlongo, M.I. Untargeted metabolite profiling to elucidate rhizosphere and leaf metabolome changes of wheat cultivars (Triticum aestivum L.) treated with the plant growth-promoting rhizobacteria (T22) and Bacillus subtilis. *Front. Microbiol.* **2022**, *13*, 971836. [CrossRef]

8. Swain, H.; Adak, T.; Mukherjee, A.K.; Sarangi, S.; Samal, P.; Khandual, A.; Jena, R.; Bhattacharyya, P.; Naik, S.K.; Mehetre, S.T.; et al. Seed biopriming with Trichoderma strains isolated from tree bark improves plant growth, antioxidative defense system in rice and enhance straw degradation capacity. *Front. Microbiol.* **2021**, *12*, 633881. [CrossRef]
9. Papadopoulou, A.; Matsi, T.; Kamou, N.; Avdouli, D.; Mellidou, I.; Karamanoli, K. Decoding the potential of a new Pseudomonas putida strain for inducing drought tolerance of tomato (Solanum lycopersicum) plants through seed biopriming. *J. Plant Physiol.* **2022**, *271*, 153658. [CrossRef]
10. Tyśkiewicz, R.; Nowak, A.; Ozimek, E.; Jaroszuk-Ściseł, J. Trichoderma: The current status of its application in agriculture for the biocontrol of fungal phytopathogens and stimulation of plant growth. *Int. J. Mol. Sci.* **2022**, *23*, 2329. [CrossRef]
11. Chakraborti, S.; Bera, K.; Sadhukhan, S.; Dutta, P. Bio-priming of seeds: Plant stress management and its underlying cellular, biochemical and molecular mechanisms. *Plant Stress* **2022**, *3*, 100052. [CrossRef]
12. Harman, G.E.; Uphoff, N. Symbiotic root-endophytic soil microbes improve crop productivity and provide environmental benefits. *Scientifica* **2019**, *2019*, 9106395. [CrossRef]
13. Yedidia, I.; Benhamou, N.; Kapulnik, Y.; Chet, I. Induction and accumulation of PR proteins activity during early stages of root colonization by the mycoparasite Trichoderma harzianum strain T-203. *Plant Physiol. Biochem.* **2000**, *38*, 863–873. [CrossRef]
14. Cai, F.; Druzhinina, I.S. In honor of John Bissett: Authoritative guidelines on molecular identification of Trichoderma. *Fungal Divers.* **2021**, *107*, 1–69. [CrossRef]
15. Persoon, C.H. Disposita methodical fungorum. *Romers Neues Mag Bot.* **1794**, *1*, 81–128.
16. Weindling, R. Trichoderma lignorum as a parasite of other soil fungi. *Phytopathology* **1932**, *22*, 837–845.
17. Weindling, R. Studies on a lethal principle effective in the parasitic action of Trichoderma lignorum on Rhizoctonia solani and other soil fungi. *Phytopathology* **1934**, *24*, 1153–1179.
18. Howell, C.R. Mechanisms employed by Trichoderma species in the biological control of plant diseases: The history and evolution of current concepts. *Plant Dis.* **2003**, *87*, 4–10. [CrossRef] [PubMed]
19. Zin, N.A.; Badaluddin, N.A. Biological functions of *Trichoderma* spp. for agriculture applications. *Ann. Agric. Sci.* **2020**, *65*, 168–178. [CrossRef]
20. Harman, G.E.; Howell, C.R.; Viterbo, A.; Chet, I.; Lorito, M. Trichoderma species—Opportunistic, avirulent plant symbionts. *Nat. Rev. Microbiol.* **2004**, *2*, 43–56. [CrossRef]
21. Sood, M.; Kapoor, D.; Kumar, V.; Sheteiwy, M.S.; Ramakrishnan, M.; Landi, M.; Araniti, F.; Sharma, A. Trichoderma: The "secrets" of a multitalented biocontrol agent. *Plants* **2020**, *9*, 762. [CrossRef]
22. Woo, S.L.; Scala, M.; Ruocco, M.; Lorito, M. The molecular biology of the interactions between *Trichoderma* spp., phytopathogenic fungi, and plants. *Phytopathology* **2006**, *96*, 181–185. [CrossRef]
23. Mona, S.A.; Hashem, A.; Abd_Allah, E.F.; Alqarawi, A.A.; Soliman, D.W.K.; Wirth, S.; Egamberdieva, D. Increased resistance of drought by Trichoderma harzianum fungal treatment correlates with increased secondary metabolites and proline content. *J. Integr. Agric.* **2017**, *16*, 1751–1757. [CrossRef]
24. Cornejo-Ríos, K.; Osorno-Suárez, M.d.P.; Hernández-León, S.; Reyes-Santamaría, M.I.; Juárez-Díaz, J.A.; Pérez-España, V.H.; Peláez-Acero, A.; Madariaga-Navarrete, A.; Saucedo-García, M. Impact of Trichoderma asperellum on chilling and drought stress in tomato (Solanum lycopersicum). *Horticulturae* **2021**, *7*, 385. [CrossRef]
25. Bashyal, B.M.; Parmar, P.; Zaidi, N.W.; Aggarwal, R. Molecular programming of drought-challenged Trichoderma harzianum-bioprimed rice (Oryza sativa L.). *Front. Microbiol.* **2021**, *12*, 655165. [CrossRef]
26. Ghorbanpour, A.; Salimi, A.; Ghanbary, M.A.T.; Pirdashti, H.; Dehestani, A. The effect of Trichoderma harzianum in mitigating low temperature stress in tomato (Solanum lycopersicum L.) plants. *Sci. Hortic.* **2018**, *230*, 134–141. [CrossRef]
27. Sánchez-Montesinos, B.; Diánez, F.; Moreno-Gavira, A.; Gea, F.J.; Santos, M. Plant growth promotion and biocontrol of Pythium ultimum by saline tolerant Trichoderma isolates under salinity stress. *Int. J. Environ. Res. Public Health* **2019**, *16*, 2053. [CrossRef] [PubMed]
28. Abd El-Baki, G.K.; Mostafa, D. The potentiality of Trichoderma harzianum in alleviation the adverse effects of salinity in faba bean plants. *Acta Biol. Hung.* **2014**, *65*, 451–468. [CrossRef] [PubMed]
29. Maurya, S.; Rashk-E-Eram; Naik, S.K.; Choudhary, J.S.; Kumar, S. Heavy metals scavenging potential of Trichoderma asperellum and Hypocrea nigricans isolated from acid soil of Jharkhand. *Indian J. Microbiol.* **2019**, *59*, 27–38. [CrossRef]
30. Sun, H.; Wu, L.; Hao, Y.; Liu, C.; Pan, L.; Zhu, Z. Tolerance mechanism of Trichoderma asperellum to Pb^{2+}: Response changes of related active ingredients under Pb^{2+} stress. *RSC Adv.* **2020**, *10*, 5202–5211. [CrossRef] [PubMed]
31. Mauch-Mani, B.; Baccelli, I.; Luna, E.; Flors, V. Defense priming: An adaptive part of induced resistance. *Annu. Rev. Plant Biol.* **2017**, *68*, 485–512. [CrossRef] [PubMed]
32. Llorens, E.; González-Hernández, A.I.; Scalschi, L.; Fernández-Crespo, E.; Camañes, G.; Vicedo, B.; García-Agustín, P. Chapter 1—Priming mediated stress and cross-stress tolerance in plants: Concepts and opportunities. In *Priming-Mediated Stress and Cross-Stress Tolerance in Crop Plants*; Hossain, M.A., Liu, F., Burritt, D.J., Fujita, M., Huang, B., Eds.; Academic Press: Cambridge, MA, USA, 2020. [CrossRef]
33. Harman, G.E.; Kubicek, C.P. Trichoderma and Gliocladium. In *Enzymes, Biological Control and Commercial Applications*; Harman, G.E., Ed.; Kubicek, Crc Pr: London, UK, 1998. [CrossRef]

34. Szczałba, M.; Kopta, T.; Gąstoł, M.; Sękara, A. Comprehensive insight into arbuscular mycorrhizal fungi, *Trichoderma* spp. and plant multilevel interactions with emphasis on biostimulation of horticultural crops. *J. Appl. Microbiol.* **2019**, *127*, 630–647. [CrossRef] [PubMed]
35. Brotman, Y.; Landau, U.; Cuadros-Inostroza, Á.; Takayuki, T.; Fernie, A.R.; Chet, I.; Viterbo, A.; Willmitzer, L. Trichoderma-plant root colonization: Escaping early plant defense responses and activation of the antioxidant machinery for saline stress tolerance. *PLoS Pathog.* **2013**, *9*, e1003221. [CrossRef]
36. Hermosa, R.; Viterbo, A.; Chet, I.; Monte, E. Plant-beneficial effects of Trichoderma and of its genes. *Microbiol. Read Engl.* **2012**, *158*, 17–25. [CrossRef]
37. Ausubel, F.M. Are innate immune signaling pathways in plants and animals conserved? *Nat. Immunol.* **2005**, *6*, 973–979. [CrossRef]
38. Morán-Diez, M.E.; Martínez de Alba, Á.E.; Rubio, M.B.; Hermosa, R.; Monte, E. Trichoderma and the plant heritable priming responses. *J. Fungi* **2021**, *7*, 318. [CrossRef] [PubMed]
39. Shoresh, M.; Harman, G.E.; Mastouri, F. Induced systemic resistance and plant responses to fungal biocontrol agents. *Annu. Rev. Phytopathol.* **2010**, *48*, 21–43. [CrossRef] [PubMed]
40. Neer, E.J. Heterotrimeric G proteins: Organizers of transmembrane signals. *Cell* **1995**, *80*, 249–257. [CrossRef] [PubMed]
41. Wu, T.Y.; Urano, D. Genetic and systematic approaches toward G protein-coupled abiotic stress signaling in plants. *Front. Plant Sci.* **2018**, *9*, 1378. [CrossRef] [PubMed]
42. Zhang, H.; Xie, P.; Xu, X.; Xie, Q.; Yu, F. Heterotrimeric G protein signalling in plant biotic and abiotic stress response. *Plant Biol.* **2021**, *23*, 20–30. [CrossRef]
43. Patel, J.S.; Sarma, B.K.; Singh, H.B.; Upadhyay, R.S.; Kharwar, R.N.; Ahmed, M. Pseudomonas fluorescens and Trichoderma asperellum enhance expression of Gα subunits of the pea heterotrimeric G-protein during Erysiphe pisi infection. *Front. Plant Sci.* **2016**, *6*, 1206. [CrossRef]
44. Aharon, G.S.; Gelli, A.; Snedden, W.A.; Blumwald, E. Activation of a plant plasma membrane Ca^{2+} channel by TGα1, a heterotrimeric G protein α-subunit homologue. *FEBS Lett.* **1998**, *424*, 17–21. [CrossRef]
45. Zhu, R.; Dong, X.; Hao, W.; Gao, W.; Zhang, W.; Xia, S.; Liu, T.; Shang, Z. Heterotrimeric G protein-regulated Ca2+ influx and PIN2 asymmetric distribution are involved in Arabidopsis thaliana roots' avoidance response to extracellular ATP. *Front. Plant Sci.* **2017**, *8*, 1522. [CrossRef]
46. Zhang, W.; Jeon, B.W.; Assmann, S.M. Heterotrimeric G-protein regulation of ROS signalling and calcium currents in Arabidopsis guard cells. *J. Exp. Bot.* **2011**, *62*, 2371–2379. [CrossRef]
47. Aldon, D.; Mbengue, M.; Mazars, C.; Galaud, J.P. Calcium signalling in plant biotic interactions. *Int. J. Mol. Sci.* **2018**, *19*, 665. [CrossRef]
48. Xu, T.; Niu, J.; Jiang, Z. Sensing mechanisms: Calcium signaling mediated abiotic stress in plants. *Front. Plant Sci.* **2022**, *13*, 925863. [CrossRef]
49. Navazio, L.; Baldan, B.; Moscatiello, R.; Zuppini, A.; Woo, S.L.; Mariani, P.; Lorito, M. Calcium-mediated perception and defense responses activated in plant cells by metabolite mixtures secreted by the biocontrol fungus Trichoderma atroviride. *BMC Plant Biol.* **2007**, *7*, 41. [CrossRef] [PubMed]
50. Ruocco, M.; Lanzuise, S.; Lombardi, N.; Woo, S.L.; Vinale, F.; Marra, R.; Varlese, R.; Manganiello, G.; Pascale, A.; Scala, V.; et al. Multiple roles and effects of a novel Trichoderma hydrophobin. *Mol. Plant-Microbe Interact.* **2015**, *28*, 167–179. [CrossRef] [PubMed]
51. Whiteford, J.R.; Spanu, P.D. Hydrophobins and the interactions between fungi and plants. *Mol. Plant Pathol.* **2002**, *3*, 391–400. [CrossRef] [PubMed]
52. Moscatiello, R.; Sello, S.; Ruocco, M.; Barbulova, A.; Cortese, E.; Nigris, S.; Baldan, B.; Chiurazzi, M.; Mariani, P.; Lorito, M.; et al. The hydrophobin HYTLO1 secreted by the biocontrol fungus Trichoderma longibrachiatum triggers a NAADP-mediated calcium signalling pathway in Lotus japonicus. *Int. J. Mol. Sci.* **2018**, *19*, 2596. [CrossRef]
53. Wojtaszek, P. Oxidative burst: An early plant response to pathogen infection. *Biochem. J.* **1997**, *322*, 681–692. [CrossRef]
54. Qu, Y.; Yan, M.; Zhang, Q. Functional regulation of plant NADPH oxidase and its role in signaling. *Plant Signal. Behav.* **2017**, *12*, e1356970. [CrossRef]
55. Suzuki, N.; Miller, G.; Morales, J.; Shulaev, V.; Torres, M.A.; Mittler, R. Respiratory burst oxidases: The engines of ROS signaling. *Curr. Opin. Plant Biol.* **2011**, *14*, 691–699. [CrossRef]
56. Marino, D.; Dunand, C.; Puppo, A.; Pauly, N. A burst of plant NADPH oxidases. *Trends Plant Sci.* **2012**, *17*, 9–15. [CrossRef]
57. Kurusu, T.; Kuchitsu, K.; Tada, Y. Plant signaling networks involving Ca^{2+} and Rboh/Nox-mediated ROS production under salinity stress. *Front. Plant Sci.* **2015**, *6*, 427. [CrossRef]
58. Evans, M.; Choi, W.G.; Gilroy, S.; Morris, R.J. A ROS-assisted calcium wave dependent on the AtRBOHD NADPH oxidase and TPC1 cation channel propagates the systemic response to salt stress. *Plant Physiol.* **2016**, *171*, 1771–1784. [CrossRef] [PubMed]
59. Mazars, C.; Thuleau, P.; Lamotte, O.; Bourque, S. Cross-talk between ROS and calcium in regulation of nuclear activities. *Mol. Plant* **2010**, *3*, 706–718. [CrossRef] [PubMed]
60. Saravanakumar, K.; Fan, L.; Fu, K.; Yu, C.; Wang, M.; Xia, H.; Sun, J.; Li, Y.; Chen, J. Cellulase from Trichoderma harzianum interacts with roots and triggers induced systemic resistance to foliar disease in maize. *Sci. Rep.* **2016**, *6*, 35543. [CrossRef]

61. Guo, K.; Sui, Y.; Li, Z.; Huang, Y.; Zhang, H.; Wang, W. Colonization of Trichoderma viride Tv-1511 in peppermint (Mentha × piperita L.) roots promotes essential oil production by triggering ROS-mediated MAPK activation. *Plant Physiol. Biochem.* **2020**, *151*, 705–718. [CrossRef] [PubMed]
62. Mukherjee, P.K.; Horwitz, B.A.; Herrera-Estrella, A.; Schmoll, M.; Kenerley, C.M. Trichoderma research in the genome era. *Annu. Rev. Phytopathol.* **2013**, *51*, 105–129. [CrossRef]
63. Chacón, M.R.; Rodríguez-Galán, O.; Benítez, T.; Sousa, S.; Rey, M.; Llobell, A.; Delgado-Jarana, J. Microscopic and transcriptome analyses of early colonization of tomato roots by Trichoderma harzianum. *Int. Microbiol.* **2007**, *10*, 19–27.
64. Ruano-Rosa, D.; Prieto, P.; Rincón, A.M.; Gómez-Rodríguez, M.V.; Valderrama, R.; Barroso, J.B.; Mercado-Blanco, J. Fate of Trichoderma harzianum in the olive rhizosphere: Time course of the root colonization process and interaction with the fungal pathogen Verticillium dahliae. *BioControl* **2016**, *61*, 269–282. [CrossRef]
65. Alonso-Ramírez, A.; Poveda, J.; Martín, I.; Hermosa, R.; Monte, E.; Nicolás, C. Salicylic acid prevents Trichoderma harzianum from entering the vascular system of roots. *Mol. Plant Pathol.* **2014**, *15*, 823–831. [CrossRef]
66. De Palma, M.; Salzano, M.; Villano, C.; Aversano, R.; Lorito, M.; Ruocco, M.; Docimo, T.; Piccinelli, A.L.; D'Agostino, N.; Tucci, M. Transcriptome reprogramming, epigenetic modifications and alternative splicing orchestrate the tomato root response to the beneficial fungus Trichoderma harzianum. *Hortic. Res.* **2019**, *6*, 5. [CrossRef]
67. Jogaiah, S.; Abdelrahman, M.; Tran, L.S.P.; Ito, S.I. Different mechanisms of Trichoderma virens-mediated resistance in tomato against Fusarium wilt involve the jasmonic and salicylic acid pathways. *Mol. Plant Pathol.* **2018**, *19*, 870–882. [CrossRef]
68. Basińska-Barczak, A.; Błaszczyk, L.; Szentner, K. Plant cell wall changes in common wheat roots as a result of their interaction with beneficial fungi of Trichoderma. *Cells* **2020**, *9*, 2319. [CrossRef] [PubMed]
69. Jones, J.D.G.; Dangl, J.L. The plant immune system. *Nature* **2006**, *444*, 323–329. [CrossRef] [PubMed]
70. Spoel, S.H.; Dong, X. How do plants achieve immunity? Defence without specialized immune cells. *Nat. Rev. Immunol.* **2012**, *12*, 89–100. [CrossRef] [PubMed]
71. Cui, H.; Tsuda, K.; Parker, J.E. Effector-triggered immunity: From pathogen perception to robust defense. *Annu. Rev. Plant Biol.* **2015**, *66*, 487–511. [CrossRef] [PubMed]
72. Bigeard, J.; Colcombet, J.; Hirt, H. Signaling mechanisms in pattern-triggered immunity (PTI). *Mol. Plant* **2015**, *8*, 521–539. [CrossRef]
73. Su, J.; Yang, L.; Zhu, Q.; Wu, H.; He, Y.; Liu, Y.; Xu, J.; Jiang, D.; Zhang, S. Active photosynthetic inhibition mediated by MPK3/MPK6 is critical to effector-triggered immunity. *PLoS Biol.* **2018**, *16*, e2004122. [CrossRef]
74. Kuć, J. Induced immunity to plant disease. *Bioscience* **1982**, *32*, 854–860. [CrossRef]
75. Choudhary, D.K.; Prakash, A.; Johri, B.N. Induced systemic resistance (ISR) in plants: Mechanism of action. *Indian J. Microbiol.* **2007**, *47*, 289–297. [CrossRef]
76. Peng, K.C.; Lin, C.C.; Liao, C.F.; Yu, H.C.; Lo, C.T.; Yang, H.H.; Lin, K.-C. Expression of L-amino acid oxidase of Trichoderma harzianum in tobacco confers resistance to Sclerotinia sclerotiorum and Botrytis cinerea. *Plant Sci.* **2021**, *303*, 110772. [CrossRef]
77. Alizadeh, H.; Behboudi, K.; Ahmadzadeh, M.; Javan-Nikkhah, M.; Zamioudis, C.; Pieterse, C.M.J.; Bakker, P.A. Induced systemic resistance in cucumber and Arabidopsis thaliana by the combination of Trichoderma harzianum Tr6 and Pseudomonas sp. Ps14. *Biol. Control.* **2013**, *65*, 14–23. [CrossRef]
78. Bigirimana, J.; Meyer, G.d.; Poppe, J.; Hoefte, M.; Elad, Y. Induction of systemic resistance on bean (Phaseolus vulgaris) by Trichoderma harzianum. *Med. Fac. Landbouw. Univ. Gent* **1997**, *62*, 1001–1007.
79. Martínez-Medina, A.; Fernandez, I.; Lok, G.B.; Pozo, M.J.; Pieterse, C.M.J.; Van Wees, S.C.M. Shifting from priming of salicylic acid- to jasmonic acid-regulated defences by Trichoderma protects tomato against the root knot nematode Meloidogyne incognita. *New Phytol.* **2017**, *213*, 1363–1377. [CrossRef] [PubMed]
80. Martinez-Medina, A.; Fernandez, I.; Sánchez-Guzmán, M.; Jung, S.; Pascual, J.; Pozo, M. Deciphering the hormonal signalling network behind the systemic resistance induced by Trichoderma harzianum in tomato. *Front. Plant Sci.* **2013**, *4*, 206. [CrossRef] [PubMed]
81. MAPK Group. Mitogen-activated protein kinase cascades in plants: A new nomenclature. *Trends Plant Sci.* **2002**, *7*, 301–308. [CrossRef] [PubMed]
82. Jonak, C.; Ökrész, L.; Bögre, L.; Hirt, H. Complexity, cross talk and integration of plant MAP kinase signalling. *Curr. Opin. Plant Biol.* **2002**, *5*, 415–424. [CrossRef]
83. Taj, G.; Agarwal, P.; Grant, M.; Kumar, A. MAPK machinery in plants. *Plant Signal. Behav.* **2010**, *5*, 1370–1378. [CrossRef]
84. Asai, T.; Tena, G.; Plotnikova, J.; Willmann, M.R.; Chiu, W.L.; Gomez-Gomez, L.; Boller, T.; Ausubel, F.M.; Sheen, J. MAP kinase signalling cascade in Arabidopsis innate immunity. *Nature* **2002**, *415*, 977–983. [CrossRef]
85. Zipfel, C.; Kunze, G.; Chinchilla, D.; Caniard, A.; Jones, J.D.G.; Boller, T.; Felix, G. Perception of the bacterial PAMP EF-Tu by the receptor EFR restricts Agrobacterium-mediated transformation. *Cell* **2006**, *125*, 749–760. [CrossRef]
86. Meng, X.; Zhang, S. MAPK Cascades in plant disease resistance signaling. *Annu. Rev. Phytopathol.* **2013**, *51*, 245–266. [CrossRef]
87. Takahashi, F.; Yoshida, R.; Ichimura, K.; Mizoguchi, T.; Seo, S.; Yonezawa, M.; Maruyama, K.; Yamaguchi-Shinozaki, K.; Shinozaki, K. The mitogen-activated protein kinase cascade MKK3–MPK6 is an important part of the jasmonate signal transduction pathway in Arabidopsis. *Plant Cell* **2007**, *19*, 805–818. [CrossRef]

88. Tsuda, K.; Mine, A.; Bethke, G.; Igarashi, D.; Botanga, C.J.; Tsuda, Y.; Glazebrook, J.; Sato, M.; Katagiri, F. Dual regulation of gene expression mediated by extended MAPK activation and salicylic acid contributes to robust innate immunity in Arabidopsis thaliana. *PLoS Genet.* **2013**, *9*, e1004015. [CrossRef] [PubMed]
89. Seo, S.; Katou, S.; Seto, H.; Gomi, K.; Ohashi, Y. The mitogen-activated protein kinases WIPK and SIPK regulate the levels of jasmonic and salicylic acids in wounded tobacco plants. *Plant J. Cell Mol. Biol.* **2007**, *49*, 899–909. [CrossRef] [PubMed]
90. Liu, Y.; Jin, H.; Yang, K.Y.; Kim, C.Y.; Baker, B.; Zhang, S. Interaction between two mitogen-activated protein kinases during tobacco defense signaling. *Plant J. Cell Mol. Biol.* **2003**, *34*, 149–160. [CrossRef] [PubMed]
91. Suzuki, K.; Yano, A.; Shinshi, H. Slow and prolonged activation of the p47 protein kinase during hypersensitive cell death in a culture of tobacco cells. *Plant Physiol.* **1999**, *119*, 1465–1472. [CrossRef] [PubMed]
92. Tsuzuki, C.; Hachisu, M.; Iwabe, R.; Nakayama, Y.; Nonaga, Y.; Sukegawa, S.; Horito, S.; Arimura, G.-I. An amino acid ester of menthol elicits defense responses in plants. *Plant Mol. Biol.* **2022**, *109*, 523–531. [CrossRef]
93. Shoresh, M.; Gal-On, A.; Leibman, D.; Chet, I. Characterization of a mitogen-activated protein kinase gene from Cucumber required for Trichoderma-conferred plant resistance. *Plant Physiol.* **2006**, *142*, 1169–1179. [CrossRef]
94. Contreras-Cornejo, H.A.; Macías-Rodríguez, L.; Alfaro-Cuevas, R.; López-Bucio, J. *Trichoderma* spp. improve growth of Arabidopsis seedlings under salt stress through enhanced root development, osmolite production, and Na+ elimination through root exudates. *Mol. Plant Microbe Interact.* **2014**, *27*, 503–514. [CrossRef]
95. Contreras-Cornejo, H.A.; López-Bucio, J.S.; Méndez-Bravo, A.; Macías-Rodríguez, L.; Ramos-Vega, M.; Guevara-García, Á.A.; López-Bucio, J. Mitogen-activated protein kinase 6 and ethylene and auxin signaling pathways are involved in Arabidopsis root-system architecture alterations by Trichoderma atroviride. *Mol. Plant Microbe Interact.* **2015**, *28*, 701–710. [CrossRef]
96. Mathys, J.; De Cremer, K.; Timmermans, P.; Van Kerkhove, S.; Lievens, B.; Vanhaecke, M.; Cammue, B.P.A.; De Coninck, B. Genome-wide characterization of ISR induced in Arabidopsis thaliana by Trichoderma hamatum T382 against infection. *Front. Plant Sci.* **2012**, *3*, 108. [CrossRef]
97. Eschen-Lippold, L.; Bethke, G.; Palm-Forster, M.A.T.; Pecher, P.; Bauer, N.; Glazebrook, J.; Scheel, D.; Lee, J. MPK11—A fourth elicitor-responsive mitogen-activated protein kinase in Arabidopsis thaliana. *Plant Signal. Behav.* **2012**, *7*, 1203–1205. [CrossRef]
98. Bethke, G.; Pecher, P.; Eschen-Lippold, L.; Tsuda, K.; Katagiri, F.; Glazebrook, J.; Scheel, D.; Lee, J. Activation of the Arabidopsis thaliana mitogen-activated protein kinase MPK11 by the flagellin-derived elicitor peptide, flg22. *Mol. Plant-Microbe Interact.* **2012**, *25*, 471–480. [CrossRef] [PubMed]
99. Lassowskat, I.; Böttcher, C.; Eschen-Lippold, L.; Scheel, D.; Lee., J. Sustained mitogen-activated protein kinase activation reprograms defense metabolism and phosphoprotein profile in Arabidopsis thaliana. *Front. Plant Sci.* **2014**, *5*, 554. [CrossRef] [PubMed]
100. Shigenaga, A.M.; Berens, M.L.; Tsuda, K.; Argueso, C.T. Towards engineering of hormonal crosstalk in plant immunity. *Curr. Opin. Plant Biol.* **2017**, *38*, 164–172. [CrossRef] [PubMed]
101. Bari, R.; Jones, J.D.G. Role of plant hormones in plant defence responses. *Plant Mol. Biol.* **2009**, *69*, 473–488. [CrossRef] [PubMed]
102. Illescas, M.; Pedrero-Méndez, A.; Pitorini-Bovolini, M.; Hermosa, R.; Monte, E. Phytohormone production profiles in Trichoderma species and their relationship to wheat plant responses to water stress. *Pathogens* **2021**, *10*, 991. [CrossRef]
103. Jaroszuk-Ściseł, J.; Tyśkiewicz, R.; Nowak, A.; Ozimek, E.; Majewska, M.; Hanaka, A.; Tyśkiewicz, K.; Pawlik, A.; Janusz, G. Phytohormones (Auxin, Gibberellin) and ACC deaminase in vitro synthesized by the mycoparasitic Trichoderma DEMTkZ3A0 strain and changes in the level of auxin and plant resistance markers in wheat seedlings inoculated with this strain conidia. *Int. J. Mol. Sci.* **2019**, *20*, 4923. [CrossRef] [PubMed]
104. Kamalov, L.S.; Turgunov, K.K.; Aripova, S.F.; Abdilalimov, O. Gibberillin A-3 from the microscopic fungus Trichoderma harzianum. *Chem. Nat. Compd.* **2018**, *54*, 421–422. [CrossRef]
105. Pérez, E.; Rubio, M.B.; Cardoza, R.E.; Gutiérrez, S.; Bettiol, W.; Monte, E.; Hermosa, R. The importance of chorismate mutase in the biocontrol potential of Trichoderma parareesei. *Front. Microbiol.* **2015**, *6*, 1181. [CrossRef]
106. Vos, C.M.F.; De Cremer, K.; Cammue, B.P.A.; De Coninck, B. The toolbox of *Trichoderma* spp. in the biocontrol of Botrytis cinerea disease. *Mol. Plant Pathol.* **2015**, *16*, 400–412. [CrossRef]
107. Pieterse, C.M.J.; Van der Does, D.; Zamioudis, C.; Leon-Reyes, A.; Van Wees, S.C.M. Hormonal modulation of plant immunity. *Annu. Rev. Cell Dev. Biol.* **2012**, *28*, 489–521. [CrossRef]
108. López-Ráez, J.A.; Verhage, A.; Fernández, I.; García, J.M.; Azcón-Aguilar, C.; Flors, V.; Pozo, M.J. Hormonal and transcriptional profiles highlight common and differential host responses to arbuscular mycorrhizal fungi and the regulation of the oxylipin pathway. *J. Exp. Bot.* **2010**, *61*, 2589–2601. [CrossRef] [PubMed]
109. Pozo, M.J.; Azcón-Aguilar, C. Unraveling mycorrhiza-induced resistance. *Curr. Opin. Plant Biol.* **2007**, *10*, 393–398. [CrossRef] [PubMed]
110. Jacobs, S.; Zechmann, B.; Molitor, A.; Trujillo, M.; Petutschnig, E.; Lipka, V.; Kogel, K.-H.; Schäfer, P. Broad-spectrum suppression of innate immunity is required for colonization of Arabidopsis roots by the fungus Piriformospora indica. *Plant Physiol.* **2011**, *156*, 726–740. [CrossRef] [PubMed]
111. Peleg-Grossman, S.; Golani, Y.; Kaye, Y.; Melamed-Book, N.; Levine, A. NPR1 protein regulates pathogenic and symbiotic interactions between Rhizobium and legumes and non-legumes. *PLoS ONE* **2009**, *4*, e8399. [CrossRef] [PubMed]
112. Alfiky, A.; Weisskopf, L. Deciphering Trichoderma–plant–pathogen interactions for better development of biocontrol applications. *J. Fungi* **2021**, *7*, 61. [CrossRef] [PubMed]

113. Morán-Diez, E.; Rubio, B.; Domínguez, S.; Hermosa, R.; Monte, E.; Nicolás, C. Transcriptomic response of Arabidopsis thaliana after 24 h incubation with the biocontrol fungus Trichoderma harzianum. *J. Plant Physiol.* **2012**, *169*, 614–620. [CrossRef]
114. Yuan, M.; Huang, Y.; Ge, W.; Jia, Z.; Song, S.; Zhang, L.; Huang, Y. Involvement of jasmonic acid, ethylene and salicylic acid signaling pathways behind the systemic resistance induced by Trichoderma longibrachiatum H9 in cucumber. *BMC Genom.* **2019**, *20*, 144. [CrossRef]
115. Mayo, S.; Cominelli, E.; Sparvoli, F.; González-López, O.; Rodríguez-González, A.; Gutiérrez, S.; Casquero, P.A. Development of a qPCR strategy to select bean genes involved in plant defense response and regulated by the Trichoderma velutinum–Rhizoctonia solani interaction. *Front. Plant Sci.* **2016**, *7*, 1109. [CrossRef]
116. Agostini, R.B.; Postigo, A.; Rius, S.P.; Rech, G.E.; Campos Bermudez, V.A.; Vargas, W.A. Long-lasting primed state in maize plants: Salicylic acid and steroid signaling pathways as key players in the early activation of immune responses in silks. *Mol. Plant Microbe Interact.* **2019**, *32*, 95–106. [CrossRef]
117. Kottb, M.; Gigolashvili, T.; Großkinsky, D.K.; Piechulla, B. Trichoderma volatiles effecting Arabidopsis: From inhibition to protection against phytopathogenic fungi. *Front. Microbiol.* **2015**, *6*, 995. [CrossRef]
118. Senaratna, T.; Touchell, D.; Bunn, E.; Dixon, K. Acetyl salicylic acid (Aspirin) and salicylic acid induce multiple stress tolerance in bean and tomato plants. *Plant Growth Regul.* **2000**, *30*, 157–161. [CrossRef]
119. Azooz, M.M.; Youssef, A.M.; Ahmad, P. Evaluation of salicylic acid (SA) application on growth, osmotic solutes and antioxidant enzyme activities on broad bean seedlings grown under diluted seawater. *Int. J. Plant Physiol. Biochem.* **2011**, *3*, 253–264. [CrossRef]
120. Ahmad, P.; Nabi, G.; Ashraf, M. Cadmium-induced oxidative damage in mustard [Brassica juncea (L.) Czern. & Coss.] plants can be alleviated by salicylic acid. *South Afr. J. Bot.* **2011**, *77*, 36–44. [CrossRef]
121. Gunes, A.; Inal, A.; Alpaslan, M.; Eraslan, F.; Bagci, E.G.; Cicek, N. Salicylic acid induced changes on some physiological parameters symptomatic for oxidative stress and mineral nutrition in maize (Zea mays L.) grown under salinity. *J. Plant Physiol.* **2007**, *164*, 728–736. [CrossRef]
122. Shakirova, F.M.; Sakhabutdinova, A.R.; Bezrukova, M.V.; Fatkhutdinova, R.A.; Fatkhutdinova, D.R. Changes in the hormonal status of wheat seedlings induced by salicylic acid and salinity. *Plant Sci.* **2003**, *164*, 317–322. [CrossRef]
123. Dat, J.F.; Foyer, C.H.; Scott, I.M. Changes in salicylic acid and antioxidants during induced thermotolerance in mustard seedlings. *Plant Physiol.* **1998**, *118*, 1455–1461. [CrossRef]
124. Moustafa-Farag, M.; Mohamed, H.I.; Mahmoud, A.; Elkelish, A.; Misra, A.N.; Guy, K.M.; Kamran, M.; Ai, S.; Zhang, M. Salicylic acid stimulates antioxidant defense and osmolyte metabolism to alleviate oxidative stress in watermelons under excess boron. *Plants* **2020**, *9*, 724. [CrossRef]
125. Huang, H.; Liu, B.; Liu, L.; Song, S. Jasmonate action in plant growth and development. *J. Exp. Bot.* **2017**, *68*, 1349–1359. [CrossRef]
126. Wang, Y.; Mostafa, S.; Zeng, W.; Jin, B. Function and mechanism of jasmonic acid in plant responses to abiotic and biotic stresses. *Int. J. Mol. Sci.* **2021**, *22*, 8568. [CrossRef]
127. Wasternack, C.; Hause, B. Jasmonates: Biosynthesis, perception, signal transduction and action in plant stress response, growth and development. An update to the 2007 review in Annals of Botany. *Ann. Bot.* **2013**, *111*, 1021–1058. [CrossRef]
128. Howe, G.A.; Jander, G. Plant immunity to insect herbivores. *Annu. Rev. Plant Biol.* **2008**, *59*, 41–66. [CrossRef]
129. Contreras-Cornejo, H.A.; Macías-Rodríguez, L.; Beltrán-Peña, E.; Herrera-Estrella, A.; López-Bucio, J. Trichoderma-induced plant immunity likely involves both hormonal- and camalexin-dependent mechanisms in Arabidopsis thaliana and confers resistance against necrotrophic fungi Botrytis cinerea. *Plant Signal. Behav.* **2011**, *6*, 1554–1563. [CrossRef] [PubMed]
130. Sáenz-Mata, J.; Jiménez-Bremont, J.F. HR4 gene is induced in the Arabidopsis-Trichoderma atroviride beneficial interaction. *Int. J. Mol. Sci.* **2012**, *13*, 9110–9928. [CrossRef] [PubMed]
131. Bakshi, M.; Oelmüller, R. WRKY transcription factors: Jack of many trades in plants. *Plant Signal. Behav.* **2014**, *9*, e27700. [CrossRef] [PubMed]
132. Chen, X.; Li, C.; Wang, H.; Guo, Z. WRKY transcription factors: Evolution, binding, and action. *Phytopathol. Res.* **2019**, *1*, 13. [CrossRef]
133. Bai, Y.; Sunarti, S.; Kissoudis, C.; Visser, R.G.F.; van der Linden, C.G. The role of tomato WRKY genes in plant responses to combined abiotic and biotic stresses. *Front. Plant Sci.* **2018**, *9*, 801. [CrossRef]
134. Sáenz-Mata, J.; Salazar-Badillo, F.B.; Jiménez-Bremont, J.F. Transcriptional regulation of Arabidopsis thaliana WRKY genes under interaction with beneficial fungus Trichoderma atroviride. *Acta Physiol. Plant.* **2014**, *36*, 1085–1093. [CrossRef]
135. Xu, X.; Chen, C.; Fan, B.; Chen, Z. Physical and functional interactions between pathogen-induced Arabidopsis WRKY18, WRKY40, and WRKY60 transcription factors. *Plant Cell* **2006**, *18*, 1310–1326. [CrossRef]
136. Agrios, G.N. *Plant Pathology*, 5th ed.; Elsevier Academic Press: San Diego, CA, USA, 2005.
137. Van Loon, L.C.; Van Strien, E.A. The families of pathogenesis-related proteins, their activities, and comparative analysis of PR-1 type proteins. *Physiol. Mol. Plant Pathol.* **1999**, *55*, 85–97. [CrossRef]
138. Ebrahim, S.; Kalidindi, U.; Singh, B. Pathogenesis related (PR) proteins in plant defense mechanism. In *Science against Microbial Pathogens: Communicatiog Research and Tecnological Advances*; Méndez-Vilas, A., Ed.; Formatex Researc Center: Badajoz, Spain, 2011; pp. 1043–1054.
139. Tucci, M.; Ruocco, M.; De Masi, L.; De Palma, M.; Lorito, M. The beneficial effect of *Trichoderma* spp. on tomato is modulated by the plant genotype. *Mol. Plant Pathol.* **2011**, *12*, 341–354. [CrossRef]

140. Alfano, G.; Ivey, M.L.L.; Cakir, C.; Bos, J.I.B.; Miller, S.A.; Madden, L.V.; Kamoun, S.; Hoitink, H.A.J. Systemic modulation of gene expression in tomato by Trichoderma hamatum 382. *Phytopathology* **2007**, *97*, 429–437. [CrossRef] [PubMed]
141. Aamir, M.; Kashyap, S.P.; Zehra, A.; Dubey, M.K.; Singh, V.K.; Ansari, W.A.; Upadhyay, R.S.; Singh, S. Trichoderma erinaceum bio-priming modulates the WRKYs defense programming in tomato against the Fusarium oxysporum f. sp. lycopersici (Fol) challenged condition. *Front. Plant Sci.* **2019**, *10*, 911. [CrossRef]
142. Perazzolli, M.; Moretto, M.; Fontana, P.; Ferrarini, A.; Velasco, R.; Moser, C.; Delledonne, M.; Pertot, I. Downy mildew resistance induced by Trichoderma harzianum T39 in susceptible grapevines partially mimics transcriptional changes of resistant genotypes. *BMC Genom.* **2012**, *13*, 660. [CrossRef] [PubMed]
143. Shoresh, M.; Yedidia, I.; Chet, I. Involvement of jasmonic acid/ethylene signaling pathway in the systemic resistance induced in Cucumber by Trichoderma asperellum T203. *Phytopathology* **2005**, *95*, 76–84. [CrossRef] [PubMed]
144. Hoth, S.; Morgante, M.; Sanchez, J.P.; Hanafey, M.K.; Tingey, S.V.; Chua, N.H. Genome-wide gene expression profiling in Arabidopsis thaliana reveals new targets of abscisic acid and largely impaired gene regulation in the abi1-1 mutant. *J. Cell Sci.* **2002**, *115*, 4891–4900. [CrossRef]
145. Rubio, M.B.; Quijada, N.M.; Pérez, E.; Domínguez, S.; Monte, E.; Hermosa, R. Identifying beneficial qualities of Trichoderma parareesei for plants. *Appl. Environ. Microbiol.* **2014**, *80*, 1864–1873. [CrossRef]
146. Yedidia, I.; Shoresh, M.; Kerem, Z.; Benhamou, N.; Kapulnik, Y. Concomitant induction of systemic resistance to Pseudomonas syringae pv. lachrymans in Cucumber by Trichoderma asperellum (T-203) and accumulation of phytoalexins. *Appl. Environ. Microbiol.* **2003**, *69*, 7343–7353. [CrossRef]
147. Morán-Diez, M.E.; Tranque, E.; Bettiol, W.; Monte, E.; Hermosa, R. Differential response of tomato plants to the application of three Trichoderma species when evaluating the control of Pseudomonas syringae populations. *Plants* **2020**, *9*, 626. [CrossRef]
148. Hasanuzzaman, M.; Parvin, K.; Bardhan, K.; Nahar, K.; Anee, T.I.; Masud, A.A.C.; Fotopoulos, V. Biostimulants for the regulation of reactive oxygen species metabolism in plants under abiotic stress. *Cells* **2021**, *10*, 2537. [CrossRef]
149. Mastouri, F.; Björkman, T.; Harman, G.E. Trichoderma harzianum enhances antioxidant defense of tomato seedlings and resistance to water deficit. *Mol. Plant Microbe Interact.* **2012**, *25*, 1264–1271. [CrossRef]
150. Noctor, G.; Foyer, C.H. Ascorbate and glutathione: Keeping active oxygen under control. *Annu. Rev. Plant Physiol. Plant Mol. Biol.* **1998**, *49*, 249–279. [CrossRef] [PubMed]
151. de Pinto, M.C.; Francis, D.; De Gara, L. The redox state of the ascorbate-dehydroascorbate pair as a specific sensor of cell division in tobacco BY-2 cells. *Protoplasma* **1999**, *209*, 90–97. [CrossRef] [PubMed]
152. Chen, S.C.; Ren, J.J.; Zhao, H.J.; Wang, X.L.; Wang, T.H.; Jin, S.D.; Wang, Z.-H.; Li, C.-Y.; Liu, A.-R.; Lin, X.-M.; et al. Trichoderma harzianum improves defense against Fusarium oxysporum by regulating ROS and RNS metabolism, redox balance, and energy flow in Cucumber roots. *Phytopathology* **2019**, *109*, 972–982. [CrossRef] [PubMed]
153. Fu, J.; Liu, Z.; Li, Z.; Wang, Y.; Yang, K. Alleviation of the effects of saline-alkaline stress on maize seedlings by regulation of active oxygen metabolism by Trichoderma asperellum. *PLoS ONE* **2017**, *12*, e0179617. [CrossRef] [PubMed]
154. Hashem, A.; Abd_Allah, E.F.; Alqarawi, A.A.; Al Huqail, A.A.; Egamberdieva, D. Alleviation of abiotic salt stress in Ochradenus baccatus (Del.) by Trichoderma hamatum (Bonord.) Bainier. *J. Plant Interact.* **2014**, *9*, 857–868. [CrossRef]
155. Yasmeen, R.; Siddiqui, Z.S. Physiological responses of crop plants against Trichoderma harzianum in saline environment. *Acta Bot. Croat.* **2017**, *76*, 154–162. [CrossRef]
156. Mironenka, J.; Różalska, S.; Bernat, P. Potential of Trichoderma harzianum and its metabolites to protect wheat seedlings against Fusarium culmorum and 2,4-D. *Int. J. Mol. Sci.* **2021**, *22*, 13058. [CrossRef]
157. Zhang, S.; Gan, Y.; Xu, B. Application of Plant-Growth-Promoting fungi Trichoderma longibrachiatum T6 enhances tolerance of wheat to salt stress through improvement of antioxidative defense system and gene expression. *Front. Plant Sci.* **2016**, *7*, 1405. [CrossRef]
158. El-Esawi, M.A.; Elansary, H.O.; El-Shanhorey, N.A.; Abdel-Hamid, A.M.E.; Ali, H.M.; Elshikh, M.S. Salicylic acid-regulated antioxidant mechanisms and gene expression enhance rosemary performance under saline conditions. *Front. Physiol.* **2017**, *8*, 716. [CrossRef]
159. Saidi, I.; Yousfi, N.; Borgi, M.A. Salicylic acid improves the antioxidant ability against arsenic-induced oxidative stress in sunflower (Helianthus annuus) seedling. *J. Plant Nutr.* **2017**, *40*, 2326–2335. [CrossRef]
160. Torun, H.; Novák, O.; Mikulík, J.; Pěnčík, A.; Strnad, M.; Ayaz, F.A. Timing-dependent effects of salicylic acid treatment on phytohormonal changes, ROS regulation, and antioxidant defense in salinized barley (Hordeum vulgare L.). *Sci. Rep.* **2020**, *10*, 13886. [CrossRef] [PubMed]
161. Serrano, I.; Audran, C.; Rivas, S. Chloroplasts at work during plant innate immunity. *J. Exp. Bot.* **2016**, *67*, 3845–3854. [CrossRef] [PubMed]
162. Şesan, T.E.; Oancea, A.O.; Ştefan, L.M.; Mănoiu, V.S.; Ghiurea, M.; Răut, I.; Constantinescu-Aruxandei, D.; Toma, A.; Savin, S.; Bira, A.F.; et al. Effects of foliar treatment with a Trichoderma plant biostimulant consortium on Passiflora caerulea L. *Yield and quality. Microorg.* **2020**, *8*, 123. [CrossRef] [PubMed]
163. Harman, G.E.; Doni, F.; Khadka, R.B.; Uphoff, N. Endophytic strains of Trichoderma increase plants' photosynthetic capability. *J. Appl. Microbiol.* **2021**, *130*, 529–546. [CrossRef] [PubMed]
164. Andrzejak, R.; Janowska, B. Flowering, nutritional status, and content of chloroplast pigments in leaves of Gladiolus hybridus L. 'Advances Red' after application of *Trichoderma* spp. *Sustainability* **2022**, *14*, 4576. [CrossRef]

165. Song, Y.; Feng, L.; Alyafei, M.A.M.; Jaleel, A.; Ren, M. Function of chloroplasts in plant stress responses. *Int. J. Mol. Sci.* **2021**, *22*, 13464. [CrossRef]
166. Li, M.; Kim, C. Chloroplast ROS and stress signaling. *Plant Commun.* **2021**, *3*, 100264. [CrossRef]
167. Fu, J.; Wang, Y.-F.; Liu, Z.-H.; Li, Z.-T.; Yang, K.-J. Trichoderma asperellum alleviates the effects of saline-alkaline stress on maize seedlings via the regulation of photosynthesis and nitrogen metabolism. *Plant Growth Regul.* **2018**, *85*, 363–374. [CrossRef]
168. Metwally, R.A.; Al-Amri, S. Individual and interactive role of Trichoderma viride and arbuscular mycorrhizal fungi on growth and pigment content of onion plants. *Lett. Appl. Microbiol.* **2020**, *70*, 79–86. [CrossRef]
169. Abdel-Fattah, G.M.; Shabana, Y.M.; Ismail, A.E.; Rashad, Y.M. Trichoderma harzianum: A biocontrol agent against Bipolaris oryzae. *Mycopathologia* **2007**, *164*, 81–89. [CrossRef]
170. da Silva, L.R.; Valadares-Inglis, M.C.; Peixoto, G.H.; de Luccas, B.E.; Muniz, P.H.; Magalhães, D.M.; Moraes, M.C.B.; de Mello, S.C.M. Volatile organic compounds emitted by Trichoderma azevedoi promote the growth of lettuce plants and delay the symptoms of white mold. *Biol. Control.* **2021**, *152*, 104447. [CrossRef]
171. Hosseinzeynali, A.; Abbaszadeh Dahaji, P.; Alaei, H.; Hosseinifard, J.; Akhgar, A. Effect of Trichoderma on growth and nutrition of Pistachio trees under common garden condition. *J. Soil Biol.* **2021**, *8*, 115–129.
172. Stahl, W.; Sies, H. Antioxidant activity of carotenoids. *Mol. Asp. Med.* **2003**, *24*, 345–351. [CrossRef] [PubMed]
173. Yano, A.; Suzuki, K.; Uchimiya, H.; Shinshi, H. Induction of hypersensitive cell death by a fungal protein in cultures of tobacco cells. *Mol. Plant-Microbe Interact.* **1998**, *11*, 115–123. [CrossRef]

MDPI
St. Alban-Anlage 66
4052 Basel
Switzerland
www.mdpi.com

Horticulturae Editorial Office
E-mail: horticulturae@mdpi.com
www.mdpi.com/journal/horticulturae

www.ingramcontent.com/pod-product-compliance
Lightning Source LLC
Chambersburg PA
CBHW061844210926
51850CB00006B/969
* 9 7 8 3 0 3 6 5 9 4 4 4 6 *